ADVANCED MAGNETIC NANOSTRUCTURES

ADVANCED MAGNETIC NANOSTRUCTURES

Edited by

David Sellmyer
University of Nebraska
Lincoln, Nebraska
USA

Ralph Skomski
University of Nebraska
Lincoln, Nebraska
USA

Springer

David Sellmyer
University of Nebraska
Lincoln, Nebraska, USA

Ralph Skomski
University of Nebraska
Lincoln, Nebraska, USA

Advanced Magnetic Nanostructures

ISBN 10: 0-387-23316-4 (e-book)

ISBN 13: 978-1-4419-3603-5

ISBN 13: 978-0-387-23316-1 (e-book)

Printed on acid-free paper.

9 8 7 6 5 4 3 2 1

springeronline.com

Contents

PREFACE

A key trend in modern science and technology is the exploitation of phenomena occurring on length scales between 1 nm and 1000 nm. This nanotechnology or nanoscience approach has lead to the emergence of fields such as nanobiology, nanoelectronics, and nanochemistry. An important and—in many respects—pivotal area is *nanomagnetism*. From early precursors in the first half of the 20th century to recent developments, magnetic nanostructures are interesting scientific objects with many present and emerging applications, including permanent magnets, soft magnets, magnetic recording media, sensors, and structures and materials for spin electronics. A key advantage of artifical magnetic nanostructures is their ability to surpass the performance of naturally occurring magnetic compounds. Examples are nanostructured permanent and soft magnets.

Magnetic nanostructures can be produced in a variety of geometries, such as nanoparticles, nanowires, dots and antidots, particulate thin films, nanotubes, nanojunctions, and nanorings. In addition, much progress has recently been made towards tuning the chemistry and crystallographic microstructure for a given geometry. For example, nanotubes can be produced as soft- and hard-magnetic structures.

A fascinating aspect of nanomagnetism is that the involved physics goes beyond a mixture of atomic-scale and macroscopic effects. The main competition between electrostatic interactions, such as exchange, and relativistic corrections, such as spin-orbit coupling, is organized on a length scale of a few nanometers. This gives rise to a variety of zero- and finite-temperature phenomena governing the static and dynamic behavior of the structures.

This book is devoted to the fabrication, characterization, experimental investigation, theoretical understanding, and utilization of advanced magnetic nanostructures. Focus is on various types of 'bottom-up' and 'top-down' artificial nanostructures, as contrasted to naturally occurring magnetic nanostructures, such as iron-oxide inclusions in magnetic rocks, and to structures such as perfect thin films.

Chapter 1 is an introduction into some basic concepts of interest to more than one chapter, such as the definitions of basic magnetic quantities. Chapters 2-4 are devoted to the theory of magnetic nanostructures, §5 deals with the characterization of the structures, and §6-10 are devoted to specific systems. Applications of advanced magnetic nanostructures are discussed in §11-15 and, finally, the appendix lists and briefly discusses magnetic properties of typical starting materials.

The book project has been supported by NSF MRSEC, by the W. M. Keck Foundation, and by various agencies and bodies acknowledged in the individual chapters. Thanks are due to the chapter authors for their enthusiasm and their timely contributions, and to our colleagues Prof. R. D. Kirby, Prof. S.-H. Liou, Dr. Y. Sui, and Dr. J. Zhou for numerous discussions. Finally, we are grateful to S. Krupicka and V. Skomski for their help in preparing the manuscript.

Lincoln, September 2005 D. J. S. and R. S.

Authors

M. J. Bonder
Department of Physics and Astronomy
University of Delaware
Newark, DE 19716, USA

S. Cardoso
INESC Microsystems and Nano-technologies, Rua Alves Redol 9 and Physics Department, Instituto Superior Tecnico, Avenida Rovisco Pais, 1000 Lisbon, Portugal

S. van Dijken
SFI Trinity Nanoscience Laboratory
Physics Department, Trinity College
Dublin 2, Ireland

R. Dittrich
Solid State Physics
Vienna University of Technology
A-1040 Vienna, Austria

R. E. Dunin-Borkowski
Department of Materials Science and Metallurgy
University of Cambridge
Cambridge, CB2 3QZ,
United Kingdom

O. Ertl
Solid State Physics
Vienna University of Technology
A-1040 Vienna, Austria

H. Ferreira
INESC Microsystems and Nano-technologies, Rua Alves Redol 9 and Physics Department, Instituto Superior Tecnico, Avenida Rovisco Pais, 1000 Lisbon, Portugal

R. Ferreira
INESC Microsystems and Nano-technologies, Rua Alves Redol 9 and Physics Department, Instituto Superior Tecnico, Avenida Rovisco Pais, 1000 Lisbon, Portugal

J. Fidler
Solid State Physics
Vienna University of Technology
A-1040 Vienna, Austria

P. P. Freitas
INESC Microsystems and Nano-technologies, Rua Alves Redol 9 and Physics Department, Instituto Superior Tecnico, Avenida Rovisco Pais, 1000 Lisbon, Portugal

D. Givord
Laboratoire Louis Néel (CNRS)
25 avenue des Martyrs, BP 166,
38042 Grenoble Cedex 9, France

J. Gregg
Clarendon Laboratory
University of Oxford
Parks Road OX1 3PU, UK

G. C. Hadjipanayis
Department of Physics and Astronomy
University of Delaware
Newark, DE 19716, USA

G. Herzer
Vacuumschmelze GmbH & Co. KG
D-63450 Hanau, Germany

G. Hrkac
Solid State Physics
Vienna University of Technology
A-1040 Vienna, Austria

Y. Huang
Department of Physics and Astronomy
University of Delaware
Newark, DE 19716, USA

S. S. Jaswal
Dept. of Physics and Astronomy
University of Nebraska
Lincoln,NE 68588, USA

A. Kashyap
LNM Institute of Information Technology
Jaipur - 302015, India

M. Kirschner
Solid State Physics
Vienna University of Technology
A-1040 Vienna, Austria

R. H. Kraus, Jr.
Biophysics Group
Los Alamos National Laboratory
Los Alamos, NM 87545, USA

V. Labhasetwar
Department of Pharmaceutical Sciences
College of Pharmacy
986025 Nebraska Medical Center
Omaha, NE 68198-6025, USA

D. L. Leslie-Pelecky
Department of Physics and Astronomy
Center for Materials Research & Analysis
University of Nebraska
Lincoln, NE 68588-0111, USA

J. C. Lodder
Systems & Materials for Information Storage, MESA and Institute for Nanotechnology,
University of Twente, PO Box 217,
7500AE Enschede, The Netherlands

M. R. McCartney
Department of Physics and Astronomy and Center for Solid State Science
Arizona State University
Tempe AZ 85287-1504, USA

T. McDaniel
Seagate Research
1251 Waterfront Place
Pittsburgh, PA 15222 -4215, USA

S. Rivoirard
CRETA (CNRS)
25 avenue des Martyrs
BP 16
38042 Grenoble Cedex 9, France

R. Sabirianov
Department of Physics
University of Nebraska
Omaha, NE 68182, USA

T. Schrefl
Department of Engineering Materials
University of Sheffield
Sheffield, S1 3JD, UK

D. J. Sellmyer
Center for Materials Research and Analysis and Department of Physics and Astronomy
University of Nebraska
Lincoln, NE 68588, USA

R. Skomski
Center for Materials Research and Analysis and Department of Physics and Astronomy
University of Nebraska
Lincoln, NE 68588, USA

D. J. Smith
Department of Physics and Astronomy and Center for Solid State Science
Arizona State University
Tempe AZ 85287-1504, USA

D. Suess
Solid State Physics
Vienna University of Technology
A-1040 Vienna, Austria

S. Sun
Department of Chemistry,
Brown University
Providence, RI 02912, USA

K. Suzuki
School of Physics and Materials Engineering
Monash University
Clayton, VIC 3800, Australia

D. Weller
Seagate Recording Media Operations
47010 Kato Road
Fremont, CA 94538, USA

W. Wernsdorfer
Laboratoire Louis Néel, CNRS
BP166
38042 Grenoble Cedex 9, France

Y. F. Xu
Center for Materials Research and Analysis and Department of Physics and Astronomy
University of Nebraska
Lincoln, NE 68588, USA

M. L. Yan
Center for Materials Research and Analysis and Department of Physics and Astronomy
University of Nebraska
Lincoln, NE 68588, USA

J. Zhou
Center for Materials Research and Analysis and Department of Physics and Astronomy
University of Nebraska
Lincoln, NE 68588, USA

R. Skomski
Center for Materials Research and Analysis and Department of Physics and Astronomy
University of Nebraska
Lincoln, NE 68588, USA

D. J. Smith
Department of Physics and Astronomy and Center for Solid State Science
Arizona State University
Tempe AZ 85287-1504, USA

D. Suess
Solid State Physics
Vienna University of Technology
A-1040 Vienna, Austria

S. Sun
Department of Chemistry,
Brown University
Providence, RI 02912, USA

K. Suzuki
School of Physics and Materials Engineering
Monash University
Clayton, VIC 3800, Australia

D. Weller
Seagate Recording Media Operations
47010 Kato Road
Fremont, CA 94538, USA

W. Wernsdorfer
Laboratoire Louis Néel, CNRS
BP166
38042 Grenoble Cedex 9, France

Y. F. Xu
Center for Materials Research and Analysis and Department of Physics and Astronomy
University of Nebraska
Lincoln, NE 68588, USA

M. L. Yan
Center for Materials Research and Analysis and Department of Physics and Astronomy
University of Nebraska
Lincoln, NE 68588, USA

J. Zhou
Center for Materials Research and Analysis and Department of Physics and Astronomy
University of Nebraska
Lincoln, NE 68588, USA

Chapter 1

INTRODUCTION

The nanostructures considered in this book are magnetic and characterized by *structural* length scales ranging from a few interatomic distances to about one micrometer. The basic length unit is the *nanometer* (1 nm = 10^{-3} μm = 10^{-9} m), corresponding to about four interatomic Fe-Fe distances. Magnetic nanostructures pose experimental challenges, exhibit interesting physical phenomena, and have many present or potential applications. An important aspect is that structural lengths affect, but only partly determine, the magnetic length scales encountered in the structures. Examples are domains in semihard nanoparticles, where both the domain size and the domain-wall thickness may be smaller than the particle size, and polycrystalline soft-magnetic nanostructures, where the magnetic correlation length is much larger than the crystallite size.

Due to rapid progress in the fabrication and processing of nanostructures, it is now possible to realize a broad variety of geometries, crystalline textures, and chemistries. For a given geometry, the structures can be fabricated using a variety of magnetic materials (compounds), with different local magnetic properties and crystalline textures. The Appendix presents some magnetic materials of interest in nanomagnetism.

1. BASIC DEFINITIONS AND UNITS

The magnetic moment **m** of the atoms in a nanostructure nearly exclusively originates from the electrons in the partially filled inner shells of transition or rare-earth metals. There are both spin (S) and orbital (L) contributions, but since L is much smaller than S in most iron-series transition-metal magnets, the magnetic moment is often equated with the spin polarization. The situation is similar to that encountered in bulk magnets, although both S and L may be modified at surfaces and interfaces (Ch. 2). As in infinite solids, nuclear moments are much smaller than electron moments and can be ignored safely for most applications.

The magnetic moment of an atom is created by intra-atomic or Hund's-rules exchange, which favors parallel spin alignment on an atomic scale. In addition, ferromagnetism requires *inter*atomic exchange, to ensure parallel alignment of the moments of neighboring atoms. The resulting net moment gives rise to the magnetization $\mathbf{M} = \mathbf{J}/\mu_o$ which is defined as the moment per unit volume. In bulk ferromagnets, the competition between interatomic exchange and thermal disorder leads to the vanishing of the spontaneous magnetization $M_s = |\mathbf{M}(\mathbf{r})|$ at a well-defined sharp Curie temperature T_C. However, the existence of a Curie-point singularity is limited to infinite bodies, and in nanostructures, the concepts of magnetic phase transitions must be reevaluated (Ch. 3).

Spin-orbit coupling in combination with local crystal-field interactions gives rise to magnetic anisotropy. In the simplest case, magnetic anisotropy is of the uniaxial type and described by the lowest-order anisotropy constant K_1. This constant is equal to the energy density necessary to turn the magnetization from the easy to a hard magnetization axis. In addition, there are shape-anisotropy contributions of magnetostatic origin. In nanostructures, surface, interface, and shape anisotropy contributions are often important, particularly in cubic materials, where the lowest-order bulk anisotropy is zero.

An applied magnetic field **H** changes the magnetization by rotating the local moment. Since the magnetic anisotropy yields energy barriers between different magnetization states or spin configurations $\mathbf{M}(\mathbf{r})$, the field dependence of the net magnetization exhibits *hysteresis*. Important hysteretic properties are remanence M_r, defined as the zero-field magnetization after saturation in a strong magnetic field, and the coercivity H_c. The latter is defined as the reverse magnetic field at which the volume-averaged magnetization of an initially saturated magnet reaches zero. Some other hysteretic properties of specific nanostructuring offers many tools to tune hysteretic properties. For example, the coercivity of advanced magnetic nanostructures varies from about 1 μT to several T. Analytical and numerical aspects of the hysteresis of magnetic nanostructures will be treated in Chs. 3 and 4, respectively.

The structural length scales of nanomagnets are intermediate between interatomic and macroscopic distances, but nanomagnetism cannot be reduced to a mixture of atomic-scale and macroscopic phenomena. For example, most extrinsic properties are realized on a nanoscale, and nanostructuring is used to produce optimized hard, soft, information-storage and sensor materials. A dynamic aspect of this interplay between atomic (or intrinsic) and hysteretic (or extrinsic) phenomena is that equilibration times vary from less than 1 ns to millions of years. This determines, for example, the magnetic switching time of spin-electronics structures and the lifetime of information stored in magnetic recording media. This is related to the thermal instability of the magnetization direction known as superparamagnetism (Bean and Livingston 1959). This effect occurs in very small particles and is strongly temperature dependent (Ch. 3).

In this book, the preferred *length unit* is 1 nm (10^{-9} m), but 1 Å = 0.1 and 1 μm = 1000 nm are also used, particularly for submicron features having sizes of several 100 nm. Recording densities are also measured in bytes per square inch or bytes per square centimeter (1 in^{-2} = 6.452 cm^{-2}). Both SI and Gaussian units will be used for *magnetic quantities*, with explicit conversions occasionally included in square brackets. This also includes the parallel or alternate use of the SI unit tesla for coercivity and magnetization, as compared to the correct but cumbersome unit A/m. The latter corresponds to $B = \mu_o(M + H)$, and the former is obtained by incorporating μ_o into M and H, so that one actually considers $\mu_o M$ and $\mu_o H$. In the Gaussian system, multiplying the magnetization (emu/cm^3) by the dimensionless number 4π changes the unit to gauss. (This is similar to measuring the perimeter of an island in miles and its diameter in feet.) Some everyday numerical *conversion rules* are: (i) 1 T = 10 kOe ≈ 0.8 MA/m (coercivity), (ii) 1 emu/cm^3 = 1 kA/m and 1 T = 10 kG ≈ 0.8 MA/m (magnetization), (iii) 1 T^2 = 100 MGOe ≈ 800 kJ/m^3 (energy product). Finally, an SI susceptibility of 1 corresponds to a Gaussian susceptibility of 4π. In the above conversions, we have used the numerical relation that $10/4\pi = 0.7958 \approx 0.8$.

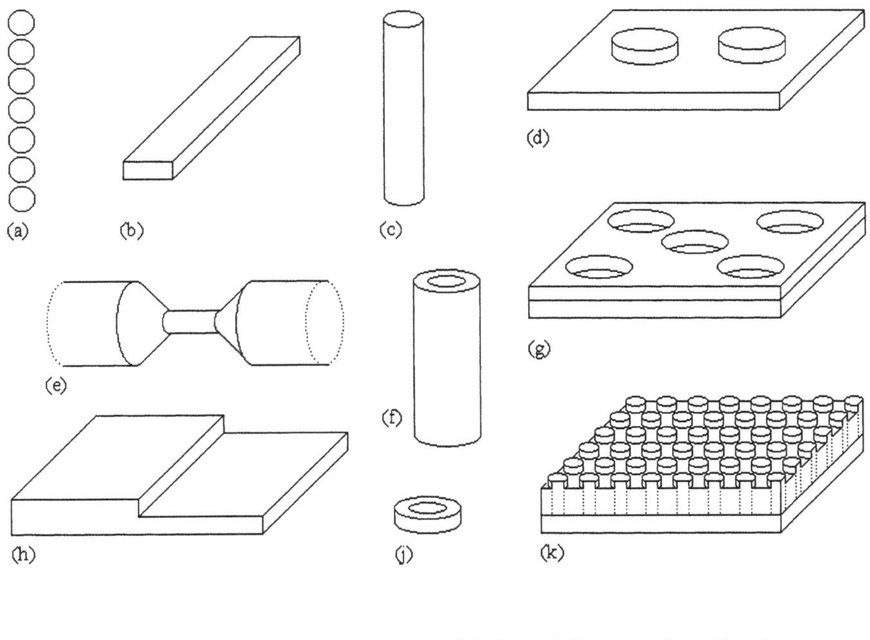

Figure 1. Typical nanostructure geometries: (a) chain of fine particles, (b) striped nanowire, (c) cylindrical nanowire, (d) nanodots, (e) nanojunction, (f) nanotube, (g) antidots, (h) vicinal surface step, (j) nanoring, and (k) patterned thin film. Note that the figures can consist of multilayered and granular nanocomposites.

2. MAGNETIC NANOSTRUCTURES

Magnetic nanostructures can be produced in a wide range of geometries. Figure 1 shows some examples. In combination with specific choices of magnetic materials for the structures – or for parts of the structures – this versatility is a major reason for interest in magnetic nanostructures. Several chapters of this book deal with the fabrication, investigation, and application of individual geometries, such as nanowires and patterned thin films. The following paragraphs briefly characterize typical geometries, mention some systems of practical or scientific interest, and provide links to the individual chapters of this book. (For references, see the Chapters 2 to 15 and the further-reading section below.)

2.1. Nanoparticles, Clusters, and Molecular Magnets

Small magnetic particles exist in nature or are produced artificially (Chs. 6 to 8). Nanoparticles have sizes ranging from a few nanometers to submicron dimensions (Ch. 7), whereas molecular magnets (Ch. 6) contain a few magnetic atoms in well-defined atomic environments. Clusters are intermediate structures, with less well-defined atomic environments but exhibiting atomic features such as facets (Ch. 8).

Examples of naturally occurring nanoparticles are magnetite (Fe_3O_4) nanoparticles precipitated in bacteria, insects and higher animals, and magnetite and other oxide particles responsible for rock magnetism. Nanobiomagnetics is concerned not only with questions such as the role of magnetite particles for horizontal and vertical orientation of animals but also with important medical issues, such as local drug administration and cancer diagnosis (Ch. 15). The small remanent magnetization of magnetic rocks, first analyzed by Néel in the 1940's, is exploited, for example, in archeomagnetic dating and to monitor changes in the Earth's magnetic field. Small oxide particles, less than 10 nm in diameter, are observed in gels having the nominal composition $FeO(OH)\cdot nH_2O$. Fine particles are also encountered in meteorites.

Some artificially produced magnetic nanoparticle structures are Fe in Al_2O_3 and so-called 'elongated single-domain particles'. Interesting applications of small particles are stable colloidal suspensions known as *ferrofluids*. A variety of materials can be used, such as Fe_3O_4, $BaFe_{12}O_{19}$, Fe, Co, and Ni, and a typical particle size is 10 nm. Most ferrofluids are based on hydrocarbons or other organic liquids, whereas water-based ferrofluids are more difficult to produce. They are used as liquids in bearings and to monitor magnetic fields and domain configurations.

2.2. Granular Nanostructures

Embedded clusters, granular materials, and other bulk nanostructures are of great importance in nanoscience. The structural correlation lengths of typical nanocomposite materials range from about 1 nm in x-ray amorphous structures to several 100 nanometers in submicron structures. Magnetic glasses and atomic-scale defect structures are beyond the scope of nanomagnetics, but they are of indirect interest as limiting cases and because nanomagnetic phenomena have their quantum-mechanical origin in atomic-scale magnetism.

Depending on grain size and microchemistry, granular nanostructures are used for example as permanent magnets (Nd-Fe-B), soft magnets (Fe-Cu-Nb-Si-B), and magnetoresistive materials (Co-Ag). There are two types of exchange-coupled permanent magnets: isotropic magnets, which exhibit random anisotropy and remanence enhancement, and oriented hard-soft composites, which utilize exchange coupling of a soft phase with a high magnetization to a hard skeleton.

Closely related systems with many potential applications are magnetic clusters deposited in a matrix. For example, the narrow size distribution of 10-20% makes this material interesting as a granular media for magnetic recording. A well-known *soft-magnetic* nanocomposite is the 'Yoshizawa' alloy $Fe_{73.5}Si_{13.5}B_9Cu_1Nb_3$, which consists of DO_3-structured Fe_3Si grains embedded in an amorphous matrix.

2.3. Particle Arrays and Functional Components

Two-dimensional arrays of nanoparticles are of interest as scientific model systems and have many present or future applications. For example, advanced magnetic recording media can be characterized as a complex array of magnetic particles, and interest in dot arrays has been sparked by the search for ever-increasing storage densities in magnetic recording. In very small dots, quantum-mechanical effects are no longer negligible, and there are phenomena such as quantum-well states. These effects are of interest in quantum computing and spin electronics.

Most easily produced and investigated are submicron dots made from iron-series transition metals, such as Ni, but it is also possible to use metallic alloys, such as Permalloy, and to reduce the dot size to less than 100 nm. The dots may form square or hexagonal arrays, or structures such as corrals. Among the investigated phenomena are the properties of individual dots and interdot interactions. A related class of nanostructures are *antidots*, that is, holes in a film rather than dots on a film. Potential applications include magnetic recording, sensors, magnetic and quantum computing, micron- and submicron-size mechanical devices, short-wavelength optics, and spin electronics. Other 'functional' building blocks are, for example, nanojunctions, spin valves, and tips for magnetic-force microscopy (MFM tips).

2.4. Nanowires

There is a smooth transition from elongated dots and thin-film patches to nanowires. Magnetic nanowires have present or potential applications in many areas of advanced nanotechnology, including patterned magnetic media, magnetic devices, and materials for microwave technology. There a various methods to produce nanowires, such as deposition on vicinal surfaces and electrodeposition, including electrodeposition into porous alumina templates (Section 3).

Much of the early work on magnetic nanowire arrays was concerned with exploratory issues, such as establishing an easy axis for typical preparation conditions, the essential involvement of shape anisotropy, as opposed to magnetocrystalline anisotropy, and the description of magnetostatic inter-actions between wires. More recently, attention has shifted towards the understanding of magnetization processes, such as the transition from curling-type to quasi-coherent nucleation, the influence of deposition-dependent polycrystallinity of typical transition-metal nanowires. Some other interesting phenomena are magnetic-mode localization, as evident *e.g.* from experimental activation volumes, spin-waves, and current-induced magnetization reversal.

2.5. Magnetic Thin Films and Multilayers

Magnetic thin films and multilayers can be classified as magnetic nanostructures, too, but it is common to treat homogeneous thin films and multilayers as a separate branch of magnetism, intermediate between nanomagnetism and surface magnetism. However, many recently developed and investigated nanostructures are thin-film nanostructures. Examples are self-assembled thin-film nanostructures (Ch. 9), patterned nanomagnetic thin films (Ch. 10), hard-magnetic thin-film nanostructures and thin films for magnetic recording (Ch. 11).

Semihard thin films are used in magnetic recording media and have, more recently, attracted attention as tools for magnetic information processing. In addition, on a length scale of a few interatomic distances, there is a variety of interesting thin-film effects, such as vicinal and interface anisotropies, moment modifications at surfaces and interfaces, thickness-dependent domain-wall and coercive phenomena, interlayer exchange-coupling, and finite-temperature magnetic ordering. A specific example is the nanoscale exchange-coupling or 'exchange-spring' effects in multilayers.

3. FABRICATION AND CHARACTERIZATION

The broad variety of magnetic nanostructures corresponds to a diverse range of processing methods. The suitability of individual methods depends on the length scale and geometry of the nanostructures. In addition, each method is usually restricted to a relatively narrow class of magnetic materials.

Granular nanostructures are produced by methods such as mechanical alloying and chemical reactions. A traditional though somewhat cumbersome method to fabricate *nanoscale particle arrays* of magnetic, dots, and wires on thin films is nanolithography. Other examples are molecular-beam epitaxy, the use of STMs, and chemical vapor deposition. The call for well-characterized large-area arrays of nanoparticles has stimulated the search for advanced production methods such as laser-interference lithography (LIL), where laser-intensity maxima effect a local transformation of a nonferromagnetic material into ferromagnetic islands. Another development is the use of focused ion-beam milling (FIB) to create small particles and particle arrays with well-defined properties.

Thin-film *nanowires* are comparatively easily obtained by depositing magnetic materials on vicinal surfaces and by exploiting structural anisotropies of the substrate. They can be produced with thicknesses down to one or two monolayers. Electrodeposition of magnetic materials into porous alumina may be used to produce regular wire arrays (see Sellmyer *et al.* 2001 and references therein). Other ways of fabricating cylindrical nanowires include deposition into molecular sieves, track-etched polymer membranes, and mica templates. By electrodeposition into porous anodic alumina it is now possible to produce hexagonal Fe, Co, and Ni nanowire arrays with diameters ranging from 4 to 200 nm, and lengths of up to about 1 μm, and variable center-to-center spacings of the order of 50 nm. The resulting materials are of interest as magnetic recording media, for optical and microwave applications, and as electroluminescent display devices. Aside from the above-mentioned iron-series transition-metal elements, there is interest in depositing alloys and multilayers, such as Fe/Pt, into porous templates.

The structural and magnetic characterization of magnetic nanostructures is the main focus of Ch. 5 and of various sections and subsections throughout the book. Structural correlation lengths can be probed for example by X-ray diffraction, small-angle neutron scattering (SANS) and electron microscopy. Magnetic measurements are performed with the methods known from bulk and surface magnetism, although some techniques must be adjusted to the small signals from certain structures. Examples are vibrating sample magnetometry (VSM), magneto-optical Kerr effect (MOKE) measurements, and SQUID magnetometry. Some methods, such as magnetic-force magnetometry, are nanospecific and presently being applied to measure hysteresis loops of nanoscale magnetic particles.

4. APPLICATIONS

Magnetic nanostructures are used in the form of traditional magnetic materials, such as hard and soft magnets, and in specific functional structures, such as sensors. Hard or *permanent magnets* are used, for example, in electromotors, hard-disk drives, loudspeakers, windshield wipers, locks, refrigerator magnets, and microphones. Some applications, such as toys, do not usually require high-performance magnets, but hard-disk drives and other high-tech applications require highly sophisticated rare-earth permanent magnets with well-defined nanostructures (Ch. 12). Compared to the highly anisotropic hard magnets, *soft magnets* exhibit very low magnetic anisotropy. They are widely used for flux guidance in permanent-magnet and other systems, in transformer cores, and for high-frequency and microwave applications, and in recording heads. In advanced soft-magnetic materials, nanostructuring is used to reduce magnetic losses by controlling anisotropy, eddy-current losses, and other properties (Ch. 13).

A key application and driving force of magnetic nanotechnology is *magnetic recording media*. They are used not only for audio-visual technology, for example in audio and video tapes, but also in computer technology, for example in hard-disks (Ch. 11). A remarkable increase in areal density of many orders of magnitude in the last two decades has relied heavily on nanostructuring of media and read and write heads.

Artificial nanostructuring is a way of creating completely new technologies. One area is spin electronics, and various types of nanostructures, such as multilayers and nanojunctions, are being used or investigated in this context (Ch. 14). The magnetoresistance of metallic thin films, granular nanostructures and magnetic oxides are exploited in sensors, and a problem of current interest is spin injection into nonferromagnetic metals and magnetic semiconductors. Other recent developments are magnetic nanostructures for quantum computing, multiferroics (where nanoscale effects are exploited to synergize electric and magnetic degrees of freedom), and nanoparticle ferrofluids for cancer treatment, guided by a magnet and delivering high local doses of drugs or radiation. Nanoscale effects are also exploited in micro-electromechanical systems (MEMS) and magnetic-force nanotips made from CoPt.

Further Reading

General Magnetism

M. J. Aitken, "Archaeological Dating using Physical Phenomena", Rep. Prog. Phys. 62, 1333 (1999).

N. W. Ashcroft and N. D. Mermin, "Solid State Physics", Saunders, Philadelphia 1976.

C P. Bean and J. D. Livingston, "Superparamagnetism", J. Appl. Phys. 30, 120S (1959).

S. Chikazumi, "Physics of Ferromagnetism", Oxford University Press, New York 1997.

D. Craik, "Magnetism: Principles and Applications", Wiley, New York 1995.

J. L. Dormann and D. Fiorani (Eds.), "Studies of Magnetic Properties of Fine Particles and their Relevance to Materials Science", Elsevier, Amsterdam 1992.

J. E. Evetts (Ed.), "Concise Encyclopedia of Magnetic and Superconducting Materials", Pergamon, Oxford 1992.

R. C. O'Handley, "Modern Magnetic Materials, Principles and Applications", John Wiley and Sons, New York 2000.

K. Moorjani and J. M. D. Coey, "Magnetic Glasses", Elsevier, Amsterdam 1984.

A. Hubert and R. Schäfer, "Magnetic Domains", Springer, Berlin 1998.

R. Skomski and J. M. D. Coey, "Permanent Magnetism", Institute of Physics, Bristol 1999.

J. M. Yeomans, "Statistical Mechanics of Phase Transitions", University Press, Oxford 1992.

Nanoscale Magnetic Phenomena

M. Bander and D. L. Mills, "Ferromagnetism of Ultrathin Films", Phys. Rev. B **38**, 12015 (1988).

D. J. Dunlop, "Developments in Rock Magnetism", Rep. Prog. Phys. **53**, 707 (1990).

J. D. Livingston and C. P. Bean, "Anisotropy of Superparamagnetic Particles as Measured by Toque and Resonance", J. Appl. Phys. **30**, 118S (1959).

A. Michels, J. Weissmüller, U. Erb, and J. G. Barker, "Measurement of a Magnetic-Field Dependent Correlation Length in Nanocrystalline Ni Using Small-Angle Neutron Scattering", phys. stat. sol. (a) **189**, 509 (2002).

D. Sander, R. Skomski, C. Schmidthals, A. Enders, and J. Kirschner, "Film Stress and Domain Wall Pinning in Sesquilayer Iron Films on W(110)", Phys. Rev. Lett. **77**, 2566 (1996).

R. Skomski, "Nanomagnetics", J. Phys.: Condens. Matter **15**, R841 (2003).

M. Ziese and M. J. Thornton (Eds.), "Spin Electronics", Springer, Berlin 2001.

Specific Magnetic Nanostructures

J. A. C. Bland and B. Heinrich (Eds.), "Ultrathin Magnetic Structures I", Springer, Berlin 1994.

R. Coehoorn, D. B. de Mooij, and C. de Waard, "Meltspun Permanent Magnet Materials Containing Fe_3B as the Main Phase", J. Magn. Magn. Mater. **80**, 101 (1989).

R. P. Cowburn, A. O. Adeyeye, and J. A. C. Bland, "Magnetic Domain Formation in Lithographically Defined Antidot Permalloy Arrays", Appl. Phys. Lett. **70**, 2309 (1997).

A. D. Kent, S. von Molnár, S. Gider, and D. D. Awschalom, "Properties and Measurement of Scanning Tunneling Microscope Fabricated Ferromagnetic Particle Arrays", J. Appl. Phys. **76**, 6656 (1994).

R. M. H. New, R. F. W. Pease, and R. L. White, "Lithographically Patterned Single-Domain Cobalt Islands for High-Density Magnetic Recording", J. Magn. Magn. Mater. **155**, 140 (1996).

D. J. Sellmyer, C. P. Luo, Y. Qiang, and J. P. Liu, "Magnetism of Nanophase Composite Films", in: "Handbook of Thin Film Materials, vol. 5: Nanomaterials and Magnetic Thin Films", Ed. H. S. Nalwa, Academic Press, San Diego 2002, p. 337-374.

D. J. Sellmyer, M. Zheng, and R. Skomski, "Magnetism of Fe, Co and Ni Nanowires in Self-Assembled Arrays", J. Phys.: Condens. Matter **13**, R433 (2001).

J. Shen, R. Skomski, M. Klaua, H. Jenniches, S. S. Manoharan, and J. Kirschner, "Magnetism in One Dimension: Fe on Cu(111)", Phys. Rev. B **56**, 2340 (1997).

R. Skomski and J. M. D. Coey, "Giant Energy Product in Nanostructured Two-Phase Magnets", Phys. Rev. B **48**, 15812 (1993).

Y. C. Sui, R. Skomski, K. D. Sorge, and D. J. Sellmyer, "Nanotube Magnetism", Appl. Phys. Lett. **84**, 1525 (2004).

Y.Yoshizawa, S. Oguma, and K. Yamauchi, "New Fe-Based Soft Magnetic Alloys Composed of Ultrafine Grain Structure", J. Appl. Phys. **64**, 6044 (1988).

Ch.-T. Yu, D.-Q. Li, J. Pearson, and S. D. Bader, "Self-Assembled Metallic Dots and Antidots: Epitaxial Co on Ru(0001)", Appl. Phys. Lett. **78**, 1228 (2001).

G. Zangari and D. N. Lambeth, "Porous Aluminum Oxide Templates for Nanometer-Size Cobalt Arrays", IEEE Trans. Magn. **33**, 3010 (1997).

M. Zheng, M. Yu, Y. Liu, R. Skomski, S. H. Liou, D. J. Sellmyer, V. N. Petryakov, Y. K. Verevkin, N. I. Polushkin, and N. N. Salashchenko, "Magnetic Nanodot Arrays Produced by Direct Laser Interference Lithography", Appl. Phys. Lett. **79**, 2606 (2001).

Chapter 2

SPIN-POLARIZED ELECTRONIC STRUCTURE

A. Kashyap
IFW Dresden
P.O. Box 27001
D-0117, Dresden, Germany

R. Sabirianov
Department of Physics
University of Nebraska
Omaha, NE 68182, USA

S. S. Jaswal
Department of Physics and Astronomy
University of Nebraska
Lincoln, NE 68588, USA

Abstract This chapter is devoted to the electronic structure of nanoscale metallic magnets. After an introduction to methods of electronic structure calculations, we review how recent trends translate into the description of magnetic nanostructures. Among the considered structures are nanowires, small particles, surfaces and interfaces, and multilayers, and emphasis is on magnetic properties such as moment and magnetization, interatomic exchange, and anisotropy.

1. INTRODUCTION

Nanostructures open new possibilities to tailor the mechanical, chemical, magnetic and electronic properties of materials and, at present, there is strong demand for basic understanding of new phenomena that nano-structures may exhibit. Nanomagnetic objects are different from both atoms and bulk materials, thereby providing an interface between physics, chemistry, material sciences, engineering and biology. For example, the length scale of typical nanostructures allows a direct use in many systems, including the human body. Some phenomena described by first principles are surface-and interface magnetic anisotropies, reduced or enhanced magnetic moments at particle or film surfaces, and optical properties, including magnetic dichroism. In addition, parameters obtained from first-principle calculations serve as an input in numerical micromagnetic simulations (Ch. 4).

Here we focus on iron-series transition-metal elements and their alloys. Magnetic oxides and rare-earth compounds can also be used in nanomagnets, but their magnetism goes beyond the scope of this review. For example, the large magnetocrystalline anisotropy of rare-earth transition-metal intermetallics, which are used in nanostructured permanent magnets, largely reflects the Hund's-rules ground state of tripositive rare-earth ions (Ch. 3).

Magnetic clusters containing a few atoms have been investigated for many decades, both theoretically and experimentally [1]. The electronic structure of atomic clusters was traditionally treated by real-space methods based on various approaches from tight-binding (TB) to density-functional methods [2, 3]. The methods of calculations of electronic structure of bulk materials use periodic boundary condition and k-space representation. They are very well developed and applied to broad variety of materials [4]. Nano-structures include 100-100000 atoms and do not exhibit periodicity, so that their first-principle study is a very challenging problem. Direct atomistic treatment of real nanodevices is very difficult due to the large number of atoms. The free-electron model, tight binding approaches and order O(N) density functional approach have all been applied to the nanoscale geometries to study their electronic properties. The choice of an approach depends on the degree of compromise between numerical and conceptual tractability on the one hand and quantitative accuracy on the other.

2. METHODS OF ELECTRONIC-STRUCTURE CALCULATIONS

Electronic structure methods for studies of nanostructures can be divided broadly into supercell methods and real-space methods. Supercell methods use standard k-space electronic structure techniques separating periodically repeated nanostructures by distances large enough to neglect their interactions. Direct space methods do not need to use periodic boundary conditions. Various electronic structure methods are developed and applied using both approaches. In this section we will shortly discuss few popular but powerful electronic structure methods: the pseudopotential method, linear muffin-tin orbital and related methods, and tight-binding methods.

2.1. Linear Muffin-Tin Orbital and Related Methods

Basis-set methods may be divided into two categories, depending on whether they (i) use wave function expansions in some set of fixed basis functions like atomic orbitals, Gaussian and plane waves, or (ii) expand the wave function in a set of energy and potential dependent partial waves as done in Korringa-Kohn-Rostoker (KKR) and the augmented plane wave (APW) methods. The KKR and APW methods require computational efforts which, despite attempts to improve the efficiency, are barely feasible and limited to sp-like valence and conduction electrons. Computationally, this can be remedied by the addition of localized orbitals to the plane wave basis set. But such a hybrid scheme is neither elegant nor in accordance with the chemical and physical intuition based on the smooth trends observed through the periodic table. It is necessary to use the self-consistent methods for computing one electron eigenvalues and eigenvectors with speed and accuracy. The so-called linear methods of band theory satisfy the requirements rather well. This is true for the linear-muffin-tin-orbital method (LMTO) [5-8]. This method is linearized version of the KKR method. Almost identical with the solid state LMTO method is the augmented spherical wave (ASW) method of Williams *et al.* [9].

The LMTO has also been extended to treating crystalline impurities with the Green's function technique by Koeing *et al.* [10], and Gunnarson *et al.* [11] have used it for both metal and semiconductor hosts. Harris [12], Casula and Herman [13] and Springborg *et al.* [14, 15] have developed the LMTO method for clusters and molecules. For surfaces and thin films, LMTO techniques have been devised by Fuziwara [16] and Fernando *et al.* [17]. It was also discovered that the conventional solid-state LMTO basis set can be transformed exactly into orthogonal [6, 7], tight binding (TB) [6] and minimal [100] basis sets, which simplifies and generalizes the solid-state LMTO method considerably.

The use of the atomic-sphere approximation (ASA) [18], where the single-electron potential is modeled as a superposition of spherical potentials inside overlapping space-filling spheres, makes LMTO methods computationally very efficient. Where this approximation is applicable, the LMTO-ASA method is presumably the most efficient procedure available for solving the density functional equations to a reasonably high degree of accuracy. However, a full-potential treatment going beyond the ASA is needed for many systems, including surfaces and impurities, and for total-energy changes associated with phenomena such as phonon distortions and atomic relaxations etc. A number of full potential methods have been developed, namely, FPLAPW [19, 20], FPLMTO [21] and FPLO [22]. A real-space version of the full-potential approach was implemented by Beck [23].

2.2. Plane-Wave Pseudopotential Methods

The pseudopotential method relies on the separation (in both energy and space) of electrons into core and valence electrons and implies that most physical and chemical properties of materials are determined by valence electrons in the interstitial region. One can therefore combine the full ionic potential with that of the core electrons to give an effective potential (called the pseudopotential), which acts on the valence electrons only. On top of this, one can also remove the rapid oscillations of the valence wavefunctions inside the core region such that the resulting wavefunction and potential are smooth.

Beyond a chosen cutoff radius, the all-electron and pseudofunctions (potential and wavefunction) are identical, while inside the core region both the pseudopotential and pseudowavefunction are smoothly varying. After the construction of these pseudofunctions for a single atom and ensuring that their scattering properties are almost identical to those of the all-electron functions, they can be used in any chemical environment.

The pseudopotential method has various advantages. Eliminating the core electrons from the problem reduces the number of particles that must be considered in the Kohn-Sham (KS) equations for the effective one-electron potential. For example, a pseudopotential calculation for bulk silicon (with 10 core and 4 valence electrons) requires the calculation of 4 occupied bands at each k-point, while an all-electron approach would require the calculation of 14 occupied bands. More importantly, the smooth spatial variation of the pseudopotential and pseudowavefunction allows the use of computationally convenient and unbiased basis, such as plane wave basis sets or grids in space.

The plane-wave pseudopotential method has proven to be an excellent computational tool for solving large scale electronic structure problems [24, 25]. Notable strengths of the method are the ability to use fast Fourier transform to update the Kohn-Sham equations, the lack of dependence on the basis on atom positions, and the clear control of convergence with the cutoff energy determined by the shortest wavelength mode. However, the method encounters difficulties in treating widely varying length scales. This issue is especially relevant for surfaces, clusters, and the hard pseudopotentials of first row elements or transition metals, which vary rapidly near the nucleus. A real-space version of the pseudopotential method was developed by Chelikowsky [26].

2.3. Tight-Binding Methods

The traditional approach to obtain the electronic structure of a periodic solid with *ab initio* pseudopotentials has been to solve the Kohn-Sham (KS) equations in momentum space using a plane-wave basis set. There are many advantages of plane-wave basis sets. For example, due to their completeness and orthonormality, they allow convenient and unbiased representation of the charge density and wavefunctions for calculating operator matrix elements. For non-periodic and localized systems, such as defects, surfaces, clusters, quantum dots, using a plane wave basis set requires an artificial periodicity in order to implement standard electronic structure algorithms. This artificial periodicity is introduced through the use of large unit cells called *supercells*.

While the supercell approach works well for localized systems, it is typically necessary to consider a very large supercell. This results in a plane-wave basis replicating not only the relevant electronic states but also vacuum regions imposed by the supercell. A much more efficient method to implement for investigating the electronic structures of localized systems is to use real space methods such as the recursion methods [27] and the moments methods [28]. These methods do not require symmetry and their cost grows linearly with the number of inequivalent atoms being considered. For these reasons, real space methods are very useful for a description of the electronic properties of complex systems, for which the usual k-space methods are either inapplicable or extremely costly.

Real-space methods are generally impractical, but they are extremely efficient when the system in consideration is well described by a tight-binding (TB) Hamiltonian. For a long time, parameterized TB Hamiltonians in conjunction with real space methods have been used to study the properties of complex transition metal alloys. In some cases, simple model TB Hamiltonians are very useful to stress the essential physics of the phenomenon in question and even to give the correct trends found in real systems. But in other cases, the effects are too subtle and the results obtained by the use of these parameterized Hamiltonians can be misleading. Even when trends can be well described, to obtain quantitative results a more exact Hamiltonians is usually needed. This has been a major problem regarding real-space electronic structure calculations. Usually, the involved parameters needed to build the TB Hamiltonian, such as hopping integrals, are obtained from a linear combination of atomic orbitals (LCAO) fit to more exact k-space calculations or are adjusted to fit experimental results [29].

When one considers complex systems, for which no exact results are available, parameters are usually obtained from a simple similar system and then transferred to the more complex system in question using the Slater-Koster rules [30] and some other empirical formulas which are known to be roughly obeyed. The results were often encouraging, but the lack of a solid theoretical background to justify the procedure left some fundamental questions unanswered. For example, how could one simulate the crystal-field splitting which depends on local environment and can not be transferred from one calculation to another, and how do we treat the wave function? The solution to these problems is not unique and several interesting points of view have developed. However, major progress towards obtaining a TB-Hamiltonian based on a solid theoretical understanding of the problem came in 1984 with the advent of the LMTO-ASA-TB formalism [31]. This gave rise to a new era regarding the quantitative description of the electronic structure of complex systems by using real-space methods.

3. MAGNETIC PROPERTIES

In this chapter we describe how basic or 'intrinsic' magnetic properties are calculated from first principles. Technological applications, for example in magnetic recording, tend to require a large magnetization, thermal stability, and control over the coercivity. The reversal process is described usually by micromagnetic modeling which takes into account dipole-dipole interactions. The description of magnetization reversal is beyond the scope of our paper. For details of micromagnetic modeling, readers are referred to Chs. 3 and 4 and to the specialized literature [31].

Magnetic properties can be understood and fully described from microscopic considerations using ab-initio electronic-structure methods. The local magnetic moment in combination with intra-atomic exchange defines the magnetization, interatomic exchange describes the thermal stability, and the coercivity reflects the magnetic anisotropy of the system. These interactions have different energy scales. The intra-atomic exchange is of the order of 2 eV, interatomic exchange of the order of 10 meV, and anisotropy of the order of 1 meV. However, all of them are extremely important to describe the behavior of a magnet.

The description of nanomagnets requires new approaches. First, nanostructures are not periodic and tend to have large surface-to-volume ratios. Because of this the magnetization is not uniform across the nanostructure, local magnetic moments differ from site to site, exchange coupling varies throughout the nanostructure, and the anisotropy can be quite different from bulk or surface anisotropies. Second, it is hard to define properties in the similar fashion as in the bulk or as in case of molecules.

3.1. Magnetic Moment

The magnetic moment is given by the difference between the filling of majority and minority spin bands, which is directly related to the ferromagnetic exchange splitting between the bands. There is also an orbital-moment contribution to the moment, but in 3d-based magnets, it is often small due to quenching (Ch. 3). Total energy minimization over all occupied band states yields crystal structure and magnetic ordering. Magnetic coupling and magnetic moment are sensitive functions of the interatomic spacing and the atomic volume, respectively. Both are dominated by the Coulomb and by exchange interactions at small distances. The moment is determined primarily by intraatomic exchange interactions and the Curie temperature by the interatomic exchange interactions.

The most common approach to calculate the magnetization from first principles is to use the local density formalism, where the spin dependent part of the total energy is approximated by an expression derived for the electron gas [32]. The corresponding exchange-correlation potential depends only on the majority and minority spin densities. Different magnetic structures, such as paramagnetic, ferromagnetic and various antiferromagnetic configurations, exhibit different total energies, and the lowest energy configuration obtained by local density theory represents the observed magnetic state in most cases. Room-temperature ferromagnetism occurs in rather few elements, Fe, Co, Ni, Gd, and several other rare earths are ferromagnetic at low temperatures. However, many alloys are ferromagnetic. The stability of the ferromagnetism of itinerant magnets can be explained by the *Stoner criterion*, which takes the DOS at the Fermi level, $D(E_F)$, and an atomic exchange integral I as input [33-35]. The transition from paramagnetism to ferromagnetism is favourable for $D(E_F)\, I > 1$. Then the system can lower its energy by bringing enough majority spin electrons down in energy by opening up the ferromagnetic exchange splitting. The Stoner criterion explains why Fe, Co and Ni are singled out for ferromagnetism. Several other elements are close to fulfilling the criterion, for example Pd. Since the DOS is different for different atomic sites, nanostructures may exhibit different local moments and complicated magnetic transitions.

With increasing atomic volume, one approaches the free atom limit where Hund's first rule postulates maximum spin, so that the individual spins of the electrons in a shell are aligned parallel. More generally, Pauli's exclusion principle implies that electrons with parallel spins have different spatial wavefunctions, reduces the Coulomb repulsion and is seen as exchange interaction. When the atoms are squeezed into a solid, some of the electrons are forced into common spatial wavefunctions, with antiparallel spins and reduction of the overall magnetic moment. At surfaces and interfaces, the reduced coordination reverses this effect, and a part of the atomic moment is recovered.

3.2. Anisotropy

Magnetic anisotropy is one of the most important properties of a magnetic material. As mentioned in Ch. 1, it is the energy necessary to deflect the magnetic moment in a single crystal from the easy to a hard direction. The main contribution to the anisotropy is *magnetocrystalline anisotropy*. The easy and hard directions arise from the interaction of the spin magnetic moment with the crystal lattice that is due to spin-orbit coupling. In crystals, the magnetocrystalline anisotropy energy is given by a series expansion in terms of the angles between the direction of magnetization and the crystal axes (Ch. 4). It is often sufficient to represent the anisotropy energy in an arbitrary direction by just the first two terms in the series expansion. These two terms each have an empirical constant associated with them called the first- and second order uniaxial anisotropy constants K_1 and K_2, respectively. In cubic crystals, the lowest-order term is the fourth-order term, also known as the 'cubic' K_1.

There is also some shape anisotropy due to magnetostatic dipole interactions (Ch. 3). This anisotropy is important in some nanomagnets, for example in elongated nanoparticles, but unrelated to the electronic structure.

A contribution caused by spin-orbit coupling and closely related to magnetocrystalline anisotropy is *magnetoelastic anisotropy*. Mechanical stress creates a strain which amounts to a lattice distortion and yields a correction to the magnetocrystalline anisotropy. Surface anisotropy is a manifestation of magnetocrystalline anisotropy, too (sections below and Ch. 3).

Starting with Brooks (1940), calculations of the magnetocrystalline anisotropy energy (MAE) of itinerant magnets have developed in two different approaches have been developed since the pioneering studies of Takayama, Bohnen, and Fulde [36] and of Gay and Richter [37]. The first approach uses perturbation theory within a semi-empirical tight-binding framework, justified by the argument that the MAE is small compared to a characteristic bandwidth [36, 38, 39, 40]. The second approach relies on *ab initio* spin-polarized total-energy calculations, with spin-orbit coupling included either self-consistently within the scalar-relativistic approximation [41, 42] or as a final perturbation to a calculation neglecting spin-orbit coupling [37, 43-45] and using the force theorem [46, 47]. The main difficulty with this approach arises from the fact that a change in the direction of the magnetic moment changes the occupation of the eigenstates only in a narrow region of the Brillouin zone, hence the *k*-space integrations for the total energy are only very slowly convergent. To avoid the necessity to use a huge number of *k*-points, Wang, Wu, and Freeman [45] introduced a state tracking procedure, using information on the change of the band structure with increasing spin-orbit interaction to extrapolate the Brillouin-zone integrals.

The description of the magnetic anisotropy of nanoparticles is more complicated than that of bulk systems, partially because shape and surface anisotropies (Ch. 3) are of the same order of magnitude as the magnetocrystalline anisotropy. The ratio of atoms on the surface to the number of atoms in the core of a nanoparticle is very large. This changes the electronic structure of the core atoms as well and its anisotropy may be quite different from the bulk systems. Surface anisotropy may cause drastic change in magnetization and reverse properties of nanostructures. Even in a naive magnetostatic approach, surface anisotropy causes the renormalization of switching parameters [48].

3.3. Spin Structure and Noncollinearity

The term spin structure refers to the ferromagnetic, antiferromagnetic or ferrimagnetic alignment of the atomic spins. In ferromagnets, all spins are parallel and the atomic moments add, whereas ferromagnets and antiferromagnets are characterized by two or more sublattices with opposite moments. This leads to a reduction or absence of a net magnetic moment. Competing exchange interactions in periodic crystals and in disordered magnets give rise to noncollinear spin arrangements. Examples are helimagnetic order in perfect crystals, which is caused by the competing interactions between next and more distant neighbors, and spin-glass behaviour in magnets with atomic scale disorder [49]. Deviations from parallel or antiparallel spin alignment may also occur at surfaces and interfaces. Further, it is important to distinguish between zero temperature magnetism and finite temperature magnetic order.

It is difficult to calculate finite-temperature magnetic properties from first principles, particularly Curie and Néel temperatures. In the atomic limit, for example using the Heisenberg model, one assumes a magnetic moment on each atom that persists beyond Curie temperature. In the paramagnetic phase, the magnetization vanishes by orienting the moments at random. The other extreme, that is the Stoner model of band like ferromagnetism, assumes that the magnetic moment gradually decreases at elevated temperature and vanishes at the Curie temperature. It yields Curie temperatures that are several times higher than observed [35]. The basic drawback is the assumption that the magnetic order disappears by spin flip excitations across the Stoner gap. They cost more energy than is actually needed to disorder magnetic moments. More realistic models contain small domains with local order whose orientations are starting to fluctuate when approaching the Curie temperature (Ch. 3). Some degree of local correlation remains even above T_C. Such a picture brings in spin wave excitations (magnons) which are not contained in the ground state energy bands. The efforts to model the spin structure just below and above the Curie point from first principles give qualitatively correct Curie temperatures [50], even if the method used in the paper is more appropriate to very weak itinerant ferromagnets such as $ZrZn_2$ (Ch. 3)

3.4. Interatomic Exchange

Magnetic order occurs because the magnetic moments of neighbouring atoms couple by interatomic exchange. As the intra-atomic exchange responsible for the magnetic moment, it reflects the overlap of electrons in combination with Pauli's exclusion principle. Depending on the positive or negative sign of the exchange, the interaction between the ions will force the individual moments into parallel (ferromagnetic) or antiparallel (antiferromagnetic) alignment. The direct exchange between overlapping orbitals was proposed first. Later other types of interatomic exchange were identified such as an indirect or RKKY-type exchange, superexchange, double exchange, anisotropic exchange, and higher-order exchange (such as biquadratic). All of them may be present in a system and responsible for a variety of effects governing the magnetic structure and finite-temperature behaviour of magnet.

Direct exchange operates between moments which are close enough to have sufficient overlap of their wavefunctions. It gives a strong but short range coupling which decreases rapidly as the ions are separated. Direct exchange as originally envisaged by Heisenberg, is always ferromagnetic, but there is a negative hopping contribution. When the atoms are very close together, the electrons minimize the single-electron interaction by spending a considerable time between the nuclei. Since the electrons are then at the same place in space at the same time, Pauli's exclusion principle requires them to possess opposite spins. This gives rise to antiparallel alignment, corresponding and therefore negative or antiferromagnetic exchange. Note that the direct intra-atomic exchange between electrons in the same atom J is positive, leading to Hund's first rule.

Indirect exchange couples moments over relatively large distances. It is the dominant exchange interaction in metals where there is little or no direct overlap between neighbouring magnetic electrons. It therefore acts through an intermediary which in metals is the conduction electrons (itinerant electrons). This type of exchange is also known as the Ruderman, Kittel, Kasuya and Yoshida (RKKY) interaction, especially when mediated by free electrons. This spin polarization in the itinerant electrons is felt by the moments of neighboring magnetic ions within range, leading to an indirect coupling. Since RKKY exchange oscillates from positive to negative as the separation of the ion changes, the magnetic coupling can be ferromagnetic or antiferromagnetic. In rare-earth metals whose magnetic electrons in the 4f shell are shielded by the 5s and 5p electrons, direct exchange is rather weak and insignificant and indirect exchange via the conduction electrons gives rise to magnetic order in these materials.

A method to calculate J_{ij}, based on the local approximation to spin density functional theory has been developed by Liechtenstein *et al.* [51, 52]. Using spherical charge and spin densities and a local force theorem, expression for J_{ij} is

$$J_{ij} = \frac{1}{4\pi} \sum_{LL'} \mathrm{Im} \int_{-\infty}^{\varepsilon_F} d\varepsilon\, \Delta^i_l(\varepsilon) T^{ij\uparrow}_{LL'}(\varepsilon) \Delta^j_{l'}(\varepsilon) T^{ij\downarrow}_{LL'}(\varepsilon) \quad (1)$$

Here $T^{ij\sigma}_{LL'}(\varepsilon)$ is the scattering path operator in the site (i,j) representation for different spin projections ($\sigma = \uparrow,\downarrow$), and $\Delta^i_l(\varepsilon) = t^{-1}_{i\uparrow} - t^{-1}_{i\downarrow}$ is the difference of the inverse single-site scattering matrices. The total exchange of a given site with all sites ($J_o = \sum_{i\neq 0} J_{oi}$) can also be calculated from the relation

$$J_0 = \frac{1}{4\pi} \sum_{LL'} \mathrm{Im} \int_{-\infty}^{\varepsilon_F} d\varepsilon \{\Delta^0_l(\varepsilon)[T^{00\uparrow}_{LL'}(\varepsilon) - T^{00\downarrow}_{LL'}(\varepsilon)] + \Delta^0_l(\varepsilon) T^{00\uparrow}_{LL'} \Delta^0_{l'}(\varepsilon) T^{00\downarrow}_{LL'}(\varepsilon)\} \quad (2)$$

The parameter J_o reflects the energy change due to small-angle rotation of the moment at one site. In contrast to the J_{ij}, it is given by the site-diagonal scattering matrix (or Green function), where $i = j$. This approach has been used to study the magnetism of Y-Fe, Y-Co and $L1_0$ systems [53-55].

4. ELECTRONIC STRUCTURE OF NANOMAGNETS

Recent advantages in the preparation and characterization of thin films, surfaces and nanostructures have lead to an intriguing question: is it possible to fabricate ‘designer solids’ by controlling materials on the atomic scale, that is layer by layer, row by row, and ultimately atom by atom? Engineered molecules are common place in biochemistry, and the same idea can be apply to solids and electronic materials. Electronic properties of semi-conductor devices have been controlled by heterostructures, quantum wells and superlattices. Magnetism as a cooperative phenomenon also lends itself to manipulation in small structures, where neighboring atoms can be replaced systematically by species with stronger or weaker magnetism. In fact, a class of magnetic/nonmagnetic multilayers termed ‘spin valves’ has been introduced into magnetic storage devices (Ch. 14). The wave functions of electrons undergo changes on confining them to dimensions comparable with their wavelength. In semiconductors, confinement and quantization phenomena are visible already at dimensions greater than 200 nm [56, 57], whereas in metals they typically are seen at 1 nm. In fact, the Fermi wavelength of typical metals has atomic dimensions, but beat frequencies with the lattice can be an order of magnitude larger.

Consider the formation of low dimensional electronic states by quantization. Confining electrons to small structures causes the continuous bulk bands to split up into discrete levels, for example quantum well states in a slab. For N atomic layers in the slab there are N levels. In order to exhibit two dimensional behavior there should be only a single level within about 6 k_BT of the Fermi level. Several levels within the Fermi cutoff would already approach a three dimensional continuum. For a coarse estimate of the corresponding slab thickness, one may set the energy E of the lowest level equal to k_BT. For room temperature ($E = k_BT = 0.026$ eV), one obtains a de Broglie wavelength $h/p = h/(2mE)^{1/2} \sim 8$ nm, which is comparable with the spatial extent of the lowest quantum state. Thus, both the high electron density in magnetic metals and the requirement of room temperature operation for quantum devices point to dimensions of a few nanometers. Note, however, that Fermions are quantum-mechanical objects at any temperature below E_F/k_B.

The contribution of first-principle calculations to physics, chemistry, materials science, and, recently, geology and biology is more important than ever. The main reasons are the steady increase in computer power and the continuous progress in methodology (both efficiency and accuracy of algorithms and approximations). However, the application of these methods to investigate structural, electronic, and optical properties of nanostructures has not been straightforward due to extensive computational needs and to the new physics inherent in the nanometer and subnanometer region. One particularly useful extension to overcome the computational demand imposed by localized nanostructures has been the introduction of methods based on a real-space implementation, such as the higher-order finite-difference pseudopotential methods. In the most favorable cases, quantum mechanical formulations of the electronic structure of atomic systems scale as the cube of the number of atoms of the system. This makes it very difficult to reach system sizes larger than a few hundreds of atoms, and is therefore a huge barrier for the study of problems in nanoscale materials. The way out of this trap was found several years ago, when a number of ideas suggested the possibility of developing approximate although accurate schemes to reduce the computational cost to linear scaling. These so called $O(N)$ [58] methods have matured since those first proposals, and now constitute a viable route for studying systems with unprecedented size.

5. FIRST-PRINCIPLE STUDIES OF NANOSTRUCTURES

Recent advances in theoretical methods and high-performance computing allow for reliable first principle calculations of complex nanostructures. Nanostructured materials are characterized by a fascinating diversity of geometries, but here we restrict ourselves mainly to first-principle calculations for nanoparticles and clusters, nanowires and nanocontacts. Nanoscale multilayers are also discussed very briefly, although multilayers are often considered as a subfield of thin-film physics rather than nanoscience. We also ignore nanotubes, because most of the work in this direction has been done on nonmagnetic carbon nanotubes.

5.1. Nanoparticles and Clusters

In magnetic recording, a bit of information corresponds to a small area or volume of ferromagnetic material with a well-defined and as high as possible magnetic moment. Furthermore, below a given size the grains tend to become superparamagnetic at room temperature, and the stored information would become unstable. It is known, both experimentally and theoretically, that the magnetic moments and the magnetic anisotropy energy per atom are often enhanced in small metallic clusters, as compared to the bulk crystalline state. Nanosize transition-metal particles, between the molecular and metallic states, are of particular interest in this context due to their intrinsically high magnetic moment. The properties of these clusters depend strongly on its surrounding. We will compare magnetic properties of free-standing clusters, clusters supported on substrate, and clusters embedded in the media.

Studies of small free-standing FeN, CoN and NiN clusters have revealed that their magnetic moments are significantly larger than the corresponding bulk magnetizations [59-65]. Non-vanishing magnetic moments have even been observed in small clusters of some 4d transition metals, for example Pd, Ru and Rh, which are non-magnetic in the solid state [66-70]. For example, Kumar *et al.* performed ab initio calculations on the growth and magnetic behavior of Pd clusters having up to 147 atoms and found an icosahedral growth and a ferromagnetic behavior in these clusters [68].

The magnetization of free clusters of elemental metals exhibits an oscillatory behavior with increasing cluster size. The origin of these oscillations is the successive addition of atomic shells with increasing cluster size. The variation of the magnetic moment in the outermost shell is due to the changing coordination number of surface atoms. As a rule, surface atoms with lower coordination (lower number of nearest neighbors) have larger moments, and the corresponding theory of magic numbers seems to explain experimental observations.

Using a tight-binding Hubbard Hamiltonian in the unrestricted Hartree-Fock approximation, Pastor *et al.* [71] have shown the size and structural dependence of magnetic properties of free Cr_n ,Fe_n , and Ni_n ($n < 15$) clusters, and also the exchange interaction and local environments effects in Fe_n clusters [72]. The finite-temperature magnetism of small clusters, which is remarkably different from that of bulk systems, has also been studied in terms of a similar approach by taking into account both electronic and structural excitations [73]. A great advantage of the TB methods is that they easily can be combined with molecular dynamics calculations, enabling the investigation of relaxation effects which proved to be important in determining the magnetic moments and the MAE of transition metal clusters [74-76].

As mentioned above, the anisotropy of the free standing clusters is very complicated. Experimental studies predict a very complicated switching behavior, suggesting a strong competition between different contributions to the anisotropy energy [77, 78]. The main reason is that different pairs of spin-orbit split levels yield anisotropy contributions of opposite sign, so that the net anisotropy is the sum of many positive and negative contributions. Phenomenological studies complement this point and yield good agreement with experiment [79, 80].

The supercell-based studies of supported clusters show that the magnetic moment tends to increase as well [81]. For example, small Co nanoclusters may exhibit a large orbital moment and large anisotropy on Pt surface [82]. The pros and cons of large orbital moments in applications of nanostructures was discussed by Andriotis *et al.* [83]. Magnetic-nanoparticle patterned media are extensively studied as a potential breakthrough technology for high-density storage of information. The key requirement is that the individual nanoparticles exhibit both a large magnetic moment and a high anisotropy. More generally, in the last decades, theoretical and experimental studies have shown that the reduction of size and dimensionality can give rise to a large range of novel materials. This has opened the possibility of generating specifically designed nanomaterials with tailored properties. Understanding the conditions for the development of such properties is crucial in order to be able to properly adjust and control these parameters, in particular for magnetic applications.

The embedding technique based on the Korringa-Kohn-Rostoker (KKR) Green-function method in the local spin-density approximation (LSDA) has been applied to the magnetism of transition metal adatoms and clusters deposited on surfaces. The main feature of this approach is that the interaction between adatoms and host surface atoms can be analyzed within first-principles electronic structure calculations [84, 85]. An accurate calculation of the total energy in terms of full potential or full charge density schemes made possible the investigations of the energetics of adatoms

[86-88]. As compared to TB methods an obvious drawback of the embedded KKR technique is that, with respect to computational limitations, the number of the atoms in the cluster is restricted to about less than 100. Furthermore, the inclusion of structural relaxations is exceedingly difficult. On the level of a fully relativistic spin-polarized electron theory, recently, strongly enhanced orbital magnetism and MAE of adatoms and small clusters on Ag and Au (100) surfaces have been reported [89, 90].

Spin polarized electronic structure calculations of the Co clusters embedded in Cu matrix have been reported by Qiang *et al.* [91]. The authors found that the magnetization of the embedded Co clusters in Cu matrix is somewhat lower than the magnetization of the bulk Co, i.e. the effect is opposite to the free cluster case. This is caused by the slight suppression of the moment of cobalt sites in the interface region. The magnetization oscillations are similar to the free case in theory if shell by shell filling of the clusters is expected. However, this condition is not satisfied in case of embedded clusters [92] and magic number theory does not work. Moreover, there is always some interdiffusion on the interface between atoms of the matrix and atoms of the cluster in case of larger clusters. Frequently it creates a magnetically dead layer at the interface.

The exchange interactions between metallic clusters in metallic media exhibit oscillatory dependence on the intercluster distance like it happens in Co clusters embedded in the Cu matrix (see also Ch. 3). The calculated inter-cluster exchange interactions shows both ferromagnetic and aniferromagnetic exchange between clusters, and the values of exchange at short distances is of the order of interatomic exchange in the bulk cobalt. The spin-glass type behavior in this and similar systems arises from frustration. This is very likely responsible for the lowering of the magnetization with the increase of cobalt concentration from 10 to 50% [91].

Nanostructured permanent magnets use the idea of mixing soft magnetic materials and hard magnetic materials in a composite structure. The goal is to combine a large magnetocrystalline anisotropy of hard magnetic materials and large magnetization of the soft magnetic materials in order to obtain large energy product $(BH)_{max}$, which requires large magnetization and large coercivity. This idea was first proposed by Kneller and Hawig [93], by estimating the effect in multilayers. Micromagnetic calculations based on parameters obtained from first principles show that FePt/Fe nanostructures may exhibit a $(BH)_{max}$ as high as 90 MGOe [54]. These changes are experimental observed in various systems [94], and Section 2.3 in the following chapter discusses some systems.

The properties of embedded nanoclusters can be controlled by the properties of the matrix. For example, the matrix can exert uniaxial or biaxial strain on the embedded nanoclusters. The corresponding strain can modify the properties of nanostructure considerably. For example, Zeng *et al.* shows that applied stress can modify the Curie temperature and anisotropy of FePt [95].

5.2. Nanowires

The progress in experimental techniques has made it possible to synthesize stable ultrathin metal nanowires with diameter down to atomic sizes and of high aspect ratios [96-99]. An example is single-crystalline silver nanowires having a width of 0.4 nm width and deposited into the pores of an organic template [99]. Kondo *et al.* [100] have observed novel helical multishell structures in the suspended ultrathin gold and platinum nanowires. This is a relatively mature field of studies with thousands of articles published in the last few years, and there are various review articles available [101, 102]

While magnetic properties of free-standing nanoclusters have been extensively studied experimentally, studies of magnetic nanowires are largely limited to theoretical calculations. One exception are wires freed from templates, where mechanical stability excludes very thin diameters. Decades ago, Weinert and Freeman [103] investigated the electronic structures and magnetism of the linear Fe and Ni chains by using the self-consistent full-potential linearized augmented plane wave (FLAPW) method. They predicted large magnetic moments of 3.3 μ_B and 1.1 μ_B for the Fe and the Ni chains, respectively. Free-standing Co atomic chains show an enhanced magnetization of 2.3 μ_B as well. [104]

It is interesting that first-principle calculations predict ferromagnetism in monatomic nanowires of the nonmagnetic transition metals Ru, Rh, and Pd, with respective mean-field moments of 1.1, 0.3, and 0.7 μ_B at the equilibrium bond length [105]. An analysis of the band structures indicates that the nanocontact superparamagnetic state suggested by calculations should affect the ballistic conductance between tips made of Ru, Rh or Pd, leading to possible temperature and magnetic-field dependent conductance.
One method to produce nanowires is the growth on the steps of the crystalline surface cut with the high Miller index [106]. Spišák and Hafner [107] investigated the position dependent magnetic moments of the Fe nanowire on a vicinal Cu(117) by using the pseudopotential method. From the total energy comparison, they found that the system of the Fe nanowire placed on the inner corner site of the Cu(117) step is most stable with a magnetic moment of 2.80 μ_B/atom, while the system of the Fe nanowire placed on the top of the terrace (formed by (001) plane) of Cu(117) is unstable with a magnetic moment of 2.96 μ_B/atom. Jin *et al.* [108] investigated the electronic structures and magnetism of an Fe nanowire along the [010] direction on a Cu(001) surface (Fe [010] chain) by using the FLAPW method and found large magnetic moments of the Fe atoms: 3.11 and 3.39 μ_B/atom for the Fe [010] chain and the free-standing nanowires (free-

Fe[010] chain), respectively. Since free-standing nanowires of elemental metals are one-dimensional objects, they exhibit magnetic properties similar to free standing clusters, i.e. its magnetic moment increases especially for the atoms at surface. The lower coordinated atoms have higher moment, while atoms inside of the nanostructure have magnetic moment close to the bulk values.

Recently, investigations of the electronic structures and magnetism of an Fe chain along the [110] direction on a Cu(001) were performed by using the FLAPW method within the generalized gradient approximation [109]. Hong *et al.* [104] have calculated properties of Co atomic wires on the Cu(001) surface. They show that magnetic moment of atomic size wires can be affected by the substrate. The magnetocrystalline anisotropy of Co wires supported on Cu substrate has opposite sign from the free standing case, i.e. the magnetization is along the wire on the Cu surface and perpendicular in free-standing case. Gambardella *et al.* [110] investigated a quasi-one-dimensional Co chain at the Pt(111) step edge. The symmetry breaking at the step leads to an easy magnetization axis at an odd angle of about 20 degrees towards the Pt step. The orbital and spin moment may have different orientation in this system. This reflects the specific character of the crystal-field interaction in these systems.

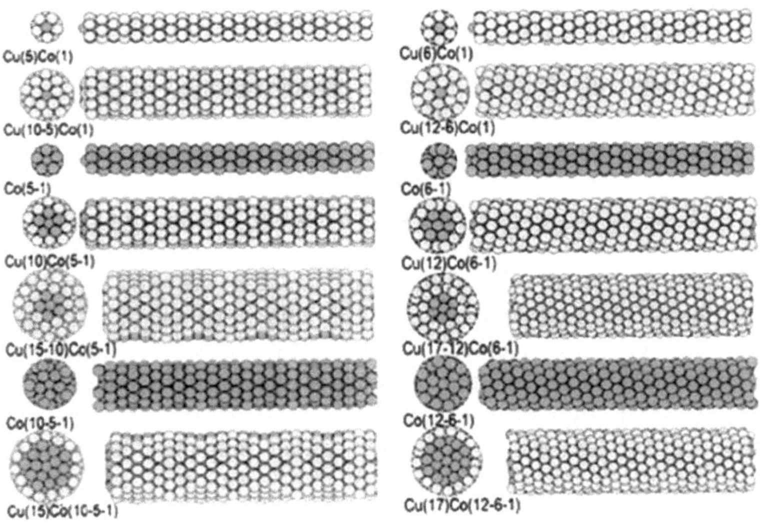

Figure 1. Optimized structures of bare and Cu-coated Co nanowires (dark balls: Co atoms; bright balls, Cu atoms). Left part is for the centered hexagonal multishell structural pattern and right one for the centered pentagonal one. The indexes of n-n1-n2-n3 in parenthesis are used to characterize each structure [111].

It is difficult to tackle the problem of embedded nanowires. An example is Co nanowires fully coated with atomic Cu shells. Wang *et al.* [111] find that Cu atoms occupy the surface, while Co atoms prefer to occupy the interior of the nanowires to form the perfect-coated structures. The coated Cu atomic layer leads to a large variation of the magnetic moment of the Co nanowires, depending on the structure and the thickness of Cu layers. A single row of Co atoms in the center of nanowire is found to be nonmagnetic when coated with two Cu layers, while all other investigated Co nanowires (Fig. 1) remain magnetic, albeit with reduced magnetic moments.

The transport properties of nanowires are of technological importance and have attracted significant attention in the recent years. Band structure gives simple solution to the analysis of the ballistic transport of periodic nanowires because the number of the bands crossing the Fermi surface is equal the number of quantum of conductance. However, the situation in nanocontacts is more complicated [112].

5.3. Nanocontacts

The understanding of spin transport in order to use it in electronics devices has been a field of intense research for the past few years. Interesting effects were reported in electrodeposited Ni nanocontacts, where resistance changes by a huge value on application of a small magnetic field at room temperature [113, 114]. This has followed after effort was focused on the magneto-resistance effect of electrodeposited nanocontacts [115-117]. Smaller magnetoresistance ratios, but interesting from the point of view of applications, were reported also for Co, Fe and Fe_3O_4 nanocontacts [118].

There have been several and partially exclusive attempts to explain the experimental data. One possibility considered is the scattering of the electrons in a domain wall [117-119]. The width of a domain wall in a nanocontact is predicted to be of the order of the nanocontact size [120] and such a reduction of the domain wall width is likely to enhance the magneto-resistance. However, this effect is very small, unless the domain walls contain atomically sharp feature, especially layers with reduced exchange [102]. Magnetostriction effects could also lead to an increased magneto-resistance [118], but most probably cannot explain the experiments. Alternatively, a thin domain wall pinned in a magnetic dead layer together with the possibility of 100% polarization of the Ni d electrons was also proposed as a possible explanation [114, 115]. Using a free-electron model and a linear response theory Zhuravlev *et al.* [121] investigated a spin-dependent electronic transport in magnetic nanocontacts in the ballistic regime of conduction. In particular, they show that the presence of a nonmagnetic region in the nanocontact separating two ferromagnetic electrodes can lead to a spin blockade resulting in very large values of magnetoresistance. Other possible explanations are a large magnetostriction in the experimental samples, or the presence of the defects in the constriction. The latter may result in the resonant tunneling and provide large MR.

In order to describe the nanocontacts theoretically, several methods have been developed by various groups. Recently, Burton *et al.* [122] investigated the electrical resistance of a constrained DW in a nanocontact using micromagnetic modeling and ballistic conductance calculations. They found that for electrodes with significantly different magnetic anisotropy, the DW can be trapped by the nanoconstriction (domain-wall pinning). They showed that the DW width and, therefore, the conductance can be controlled by an applied magnetic field. The first-principle approach is based on the tight-binding (TB) formulation [123-125]. An important conclusion coming from TB models is that conductance of the single-atomic contacts is proportional to the number of valence orbitals available at the Fermi energy [124, 126]. Ab-initio density functional (DF) calculations with jellium electrodes have been proposed by Lang *et al.* [127-130] and Kobayashi *et al.* [131-133], which addressed single-atom contacts [128,134], atomic chains of Al [127, 131], Na [128, 133], and C [130].

The formation mechanisms of atomic chains made from different type of elements such as Ni, Pd, Pt, Cu, Ag and Au were studied by means of molecular dynamic simulations [135]. Also, Mehrez *et al.* [136] and Brandbyge *et al.* [137] presented fully self-consistent DF calculations of the conductance of atomic contacts treating the electronic structure of the whole system (electrodes and the constriction region) on the same footing. The ballistic conductance through Ni, Co, and Fe nanocontacts within a semiempirical tight-binding model proposed by Velev and Butler [138] shows that the interplay of the contact and the domain-wall resistance can produce very large giant magnetoresistance ratios. However, the very giant magnetoresistance is limited to geometries in which the nanocontacts are very narrow and have very small aspect ratios, similar to the predictions in [102]. Solanki *et al.* [112] calculated the electronic structure and conductance of atomic-size Ni contacts using a real-space tight-binding LMTO method and recursion technique within the frame work of density functional theory. They used the Landauer–Buttiker approach to calculate the conductance. Their results of spin-dependent conductance as a function of energy show ballistic bulk-like behavior. The appreciable dependence of the conductance on the structure of nanocontact is reported in their work. Figure 2 shows an example of nanocontact modeling. Similar conclusion was made by Emberly *et al.* considering carbon nanowires [139].

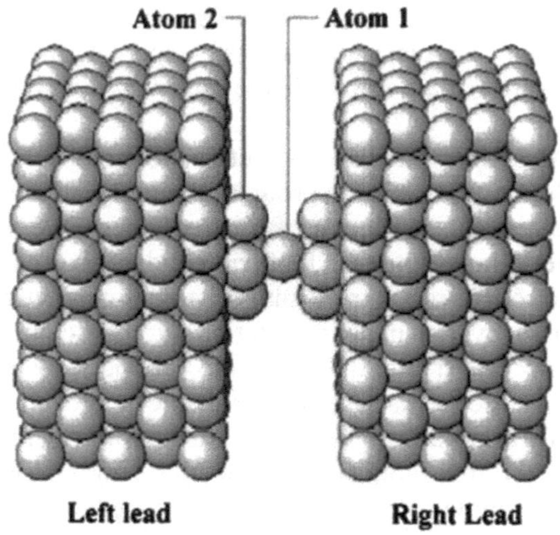

Figure 2. Geometry of a Ni nanocontact used in calculations. Atom of type 1 and atoms of type 2 occupy two non-equivalent sites within the nanocontact region [112].

We should notice that there are several research groups which could not reproduce giant MR in the nanocontacts experimentally.

5.4. Multilayers, Surface, and Interfaces

Magnetic thin films and multilayers have attracted much interest in the context of nanomagnetism, although they are often considered as a separate branch of condensed matter physics [140]. Ultrathin magnetic films, epitaxially grown on nonmagnetic substrates, are prototype systems for investigating magnetism in two dimensions, and the possibility to produce films and multilayers with perpendicular magnetic anisotropy for magnetooptic recording applications is of technological importance [141, 142]. The topic of perpendicular magnetic anisotropy is very challenging, since the magnetostatic dipolar interactions always prefer in-plane orientation of the easy axis of magnetization. Since the pioneering experimental work by Gradmann [143] and the theoretical predictions of Gay and Richter [38] much effort has been spent investigating Fe, Co, and Ni films grown on noble-metals substrates. Among the heteroepitaxial systems that show strong perpendicular magnetic anisotropy (PMA) are Co/Au(111) thin films and superlattices [144], Ni/Cu(001) [145-147] and Ni/Cu(111) [146] thin films, Co/Au(111) thin films and multilayers [144, 148], bcc Fe/Ag(001) [149, 150] and fcc Fe/Cu(001) [151, 152] thin films. Sabiryanov and Jaswal carried out first-principles studies of exchange-coupled hard-soft multilayers predicting very large energy products needed for hard-magnet applications [153].

The discovery of the giant-magnetoresistance (GMR) effect [154] has further renewed interest in the theory of transport in layered and granular systems. Several approaches to transport in layered systems are currently being used. The phenomenon of GMR is promising for its applications, such as in nonvolatile magnetic random-access memories in the information technology industry or as reading heads and various kinds of sensors in the recording and car industries, respectively. The GMR effect has been observed mostly in the diffusive transport regime in which the mean free path is much smaller than the dimension of the so-called active part of the multilayer system. In the ballistic regime, the mean free path is larger than the dimension of the active part of the multilayer system. Transport in multilayers has been studied extensively and review on the subject [155] covers the topic thoroughly.

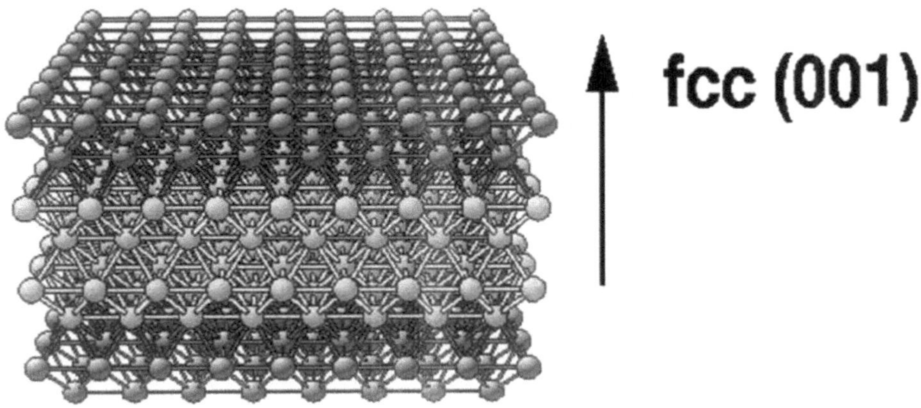

Figure 3. The modeling of layered structures for the calculational investigation [160].

The magnetocrystalline anisotropy of multilayered systems has been also calculated by many groups [156-158]. Újfalussy *et al.* [159] performed fully relativistic spin-polarized local spin density calculations for Co monolayer on Au(111). They obtained an enhancement of perpendicular magnetic anisotropy as a function of the Au coverage. They showed the close relationship between the anisotropies of orbital magnetic moments and the anisotropy energies, thus interpreting their results in terms of familiar perturbation theory. Lorentz and Hafner [157] presented calculations of the magnetic structure and of the uniaxial and planar anisotropies of thin films of Fe on Cu(001) substrates. They have used the real space recursion calculations using TB-LMTO to calculate the anisotropy. Their results for Fe monolayers reveal the importance of the surface and interface anisotropies, including the effect of nonmagnetic coverages. Essentially, this is a crystal-field effect, similar to that in bulk compounds consisting of atomic layers with different magnetic properties. Stepanyuk and Hergert have [160] performed molecular dynamics simulations of Co nanostructures on Cu(001) and Cu(111) surfaces. They first performed ab initio calculations of the electronic structure in the framework of density functional theory and then used the results for dynamical simulations. Figure 3 shows a layered structure investigated in [160].

6. SUMMARY

One of the goals of computational nanoscience is to calculate physical and chemical properties from first principles. This requires the knowledge of the electronic structures of the materials system in question. The density-functional theory (DFT) makes a huge step towards this goal, by reducing a highly complex many-electron problem to that of noninteracting particles under the influence of an effective potential. The adiabatic approximation allows one to separate the ionic degrees of freedom from those of the electrons. However, in order to apply DFT in practice one has to resort to approximations for electron exchange and correlation such as the local-density approximation (LDA) or the generalized-gradient approximation (GGA). In this chapter, we have outlined some basic principles and some important examples of this approach, representing a very large and growing body of experimental and theoretical work on magnetic nanostructures such as clusters, nanowires, nanocontacts, and multilayers.

References

[1] P. Jena, S. N. Khanna, and B. K.Rao, Materials Science Forum **232,** 1 (1996).

[2] B. Delley, D. E. Ellis, and A. J. Freeman, J. Magn. Magn. Mater. **30,** 71 (1982).

[3] D. P. Pappas, A. P. Popov, A. N. Anisimov, B. V. Reddy, S. N. Khanna, Phys. Rev. Lett. **76**, 4332 (1996).

[4] Primer in Density Functional Theory (Lecture Notes in Physics **620**; Computational Materials Science. From Basic Principles to Material Properties; Lecture Notes in Phys. Vol. **642**).

[5] O. K. Andersen, Phys. Rev. B **12**, 3060 (1975).

[6] O. K. Andersen and O. Jepsen, Phys. Rev. Lett. **53**, 2571 (1984).

[7] O. K. Andersen, Euro Phys. News **12**, 1 (1981).

[8] H. L. Skriver, "The LMTO Method", Springer, Berlin 1984.

[9] A. R. Williams, J. Kubler, and C. D. Gelatt, Phys. Rev. B **19**, 6094 (1979).

[10] C. Koenig, N. Stefanou, and J. M. Koch, Phys. Rev. B **33**, 5307 (1986).

[11] O. Gunnarson, O. Jepsen, and O. K. Andersen, Phys. Rev. B **27**, 7144 (1983).

[12] J. Harris, in “The Electronic Structure of Complex Systems”, Eds W. Temmerman and P. Phariseau, Plenum, New York 1984.
[13] F. Casula and F. Herman, J Chem. Phys. **78**, 858 (1983).
[14] M. Springborg and O.K. Andersen, J. Chem. Phys. **87**, 7125 (1987).
[15] M. Springborg and R.O. Jones, Phys. Rev. Lett. **57**, 1145 (1986).
[16] T. Fuziwara, J. Phys. F **16**, 869 (1986).
[17] G. F. Fernando, B. R. Cooper, M. V. Ramana, H. Krakauer, and C. Q. Ma, Phys. Rev. Lett. **56**, 2299 (1986).
[18] D. Glotzel, B. Segall, and O. K.Andersen, Solid State Commun. **36**, 403 (1980).
[19] P. Blaha, K. Schwarz, and P. Sorantin, Computer Physics Communications **59**, 399 (1990). See also wien2k package.
[20] H. J. F. Jansen and A. J. Freeman, Phys. Rev. B **30**, 561 (1984).
[21] M. Methfessel and M. van Schilfgaarde, Inter. J. of Modern Physics B.**7**, 262 (1993).
[22] K. Koepernik and H. Eschrig, Phys. Rev. B **59**, 1743 (1999).
[23] J. Wang and T. L. Beck, Journal of Chemical Physics **112**, 9223 (2000).
[24] G. Kresse and J. Furthmüller, Phys. Rev. B **54**, 11169 (1996).
[25] G.-M. Rignanese, X. Gonze, A. Pasquarello, "Ab-Initio Calculations of the Structural, Electronic and Dynamical Properties of High-k Dielectrics", Ed. M. Houssa, Institute of Physics Publishing, 2004, pp 431.
[26] M. M. G. Alemany, M. Jain, L. Kronik, and J. R. Chelikowsky, Phys. Rev. B **69**, 75101 (2004).
[27] R. Haydock, in “Solid State Physics: Advances in Research and Applications”, Eds H. Ehrenreich, F. Seitz, D. Turnbull, Acedemic, New York 1980, Vol. 35, p 216.
[28] J. Gaspard and F. Cyrot-Lackmann, J. Phys C **6**, 3077 (1973).
[29] W. A. Harrison, “Electronic Structure and Properties of Solids”, Freeman, San Fransisco 1980.
[30] J. C. Slater and G. F. Koster, Phys. Rev. **94**, 1498 (1954).
[31] J. Fidler and T. Schrefl, Journal of Physics D **33**, R135 (2000).
[32] U. von Barth and L. Hedin, J. Phys. C **5**, 1629 (1972).
[33] O. Gunnarson, J. Phys F **6**, 587 (1976).
[34] J. F. Janak, Phys. Rev. B **16**, 255 (1977).
[35] D. G. Pettifor, J. Magn. Magn. Mater. **15,** 847 (1980).
[36] H. Takayama, K. P. Bohnen, and P. Fulde, Phys. Rev. B **14**, 2287 (1976).
[37] J. G. Gay and R. Richter, Phys. Rev. Lett. **56**, 2728 (1986).
[38] P. Bruno, Phys. Rev. B **39**, 865 (1989).
[39] S. Pick and Dreyssee, Phys. Rev. B, **48**, 13588 (1993).
[40] M. Cinal, D. M. Edwards, and J. Mathon, Phys. Rev. B **50**, 3754 (1994).
[41] H. Victora and J. M. McLaren, Phys. Rev. B **47**, 11 583 (1993).
[42] J. Trygg, B. Johanson, O. Erikson, and J. M. Wills, Phys. Rev.Lett. **75**, 2871 (1995).
[43] G. H. O. Daalderop, P. J. Kelly, and M. F. H. Schuurmans, Phys.Rev. B **41**, 11 919 (1990); **42**, 7270 (1990).
[44] C. Li, A. J. Freeman, H. J. F. Jansen, and C. L. Fu, Phys. Rev. B **42**, 5433 (1990).
[45] D. S. Wang, R. Wu, and A. J. Freeman, Phys. Rev. Lett. **70**, 869 (1993).
[46] A. R. Mackintosh and O. K. Andersen, in “Electrons at the Fermi Surface, Ed. M. Sprinford, Cambridge University Press, Cambridge 1980, Chap. 53.

[47] V. Heine, in "Solid State Physics—Advances in Research and Applications", Eds. H. Ehrenreich, D. Turnbull, and F. Seitz, Academic Press, New York 1980, Vol. **35**, p. 114.
[48] A Aharoni, J. Magn. Magn. Mater. **197**, 786 (1999).
[49] K. Moorjani and J. M. D. Coey, "Magnetic Glasses", Elsevier, Amsterdam 1984.
[50] M. Uhl and J. Kubler, Phys. Rev. Lett. **77**, 334 (1996).
[51] A. I. Liechtenstein, M. I. Katsnelson, V. P. Antropov, and V. A. Gubanov, J. Magn. Magn. Mater. **67**, 65 (1987).
[52] V. A. Gubanov, A. I. Liechtenstein, and A. V. Postnikov, "Magnetism and the Electronic Structure of Crystals", Springer, Berlin 1992.
[53] R. F. Sabirianov and S. S.Jaswal, Phys. Rev B **57**, 7767 (1998).
[54] A. Kashyap, R. Skomski, R. F. Sabiryanov, S. S. Jaswal, and D. J. Sellmyer, IEEE Trans. Magn. **39**, 2908 (2003).
[55] A. Kashyap, R. Skomski, A. K.Solanki, Y. F. Xu, and D.J. Sellmyer, J. Appl. Phys. **95**, 7480 (2004).
[56] W. Hansen, J.P. Kottaus, and U. Merkt, Semiconductor Semimetals **35**, 279 (1992).
[57] M. Kastner, Phys. Today, Jan 24 (1993).
[58] P. Ordejon, Comp. Mat. Sci. **12**, 157 (1998).
[59] I. M. L. Billas, A. Chatelain, and W. A. de Heer, Science **265,** 1682 (1994).
[60] S. E. Apsel, J. W. Emmert, J. Deng, and L. A. Bloomfield, Phys. Rev. Lett.**76,** 1441 (1996).
[61] M. Respaud, M. Goiran, and J. M. Broto, Phys. Rev. B **59,** R3934 (1999).
[62] K. Lee, J. Callaway, and S. Dhar, Phys. Rev. B **30,** 1724 (1985).
[63] C. Jamorski, A. Martinez, M. Castro, and D. R. Salahub, Phys. Rev. B **55**, 10905 (1997).
[64] Y. Jinlong, F. Toigo, and W. Kelin, Phys. Rev. B **50**, 7915 (1994).
[65. G. M. Pastor, "Atomic Clusters and Nanoparticles", NATO ASI Series, Les Houches Session LXXIII, Eds. C. Guet, P. Hobza, F. Spiegelman, and F. David, Springer, Berlin 2001, p 335.
[66] V. Kumar and Y. Kawazoe, Phys. Rev. B **66**, 144413 (1994).
[67] A. J. Cox, J. G. Louderback, and L. A. Bloomfield, Phys. Rev. Lett. **71**, 923 (1993).
[68] A. J. Cox, J. G. Louderback, S. E. Apsel, and L. A. Bloomfield, Phys. Rev. B **49**, 12295 (1994).
[69] B. V. Reddy, S. N. Khanna, and B. I. Dunlap, Phys. Rev. Lett. **70**, 3323 (1993).
[70] G. Ganteför and W. Eberhardt, Phys. Rev. Lett. **76**, 4975 (1996).
[71] G. M. Pastor, J. Dorantes-Dávila, and K. H. Bennemann, Phys. Rev. B **40**, 7642 (1989).
[72] J. Dorantes- Dávila, H. Dreysse and G. M. Pator, Phys. Rev. B **46**, 10432 (1992).
[73] F. Lopez-Urias, G. M. Pastor, and K. H. Bennemann, J. Appl. Phys. **87**, 4909 (2000).
[74] B. Piveteau, M. C. Desjonquéres, A. M. Oles, and D. Spanjaard, Phys. Rev. B **53**, 9251 (1996).
[75] J. L. Rodriguez-López, F. Aguilera-Granja, A. Vega, and J. A. Alonso, Eur. Phys. J. D **6**, 235 (1999).
[76] R. Guirado-López, M. C. Desjonquéres, and D. Spanjaard, Phys. Rev. B **62**, 13188 (2000).

[77] M. Jamet, M. Negrier, V. Dupuis, J. Tuaillon-Combes, P. Melinon, A. Perez, W. Wernsdorfer, B. Barbara, and B. Baguenard, J. Magn. Magn. Mater. **237**, 293 (2001).
[78] M. Jamet, W. Wernsdorfer, C. Thirion, D. Mailly, V. Dupuis, P. Melinon, and A. Perez, Phys. Rev. Lett. **86**, 4676 (2001).
[79] A. Thiaville, Phys. Rev. B **61**, 12221 (2000).
[80] R. Dittrich, A. Thiaville, J. Miltat, and T. Schrefl, J. of Appl. Phys. **93**, 7891 (2003).
[81] R. F. Sabiryanov, M. I. Larsson, K. Cho, W. D. Nix, B. M. Clemens, Phys. Rev. B **67**, 125412 (2003).
[82] P. Gambardella, S. Rusponi, M. Veronese, S. S. Dhesi, C. Grazioli, A. Dallmeyer, I. Cabria, R. Zeller, P. H. Dederichs, K. Kern, C. Carbone, and H. Brune, Science **300**, 1130 (2003).
[83] A. N Andriotis and M. Menon, Phys. Rev. Lett. **93**, 026402/1 (2004).
[84] P. Lang, V. S. Stepanyuk, K. Wildberger, R. Zeller, and P. H. Dederichs, Solid State Commun. **92**, 755,1994.
[85] K. Wildberger, V. S. Stepanyuk, P. Lang, R. Zeller, and P. H. Dederichs, Phys. Rev. Lett. **75**, 509 (1995).
[86] B. Nonas, K. Wildberger, R. Zeller, and P. H. Dederichs, Phys. Rev. Lett. **80**, 4574 (1998).
[87] N. A. Levanov, V. S. Stepanyuk, W. Hergert, D. I. Bazhanov, P. H. Dederichs, A. A. Katsnelson, and C. Massobrio, Phys. Rev. B **61**, 2230 (2000).
[88] V. S. Stepanyuk and W. Hergert, Phys. Rev. B **62**, 7542 (2000).
[89] I. Cabria, B. Nonas, R. Zeller, and P. H. Dederichs, Phys. Rev. B **65**, 054414 (2002).
[90] B. Lazarovits, L. Szunyogh, and P. Weinberger, Phys. Rev. B **65**, 104441 (2002).
[91] Y. Qiang, R. Sabirianov, S. S. Jaswal, Y. Liu, H. Haberland, and D. J. Sellmyer, Phys. Rev. B **66**, 064404 (1977).
[92] R. Robles, R. C. Longo, A. Vega, C. Rey, V. Stepanyuk, and L. J. Gallego, , Phys. Rev. B **66**, 064410/1 (2002).
[93] E. F. Kneller and R. Hawig, IEEE Trans. Magn. **27**, 3588 (1991).
[94] Z. Hao, L. Jing, J. P. Liu, Z. L. Wang, and Sun Shouheng, Nature **420**, 395 (2002).
[95] H. Zeng, R. Sabirianov, O. Mryasov, K. Cho, and D. J. Sellmyer, Phys. Rev. B **66**, 184425 (2002).
[96] Y. Kondo and K. Takayanag, Phys. Rev. Lett. **79**, 3455 (1997).
[97] I. Lisiecki and A. Filankembo, Phys. Rev. B **61**, 4968 (2000).
[98] J. L. Lin, D. Y. Petrovykh, A. Kirakosian, H. Rauscher, and F. J. Himpsel, Appl. Phys. Lett. **78**, 829 (2001).
[99] B. H. Hong, S. C. Bae, C.-W. Lee, S. Jeong, K. S. Kim, Science **294**, 348 (2001).
[100] K. Kondo and K. Takayanagi, Science **289**, 606 (2000). Y. Oshima, H. Koizumi, K. Mouri, H. Hirayama, and K. Takayanagi, Phys. Rev. B **65**, 121401 (2002).
[101] A. Fert and L. Piraux, J. Magn. Magn. Mater. **200**, 338 (1999).
[102] R. Skomski, J. Phys.: Condens. Matter **15**, R841 (2003).
[103] M. Wienert and A.J. Freeman, J. Magn. Magn. Mater. **38**, 23 (1983).
[104] H. Jisang and R. Q. Wu, J. Appl. Phys. **93**, 8764 (2003).
[105] A. Delin and E. Tosatti, J. Phys.: Condens. Matter **16**, 8061 (2004).
[106] P. Gambardella, A. Dallmeyer, K. Maiti, M.C. Malagoli, S. Rusponi, P. Ohresser, W. Eberhardt, C. Carbone, and K. Kern, Phys. Rev. Lett. **93**, 077203/1 (2004).

[107] D. Spisak and J. Hafner, Phys. Rev. B **65**, 235405 (2002).
[108] Y. J. Jin, I. G. Kim, and J. I. Lee, J. Korean Phys Soc **43**, 1071 (2003).
[109] Y. J. Jin, I. G. Kim, and J. I. Lee, Phys Stat Sol. (b) **241**, 1431 (2004).
[110] P. Gambardella, A. Dallmeyer, K. Maiti, M. C. Malagoli, W. Eberhardt, K. Kern, and C. Carbone, Nature (London) **416**, 301 (2002).
[111] B. Wang, X. Chen, G. Chen, G. Wang, and J. Zhao, Solid State Commun. **129**, 25 (2004).
[112] A. K. Solanki, R. F.Sabiryanov, E. Y. Tsymbal, and S. S. Jaswal, J. Magn. Magn. Mater. **272-276**, 1730 (2004).
[113] H. D. Chopra and S. Z. Hua, Phys. Rev. B **66**, 020403(R) (2002).
[114] H. Wan, H. Cheng, and N. Garcia, arXiv:cond-mat/0207516.
[115] N. Garcia, M. Munoz, V. V. Osipov, E. V. Ponizovskaya, G. G. Qian, I. G. Saveliev, Y.-W. Zhao, J. Magn. Magn. Mater. **240**, 92 (2002).
[116] N. Garcia, M. Munoz, and Y.-W. Zhao, Phys. Rev. Lett. **82**, 2923 (1999).
[117] G. Tatara, Y.-W. Zhao, M. Munoz, and N. Garcia, Phys. Rev. Lett. **83**, 2030 (1999).
[118] J. J. Versluijs, M. A. Bari, and J. M. D. Coey, Phys. Rev. Lett. **87** 026601 (2001).
[119] L. R. Tagirov, B. P. Vodopyanov, and K. B. Efetov, Phys. Rev. B **63**, 104428 (2001).
[120] P. Bruno, Phys. Rev. Lett. 83, 2425 (1999); V. A Molyneux, V. V. Osipov, and E. V. Ponizovskaya, Phys. Rev. B **65**, 184425 (2002).
[121] M. Y. Zhuravlev, E. Y. Tsymbal, S. S. Jaswal, A. V. Vedyayev, and B. Dieny, App. Phys. Lett. **83**, 3534 (2003).
[122] J. D. Burton, A. Kashyap, M. Ye. Zhuravlev, R. Skomski, E. Y. Tsymbal, S. S. Jaswal, O. N. Mryasov, and R. W. Chantrell, App. Phys. Lett. **85**, 251 (2004).
[123] J. C. Cuevas, A. Levy Yeyati, and A. Martín-Rodero, Phys. Rev. Lett. **80**, 1066 (1998).
[124] A. Levy Yeyati, A. Martín-Rodero, and F. Flores, Phys. Rev. B **56**, 10369 (1997).
[125] M. Brandbyge, N. Kobayashi, and M. Tsukada, Phys. Rev. B **60**, 17064 (1999).
[126] E. Scheer, N. Agrait, J. Carlos Cuevas, A. Levy Yeyati, B. Ludoloph, A. Martn-Rodero, G. Rubio Bollinger, J. M. van Ruitenbeek, and C. Urbina, Nature **394**, 154 (1998).
[127] N. D. Lang, Phys. Rev. B **52**, 5335 (1995).
[128] N. D. Lang, Phys. Rev. B **55**, 9364 (1997).
[129] N. D. Lang and Ph. Avouris, Phys. Rev. Lett. **81**, 3515 (1998).
[130] N. D. Lang and Ph. Avouris, Phys. Rev. Lett. **84**, 358 (2000).
[131] N. Kobayashi, M. Brandbyge, and M. Tsukada, Jpn. J. Appl. Phys. **38**, 336 (1999).
[132] N. Kobayashi, M. Aono, M. Tsukada, Phys. Rev. B **64**, 121402 (2001).
[133] N. Kobayashi, M. Brandbyge, M. Tsukada, Phys. Rev. B **62**, 8430 (2000).
[134] N. D. Lang, Phys. Rev. B **36**, 8173 (1987); N. D. Lang, Phys. Rev. B **55**, 9364 (1997).
[135] S. R. Bahn and K. W. Jacobsen, Phys. Rev. Lett. **87**, 266101 (2001).
[136] H. Mehrez, A. Wlasenko, B. Larade, J. Taylor, P. Grütter, H. Guo, J. Wang, and C. Roland, Phys. Rev. B **65**, 195419 (2002).
[137] M. Brandbyge, J. Mozos, P.Ordejon, J. Taylor, and K. Stokbro, Phys. Rev. B **65**, 165401 (2002).
[138] J. Velev and W. H. Butler, Phys.Rev. B **69**, 094425 (2004).

[139] E. G. Emberly and G. Kirczenow, Phys. Rev. B **60**, 6028 (1999).

[140] M. N. Baibich, J. M. Broto, A. Fert, F. Nguyen Van Dau, and F. Petroff , P. Eitenne, G. Creuzet, A. Friederich, and J. Chazelas, Phys. Rev. Lett. **61**, 2472 (1988).

[141] R. Allenspach, J. Magn. Magn. Mater. **129**, 160 (1994).

[142] "Ultrathin Magnetic Structures I, II" Eds. B. Heinrich and J. A. C. Bland, Springer, Berlin 1994.

[143] U. Gradmann, Ann. Phys. **7**, 91 (1966); U. Gradmann, J. Magn. Magn. Mater. **54**, 733 (1986).

[144] R. Allenspach, M. Stampanoni, and A. Bischoff, Phys. Rev. Lett. **65**, 3344 (1990).

[145] B. Schulz and K. Baberschke, Phys. Rev. B **50**, 13467 (1994).

[146] F. Huang, M. T. Kief, G. J. Mankey, and R. F. Willis, Phys. Rev. B **49**, 3962 (1994).

[147] G. Bochi, C. A. Ballentine, H. E. Inglefield, C. V. Thomson, R. C. O'Handley, Hans J. Hug, B. Stiefel, A. Moser, and H. J. Gűntherodt, Phys. Rev. B **52**, 7311 (1995).

[148] C. H. Lee, H. Hee, F. J. Lamelas, W. Vavra, C. Uher, and R. Clarke, Phys. Rev. B **42**, 1066 (1990).

[149] B. Heinrich, K. B. Urquardt, A. S. Arrott, J. F. Cochran, K. Martle, and S. T. Purcell, Phys. Rev. Lett. **59**, 1756 (1987).

[150] Z. Q. Qiu, J. Pearson, and S. D. Bader, Phys. Rev. Lett. **70**, 1006 (1993).

[151] C. Liu, E. R. Moog, and S. D. Bader, Phys. Rev. Lett. **60**, 2422 (1988).

[152] D. P. Pappas, K. P. Kämper, and H. Hopster, Phys. Rev. Lett. **64**, 3179 (1990).

[153] R. F. Sabiryanov and S. S. Jaswal, J. Magn. Magn. Mater. **177-181**, 989 (1998); Phys. Rev. B **58**, 12071 (1998).

[154] S. van Dijken, X. Jiang, and S. S. P. Parkin, App. Phys. Lett. **83**, 951 (2003).

[155] E. Y. Tsymbal, O. N. Mryasov, and P. R. LeClair, J. Phys.: Cond. Matter **15** , R109-R142 (2003).

[156] G. H. O. Daalderop, P. J. Kelly, and M. F. H. Schuurmans, Phys. Rev. B **50**, 9989 (1994).

[157] R. Lorentz and J. Hafner, Phys. Rev. B **54**, 15937 (1996).

[158] Le Bacq, O. Eriksson, and B. Johansson, Phys. Rev. B **65**, 134430 (2002).

[159] B. Ujfalussy, L. Szunyogh, P.Bruno, and P. Weinberger, Phys. Rev. B **77**, 1805 (1996).

[160] V. S. Stepanyuk and W. Hergert, "Magnetism Structure and Interactions at the Atomic Scale", Lect. Notes Phys. 642, Springer, Berlin Heideberg 2004, 159-176.

Chapter 3

NANOMAGNETIC MODELS

Ralph Skomski and Jian Zhou
Center for Materials Research and Analysis and
Department of Physics and Astronomy
University of Nebraska
Lincoln, NE 68588, USA

Abstract The atomic-scale and mesoscopic physics of magnetic nanostructures is reviewed. Emphasis is on the description of magnetic phenomena and properties by analytical models, as contrasted to numerical approaches. Nanostructuring affects the magnetic properties on different length scales, from a few interatomic distances for intrinsic properties such as magnetization and anisotropy to more than 10 nm for extrinsic properties, such as coercivity. The consideration includes static and dynamic mechanisms, as well as nanoscale finite-temperature effects. Some explicitly discussed examples are Curie-temperature changes due to nanostructuring, the effect of narrow and constricted walls, the potential use of magnetic nanodots for finite-temperature quantum computing, and exchange-coupled hard-soft nanocomposites. The temperature dependence of extrinsic properties reflects the atomic-scale static or 'intrinsic' temperature dependence of the free-energy barriers and thermally activated dynamic or 'extrinsic' jumps over metastable free-energy barriers.

1. INTRODUCTION

Nanomagnetism involves both atomic and macroscopic features, but it cannot be reduced to a superposition of these two limits. Macroscopic interactions, as exemplified by the compass needle and the geomagnetic field, and atomic-scale magnetic phenomena, such as quantum-mechanical exchange [1-4], are necessary but not sufficient to understand nano-magnetism. Solid-state matter is made from atoms, and the corresponding length scale, Bohr's radius $a_o = 0.52$ Å, determines the range of the exchange responsible for bonding and magnetism. On the other hand, Maxwell's equations are macroscopic, that is, they do not correspond to any characteristic length. Why are the phenomena considered in this book realized on a length scale of a few nanometers, as contrasted to, for example, a few meters?

An answer is provided by the relativistic nature of many magnetic phenomena of importance in nanomagnetism [5, 6]. For example, typical domain walls have a thickness of few nanometers. This length reflects the competition between nonrelativistic exchange and relativistic magnetic anisotropy. A simple but qualitatively correct picture is obtained from the relativistic electron energy $mc^2\sqrt{1 + v^2/c^2}$, where v is the electron velocity. Expanding the energy into powers of v/c yields the rest energy mc^2, the electrostatic or 'nonrelativistic' energy $mv^2/2$, and the lowest-order relativistic correction $(\alpha/2)^2 mv^2/2$, where $\alpha = 4\pi\varepsilon_0 e^2/\hbar c \approx 1/137$ is Sommerfeld's fine-structure constant. Here we have exploited that typical electron velocities in solids are of order $v = \alpha\, c$. Respective examples of nonrelativistic and relativistic magnetic interactions are exchange, which has the character of an integral over electrostatic interactions, and spin-orbit coupling, which leads to magnetocrystalline anisotropy [7, 8]. On an atomic scale, relativistic interactions are unable to compete against atomic-scale exchange effects. For example, Heisenberg exchange may exceed 1000 K, whereas typical anisotropies are less than 1 K. However, electrostatic and relativistic contribution become comparable on length scales of order a_o/α = 7.25 nm [5, 6, 9].

In addition to the range of interactions, there is the question of interference with structural length scales. For example, Bloch wave functions, which form the basis for the band-structure theory of itinerant magnetism, require infinite crystals with perfect periodicity. How does nanostructuring interfere with this requirement? Similarly, from a thermodynamic point of view, ferromagnetism is limited to infinite crystals. It fact, the spontaneous magnetization of any finite magnet is zero, because thermal fluctuations cause the magnetization to average. This leads to the next consideration, the dependence of equilibration or averaging times on structural length scales.

There is a fundamental distinction between intrinsic and extrinsic properties. Examples of *intrinsic* properties are the spontaneous magnetization M_s, the Curie temperature T_c, and the anisotropy K_1. Intrinsic properties describe perfect crystals or surfaces, but their physical origin is atomic and involves quantum phenomena such as exchange, crystal-field interaction, interatomic hopping, and spin-orbit coupling [1, 2, 8, 10, 11]. Intrinsic properties tend to approach their bulk values on fairly small length scales. For example, 'long-range' thermodynamic fluctuations, as involved in the realization of the Curie temperature, and deviations from the Bloch character of metallic wave functions yield only small corrections when the size of the magnetic particle exceeds about 1 nm. The dynamics is characterized by fast equilibration times which means that intrinsic properties can be treated by equilibrium statistical mechanics. This makes it possible to treat intrinsic properties as local parameters. For example, $M_s(\mathbf{r})$ and $K_1(\mathbf{r})$ reflect the local chemistry, and the unit vector $\mathbf{n}(\mathbf{r})$ of the easy magnetization direction corresponds to the local c-axis orientation of the crystallites.

Extrinsic or hysteretic magnetic properties, such as the coercivity H_c and the remanence M_r, reflect the magnet's real-structure [12-16]. For example, the coercivity of technical iron doubles by adding 0.01 wt.% nitrogen [15]. Such small concentrations have little effect on the intrinsic properties but lead to inhomogeneous lattice strains that affect the propagation of magnetic domain walls and explain the observed coercivity increase. The hysteretic character of extrinsic properties means that equilibration times may be very long. At room temperature, the switching of a single atomic moment is a frequent event, but the thermally activated switching of nanoscale cooperative units, such as domain-wall segments, is very rare. This is the thermodynamic origin of hysteresis, enabling us to build permanent magnets and to store information on magnetic disks.

This chapter investigates how intrinsic and extrinsic properties are affected by nanostructuring. Emphasis is on model calculations and analytical approximations, as contrasted to Chapter 2 by Kashyap *et al.* which focuses on numerical calculations, and Ch. 4 by Schrefl *et al.*, where emphasis is on micromagnetic models and simulations. Section 2 is devoted to static properties, whereas section 3 is concerned with magnetization dynamics. Finally, section 4 presents a number of case studies.

2. MESOSCOPIC MAGNETISM

2.1. Nanoscale Spin Structure

2.1.1. Magnetic moment

The magnetic moment nearly exclusively originates from the spin and orbital moments of transition-metal electrons. The magnetic moment of iron-series transition-metal atoms in metals (Fe, Co, Ni, YCo_5) and nonmetals (Fe_3O_4, NiO) is largely given by the *spin*, and the moment, measured in μ_B, is equal to the number of unpaired spins. The orbital moment is very small, typically of the order of 0.1 μ_B, because the orbital motion of the electrons is quenched by the crystal field [16-18]. By contrast, rare-earth moments are given by Hund's rules, which predict the spin and orbital moment as a function of the number of inner-shell electrons [17]. In some cases, atoms are spin-polarized by neighboring atoms. An example of importance in nanomagnetism is $L1_0$ magnets such as FePt, where the Pt carries a magnetic moment.

The spin moment is largely determined by intra-atomic exchange. It is an electrostatic many-body effect, caused by the $1/|\mathbf{r} - \mathbf{r}'|$ Coulomb interaction between electrons located at $\mathbf{r}$ and $\mathbf{r}'$. Physically, ↓↑ electron pairs in an atomic orbital are not forbidden by the Pauli principle but are unfavorable from the point of view of Coulomb repulsion. Parallel spin alignment, ↑↑, means that the two electrons are in different orbitals, which is electrostatically favorable. However, the corresponding gain in Coulomb energy competes against an increase in one-electron energies, because one of the two electrons must occupy an excited state.

The magnetic moments of insulating transition-metal oxides and rare-earth metals are located on well-defined atomic sites. However, in Fe, Co, and Ni, as well as in many alloys, the moment is delocalized or *itinerant*. Nonmagnetic metals, or Pauli paramagnets, have two equally populated ↑ and ↓ subbands; and an applied magnetic field transfers a few electrons from the ↓ band to the ↑ band. The corresponding spin polarization is very small, of the order of 0.1%, because the Zeeman interaction is a small relativistic correction [19]. Itinerant ferromagnetism is realized by narrow bands, where the intra-atomic exchange is stronger than the band-width related gain in single-electron hybridization (Stoner criterion).

The Bloch character of itinerant wave functions means that the wave functions extend to infinity. This is not realistic for two reasons. First, magnets encountered in reality, in particular nanomagnets, cannot be considered as infinite. Second, finite-temperature excitations create spin disorder and break the Bloch symmetry of the ↑ and ↓ wave functions. The problem of nonequivalent sites can be tackled, for example, by real-space approaches [16, 20-23]. Restricting the consideration to nearest neighbors yields the correct band width, but details of the band structure, such as peaks in the density of states, are ignored. Increasing the number of neighbors improves the resolution of the density of states and makes it possible to distinguish between bulk sites and sites close to surfaces. As a consequence, magnetic moments are determined by the local atomic environment, typically without major nanoscale corrections.

2.1.2. Interatomic exchange

The spin structure of a magnetic moment is the relative orientation of the atomic magnetic moments. It includes types of zero-temperature magnetic order, such as ferromagnetism, ferrimagnetism, and antiferromagnetism, and finite temperature magnetic order. However, micromagnetic structures, such as domains and domain walls, are usually excluded from the consideration. Figure 1 shows some spin structures of interest in the present context.

To a large extent, the spin structure of bulk and nanomagnets is determined by the interatomic *Heisenberg exchange*, $J(\mathbf{R}_i - \mathbf{R}_j)\ \mathbf{S}_i \cdot \mathbf{S}_j = J_{ij}\ \mathbf{S}_i \cdot \mathbf{S}_j$. For positive and negative values of J_{ij} it favors parallel and antiparallel spin alignment, respectively. In ferromagnets, such as Fe, Co, and $Nd_2Fe_{14}B$, all spins are parallel and the atomic moments add. Ferrimagnets, such as Fe_3O_4 and $BaFe_{12}O_{19}$, and antiferromagnets, such as CoO and MnF_2, are characterized by two (or more) sublattices with opposite moments. This amounts to a ferrimagnetic reduction or antiferromagnetic absence of a net moment. Sublattice formation may be spontaneous, as in typical antiferromagnets, or imposed by the atomic composition, as in ferrimagnets [24] [25]. In metals, the interatomic exchange may be positive or negative and depends on the atomic environment, on the interatomic distance, and on the band filling.

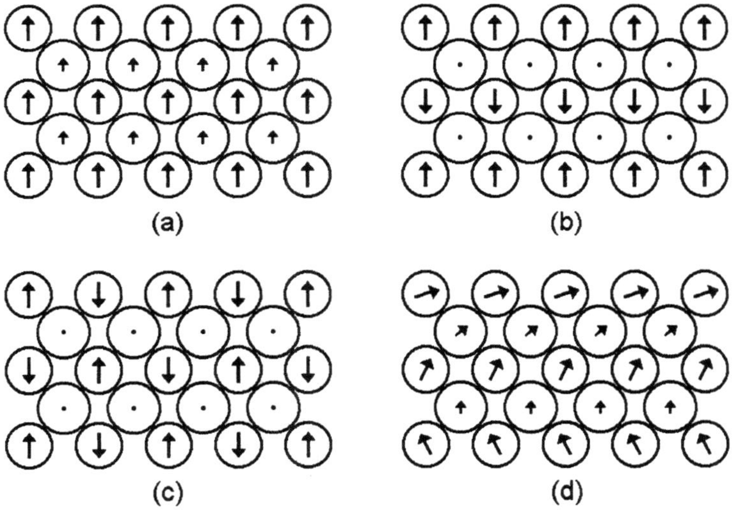

Figure 1. Spin structures (schematic): (a) ferromagnetism, (b-c) antiferromagnetism, and (d) noncollinear structure. The shown structure of the $L1_0$ type; the small atoms (with the large magnetization arrows) the iron-series transition-metal atoms, as compared to the bigger 4d/4f atoms. Examples of $L1_0$ magnets are CoPt and FePt.

A simple and asymptotically correct [26-28] model is the Ruderman-Kittel-Kasuya-Yosida or *RKKY* exchange between two localized moments in a Pauli-paramagnetic matrix. For a free-electron gas of wave-vector k_F,

$$J_{ij} = J(|\mathbf{r}_i - \mathbf{r}_j|) = J(R) \sim \cos(2k_F R)/R^3 \qquad (1)$$

The interaction is obtained by second-order perturbation theory, that is, the embedded magnetic moments lead to a mixing of one-electron wave functions. The origin of the oscillations is the sharp Fermi surface, which means that spatial features smaller than about $1/k_F$ cannot be resolved with available zero-temperature wave functions.

RKKY interactions were first considered on an atomic scale, where the oscillation period is on an Å scale. In nanostructures, the fast oscillations do not average to zero but increase with the size of the embedded clusters or nanoparticles. However, the increase is less pronounced than that of magnetostatic interactions, and for particles sizes larger than about 1 nm, the magnetostatic interactions become dominant [27, 29]. In semiconductors and semimetals, such as Sb, the low density of carriers means that k_F is small, and the period of the oscillations is nanoscale [16, 28]. This contributes to the complexity of the physics of diluted magnetic semiconductors [30, 31].

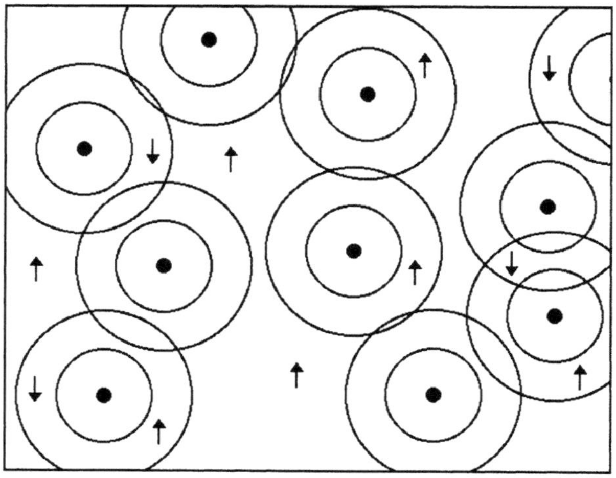

Figure 2. Carrier mediated exchange in dilute semiconductors (schematic). The mechanism is similar to RKKY interactions, but due to the essential involvement of donor or acceptor orbitals, $J(r_i, r_j)$ can no longer be written as $J(|r_i - r_j|)$.

In a strict sense, the RKKY interaction is mediated by free electrons, but there exist similar effects in other regimes, for example in the tight-binding scheme [32]. Figure 2 shows the example of a dilute magnetic semiconductor where localized impurity spins are coupled by shallow donor or acceptor carriers. The carrier orbitals have a radius of the order of 1 nm, hybridize, and yield an RKKY-type coupling. A simple case is the exchange mediated by a two weakly overlapping s-orbitals, centered at $\mathbf{R}_1$ and $\mathbf{R}_2$. The overlap leads to bonding and antibonding orbitals, and the exchange J_{ij} is obtained from the hybridized electrons by second-order perturbation theory. In terms of local electron densities $\rho(\mathbf{r}) = \psi^*(\mathbf{r})\psi(\mathbf{r})$, the exchange J_{ij} scales as $(\rho(\mathbf{r}_i - \mathbf{R}_1) - \rho(\mathbf{r}_i - \mathbf{R}_2))\,(\rho(\mathbf{r}_j - \mathbf{R}_1) - \rho(\mathbf{r}_j - \mathbf{R}_2))$. This means that the exchange is ferromagnetic if the magnetic ions are located in the same shallow s-orbital and antiferromagnetic if they are in different s-orbitals. Correlations (the Hubbard or 'Coulomb-blockade' energy of the orbitals) can be shown to reduce the ferromagnetic exchange while leaving the qualitative picture unchanged.

One effect of competing RKKY exchange is noncollinear spin structures, as illustrated in Fig. 1(d). Noncollinearity due to competing exchange is encountered, for example, in some elemental rare earths (helimagnetism), where it reflects different exchange interactions between nearest and next-nearest rare-earth layers [33]. Note that the corresponding magnetization wave vector is generally incommensurate with lattice spacing, not only in nanostructures but also in perfect crystals. Furthermore, the effect is relatively strong, with angles between neighboring atomic spins from 0 to 180°. This is in contrast to the relativistic effects considered in the next subsection.

2.1.3. Exchange stiffness

On a continuum level, the Heisenberg exchange energy of a cubic material is

$$E_{ex} = \int A(\nabla(\mathbf{M}/M_s))^2 \, dV \tag{2}$$

where A is the *exchange stiffness*. More generally, Heisenberg exchange is described by

$$E_{ex} = \iint J(\mathbf{r} - \mathbf{r}') \, \mathbf{M}(\mathbf{r}) \cdot \mathbf{M}(\mathbf{r}') \, dV dV' \tag{3}$$

A well-known derivation of Eq. (2) is in terms of magnetization angles. It assumes $\phi = 0$, so that $E_{ex} = \int A(\nabla\theta)^2 dV$, and takes into account that $\Sigma_{ij} J_{ij} \cos(\theta_i - \theta_j) \approx \Sigma_{ij} J_{ij}(1 - (\theta_i - \theta_j)^2/2)$. Using the expansion $\theta_j = \theta_i + \nabla\theta \cdot (\mathbf{r}_j - \mathbf{r}_i)$ and comparing the result with Eq. (2) then yields $A \sim \Sigma_{ij} J_{ij}(\mathbf{r}_i - \mathbf{r}_j)^2$. This result is meaningful for nearest-neighbor interactions, but it diverges for long-range interactions J_{ij}. An example is the RKKY interaction, where integration over all neighbors yields $A \sim \int 1/R^3 \, R^2 \, R^2 \, dR = \infty$. This is because $\theta_j = \theta_i + \nabla\theta \cdot (\mathbf{r}_j - \mathbf{r}_i)$ breaks down for large distances $R = |\mathbf{r}_i - \mathbf{r}_j|$.

A more general derivation of A is based on Fourier transformation, which diagonalizes $J(|\mathbf{r} - \mathbf{r}'|)$ and yields a representation in terms of $J_\mathbf{k}$. Since $\int A \, (\nabla\theta)^2 \, dV = \int J_\mathbf{k} \, \theta_\mathbf{k}^2 \, d\mathbf{k}$ and $\int A \, (\nabla\theta)^2 \, dV = - \int A \, k^2 \, \theta_\mathbf{k}^2 \, d\mathbf{k}$, A is given by the quadratic coefficient of the expansion of $J_\mathbf{k}$ with respect to $\mathbf{k}$. Putting $\mathbf{k} = k\mathbf{e}_k$, $\mathbf{R} = R \cos\theta' \, \mathbf{e}_k + R \sin\theta' \, \mathbf{e}_\perp$, and $dV = 4 \pi R^2 \sin\theta' \, d\theta' \, dR$ yields

$$J_\mathbf{k} \sim \int J(R) \frac{\sin(kR)}{kR} R^2 \, dR \tag{4}$$

Here $J_k = F(|\mathbf{k}|)$ is the Lindhard screening function [26, 34]. Note that noncollinear or incommensurate spin states then correspond to a minimum of $J(\mathbf{k})$.

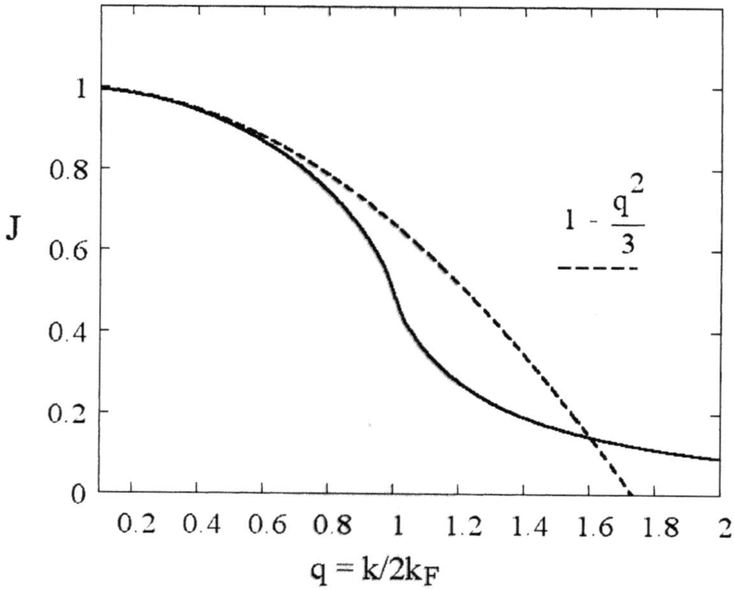

Figure 3. Exchange energy as a function of the wave vector of the magnetization inhomogenity: Lindhard function (solid line) and exchange-stiffness or continuum approximation (dashed line). For late 3d elements (Cu), $k/2k_F = 1$ corresponds to modulation wavelength of 0.23 nm

Figure 3 compares the Lindhard function (solid line) with the exchange-stiffness approximation (dashed line). We see that the exchange-stiffness approximation works well unless k is comparable to k_F.

In *noncubic* materials, A must be replaced by the 3×3 exchange-stiffness tensor $A_{\mu\nu}$, and the energy is $\Sigma_{\mu\nu}\int A_{\mu\nu}\, \partial\mathbf{M}/\partial x_\mu \cdot \partial\mathbf{M}/\partial x_\nu \, \mathrm{d}V$. Here the indices μ and ν denote the spatial coordinates x, y, and z of the bonds. The energy is anisotropic with respect to the nabla operator $\nabla_\mu = \partial/\partial_\mu$ (bond anisotropy) but isotropic with respect to the magnetization $\mathbf{M}$. By contrast, the relativistic anisotropic exchange $\Sigma_{\alpha\beta}\int A_{\alpha\beta}\, \nabla M_\alpha\, \nabla M_\beta\, \mathrm{d}V$ is isotropic with respect to ∇ but anisotropic with respect to $\mathbf{M}$.

2.1.4. *Curie temperature*

Thermal disorder competes against interatomic exchange and causes the magnetization of ferromagnets to vanish at a well-defined sharp Curie temperature T_C. In a strict sense, ferromagnetism is limited to infinite magnets, because thermal excitations in finite magnets cause the net moment to fluctuate between opposite directions. The Curie temperature is determined by the site-resolved exchange coefficients J_{ij}, and since M_S and T_c are equilibrium properties, it is sufficient to know the *partition function* $Z = \Sigma_\mu \exp(-E_\mu/k_B T)$, where the summation includes all microstates or spin-configurations μ. However, the number of terms in Z increases exponentially with the size of the magnet, and there exist exact solutions only in a few cases [35]. The simplest approximation is the *mean-field approximation*, where the interactions are mapped onto a selfconsistent field. The homogeneous nearest-neighbor Heisenberg ferromagnet has the mean-field Curie-temperature $T_c = (S+1)zJ/3k_B S$, where S is the spin quantum number and z is the number of nearest neighbors.

The mean-field model is easily generalized to two or more sublattices. This site-resolved or lattice mean-field theory includes the case of nanomagnets, which have a very large number N of non-equivalent atomic sites or 'sublattices'. Since the J_{ij} form an N × N matrix, there are N coupled algebraic equations, and T_c is given by the largest eigenvalue of the matrix J_{ij}/k_B [24, 36]. Using averaged exchange constants $\langle J_{ij} \rangle$ fails to properly account for the spatial dispersion of the exchange. An extreme example is a mixture of two ferromagnetic phases with equal volume fractions but different Curie temperatures T_1 and $T_2 > T_1$. In the above approximation, $T_c = (T_1+T_2)/2$, but in reality $T_c = T_2$ [36].

Note that mean-field theory is unable to describe the long-range correlations aspect of the problem, but the involved energy contributions are small, and the long-range features of the thermodynamics are not affected by the nanoscale effects [36].

In practice, it is difficult to distinguish the magnetism of particles or nanostructural features larger than about 1 nm from true ferromagnetism, because interatomic exchange ensures well-developed ferromagnetic correlations on a nanoscale. For example, when the radius of a particle is larger than a few interatomic distances, then the $M_s(T)$ curve is difficult to distinguish from a ferromagnet. Disordered two-phase nanostructures have a single common Curie temperature close to the Curie temperature of the phase with the strongest exchange coupling [5, 36, 37]. Similar considerations apply to multilayers [38-42] and to systems such as magnetic semiconductors [30]. Figure 4 illustrates that it is not possible to enhance the finite-temperature magnetization of a phase having a low Curie temperature by exchange-coupling it to a phase with a high bulk Curie temperature [36]. This is clear contrast to the nanoscale improvement of extrinsic properties (§2.3.5).

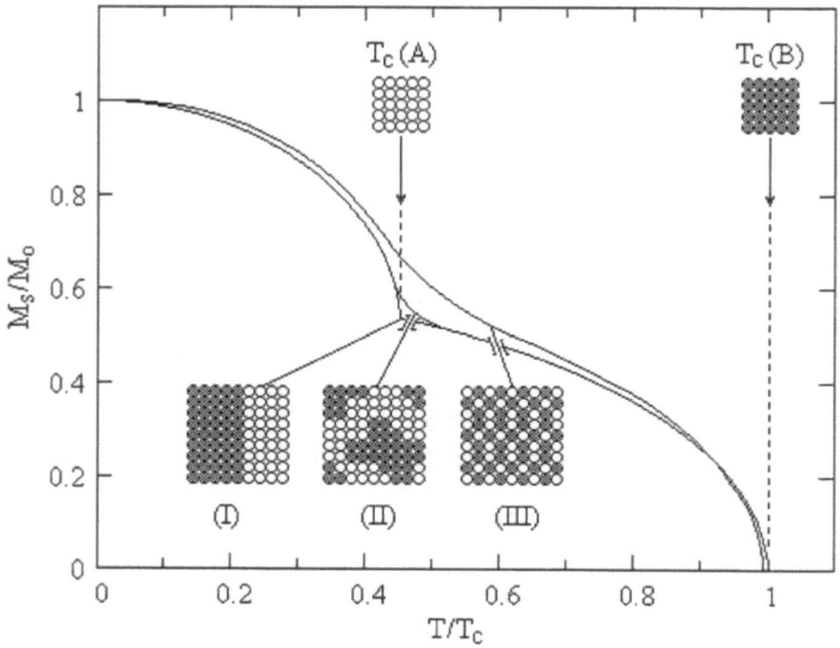

Figure 4. Spontaneous magnetization of inhomogeneous magnets: (I) macroscopic mixture, (II) nanostructure, and (III) alloy. In alloys and nanostructures, there is only one Curie temperature, although the $M_s(T)$ curves of nanostructures exhibit a two-phase like inflection whose curvature may be difficult to resolve experimentally [36].

Heisenberg interactions require well-defined atomic magnetic moments. where $\mathbf{S}^2 = S_o^2$. In insulators, $S_o^2 = S(S+1)$, whereas in metals, S_o is an expectation value and $\mathbf{S}/S_o$ has the character of a unit vector that describes the local magnetization direction. The assumption of a constant magnetic moment is often justified, because the total interatomic exchange per atom, of order 100 meV, tends to be much smaller than typical intra-atomic exchange energies of about 1000 meV [43]. Some exceptions are $L1_0$ magnets, where the 4d or 5d moments (Pd or Pt) are spin-polarized by the 3d atoms (Fe or Co) [44], and very weak itinerant ferromagnets, such as $ZrZn_2$ [16, 45]. Simplifying somewhat, the overall situation is intermediate between Heisenberg magnetism with stable local moments and a Stoner-like behavior where the moment vanishes at T_c. However, itinerant magnets such as Fe tend to be close to the Heisenberg limit (Ch. 2), which establishes the spin-fluctuation picture of finite-temperature magnetism [16, 34].

2.1.5. *Anisotropic Exchange*

Heisenberg exchange is magnetically *isotropic*, that is, coherent rotation of a magnet's spin system does not change the Heisenberg exchange energy. For example, layered structures, such as YCo_5 and $L1_0$ magnets, tend to exhibit different intra- and interlayer interactions [44, 46], but the exchange does not depend on whether the magnetization is in-plane or normal to the layers. This exchange-bond anisotropy affects the spin structure of a magnet at both zero and nonzero temperatures. For example, it is the main source of spin noncollinearities encountered in elemental rare earths [33] and in magnetoresistive materials, such as NiMnSb [47]. Figure 1 shows some spin structures.

The bond anisotropy must not be confused with the relatively weak relativistic exchange anisotropies, which involve spin-orbit coupling and depend on the angle between the magnetization and the crystal axes. Examples are the exchange interactions assumed in the Ising and XY models, the magnetocrystalline anisotropy, and the unidirectional Dzyaloshinskii-Moriya exchange. For example, the exchange anisotropies $J_{xx} - J_{zz}$ and $J_{yy} - J_{zz}$ are small corrections to the isotropic exchange $J = (J_{xx} + J_{yy} + J_{zz})/3$. Another example is the *Dzyaloshinskii-Moriya* (or DM) interaction $H_{DM} = - \frac{1}{2} \Sigma_{ij} \mathbf{D}_{ij} \cdot \mathbf{S}_i \times \mathbf{S}_j$, where the vector $\mathbf{D}_{ij} = - \mathbf{D}_{ji}$ reflects the local environment of the magnetic atom [48]. Net DM interactions require local environments with sufficiently low symmetry and occur, for example, in some crystalline materials, such as α-Fe_2O_3 (hematite), amorphous magnets, spin glasses, and magnetic nanostructures [5, 33, 48]. Micromagnetic noncollinearities, such as domain walls, also stem from relativistic effects, because they involve magnetocrystalline anisotropy, but their domain is nanoscale rather than atomic, and they are traditionally treated in the context of micromagnetism (§2.3). Compared to Heisenberg exchange, relativistic contributions are smaller by a factor of order α^2, where $\alpha = 1/137$ [5]. For example, typical DM canting angles are about 0.1°.

2.2. Magnetic Anisotropy

The dependence of the magnetic energy on the orientation of the magnetization with respect to the crystal axes is known as *magnetic anisotropy*. Permanent magnets need a high magnetic anisotropy, in order to keep the magnetization in a desired direction. Soft magnets are characterized by a very low anisotropy, whereas materials with intermediate anisotropies are used as magnetic recording media. In terms of the magnetization angles

ϕ and θ, the simplest anisotropy-energy expression for a magnet of volume V is $E_a = K_1\ V \sin^2\theta$. This anisotropy is known as lowest-order (or second-order) *uniaxial* anisotropy, and K_1 is the first uniaxial anisotropy constant. It is often convenient to express anisotropies in terms of anisotropy fields. For example, the expression $E_a = K_1 V \sin^2\theta$ yields $H_a = 2K_1/\mu_o M_s$.

For magnets of low symmetry (orthorhombic, monoclinic, and triclinic), the lowest-order anisotropy energy is

$$E_a = K_1 V \sin^2\theta + K'_1 V \sin^2\theta \cos(2\phi) \tag{5}$$

where K_1 and K'_1 are, in general, of comparable magnitude. This expression must also be used for magnets having a low-symmetry shape, such as ellipsoids with three unequal principal axes, for a variety of surface anisotropies, such as that of bcc (011) surfaces [49], and for nanoparticles with random surfaces. Equation 5 can also be written as $E_a = -\mathbf{M}\cdot\underline{\underline{\mathbf{K}}}\cdot\mathbf{M}/M_s^2$, where $\underline{\underline{\mathbf{K}}}$ is a 3×3 tensor. It obeys $\mathrm{Tr}\underline{\underline{\mathbf{K}}} = 0$, and the two independent eigenvalues of $\underline{\underline{\mathbf{K}}}$ correspond to K_1 and K'_1. *Higher-order* anisotropy expressions contain, in general, both uniaxial and planar terms. For example, $E_a/V = K_1 \sin^2\theta + K_2 \sin^4\theta$ contains second- and fourth-order uniaxial anisotrop terms and describes hexagonal and rhombohedral crystals [16, 50].

2.2.1. *Origin of anisotropy*

Figure 5 illustrates that there are two main sources of anisotropy: shape anisotropy and magnetocrystalline anisotropy. Shape anisotropy is important in magnetic nanostructures made from soft-magnetic materials, for example in Fe, Co, and Ni particles [16, 51] and in nanowires [52-55]. However, the anisotropy of most materials is of magnetocrystalline origin, reflecting the competition between electrostatic crystal-field interaction and spin-orbit coupling [7]. Note that the same mechanism is responsible for the quenching (or unquenching) of the orbital moment and for phenomena such as magnetic circular dichroism and anisotropic magnetoresistance.

For shape anisotropy, $K_1 = \mu_o(1\text{-}3D)M_s^2/4$, where D is the demagnetizing factor ($D = 0$ for long cylinders, $D = 1/3$ for spheres, and $D = 1$ for plates) [56]. It is important to note that shape anisotropy is limited to very small particles (§2.2). In large particles, shape anisotropy is destroyed by internal flux closure, indicated at the bottom of Fig. 5(a).

The crystal field [57], which contains both electrostatic and hopping contributions [58], acts on the orbits of the inner d and f electrons. That is, the electron orbits reflect the anisotropic crystalline environment, and adding spin-orbit coupling translates this anisotropic electron motion into magnetic anisotropy.

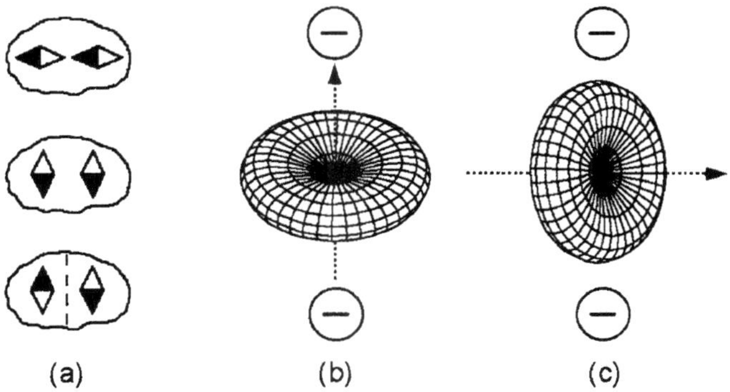

Figure 5. Physical origin of magnetic anisotropy: (a) compass-needle analogy of shape anisotropy and (b-c) magnetocrystalline anisotropy. In (b) and (c), the anisotropy energy is given by the electrostatic repulsion between the tripositive rare-earth ions and the negative crystal-field charges.

The magnitude of the magnetocrystalline anisotropy depends on the ratio of crystal-field energy and spin-orbit coupling. As a relativistic phenomenon, spin-orbit coupling is most pronounced for inner-shell electrons in heavy elements, such as rare-earth 4f electrons. This leads to a rigid or 'unquenched' coupling between spin and orbital moment, and the magnetocrystalline anisotropy is given by the relatively small electrostatic crystal-field interaction of the 4f charge clouds [59] with the crystal field [16, 60, 61]. This largely electrostatic mechanism, illustrated in Fig. 5(b) and 5(c), is responsible for the high room-temperature anisotropy of rare-earth permanent magnets, $K_1 \sim 10$ MJ/m^3 (see Appendix).

In 3d atoms, the spin-orbit coupling is much smaller than the crystal-field energy, and the magnetic anisotropy is a perturbative effect [7, 8, 16]. Typical second- and fourth-order transition-metal anisotropies are of the orders of 1 MJ/m^3 and 0.01 MJ/m^3, respectively. A manifestation of magnetocrystalline anisotropy is magnetoelastic anisotropy, where the crystal field is changed by mechanical strain [5, 16].

2.2.2. *Surface and interface anisotropy*

To realize second-order anisotropy, the atomic environment of the transition-metal atoms must have a sufficiently low symmetry [49, 62-65]. Figure 6 illustrates that this is often, but not always, the case for surface atoms. Magnetic *surface anisotropy*, first analyzed by Néel [62], is important in complicated structures and morphologies such as ultrathin transition-metal films [66], multilayers [67], rough surfaces [65], small particles [68], and surface steps [69]. In a variety of cases it has been possible to calculate surface anisotropies from first principles [64, 67, 70-72]. The same is true for some other low-geometries, such as Fe wires embedded in Cu [73] and free-standing monatomic Co wires [74]. An interesting point is that surface anisotropies easily dominate the bulk anisotropy of cubic materials. From the tables in the appendix we see that bulk anisotropies are about two orders of magnitude smaller than lowest-order anisotropies. Due to the comparatively large number N_s of surface atoms of small particles, the surface contribution dominates the bulk anisotropy in particles smaller than about 3 nm, even if one takes into account that the net surface anisotropy is not necessarily linear in N_s but tends to scale as $N_s^{1/2}$ due to random-anisotropy effects.

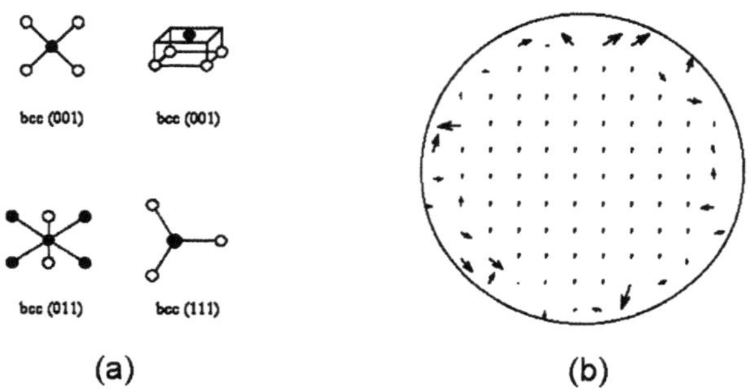

Figure 6. Surface anisotropy: (a) atomic origin and (b) realization in a nanoparticle. Lowest–order biaxial anisotropy is realized for bcc (011) but not for bcc (001) and bcc (111). The large surface-to-volume ratio of clusters leads to a comparatively strong diameter dependence of the intrinsic properties such as anisotropy [68] and magnetization.

Magnetocrystalline anisotropy is characterized by a pronounced temperature dependence [16, 61, 75-77]. For example, the leading rare-earth anisotropy contribution of permanent magnet intermetallics such as $SmCo_5$ and $Nd_2Fe_{14}B$ scales as $1/T^2$ [78]. The main reason is that typical anisotropy energies per atom are quite small, E_a ranging from less than 0.1 K to a few K. The realization of room-temperature anisotropy requires the support of the interatomic exchange field, which suppresses the switching of individual atomic spins into states with reduced anisotropy contributions [16, 79, 80].

Magnetocrystalline anisotropy is, essentially, a single-ion property, realized by embedding the atom in a metallic or nonmetallic crystalline environment [16, 58]. This must be compared to the popular Néel model [62], which ascribes anisotropy to pair interactions. Figure 7 illustrates the difference. The Néel model requires two interacting magnetic atoms (black), whereas the single-ion or crystal-field model amounts to hopping or crystal-field interactions with atoms that are not necessarily magnetic (white). The principal failure of the Néel model is seen by comparing Sm_2Fe_{17} and $Sm_2Fe_{17}N_3$, where the electronegative nitrogen is nonmagnetic but strongly affects the crystal field and changes the room-temperature anisotropy from -0.8 MJ/m^3 to 8.6 MJ/m^3 [16].

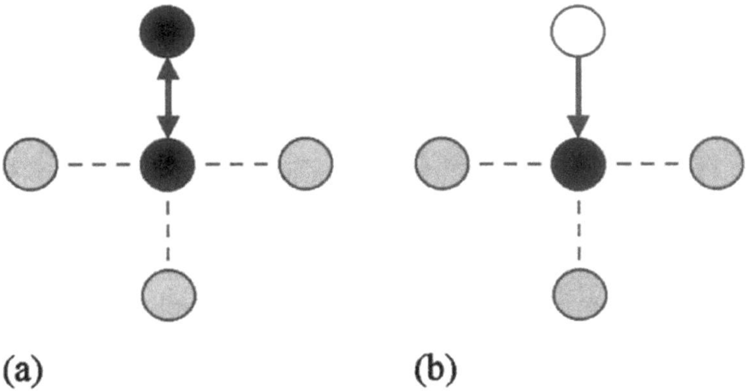

Figure 7. Models of magnetic anisotropy: (a) Néel model and (b) single-ion crystal-field model. Both models reproduce the correct symmetry, but (b) is physically more adequate for most systems.

2.2.3. Temperature dependence of anisotropy

In most materials, including nanostructures, the magnetocrystalline anisotropy is strongly decreases with increasing temperature. This is due to intra-atomic excitations. The strong temperature dependence of the leading rare-earth anisotropy contribution of hard-magnetic materials such as $SmCo_5$ and $Nd_2Fe_{14}B$ reflects intramultiplet excitations [16, 60, 61]. The excitations may be visualized as changes between Fig. 5(b) and 5(c). The figure indicates that these excitations compete against the crystal field. However, the crystal field is only one consideration; the main contribution is from the inter-sublattice exchange, which dominates thermal spin disorder

For one-sublattice magnets, such as Fe and Co, the Akulov or Callen and Callen theory [81] relates the temperature dependence of the anisotropy to the spontaneous magnetization and yields M^3 and M^{10} power laws for uniaxial and cubic magnets, respectively. This theory has become popular far beyond its range of applicability [82] but is unable to describe structures such as rare-earth transition-metal magnets [16, 60], actinide magnets [83], and $L1_0$ type compounds [44, 84].

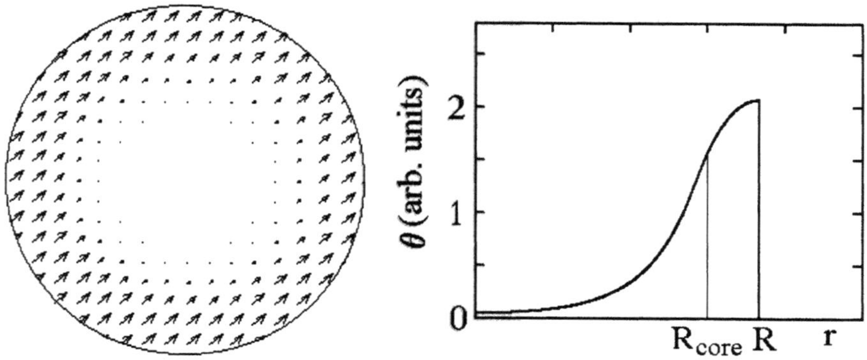

Figure 8. Nucleation mode in a small $L1_0$ particle. Due to reduced anisotropy at the surface, the reversal starts at the surface [89].

An interesting example is the finite-temperature magnetization of $L1_0$ magnets [44, 84, 85]. Elemental 4d/5d magnets, such as Pd and Pt, are exchanged-enhanced Pauli paramagnets, but in a ferromagnetic environment they are easily spin-polarized by neighboring 3d atoms [86, 87]. The 4d/5d moment contributes little to the magnetization and Curie temperature, but it plays a key role in the realization of magnetic anisotropy, which is of the order of 5 MJ/m^3 at room temperature [88].

The temperature dependence of the anisotropy is reflects the collapse of the 4d/5d moment. The result of the calculation is an M^2 law [44], as compared to the Callen-Callen prediction M^3 and to refined simulations that yield an $M^{2.08}$ dependence [84]. By comparison, for uniaxial 3d magnets, such as Co and YCo_5, $m = 3$ [81], cubic and noncubic actinide magnets exhibit $m = 1$ [83], and for cubic 3d magnets, such as Fe and Ni, $m = 10$ [81]. Finally, rare-earth transition-metal intermetallics exhibit $m \approx 0$, that is, the 4f sublattice anisotropy is largely independent of the leading 3d magnetization [16]. The exponents $m = 2$ and $m = 3$ are not very dissimilar [85], but the different physics—the crucial involvement of two sublattices—speaks in favor of $m = 2$. In fact, recent calculations by Mryasov *et al.* have yielded $m = 2.08$, amounting to a single-sublattice contribution of the order of 8%. As also pointed out in [84], the reduction of the number of 3d neighbors in magnetic nanoparticles has a very similar surface-anisotropy reduction effect.

2.3. Hysteresis of Magnetic Nanostructures

Magnetic anisotropy yields easy magnetization directions corresponding to local energy minima and energy barriers that separate the easy directions. On an atomic scale, the barriers are easily overcome by thermal fluctuations, but on nanoscale or macroscopic length scales the excitations are usually too weak to overcome the barriers. This is observed as *magnetic hysteresis*.

Zeeman and selfinteraction (demagnetization) magnetic fields, interatomic exchange, and magnetic anisotropy all contribute to the rotation, which occurs on a mesoscopic scale and has been known historically as micromagnetism [90], although nanomagnetism would be a better name to characterize the involved length scales. Magnetic nanostructures exhibit a particularly rich extrinsic behavior, but even traditional 'microstructured' magnets exploit nanometer-scale features for performance optimization [91]. For example, the best room-temperature permanent magnets are now made from Nd-Fe-B [92], but as-cast samples with the correct stoichiometry exhibit a disappointingly low coercivity unless the grain-boundary structure is optimized by a specific heat treatment.

Figure 9 shows a typical hysteresis loop and illustrates how magnetic hysteresis is realized in real space. In the example of Fig. 9, the hysteresis reflects domain-wall pinning in a small particle. This means that magnetic domains are separated by domain walls (dotted lines) whose motion is impeded by real-structure defects or 'pinning centers' in the bulk or at the surface. Aside from a few basic hysteresis mechanisms, such as pinning, coherent rotation, curling, and localized nucleation, there exist many variations and combinations. The reason is the real-structure dependence of magnetic hysteresis, which makes it necessary to consider each material or each class of materials separately.

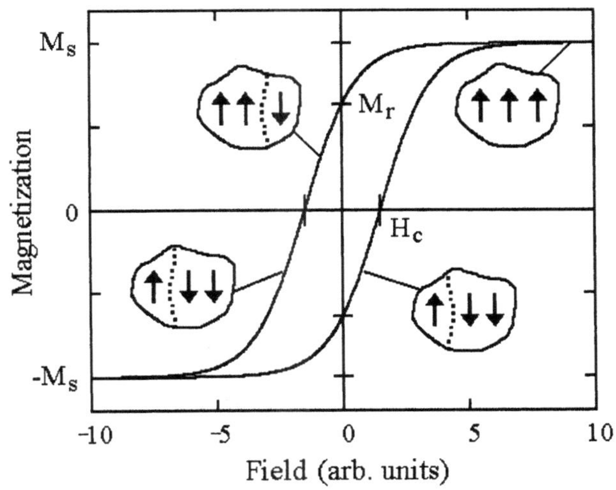

Figure 9. Magnetic hysteresis: origin and phenomenology of hysteresis. The coercivity of the particles shown in this figure is caused by domain-wall pinning at the grain-boundary phase.

2.3.1. Micromagnetic free energy

A key theoretical problem is to derive magnetization curves by simulating or modeling the magnet's nanostructure. This requires the determination of the local magnetization **M**(**r**), from which the hysteresis loop is obtained by averaging. The large strength of the intra-atomic exchange means that typical magnetization changes in magnetic solids are caused by moment *rotations* rather than by changes in the moments' magnitude. The result is the complicated nonlinear, nonequilibrium, and nonlocal problem of hysteresis. In addition, there is a strong real-structure influence. Properties related to hysteresis (extrinsic properties) are also known as micromagnetic properties [90], but this term is somewhat unfortunate, because most micromagnetic phenomena are nanoscale.

Hysteresis problems are usually treated on a *continuum* level [16, 90, 93]. Narrow-wall phenomena, which have been studied for example in rare-earth cobalt permanent magnets [94] and at grain boundaries [95, 96], involve individual atoms and atomic planes and lead to comparatively small corrections to the extrinsic behavior (§4.2). Furthermore, in contrast to the intrinsic phenomena considered in Section 2, which affect the spontaneous magnetization $M_s = |\mathbf{M}|$, micromagnetic phenomena are realized by local *rotations* of the magnetization vector. This is because typical micromagnetic energies are much smaller than the quantum-mechanical energy contributions that establish M_s.

To explain the hysteresis loop of magnetic materials one needs to trace the local magnetization $\mathbf{M}(\mathbf{r}) = M_s\mathbf{s}(\mathbf{r})$ as a function of the applied field H. This is achieved by considering the free-energy functional

$$F = \int \left\{ A\left[\nabla\left(\frac{\mathbf{M}}{M_s}\right)\right]^2 - K_1 \frac{(\mathbf{n}.\mathbf{M})^2}{M_s^2} - \mu_o\, \mathbf{M}\cdot\mathbf{H} - \frac{\mu_o}{2}\,\mathbf{M}\cdot\mathbf{H}_d(\mathbf{M}) \right\} \mathrm{d}V \qquad (6)$$

Here $M_s(\mathbf{r})$ is the spontaneous magnetization, $K_1(\mathbf{r})$ is the first uniaxial anisotropy constant, $A(\mathbf{r})$ denotes the exchange stiffness, and $\mathbf{n}(\mathbf{r})$ is the unit vector of the local anisotropy direction. $\mathbf{H}$ is the external magnetic field, and $\mathbf{H}_d$ is the magnetostatic self-interaction field. The latter can be written as

$$\mathbf{H}_d(\mathbf{r}) = \frac{1}{4\pi}\int \frac{3\,(\mathbf{r}-\mathbf{r}')\,(\mathbf{r}-\mathbf{r}').\mathbf{M}(\mathbf{r}') - |\mathbf{r}-\mathbf{r}'|^2\,\mathbf{M}(\mathbf{r}')}{|\mathbf{r}-\mathbf{r}'|^5}\, \mathrm{d}V' \qquad (7)$$

The *free*-energy character of F reflects the intrinsic or equilibrium temperature dependence of the parameters A, K_1, and M_s. Furthermore, these parameters are local parameters, because they depend on chemistry, crystal structure, and crystallite orientation.

Depending on the considered system, additional terms must be added to the micromagnetic equation. In lowest order, DM interactions amount to a random field $\Sigma_j(D_{ij,y}\mathbf{e}_x - D_{ij,x}\mathbf{e}_y)/2$ where the summation (or integration) over j includes all atomic neighbors; the resulting structure may be called a 'spin colloid'.

2.3.2. Micromagnetic length scales

Equation (6) yields not only the hysteresis loop but also the underlying micromagnetic spin structure. This includes features such as domains and domain walls. An aspect of great importance in nanomagnetism is the length scales on which these features are realized. Dimensional analysis of Eq. (6) yields two fundamental quantities, namely the wall-width parameter $\delta_o = \sqrt{A/K_1}$ and the exchange length $l_{ex} = \sqrt{A/\mu_o M_s^2}$. Other length and dimensionless parameters are derived from δ_o and l_{ex}.

The *wall-width parameter* δ_o varies from about 1 nm in extremely hard materials to several 100 nm in very soft materials. It determines the thickness $\delta_B = \pi\delta_o$ and energy $\gamma_w = 4K_1\delta_o$ of Bloch domain walls [13, 14, 97, 98], and describes the spatial response of the magnetization to local perturbations [95]. Essentially, the thickness of the walls is determined by the competition between exchange, which favors extended walls, and anisotropy, which favors narrow transition regions.

The *exchange length* l_{ex} is the length below which atomic exchange interactions dominate typical magnetostatic fields. It determines, for example, the particle cylinder radius $R_{coh} \sim 5l_{ex}$ above which curling is more favorable than coherent rotation (§2.3.3). It also yields the thickness of soft-magnetic films below which Néel walls are energetically more favorable than Bloch walls. For a broad range of ferromagnets, $l_{ex} \sim 2$ nm [16]. Note that the wall-width parameter δ_o is sometimes interpreted as an exchange length. If this were a valid consideration, then ideally soft materials, where $K_1 = 0$ and $\delta_o = \infty$, would realize exchange coupling on a truly macroscopic scale. This is at odds with experiment.

Compared to domain-wall thicknesses, the determination of domain *sizes* tends to be very complicated [13, 14, 51, 93, 99-101]. This is because domains are caused by the strongly geometry- and size-dependent magnetostatic selfinteraction. An exception is the *critical single-domain radius*

$$R_{SD} = \frac{36\sqrt{AK_1}}{\mu_o M_s^2} \tag{8}$$

of spherical particles, above which a two-domain state is more favorable than the single-domain state. This expression reflects the competition between domain-wall energy, $4\pi\sqrt{AK_1}R^2$, and gain in magnetostatic energy $\mu_o M_s^2 V/12$ [97]. In soft magnets, the critical single-domain size is only a few nanometers, but in very hard magnets it exceeds 1 μm.

It is important to emphasize that the critical single-domain radius is an *equilibrium* property. It involves the comparison of the of single-domain and multi-domain energies but is independent of the energy barriers separating the states. It determines, for example, the initial or virgin state after thermal demagnetization. By contrast, hysteresis is a *non*equilibrium phenomenon caused by energy barriers. Furthermore, equilibrium domains are qualitatively different from the nonuniform (incoherent) magnetization states occurring during magnetization reversal (§2.3.3). The popular but incorrect equating of single-domain magnetism and coherent rotation has its origin in the focus on soft and semi-hard magnets in the first half of the 20th century.

Note than typical domain-wall widths are much smaller than the domains themselves. When the size of a magnetic particle is smaller than the domain-wall width δ_B, as encountered for example in small soft-magnetic nanodots, then the distinction between domains and domain walls blurrs, and the determination of the micromagnetic spin structure requires additional considerations [102]. One example is curling-type flux-closure or 'vortex' states.

2.3.3. *Exact solutions*

In small particles, the exchange is sufficiently strong to ensure that **M(r)** is constant throughout the magnet, that is, $\nabla\mathbf{M}$ in Eq. (6) is zero. Depending on the context, this regime is called *coherent rotation*, uniform rotation, or Stoner-Wohlfarth reversal [90, 103, 104]. For perfect uniaxial ellipsoids of revolution having the symmetry axis parallel to the external field $\mathbf{H} = H\mathbf{e}_z$, the energy (Eq. (6)) then becomes

$$\frac{F}{V} = K_1 \sin^2\theta + \frac{\mu_o}{2}(1 - 3D)\, M_s^2 \sin^2\theta - \mu_o\, M_s\, H \cos\theta \qquad (9)$$

Expanding this equation into powers into powers of θ and analyzing the stability of the local free-energy minimum at $\theta = 0$ yields the Stoner-Wohlfarth coercivity

$$H_c = \frac{2K_1}{\mu_o M_s} + \frac{1}{2}\,(1\text{-}3D)\; M_s \qquad (10)$$

This coercivity is a simple example of a *nucleation field.* In micromagnetism, the term nucleation refers to the instability of the remanent state in a reverse magnetic field $H_z = -H_N$. It does not necessarily imply localization effects [16, 93], although localized nucleation is frequently encountered in practice.

Increasing the radius of the particle leads to a transition from coherent rotation to *curling.* Figure 10 compares coherent-rotation and curling nucleation modes. Curling is favorable from the point of view of magnetostatic selfinteraction, because the vortex-like mode yields some flux closure, but it costs some exchange energy, because $\nabla \mathbf{M} \neq 0$. The derivation of the curling mode involves the exchange term in Eq. (6). After some calculation [16, 90, 93, 104] one obtains the nucleation-field coercivity

$$H_c = \frac{2K_1}{\mu_o M_s} - DM_s + \frac{c(D)\,A}{\mu_o M_s R^2} \qquad (11)$$

Here the radius R refers to the two degenerate axes of the ellipsoid, and c = 8.666 for spheres (D = 1/3) and c = 6.678 for needles (D = 0). In (11), the magnetostatic contribution, $-DM_s$, is always negative, in contrast to the term $(1 - 3D)M_s/2$ in (10). This means that there is no shape anisotropy in large magnets, although the exchange term in (11) partly compensates the absence of the exchange field [105].

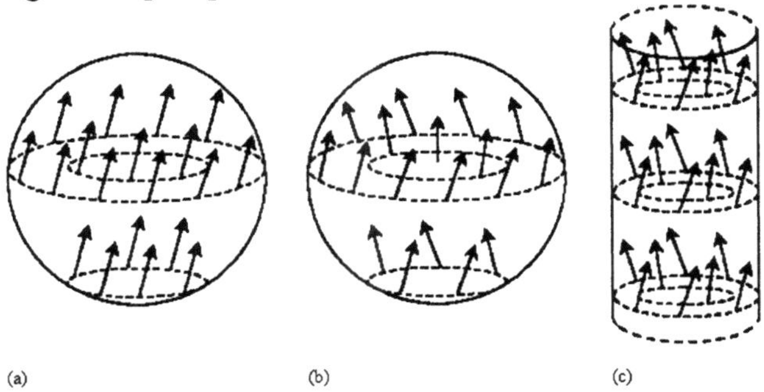

Figure 10. Nucleation modes in homogeneous magnets: (a) coherent rotation in a sphere, (b) curling in a sphere, and (c) curling in a cylinder. The arrows show the local magnetization $M = M_z\, e_z + m$, where e_z is parallel to the axis of revolution of the ellipsoid (cylinder).

Comparison of Eqs. (10) and (11) yields the radius R_{coh} for the transition from coherent rotation to curling. For $R < R_{coh}$, the exchange energy dominates, and the nucleation is realized by coherent rotation, whereas for $R > R_{coh}$ the nucleation behavior is dominated by flux closure and realized by curling. For spheres and wires (cylinders), one obtains $R_{coh} = 5.099\ l_{ex}$ and

$R_{coh} = 3.655\ l_{ex}$, respectively [16, 90, 93]. These radii are typically of the order of 10 nm. In thin films with perpendicular anisotropy ($D \approx 0$), curling occurs when the cross section of the films exceeds some value scaling as l_{ex}^2 [9]. Note that R_{coh} is anisotropy-independent, in contrast to the critical single-domain radius R_{SD}. Since $R_{coh} \ll R_{SD}$ in hard magnets, there is a broad region $R_{coh} \approx 10$ nm and $R_{SD} \approx 1$ μm where hard-magnetic single-domain particles demagnetize incoherently. The popular but incorrect equating of single-domain magnetism and coherent rotation, as epitomized by the unfortunate term 'elongated single-domain particle' (ESD), has its origin in the focus on soft and semi-hard magnets in the first half of the 20th century.

2.3.4. Localized nucleation

The Stoner-Wohlfarth approach works fairly well for very small particles, where $\nabla \mathbf{M} = 0$ is a good approximation. However, it has been known for decades that neither the Stoner-Wohlfarth theory nor the additional consideration of the curling mode account for the coercivity of real materials. For example, the coercivity of optimized permanent magnets is only 20-40% of the anisotropy field $2K_1/\mu_o M_s$, and only a part of the discrepancy can be ascribed to the curling terms in Eq. (11). The reason is that real-structure imperfections make it impossible to consider the magnets as perfect ellipsoids of revolution.

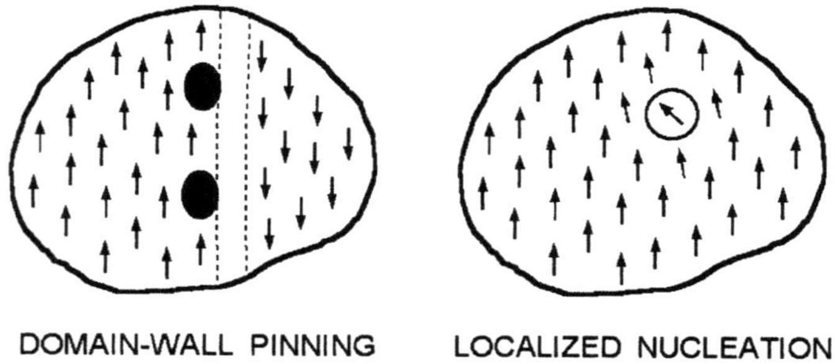

Figure 11. Pinning and nucleation. Pinning means that coercivity is created by trapping the domain wall at pronounced inhomogenities (pinning centers). In the absence of pinning centers, the coercivity is determined by the reversal field at which the original magnetization configuration becomes unstable (nucleation).

Figure 11 illustrates that there is an important distinction between domain-wall pinning and localized nucleation. Domain-wall pinning [15, 51, 99, 106], as illustrated in Fig. 11, requires relatively strong inhomogenities. For example, the high coercivity of Sm-Co-Cu-Zr permanent magnets reflects domain-wall pinning at a Cu-rich grain-boundary phase with strongly reduced anisotropy and domain-wall energy [5, 16, 77, 107, 108].

In nearly perfect magnets, pinning is unimportant, and the coercivity is determined by the stability of saturated (or nearly saturated) original magnetization state. This scenario is similar to coherent rotation and curling, except that a single nanoscale inhomogenity may initiate the magnetization reversal of a macroscopic magnet. This localized nucleation leads to strong reduction of H_N and solves Brown's paradox [90, 109], according to which the observed coercivities are often much smaller than predicted by (10).

To determine the nucleation field, we write the local magnetization as

$$\mathbf{M}(\mathbf{r}) = M_s \left(\sqrt{1 - m^2}\ \mathbf{e}_Z + \mathbf{m}(\mathbf{r})\right) \tag{12}$$

where $\mathbf{m}$ is the perpendicular magnetization component ($|\mathbf{m}| = \sin\theta$). Taking into account that $\mathbf{n} = \mathbf{e}_z$ for aligned magnets, and expanding the free energy Eq. (6) into powers of $\mathbf{m}$ yields the quadratic expression

$$F = \int [\, A\,(\nabla\mathbf{m})^2 + K_1(\mathbf{r})\,\mathbf{m}^2 - \frac{1}{2}\,\mu_o\, M_s\, H\,\mathbf{m}^2]\ \mathrm{d}\mathbf{r} \tag{13}$$

Here we have incorporated the magnetostatic selfinteraction into K_1 and H, which is a good approximation for many systems [5]. Eigenmode analysis of (13) yields the differential equation

$$A\nabla^2 m - \left(K_1(\mathbf{r}) + \frac{\mu_o}{2}\,M_s\,H\right) m = 0 \tag{14}$$

where we have assumed that A is constant throughout the magnet.

In terms of (14), imperfections appear as a modification of the local anisotropy $K_1(\mathbf{r})$ and lead to a nucleation-field and coercivity reduction [105, 110-112]. The solution of the nucleation problem is simplified by the fact that Eq. (14) has the same structure as the single-particle Schrödinger equation, $K_1(\mathbf{r})$ and H_c being the respective micromagnetic equivalents of $V(\mathbf{r})$ and E. Consider, for example, an imperfection in form of a cubic soft inclusion of volume L^3 in a hard matrix. The corresponding wave functions are particle-in-a-box states, and the nucleation field is [5]

$$H_c = \frac{6\,\pi^2 A}{\mu_o\, M_s\, L^2} \tag{15}$$

Analyzing this expression in terms of the exchange length reveals that imperfections are harmful if their size is comparable to or larger than a few nanometers.

In the past, nucleation fields such as Eq. (15) have been obtained for several cases: spherical particles in an infinitely hard matrix [110], small inclusions in a matrix of arbitrary anisotropy and exchange stiffness [105] [111], various types of multilayers [111, 113], and some core-shell and nanowire configurations [105, 114, 115]. For a discussion of the unphysical limit of very small inclusions, $L = 0$, see e.g. [5].

2.3.5. Nanostructured magnetic materials

There are various classes of magnetic materials (see Appendix), and both traditional materials and novel nanostructures exploit nanoscale phenomena [5]. An example of improving the performance of magnetic materials by nanostructuring is hard-soft permanent-magnet composites [16, 111, 116-120]. As analyzed in [16, 111], atomic-scale magnetism doesn't support substantial improvements of permanent magnets beyond existing intermetallics such as $SmCo_5$, Sm_2Co_{17}, and $Nd_2Fe_{14}B$. However, adding a soft phase to a hard phase in a suitable nanostructure can improve the permanent-magnet performance beyond that of the hard phase. In these structures, the hard phase acts as a skeleton to ensure a coercivity of the order of $H_c/2$, whereas adding the soft-magnetic high-magnetization phase enhances M_s and M_r [111]. This materials-by-design or 'metamaterials' approach makes it possible to produce materials not encountered in nature.

Note that isotropic single-phase and two-phase *permanent magnets* are comparatively easy to produce. Examples are nanocrystalline $Nd_2Fe_{14}B$/Fe_3B-Fe and $Sm_2Fe_{17}N_3$/Fe composites produced by melt-spinning [116] and mechanical alloying [121], respectively. However, the remanent magnetization M_r of randomly oriented grains with uniaxial anisotropy is only half the saturation magnetization M_s. Since the energy product of highly coercive permanent magnets scales as M_r^2, this amounts to a reduction of this key figure of merit by a factor of 4. Intergranular exchange improves the remanence by favoring parallel spin alignment in neighboring grains [116, 121-125]. However, this *remanence enhancement* is accompanied by a cooperative coercivity reduction, which limits the achievable energy product.

Magnetic nanostructures with c-axis alignment do not suffer from magnetization reduction due to random anisotropy. Examples of structures are multilayers [111, 113, 117], which are now widely associated with Kneller's concept of exchange-spring magnetism, and three-dimensional two-phase nanostructures [110, 111]. The model predictions by Skomski and Coey [16, 111] have been verified in Fe-Pt nanostructures [5, 119], but the experimental realization of high-performance multilayered [118, 120, 126-130] and granular [16, 95, 105, 111, 116, 119, 123-125, 131, 132] rare-earth transition-metal structures has remained a demanding challenge. The main reasons are difficulties in aligning the easy-axis of hard-phase grains, maintaining a uniform phase mixture on the scale of ~ 5 nm, and real-structure imperfections, which make it difficult to maintain coercivity.

3. MAGNETIZATION DYNAMICS

Magnetization processes are generally time-dependent, even if the external magnetic field is kept constant. For example, freshly magnetized permanent magnets lose a small fraction of their magnetization during the first few hours. In nanomagnets, the oscillation and relaxation times vary from less than one nanosecond to millions of years.

Atomic processes are very fast, so that intrinsic properties obey equilibrium statistics. An intermediate regime is characterized by typical magnetostatic and anisotropy energies per atom, about 0.1 meV, which correspond to times of order $\tau_o \sim 0.1$ ns. Examples are ferromagnetic resonance and related precession and damping phenomena. When energy barriers are involved, thermal excitations lead to a relatively slow relaxation governed by the Boltzmann-Arrhenius law [99, 133-137]

$$\tau = \tau_o \exp\left(\frac{E_a}{k_B T}\right) \tag{16}$$

This relation, known as the Néel-Brown law [135, 137], goes back to the 1930s [99]. Due to the exponential energy dependence, extrinsic equilibration times vary over many orders of magnitudes, from nanoseconds or milliseconds in superparamagnetic particles to decades in permanent magnets and recording media and millions of years in magnetic rocks.

3.1. Fundamental Equations

The time-dependent Schrödinger equation can, in principle, be used to predict the evolution of any physical system, but this method is not feasible in practice. First, the deterministic character of the Schrödinger equation forbids irreversible processes. Second, the many-body character of the Schrödinger equation, and the large number of degrees of freedom, such as lattice vibrations, complicate the description of real magnetic systems.

To make meaningful predictions about the relevant magnetic degrees of freedom, such as the position of a domain wall, one must treat the irrelevant atomic degrees of freedom as a heat bath, thereby introducing irreversibility. A simple classical analogue of this 'coarse graining' [5, 138-140] is a system of masses coupled by harmonic springs. The system has a recurrence time

τ_{rec} scaling as $1/\Delta\omega$, where $\Delta\omega$ is the system's smallest eigenfrequency difference. For any finite system the recurrence time is finite, but for an infinite number of degrees of freedom, corresponding to a heat bath, the solution yields a continuum of eigenvalues, $\Delta\omega = 0$ and $\tau_{rec} = \infty$.

The coarse-graining procedure simplifies the picture and provides the justification for various nonequilibrium approximations. One example is the Landau-Lifshitz equation

$$\frac{d\mathbf{M}}{dt} = \gamma \mathbf{M} \times \mathbf{H}_{eff} - \frac{1}{M_s^2 \tau_o} \mathbf{M} \times (\mathbf{M} \times \mathbf{H}_{eff}) \tag{17}$$

where γ is the gyromagnetic ratio and $\mu_o \mathbf{H}_{eff} = - \partial E/\partial \mathbf{M}(\mathbf{r})$ is a local effective field [93, 141-143]. This equation—and similar relations, such as the Gilbert and Bloch-Bloembergen equations—describe the precession of the magnetization around $\mathbf{H}_{eff}$ and its relaxation towards the local or global energy minima associated with $\mathbf{H}_{eff}$. However, they are not able to describe thermally activated jumps over energy barriers.

There are several ways of describing thermal activation. Without loss of generality, we can restrict ourselves to a single magnetic degree of freedom s, such as a single magnetization angle or projection [144]. The relevant equation is the *Langevin* equation,

$$\frac{\partial s}{\partial t} = - \frac{\Gamma_o}{k_B T} \frac{\partial E}{\partial s} + \sqrt{2\Gamma_o}\, \xi(t) \tag{18}$$

where $\xi(t)$ is a random force obeying $<\xi(t)> = 0$ and $<\xi(t)\, \xi(t')> = \delta(t - t')$. Physically, the force $- \partial E/\partial s$ drives the magnetization towards the local (free) energy minimum but competes with the random force $\xi(t)$. Figure 12 compares the two regimes describes by Eqs. (17) and (18). For the numerical realization of Landau-Lifshitz and Langevin equations see Ch. 4.

The probability distribution belonging to Eq. (18) obeys the diffusion-type magnetic *Fokker-Planck* equation [137, 140, 145]

$$\tau_o \frac{\partial P}{\partial t} = \frac{1}{k_B T} \frac{\partial}{\partial s}\left(P \frac{\partial E}{\partial s}\right) + \frac{\partial^2 P}{\partial s^2} \tag{19}$$

This equation can also be written in form of a continuity equation, $\tau_o\, \partial P/\partial t = - \partial P/\partial s$. In equilibrium, $\partial P/\partial t = 0$ and $P = Z^{-1} \exp(-E(s)/k_B T)$, justifying the magnitude of the source term in Eq. (18). For energy barriers $E_a \gg k_B T$, Eq. (19) essentially reduces to Eq. (16) [145, 146].

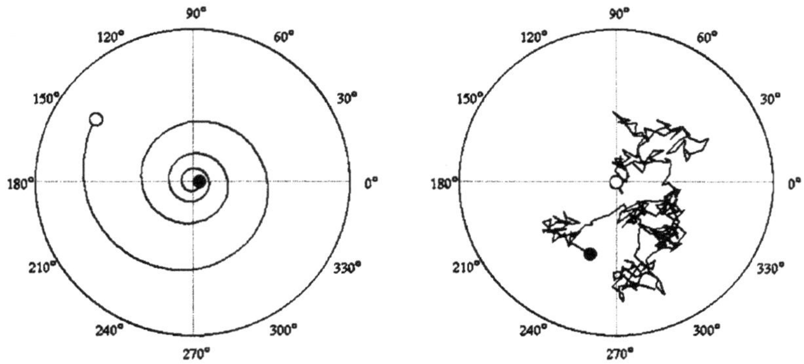

Figure 12. Magnetization dynamics of a nanoparticle: (a) damped precession and (b) random thermal motion. The curves are simulations for typical but not critical parameters, covering a time of order 0.1 ns. In both polar plots, the direction of the motion is from the white circles to the black circles.

A third and largely equivalent approach is the master or rate equation

$$\frac{\partial P(s)}{\partial t} = \int [W(s, s') P(s') - W(s', s) P(s)] ds' \qquad (20)$$

where the $W(s, s') = W(s' \rightarrow s)$ are transition rates. For example, describing the system by Fermi's golden rule yields

$$W_{ij} = \frac{2\pi}{\hbar} |\langle\Psi_i|V|\Psi_j\rangle|^2 \, \delta(E_i - E_j) \qquad (21)$$

where W_{ij} is the transition rate between two quantum states s_i and s_j. Specifying V relates the dynamics of a system to quantum mechanics, and Kramers-Moyal expansion of Eq. (20) with respect to small magnetization changes $s_i - s_j$ reproduces the diffusive Fokker-Planck equation (19).

3.2. Spin Waves in Nanostructures

Since Bloch's 1930 article on the temperature dependence of the spontaneous magnetization of ferromagnets [147], spin waves or magnons have attracted much attention in the solid-state and magnetism communities [50, 148, 149]. In particular, Bloch's spin-wave arguments indicate that there is no long-range isotropic ferromagnetism in two or fewer dimensions [10, 16, 26, 35, 48, 147, 150]. In Bloch's original approach, the long-wavelength magnetization reduction due to spin waves is proportional to the integral $\int k^{d-3}\, dk$, which exhibits a long-wavelength divergence for $d \leq 2$. However, as explained in §2.1, the behavior of experimental nanostructures is reminiscent of bulk ferromagnets. In a spin-wave picture, the energy of long-wavelength magnons is too small to interfere with the exchange energies that control moment formation and local magnetic order.

Ignoring the damping term in Eq. (17), the resonance is described by $d\mathbf{M}/dt = \gamma(\mathbf{M}\times\mathbf{H}_{eff})$. For homogeneously magnetized ellipsoids of revolution, the effective field is equal to the applied field $\mathbf{H} = H\,\mathbf{e}_z$ plus the anisotropy field $\mathbf{H}_a$, and the resonance problem is solved by the diagonalization of a 2×2 matrix. This uniform or ferromagnetic resonance (FMR) yields resonance frequencies determined by [17]

$$\omega^2 = \gamma^2 (H + H_{ax} - H_{az})(H + H_{ay} - H_{az}) \qquad (22)$$

where $\mathbf{e}_x$ and $\mathbf{e}_y$ correspond to the principal axes of the 2×2 matrix. In systems with rotational symmetry, such as perfect nanowires aligned parallel to the external magnetic field, this equation degenerates into

$$\omega = \gamma (H + 2\,K_{eff}/\mu_o M_s) \qquad (23)$$

where $K_{eff} = K_1 + \mu_o M_s^2/4$.

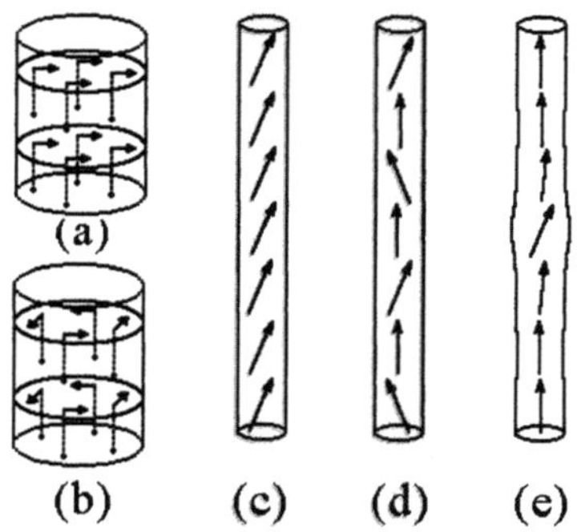

Figure 13. Spin-wave modes in nanowires: (a) is the coherent mode ($k_\perp = 0$) and (b) is of the curling type. When $k_\perp \gg 1/R$, then the perpendicular spin waves are essentially superpositions of plane waves $\exp(ik_x x)$ and $\exp(ik_y y)$. Since the diameter of typical nanowires is much larger than interatomic distance, there are many excited perpendicular modes, and the finite-temperature magnetization $M_s(T)$ is reminiscent of bulk magnets.

Recent research has largely focused on spin-wave excitations in perfect nanostructures [151-156]. In addition, real-structure effects have been considered [5, 153]. In macroscopic systems, the superposition of individual lines leads to a line broadening due to real-structure inhomogenities and field gradients. In nanostructures, the superposition of resonance lines is a poor approximation, because the modes are coupled by interatomic exchange. Since the nucleation mode is essentially an $\omega = 0$ spin-wave mode [93], the considerations of §2.3.4 carry over to the problem of spin waves in real nanostructures [5, 153].

Figure 13 shows various types of spin-wave modes in long nanowires. In very thin nanowires, where $R < R_{coh}$, curling-type modes can be ignored [55, 114], and the perpendicular magnetization components obey $M_x = M_s\, m(z) \cos(\omega t)$ and $M_y = M_s\, m(z) \sin(\omega t)$. Note that both curling and the radial spin-wave quantization in terms of Bessel functions [153] leads to very high level splittings and can be ignored in lowest order. The function $m(z)$ obeys

$$- 2A \frac{d^2 m}{dz^2} + \left(2\, K_{eff}(z) \;+\; \mu_o\, M_s\, H \;-\; \frac{\omega}{\gamma}\right) m = 0 \qquad (24)$$

For $\omega = 0$, this equation reduces to (14), whereas $dm/dz = 0$ reproduces (23). Mathematically, Eq. (24) is a well-known random-potential eigenvalue problem, which can be solved numerically or by transfer-matrix methods [5, 153].

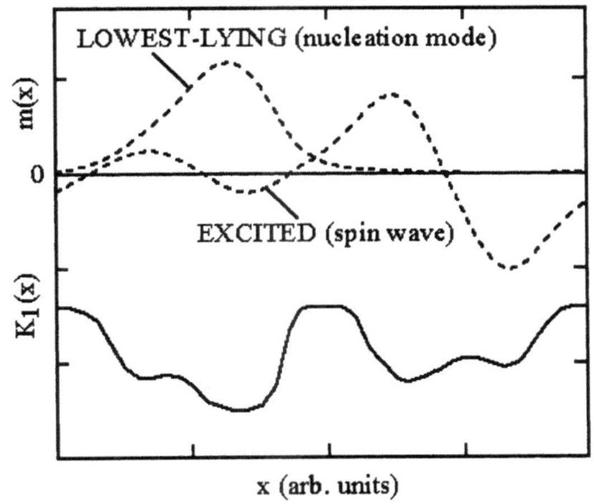

Figure 14. Localization of spin-wave modes with k vectors along the wire axis. All modes are localized, but the localization length is smallest for low-lying modes.

Figure 14 shows some examples. An interesting point is that all modes are localized [105], as one expects from the quantum-mechanical analog of a one-dimensional electron gas in a random potential [157]. Alternatively, from a localized 'tight-binding' point of view, micromagnetic delocalization can be interpreted as a tunneling through hard-magnetic regions. The localization length depends on ω and is largest for high frequencies.

A major source of real-structure inhomogenities are wire-thickness fluctuations and polycrystallinity, and geometrical features at wire ends [54, 55, 114, 158]. As expected from §2.3.4, the localization of the modes is accompanied by a coercivity reduction [54, 55, 114].

3.3. Magnetic Viscosity

The extrinsic time dependence of the magnetization is known as *magnetic viscosity*. The magnetic viscosity determines, for example, the stability of the information stored in magnetic recording media [159] and the time-dependent decay of the remanent magnetization of permanent magnets. From a thermodynamic point of view, this is the origin of hysteresis, and similar magnetic freezing processes are the superparamagnetic freezing of small particles and ferrofluids [160-162] and spin glasses [33, 48, 140, 163, 164]. A related effect is the dependence of the coercivity on the sweep rate $\mathrm{d}H/\mathrm{d}t$ of the external magnetic field [16, 99, 134]. As illustrated in Fig. 15, both effects are caused by thermal activation and involve jumps over (free) energy barriers. Thermally activated jumps yield only small corrections, because typical energy barriers in ferromagnets are much larger than k_BT.

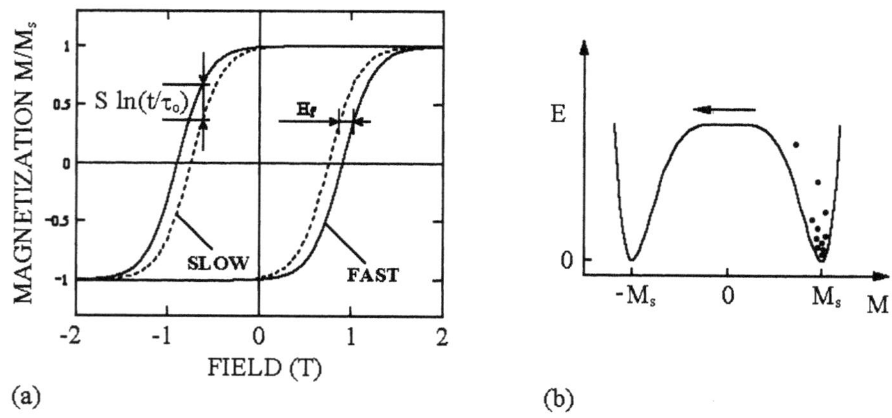

Figure 15. Dynamic hysteresis-loop effects: (a) magnetic viscosity and (b) sweep-rate dependence. The sweep-rate dependence amounts to a *fluctuation-field* [165] or sweep-rate correction to the static coercivity H_{co}.

Experiment shows that the time dependence of the magnetization is often *logarithmic* rather than exponential [16, 99, 136, 159, 166, 167]. As pointed out long ago [99], a logarithmic time dependence is obtained by averaging over exponential relaxation processes $M(t) = - M_S + 2M_S \exp(-t/\tau)$. The key assumption is that the distribution of the relaxation time τ reflects an energy-barrier barrier distribution much larger than k_BT.

In [99], a rectangular energy-barrier distribution was used, but this is an unnecessary and rather complicated assumption. In fact, introducing an energy-barrier distribution $P(E_a)$ leads to the integral

$$M(t) = - M_S + 2\, M_S \int_{-\infty}^{\infty} P(E_a)\, e^{-(t/\tau_o) \exp(-E_a/k_BT)}\, dE_a \qquad (25)$$

and to [5, 168]

$$M(t) = M(t') - 2\, k_BT\, P(0) \ln(t/t')\, M_s \qquad (26)$$

where t and t' are two arbitrary reference times. Phenomenologically, the logarithmic law is often written as $M(t) = M(t) - S\ln(t/t')$, where S is the magnetic-viscosity constant. Note that the logarithmic law breaks down for $t = \infty$. An expression with improved asymptotics is [168]

$$M(t) = 2\, M_S\, (t/\tau_O)^{-k_BT/E_O} - M_S \qquad (27)$$

However, since $(x^{\varepsilon} - 1) = \varepsilon \ln x$ for small exponents, this expression is not very different from the logarithmic law.

An important aspect of magnetic viscosity is its *field dependence*. For example, $S(H)$ tends to exhibit a maximum near the coercivity. This reflects the close relationship between the energy-barrier distribution and the irreversible part χ_{irr} of the susceptibility, and leads to $S = \chi_{iir}\, S_v$, where the viscosity parameter S_v is only weakly field-dependent [5, 133, 159, 167, 169-172].

An effect closely related to magnetic viscosity is the dependence of the coercivity on the sweep rate $\eta = dH/dt$: H_C is largest for high sweep rates, that is, for fast hysteresis-loop measurements (Fig. 15). Sweep-rate and magnetic-viscosity dynamics have the same origin, but there is a very simple way of deriving a relation for the sweep-rate dependence. Let us assume that the energy barriers exhibit a power-law dependence

$$E_a(H) = K_o\, V_o \left(1 - \frac{H}{H_o}\right)^m \qquad (28)$$

where K_o, V_o, and m are micromagnetic parameters. This law is not restricted to aligned Stoner-Wohlfarth particles, where $K_o = K_1$, $V_o = V$, and $m = 2$, but also describes a broad range of pinning and nucleation mechanisms.

Putting Eq. (28) into Eq. (16) and identifying H with reverse field at which the magnetization state escapes the local energy minimum yields [134, 140]

$$H_c = H_o\left(1 - \left(\frac{k_B T}{K_o V_o} \ln(\tau/\tau_o)\right)^{1/m}\right) \tag{29}$$

where $\tau \sim 1/\eta$. A slightly more sophisticated calculation, first published in the 1960s [134] and popularized by Sharrock more than ten years ago [173], yields a relatively unimportant factor of ln2 = 0.693, which is usually incorporated into τ_o.

An experimental approach to analyze the resulting sweep-rate dependence of the coercivity is to exploit the phenomenological relation [159]

$$H_c(\eta) = H_c(\eta_o) + \frac{k_B T}{M_s V^*} \ln\left(\frac{\eta}{\eta_o}\right) \tag{30}$$

Linearizing Eq. (30) with respect to $\ln(\eta/\eta_o) = -\ln(\tau/\tau_o)$ [16]

$$V^* = \frac{m}{2}\left(\frac{25 k_B T}{K_o V_o}\right)^{1-1/m} V_o \tag{31}$$

This equation shows that the activation volume V^* is only loosely related to the 'physical' or 'Barkhausen' volume V_o. Furthermore, V_o is not necessarily the volume of a single particle—due to cooperative and localization effects its may be smaller or larger than the particle volume.

The exponent m cannot be regarded as a fitting parameter but depends on the symmetry of the system. In most cases, $m = 3/2$ [16, 140, 158, 166, 167, 174, 175], but $m = 2$ for highly symmetric systems, such as aligned Stoner-Wohlfarth particles. In particular, the $m = 3/2$ law is realized for misaligned Stoner-Wohlfarth particles and for most domain-wall pinning mechanisms [5]. Experimental values of m tend to vary between 1.5 to 2 [136, 158]. Linear laws, where $m = 1$, are sometimes used in simplified models, but so far it hasn't been possible to derive them from physically reasonable energy landscapes [5, 16, 176]. The same is true for dependences such as $1/H - 1/H_o$ [177], where series expansion yields an $m = 1$ power law.

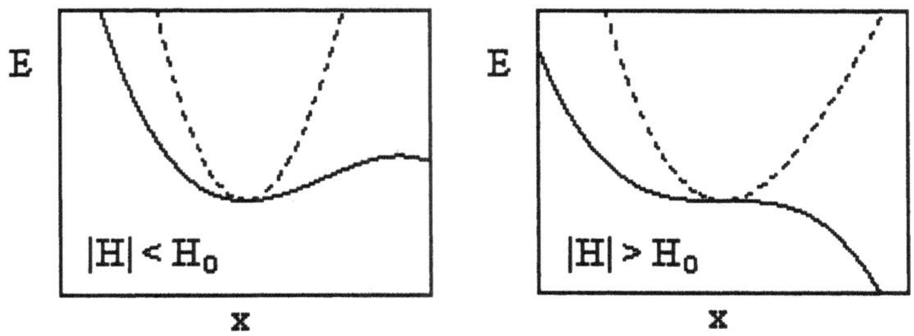

Figure 16. Multidimensionality of the coercivity problem: Only the trajectory corresponding to the lowest-lying mode contributes to the switching.

It is important to keep in mind that there are many possible magnetization-reversal modes, including modes whose energy scales as $1/H$, but only the lowest-lying one—that with the smallest nucleation or propagation field H_o—is of importance for the magnetic-viscosity and sweep-rate corrections to zero-temperature magnetization reversal [5, 37, 93, 175]. Other reversal modes are not forbidden, but typical energy-barrier differences are much larger than $k_B T$ and make the associated 'giant fluctuations' very unlikely. Fig. 16 illustrates the underlying energy landscape by schematically comparing the lowest-lying mode (solid line) with an excited mode (dashed line). The thermodynamically preferred direction is along the solid line, whereas excited modes are suppressed by a Boltzmann factor $\exp(-\Delta E_a/k_B T)$. An explicit example is shown in Fig. 17. The pinning and giant-fluctuation mechanisms (a) and (b) obey $m = 3/2$ power and $1/H$ laws, respectively. At low temperatures, magnetization reversal requires $E_a = 0$, and Eq. (29) yields some finite coercivity $H_c = H_o$. By contrast, the activation energy for the droplet configuration of Fig. 17(b), described by an $1/H$ law, amounts to the unphysical prediction of an infinite zero-temperature anisotropy.

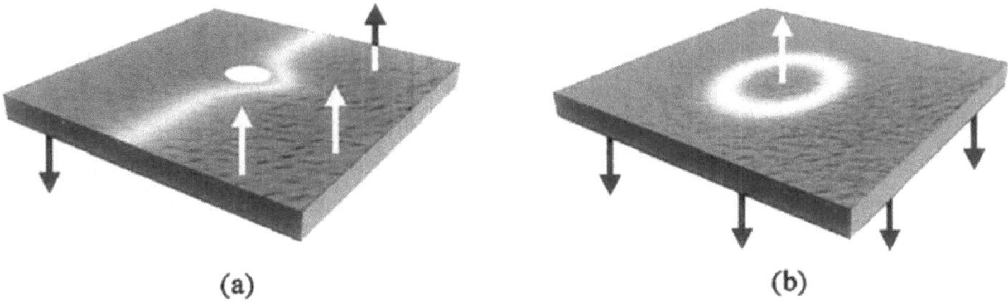

Figure 17. Coercivity mechanisms in thin films: (a) power-law type pinning and (b) $1/H$-type spontaneous reversal.

The above consideration requires energy barriers much larger than $k_B T$, as encountered in most nanostructures. An exception is very small particles, where $K_o V_o$ is comparable to $k_B T$, exhibit a rapid decay of the magnetization, which is known as *superparamagnetism* [160, 178, 179]. Defining superparamagnetism by a waiting time of τ = 100 s yields the zero-field stability condition $K_o V_o / k_B T \geq 25$, where $K_o V_o / k_B T \equiv \xi$ is referred to as the stability parameter. Thermal stability for 10 years corresponds to $\xi = 40$ [55, 159]. Other exceptions are atomic wires and monatomic thin films, where corrections to the power law of Eq. (28) must be considered.

Equation (29) amounts to a $T^{1/m}$ dependence of the coercivity [16], but this is only one consideration. In many cases, the biggest contribution to the temperature dependence of the coercivity is from the intrinsic $K_1(T)$ behavior (§2.2.3). This mechanism amounts to explicit changes in the energy landscape. The leading contribution is static, involving thermal averages $\langle K_1(T) \rangle$, but there is also a dynamic correction, with fluctuating saddle-point energies.

4. CASE STUDIES

4.1. Magnetic Localization and Cooperativity

From an atomic point of view, all nanostructures are cooperative, because interatomic exchange ensures well-defined local spin correlations. In other words, nanoscale magnetization processes involve blocks of spins rather than individual spins. The sizes of the correlated regions are nanoscale, in agreement with the analysis in §2.3, but vary from system to system [180]. The coherent-rotation and curling modes are delocalized, that is, they extend throughout the magnet. Delocalized magnetization states are favorable from the point of view of interatomic exchange, because the magnetization gradient is small. However, exchange is not the only consideration, because local variations of the magnetization cost some exchange energy but may be favorable from the point of view of local anisotropy inhomogenities.

The localization length and the coercivity reduction depend on the ratio of exchange and anisotropy energies. The stronger the anisotropy inhomogenity, the smaller the localization length L, and for a soft inclusion in a very hard matrix, as considered in §2.3, the localization length reduces to the size of the inclusion. For zero disorder, that is, for perfect ellipsoids of revolution, the localization length goes to infinity and the reversal degenerates into coherent rotation.

Cooperative effects are of great importance in advanced technology. For example, in high-density magnetic recording media they lead to the formation of interaction domains, which may improve the thermal stability but reduce the storage density. In permanent magnets the vanishing of the two-phase shoulders in hysteresis loops can be considered as a cooperative effect [180, 181], as is the above discussed low coercivity of soft-magnetic random-anisotropy magnets (Ch. 13). Similar arguments apply to finite-temperature phenomena (Fig. 18). On a local scale, the spin alignment remains largely ferromagnetic but thermal excitations and local perturbations break the cooperativity on a nanoscale.

Weakly interacting particles and grains are noncooperative. In this limit, one can use micromagnetic mean-field theories, the Preisach model [182, 183], and approaches based on Wohlfarth's remanence relation [184], such as Henkel [185], delta-M [186, 187], and delta-H [188] plots. When the interactions exceed a certain threshold, the behavior of the magnet changes from noncooperative to cooperative. In this regime, the above-mentioned approaches are no longer applicable. For example, the exchange field of strongly coupled small particles exceeds 100 T. In the mean-field approximation, this field adds to the external field and yields an unphysically high coercivity [122, 180]. In reality, the strong exchange field does not translate into a high coercivity, and cooperativity may even reduce the coercivity. This is because two strongly interacting particles behave like one particle, and there is no point in adding any strong internal interaction to the external field.

A rough criterion for the applicability of interaction-field models is obtained from the slope of the hysteresis loop at coercivity, $\chi_c = \mathrm{d}M(H_c)/\mathrm{d}H$. When M_s/χ_c is smaller than the interaction field, then the behavior of the magnet is governed by cooperative effects [180]. In terms of Fig. 14, this is not surprising, because broad switching-field distributions correspond to pronounced inhomogenities and therefore to strong localization.

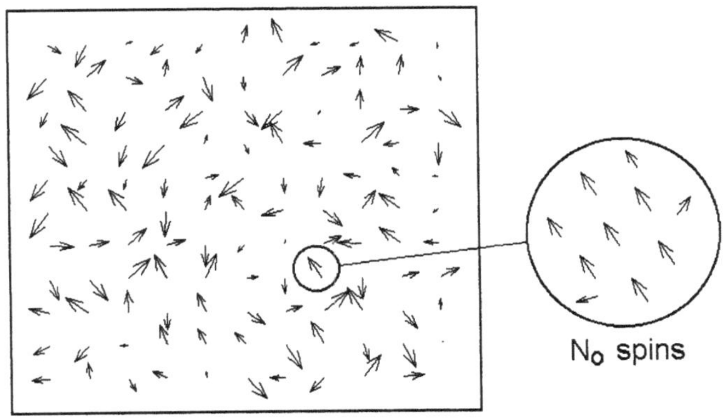

Figure 18. Cooperative spin blocks. When the size N of the particle becomes too big, then thermal activation leads to the formation of cooperative units of size N_0.

4.2. Narrow-Wall and Grain-Boundary Effects

Typical domain walls are smooth and extend over many interatomic distances. However, deviations from this continuum picture occur in very hard materials (narrow walls), at grain boundaries and in the case of geometrical constraints. Narrow-wall phenomena, which have been studied for example in rare-earth cobalt permanent magnets [189] and at grain boundaries [95, 96], involve individual atoms and atomic planes and lead to comparatively small corrections to the extrinsic behavior.

The nanoscale spin structure at grain boundaries and nanojunctions affects the performance of permanent magnets, magnetic recording media, soft magnets, and structures for spin electronics (Chs. 11-14). In permanent magnets, it is often necessary to maximize [16, 111] or minimize the intergranular exchange, depending on the desired reversal mechanism. Strong intergranular exchange reduces the coercivity of isotropic soft-magnetic nanostructures but is undesired in magnetic recording, where it negatively affects the storage density [159, 190], and the resistance of spin-electronic structures depends on the local magnetization at interfaces and at junctions [95, 96, 191]. Some aspects of grain-boundary magnetism were anticipated long ago [94, 111, 113, 192], but polycrystalline granular interfaces and constrained domain walls was first investigated have been investigated only recently.

We start from Eq. (6) and restrict ourselves to linear case of weakly textured systems, so that we can use a linear approach similar to that in §2.3.4 [5]. The corresponding equation for the local easy axis is $\mathbf{n}(\mathbf{r}) = \sqrt{1 - a^2(\mathbf{r})}\ e_z + \mathbf{a}(\mathbf{r})$, where $\mathbf{a}(\mathbf{r})$ is the transverse vector components of $\mathbf{n}$. Linearization yields [5]

$$F = \int \left[A\,(\nabla \mathbf{m})^2 + K_1\,(\mathbf{m} - \mathbf{a})^2 + \frac{1}{2}\,\mu_o\,M_s\,H\,\mathbf{m}^2] \right] \mathrm{d}\mathbf{r} \tag{32}$$

As in Eq. (32), the magnetostatic selfinteraction is incorporated into K_1 and H. To minimize F with respect to $\mathbf{m}(\mathbf{r})$ we exploit that the minimum of any functional $F = \int f\,\mathrm{d}V$ is given by the functional derivative $\delta F/\delta\mathbf{m}(\mathbf{r}) = 0$. Explicitly,

$$\frac{\delta F}{\delta \mathbf{m}(\mathbf{r})} = - \nabla \left(\frac{\partial f}{\partial\, \nabla \mathbf{m}(\mathbf{r})} \right) + \frac{\partial f}{\partial \mathbf{m}(\mathbf{r})} \tag{33}$$

so that

$$- \nabla(A\nabla\mathbf{m}) + \left(K_1 + \frac{1}{2}M_s H\right)\mathbf{m} = K_1\,\mathbf{a}(\mathbf{r}) \tag{34}$$

This equation means that the polycrystalline easy-axis disorder $\mathbf{a}(\mathbf{r})$ acts as a random inhomogenity.

The term $\nabla(A\nabla\mathbf{m})$ reflects due to local character of exchange stiffness $A(\mathbf{r})$ [111]. For sharp phase boundaries, the exchange term reduces to the boundary condition

$$\left(A(\mathrm{x})\frac{\partial m}{\partial x}\right)\Big|_{\mathrm{x}_0-\varepsilon} = \left(A(\mathrm{x})\frac{\partial m}{\partial x}\right)\Big|_{\mathrm{x}_0+\varepsilon} \tag{35}$$

Figure 19 illustrates the physical meaning of this boundary condition. A jump in A(x) leaves the magnetization continuous but yields a change in the slope of the perpendicular magnetization component m(x). The solutions of (34) are exponentially decaying in the adjacent grains and nearly linear in the interface region [95, 96].

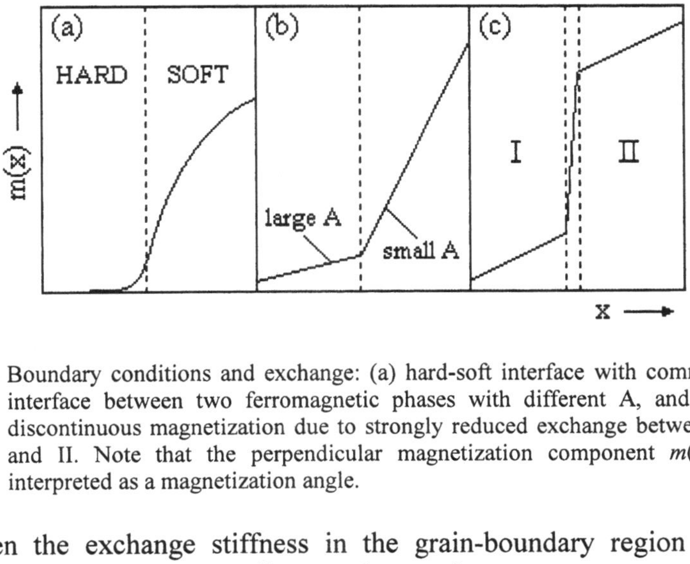

Figure 19. Boundary conditions and exchange: (a) hard-soft interface with common A, (b) interface between two ferromagnetic phases with different A, and (c) quasi-discontinuous magnetization due to strongly reduced exchange between grains I and II. Note that the perpendicular magnetization component m(x) can be interpreted as a magnetization angle.

When the exchange stiffness in the grain-boundary region is much lower than that of the two adjacent phases, then one encounters a *quasi-discontinuity* of the magnetization [95, 96], as shown in Fig. 19(c). Experimentally, reduced interface exchange reflects real-structure features such as impurity atoms diluting the interatomic exchange, oxide layers covering the grains, and interface amorphization. Putting calculated spin structures m(x) into Eq. (32) and evaluating the integral as a function of the misalignment vector $\mathbf{a}$ yields the *effective intergranular exchange*

$$J_{\mathrm{eff}} = S_\mathrm{o} \sqrt{A\,K_1}\,\frac{1}{1 + \dfrac{\pi\,D\,A}{2\,\delta_\mathrm{B} A'}} \tag{36}$$

where S_o is the interface area [96]. This equation shows that the effective exchange can never be larger than the value $S_\mathrm{o}(AK_1)^{1/2}$, which scales as Bloch-wall energy of a perfect magnetic crystal. By contrast, the widely used 'hard interface' approach, which assigns bonds of strength $J' \approx A'/a$ to each of the $N \approx S_\mathrm{o}/a^2$ adjacent pairs of atoms, yields $J_\mathrm{eff} \approx S_\mathrm{o}A'/a$ and generally overestimates the effective exchange. The explicit dependence of A on R modifies the scaling behavior to $L \sim R^{(2\text{-d})/(4\text{-d})}$ and $H_\mathrm{c} \sim R^{\mathrm{d}/(4\text{-d})}$. Note that the exponent for L changes sign for $d \leq 2$. In this case, the exchange through the grain boundary is not able to overcompensate the reduction of grain size, and there are no correlated grains with $L > R$.

In the limit of vanishing anisotropy, the situation remains similar to Fig. 19(c), because the anisotropy changes the curvature of the magnetization angle. Solving Eq. (34) for this special case yields the following relation for the effective exchange A_eff: $D/A_\mathrm{eff} = (D\text{-}t)/A + t/A^\mathrm{I}$. This is of practical importance in soft magnetism (Ch. 13).

In a *layer-resolved* or atomic analysis, the ∇ operator in Eqs. (32-35) must be replaced by magnetization-angle differences, but a comparison with the continuum solution [95, 96] reveals only minor corrections due to the discrete nature of the layers. Note, however, that the layer resolved anisotropies and exchange constants may deviate from the respective bulk values.

Using the integral $\int (\nabla\mathbf{M})^2 \mathrm{d}x \approx M_\mathrm{s}^2 \int (\nabla\mathbf{m})^2 \mathrm{d}x$ as a crude measure to gauge the spin-dependent scattering ability of an interface we find that the scattering is maximized for interface thicknesses of the order of $D = \delta_\mathrm{o}\, A'/A$. Compared to the relatively small Bloch-wall scattering, where $\int(\nabla\mathbf{m})^2\mathrm{d}x \approx 1/\delta_\mathrm{o}$, the maximum scattering is enhanced by a factor A/A' Unfortunately, strong reductions of A' are likely to negatively affect the spin injection through the boundary region, thereby reducing the magnetoresistance. Another way of enhancing the scattering is using very hard materials, where δ_o is small, but this requires large fields to switch the magnetization direction [95].

Random-anisotropy contributions, as epitomized by the **a**(**r**)-term in Eq. (34), are an important aspect of nanomagnetism, but their detailed treatment goes beyond the scope of this chapter. Some aspects of random-anisotropy magnetism will be treated in the chapter on soft magnets, and we also recommend reading or consulting the rich original and review literature [16, 48, 116, 124, 189, 193-197].

4.3. Quantum Entanglement between Magnetic Nanodots

The use of quantum bits or qubits is a promising way to meet the ever-increasing needs of information technology, with various advantages over classical information processing in areas such as factorization and cryptography [198-200]. Most systems considered at present, including magnetic structures such as spin chains and spin clusters, operate at very low temperatures, typically much smaller than 1 K [201-203]. The smallness of Bohr's magneton, $\mu_B/k_B = 0.672$ K/T, makes it difficult to exploit magnetic fields [201] at temperatures significantly above 1 K. Superconducting magnets are able to create fields much larger than 1 T (10 kOe), but they are very cumbersome and may not establish a practical alternative. Exchange anisotropy is, in principle, an alternative [201, 202], but lowest-order exchange is isotropic, and the exchange anisotropies are small and difficult-to-realize relativistic corrections to the isotropic exchange (§2.1.5). However, anisotropic magnetic nanodots may open the door for quantum information processing significantly above 4.2 K. The idea is to exploit the quantum entanglement of the dots [204].

Figure 20 shows a model of two coupled and generally nonequivalent magnetic nanodots or clusters. The dots' total spins S and S' can be written as $S = N\,S_o$ and $S' = N'\,S'_o$, where N and N' are the respective numbers of magnetic atoms per dot. The Hamiltonian of the first dot contains the Zeeman energy $- g\mu_o\mu_B H\,\hat{S}_z$ and the anisotropy term

$$H_a = - \frac{K\,N}{3\,S^2}\,(3\,\hat{S}_z^2 - \hat{S}^2) \qquad (37)$$

where K is the anisotropy energy per atom. Analog relations exist for the second dot. There are two reasons for considering nonequivalent dots. First, real nanomagnets tend to have imperfections, and it is extremely difficult to produce identical dots. Second, nonequivalent dots may exhibit unequal level spacings, which simplifies the addressing of well-defined quantum states, for example in resonance experiments.

The energy difference (level spacing) between the lowest two eigenvalues of Eq. (37), $KN(2S\text{-}1)/S^2$, determines the maximum operation temperature. We assume that the two dots are coupled by a Heisenberg-type exchange J. It may be realized, for example, by Ruderman-Kittel-Kasuya-Yosida (RKKY) interaction through a substrate or medium [27, 205], by a nanojunction, or by magnetostatic interactions. The interaction is the basis for entanglement of the dots' wave functions, which is a precondition for quantum computing.

Schrödinger's *verschränkung* or entanglement is a quantum effect without classical analog and plays a key role in quantum computing [201,

206]. In a simple two-particle interpretation, the spin of a given particle depends on that of the second spin (entangled state) or is independent of the second particle's spin (nonentangled or separable state). In the Schmidt decomposition, the two-particle wave function can be written as

$$|\Psi\rangle = \alpha|00\rangle + \beta|01\rangle + \gamma|10\rangle + \delta|11\rangle \qquad (38)$$

For example, the four maximally entangled Bell states are proportional to $|00\rangle \pm |11\rangle$ and $|01\rangle \pm |10\rangle$. To describe low-temperature entanglement, we construct a Schmidt basis from the two lowest-lying single-dot states $|0\rangle$ and $|1\rangle$, which have $|S_z| = S$ and $|S_z| = S - 1$, respectively. Note that higher excitations do not affect the entanglement of the low-lying states [204]. The right-hand side of Fig. 20 illustrates the physical meaning of this basis for ferromagnetic and antiferromagnetic couplings.

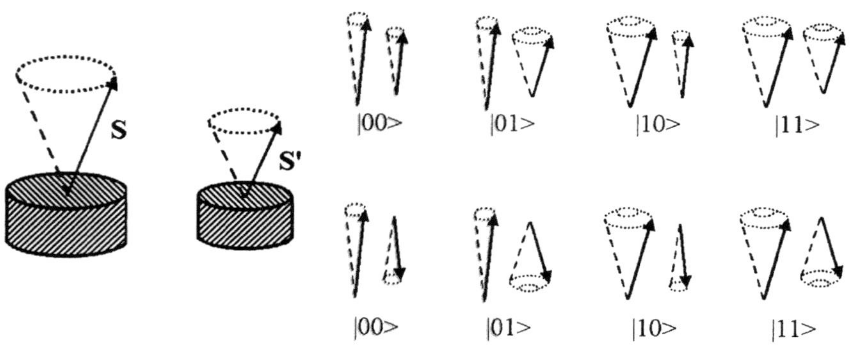

Figure 20. Quantum states of interacting nanodots. The dots (left) may be coupled ferromagnetic (top right) or antiferromagnetic (bottom right). The labels $|0\rangle$ and $|1\rangle$, denote states where $|m| = S$ and $|m| = S - 1$, respectively.

A quantitative entanglement measure is the concurrence C [206]. For nonentangled (separable) states, $C = 0$, whereas the maximally entangled Bell states exhibit $C = \pm 1$. The concurrence is obtained by evaluating the matrix elements of the total Hamiltonian and examining the lowest-lying two-dot eigenstates [204, 207]. It depends on whether the coupling is ferromagnetic (FM, $J > 0$) or antiferromagnetic (AFM, $J < 0$). The ferromagnetic ground state and the first two excited AFM states are separable, that is, $C = 0$. Nonzero entanglement is encountered in the AFM ground state and between low-lying ferromagnetic excitations [204].

Figure 21 shows C as a function of the anisotropy of the second dot. The entanglement of the low-lying FM excitations exhibits a resonant peak whose width depends on the interaction strength. Since a local magnetic field shifts the single-dot energies, a field gradient can be used to tune the entanglement. The dashed line in Fig. 21 shows that there is no peak in the antiferromagnetic case.

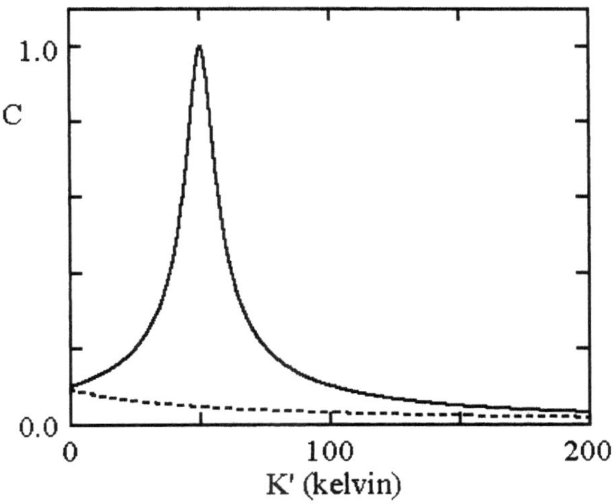

Figure 21. Entanglement as a function of the anisotropy K' of the second dot. The solid and dashed lines denote FM and AFM couplings, respectively. The parameters are $S_o = S_o' = 1$, $N = 1000$, $N' = 1100$, $K = 50$ K, and $J = 0.005$ K. For simplicity, $H = 0$ and $H' = 0$. The FM maximum is a resonance effect involving |01> and |10> states; there is no similar resonance in the AFM case.

Since the basic level splitting and the maximum operating temperature are proportional to the anisotropy, it is necessary to use dots with high magnetocrystalline anisotropy. Dots with shape anisotropy and semihard dots, such as Co, cannot be used, because they correspond to temperatures of one kelvin or less.

Much higher temperatures are achievable for $SmCo_5$ and other highly anisotropic materials. However, typical hard-magnetic materials have temperature-dependent anisotropies that are maximized at or above room temperature [16, 77]. Little work has been done to optimize anisotropies at low temperatures, although it is known that some compounds with $T_c < 300$ K have huge anisotropies of 100 to 1000 MJ/m^3 [83, 208].

In sufficiently small dots, the maximum operating temperature is proportional to the magnetocrystalline anisotropy of the dot material. An upper limit of the operation temperature is significantly above 10 K but probably below 100 K. An upper limit to the temperature above which spin-wave-like excitations and magnetic domains destroy quantum coherence is given by the energies of the lowest-lying spin-wave states. They scale as Aa^3/L^2, where A is the exchange or spin-wave stiffness of the dot material [16], a is the interatomic distance, and L is the dot size. For typical materials at temperatures significantly higher than 1 K, this quantity does not exceed a few nanometers. In bigger dots, low-lying spin-waves excitations have energies comparable to K and decoherence occurs very fast.

5. CONCLUDING REMARKS

From a scientific point of view, nanomagnetic effects are intermediate between atomic-scale magnetism and macroscopic phenomena but cannot be reduced to a mixture of the two limits. The main reason is the competing involvement of relativistic interactions, which give rise to an additional length scale of a_0/α = 7.25 nm. Intrinsic properties, such as spontaneous magnetization, Curie temperature, and magnetocrystalline anisotropy, reflect comparatively strong quantum-mechanical and spin-orbit interactions. They are realized on atomic length scales and therefore well-defined for nano-structures. However, surface and interface contributions such as magnetic interface anisotropy may be strong and outweigh bulk contributions on a nanoscale. Extrinsic properties, such as coercivity, are realized on larger length scales and exhibit a pronounced real-structure dependence.

A particularly subtle property is the Curie temperature, which describes the onset of ferromagnetism and involves long-range thermodynamic fluctuations. In a strict sense, the Curie temperature of finite bodies is zero, but the nonequilibrium character of practically encountered magnetic phenomena and the low energy differences associated with long-range thermodynamic fluctuations mean that particles or grains larger than about 1 nm., are virtually indistinguishable from true ferromagnets. For this reason, nanostructuring cannot be used to improve the Curie temperature.

It is important to distinguish between equilibrium and nonequilibrium extrinsic properties. The equilibrium behavior of magnetic nanostructures, as epitomized by the critical single-domain radius, is largely irrelevant to hysteresis. For example, in highly anisotropic rare-earth transition-metal permanent magnets, the critical single-domain size is of the order of 1 μm, but this does not mean that the magnetization reversal in such a particle is Stoner-Wohlfarth like. In fact, the reversal is localized, and typical coercivities are realized on a length scales of a few nanometers.

From a structural point of view, surfaces, interfaces, and bottlenecks (junctions) have a strong impact on nanomagnetism. One issue is that exchange at grain boundaries affects the coupling between nanograins and, indirectly, the extrinsic properties of the structures. Reduced grain-boundary exchange leads to a quasi-discontinuity of the magnetization. By contrast, anisotropy changes in the grain-boundary region have no major effect on the spin structure.

Intrinsic properties correspond to a very fast equilibrium, whereas the equilibration times of extrinsic properties cover a broad range, from a about one nanosecond in soft-magnetic resonance experiments and to millions of years in magnetic rocks. The slowest mechanism is thermally activated jumps over nanomagnetic energy barriers. This results in small low-temperature magnetic-viscosity corrections to the leading intrinsic contribution. Since the phase-space trajectories responsible for thermally activated magnetization reversal are very close to static trajectories, 'giant' thermodynamic fluctuations involving arbitrary modes can safely be ignored.

In conclusion, magnetic nanostructures exhibit various scientifically interesting and technologically important deviations from bulk and thin-film magnets. For example, there is a resonant quantum entanglement of the low-lying states of coupled hardmagnetic dots, which may be exploited in future quantum-information processing above 4.2 K. Some others examples, mentioned or elaborated here and throughout this book, are the energy-product enhancement in nanostructured two-phase magnets whose hard-magnetic performance is improved by adding a soft-magnetic phase, multilayered and granular spin-valve structures, nanostructured soft magnets, and magnetic recording media. The principles and mechanisms outlined in this chapter make it possible to realize properties not achievable in single-phase bulk and thin-film materials, and form a basis for future experimental, technological and theoretical developments in the fascinating field of nanomagnetism.

Acknowledgement

This work has benefited from discussions with P. A. Dowben, A. Kashyap, R. D. Kirby, A. Y. Istomin, A. F. Starace, D. Leslie-Pelecky, S.-H. Liou, J.-P. Liu, K. D. Sorge, and D. J. Sellmyer. The underlying research is supported by NSF MRSEC, DARPA/ARO, DOE, ONR, AFOSR, the W. M. Keck oundation, and CMRA.

References

[1] W. Heisenberg, Z. Phys. **49**, 619 (1928).
[2] F. Bloch, Z. Phys. **57**, 545 (1929).
[3] J. C. Slater, Phys. Rev. **49**, 537-545 (1936).
[4] E. C. Stoner, Proc. Roy. Soc. A**165**, 372 (1938).
[5] R. Skomski, J. Phys.: Condens. Matter **15**, R841 (2003).
[6] R. Skomski, J. Magn. Magn. Mater. **272-276**, 1476 (2004).

[7] F. Bloch and G. Gentile, Z. Phys. **70**, 395 (1931).
[8] H. Brooks, Phys. Rev. **58**, 909 (1940).
[9] R. Skomski, H.-P. Oepen, and J. Kirschner, Phys. Rev. B **58**, 3223 (1998).
[10] E. Ising, Z. Phys. **31**, 253 (1925).
[11] J. C. Slater, Rev. Mod. Phys. **25**, 199 (1953).
[12] Depending on the context, the real structure is also known as the defect structure, morphology, or metallurgical microstructure. For example, see: D. Sander, Rep. Prog. Phys. **62**, 809 (1999).
[13] F. Bloch, Z. Phys. **74**, 295 (1932).
[14] L. Landau and E. Lifshitz, Phys. Z. Sowjetunion **8**, 153 (1935).
[15] M. Kersten, Z. Phys. **44**, 63 (1943).
[16] R. Skomski and J. M. D. Coey, "Permanent Magnetism", Institute of Physics, Bristol 1999.
[17] C. Kittel, "Introduction to Solid State Physics", Wiley, New York 1986.
[18] B. T. Thole, P. Carra, F. Sette, and G. van der Laan, Phys. Rev. Lett. **68**, 1943 (1992).
[19] Both spin-orbit coupling and Zeeman interactions are obtained as terms in the Pauli expansion of the Dirac equation.
[20] F. Cyrot-Lackmann, J. Phys. Chem. Solids **29**, 1235 (1968).
[21] V. Heine, Solid State Phys. **35**, 1 (1980).
[22] A. P. Sutton, "Electronic Structure of Materials", Oxford University Press, Oxford 1993.
[23] M. C. Desjonquères and D. Spanjaard, "Concepts in Surface Physics", Springer, Berlin 1993.
[24] J. S. Smart, "Effective Field Theories of Magnetism", Saunders, Philadephia 1966.
[25] N. H. Duc, T. D. Hien, D. Givord, J. J. M. Franse, and F. R. de Boer, J. Magn. Magn. Mater. **124**, 305 (1993).
[26] N. W. Ashcroft and N. D. Mermin, "Solid State Physics", Saunders, Philadelphia 1976.
[27] R. Skomski, Europhys. Lett. **48**, 455 (1999).
[28] R. Skomski, R. F. Sabiryanov, and S. S. Jaswal, J. Appl. Phys. **87**, 5890 (2000).
[29] J. A. De Toro, M. A. López de la Torre, J. M. Riveiro, J. Bland, J. P. Goff, and M. F. Thomas, Phys. Rev. B **64**, 224421 (2001).
[30] D. J. Priour Jr., E. H. Hwang, and S. Das Sarma, Phys. Rev. Lett. **92**, 117201 (2004).
[31] J. M. D. Coey, M. Venkatesan, and C. B. Fitzgerald, Nature Materials 2, 173 (2005).
[32] D. C. Mattis, "Theory of Magnetism", Harper and Row, New York 1965.
[33] K. Moorjani and J. M. D. Coey, "Magnetic Glasses", Elsevier, Amsterdam 1984.
[34] P. Fulde, "Electron Correlations in Molecules and Solids", Springer, Berlin 1991.
[35] J. M. Yeomans, "Statistical Mechanics of Phase Transitions", University Press, Oxford 1992.
[36] R. Skomski and D. J. Sellmyer, J. Appl. Phys. **87**, 4756 (2000).
[37] R. Skomski, D. Leslie-Pelecky, R. D. Kirby, A. Kashyap, and D. J. Sellmyer, Scripta Mater. **48**, 857 (2003).
[38] H. R. Ma and C. H. Tsai, Solid State Commun. **55**, 499 (1985).
[39] W. Maciejewski and A. Duda, Solid State Commun. **64**, 557 (1987).
[40] H. K. Sy, Phys. Lett. A **120**, 203 (1987).

[41] W. Maciejewski, IEEE Trans. Magn. **26**, 213 (1990).

[42] R. W. Wang and D. L. Mills, Phys. Rev. B **46**, 11681 (1992).

[43] There are a few exceptions, such as very weak itinerant ferromagnets (for example $ZrZn_2$) and low-spin high-spin transition in fcc iron.

[44] R. Skomski, A. Kashyap, and D. J. Sellmyer, IEEE Trans. Magn. **39**, 2917 (2003).

[45] P. Mohn, "Magnetism in the Solid State", Springer, Berlin 2003.

[46] A. Kashyap, R. Skomski, R. F. Sabiryanov, S. S. Jaswal, and D. J. Sellmyer, IEEE Trans. Magn. **39**, 2908 (2003).

[47] R. Skomski and P. A. Dowben, Europhys. Lett. **58**, 544 (2002).

[48] K.-H. Fischer and A. J. Hertz, "Spin Glasses", University Press, Cambridge 1991.

[49] D. Sander, R. Skomski, C. Schmidthals, A. Enders, and J. Kirschner, Phys. Rev. Lett. **77**, 2566 (1996).

[50] M. Farle, Rep. Prog. Phys. **61**, 755 (1998).

[51] R. M. Bozorth, "Ferromagnetism", van Nostrand, Princeton, New Jersey 1951.

[52] H. Zeng, M. Zheng, R. Skomski, D. J. Sellmyer, Y. Liu, L. Menon, and S. Bandyopadhyay, J. Appl. Phys. **87**, 4718 (2000).

[53] G. T. A. Huysmans and J. C. Lodder, J. Appl. Phys. **64**, 2016 (1988).

[54] M. Zheng, R. Skomski R, Y. Liu, and D. J. Sellmyer, J. Phys.: Condens. Matter. **12**, L497 (2000).

[55] D. J. Sellmyer, M. Zheng, and R. Skomski, J. Phys.: Condens. Matter **13**, 433 (2001).

[56] J. A. Osborn, Phys. Rev. **67**, 351 (1945).

[57] H. Bethe, Ann. Physik, **3**, 133 (1929).

[58] C. J. Ballhausen, "Ligand Field Theory", McGraw-Hill, New York 1962.

[59] M. T. Hutchings, Solid State Phys. **16**, 227 (1964).

[60] J. F. Herbst, Rev. Mod. Phys. **63**, 819 (1991).

[61] J. M. D. Coey (ed.), "Rare-earth Iron Permanent Magnets", University Press, Oxford 1996.

[62] L. Néel, J. Phys. Radium **15**, 225 (1954).

[63] Y. Millev, R. Skomski, and J. Kirschner, Phys. Rev. B **58**, 6305 (1998).

[64] J. A. C. Bland and B. Heinrich (eds.), "Ultrathin Magnetic Structures I", Springer, Berlin 1994.

[65] M. T. Johnson, P. J. H. Bloemen, F. J. A. den Broeder, and J. J. de Vries, Rep. Prog. Phys. **59**, 1409 (1996).

[66] U. Gradmann, in: "Handbook of Magnetic Materials", Vol. 7, Ed. K. H. J. Buschow, Elsevier, Amsterdam 1993, p. 1.

[67] R. H. Victora and J. M. McLaren, Phys. Rev. B **47**, 11583 (1993).

[68] M. Jamet, W. Wernsdorfer, C. Thirion, D. Mailly, V. Dupuis, P. Mélinon, and A Pérez, Phys. Rev. Lett. **86**, 4676 (2001).

[69] D. S. Chuang, C. A. Ballentine, and R. C. O'Handley, Phys. Rev. B **49**, 15084 (1994).

[70] J. G. Gay and R. Richter, Phys. Rev. Lett. **56**, 2728 (1986).

[71] G. H. O. Daalderop, P. J. Kelly, and M. F. H. Schuurmans, Phys. Rev. B **42**, 7270 (1990).

[72] D.-S. Wang, R.-Q. Wu, and A. J. Freeman, Phys. Rev. B **47**, 14932 (1993).

[73] M. Eisenbach, B. L. Györffy, G. M. Stocks, and B. Újfalussy, Phys. Rev. B **65**, 144424 (2002).

[74] M. Komelj, C. Ederer, J. W. Davenport, and M. Fähnle, Phys. Rev. B **66**, 140407 (2002).

[75] J.-F. Hu, I. Kleinschroth, R. Reisser, H. Kronmüller, and Sh. Zhou, phys. stat. sol. (a) **138**, 257 (1993).

[76] E. Lectard, C. H. Allibert, and R. Ballou, J. Appl. Phys. **75**, 6277 (1994).

[77] K. Kumar, J. Appl. Phys. **63**, R13 (1988).

[78] R. Skomski, J. Appl. Phys. **83**, 6724 (1998).

[79] J. M. Cadogan, J. P. Gavigan, D. Givord, and H. S. Li, J. Phys. F: Met. Phys. **18**, 779 (1988).

[80] J. F. Liu, T. Chui, D. Dimitrov, and G. C. Hadjipanayis, Appl. Phys. Lett. **85**, 3007 (1998).

[81] E. R. Callen and H. B. Callen, Phys. Rev., **129**, 578 (1963).

[82] J. B. Staunton, S. Ostanin, S. S. A. Razee, B. L. Gyorffy, L. Szunyogh, B. Ginatempo, and E. Bruno, Phys. Rev. Lett. **93**, 257204 (2004).

[83] R. Skomski, J. Appl. Phys. **91**, 8489 (2002).

[84] O. N. Mryasov, U. Nowak, K. Guslienko, and R. Chantrell, Europhys. Lett. 69, 805 (2005); O. N. Mryasov, U. Nowak, K. Guslienko, and R. Chantrell, unpublished (2004).

[85] N.H. Hai, N. M. Dempsey, and D. Givord, IEEE Trans. Magn. **39**, 2914 (2003).

[86] G. H. O. Daalderop, P. J. Kelly, and M. F. H. Schuurmans, in: "Ultrathin magnetic structures I", Eds.: J.A.C. Bland and B. Heinrich, Springer, Berlin 1994, p. 40.

[87] G. Brown, B. Kraczek, A. Janotti, T. C. Schulthess, G. M. Stocks, and D. D. Johnson, Phys. Rev. B **68**, 052405 (2003).

[88] R. A. McCurrie, "Ferromagnetic Materials—Structure and Properties", Academic Press, London 1994.

[89] J. Zhou, R. Skomski, K. D. Sorge, and D. J. Sellmyer, Scripta Materialia (2005) (in press).

[90] W. F. Brown, "Micromagnetics", Wiley, New York 1963.

[91] The prefix 'micro' originates from the greek word □□□□ός, meaning "small" but not implying any well-defined length scale.

[92] M. Sagawa, S. Fujimura, N. Togawa, H. Yamamoto, and Y. Matsuura, J. Appl. Phys. **55**, 2083 (1984); M. Sagawa, S. Hirosawa, H. Yamamoto, S. Fujimura and Y. Matsuura, Jpn. J. Appl. Phys. **26**, 785 (1987).

[93] A. Aharoni, "Introduction to the Theory of Ferromagnetism", University Press, Oxford, 1996.

[94] H. R. Hilzinger and H. Kronmüller, phys. stat. sol. (b), **54**, 593, (1972).

[95] R. Skomski, in: "Spin Electronics", Eds.: M. Ziese and M. J. Thornton, Springer, Berlin 2001, p. 204.

[96] R. Skomski, H. Zeng, and D. J. Sellmyer, IEEE Trans. Magn. **37**, 2549 (2001).

[97] C. Kittel, Rev. Mod. Phys. **21**, 541 (1949).

[98] D. J. Craik and R. S. Tebble, Rep. Prog. Phys. **24**, 116 (1961).

[99] R. Becker and W. Döring, "Ferromagnetismus", Springer, Berlin 1939.

[100] S. Chikazumi, "Physics of Magnetism", Wiley, New York 1964.

[101] M. Hehn, K. Ounadjela, J. Bucher, F. Rousseaux, D. Decanini, B. Bartenlian, and C. Chappert, Science **272**, 1782 (1995).

[102] R. Skomski, A. Kashyap, K. D. Sorge, and D. J. Sellmyer, J. Appl. Phys. **95**, 7022 (2004).
[103] E. C. Stoner and E. P. Wohlfarth, Phil. Trans. Roy. Soc. A**240**, 599 (1948). reprinted in IEEE Trans. Magn. **MAG-27**, 3475 (1991).
[104] A. Aharoni, Rev. Mod. Phys. **34**, 227 (1962).
[105] R. Skomski, J. P. Liu, and D. J. Sellmyer, Phys. Rev. B **60**, 7359 (1999).
[106] R. Skomski, J. Zhou, A. Kashyap, and D. J. Sellmyer, IEEE Trans. Magn. **40**, 2946 (2004).
[107] J. Zhou, R. Skomski, C. Chen, G.C. Hadjipanayis, and D.J. Sellmyer, Appl. Phys. Lett. **77**, 1514 (2000).
[108] J. Zhou, A. Kashyap, Y. Liu, R. Skomski, and D. J. Sellmyer., IEEE Trans. Magn. **40**, 2940 (2004).
[109] W. F. Brown, Rev. Mod. Phys. **17**, 15 (1945).
[110] R. Skomski, phys. stat. sol. (b) **174**, K77 (1992).
[111] R. Skomski and J. M. D. Coey, Phys. Rev. B **48**, 15812 (1993).
[112] R. Skomski, J. Appl. Phys. **83**, 6503 (1998).
[113] S. Nieber and H. Kronmüller, phys. stat. sol. (b) **153**, 367 (1989).
[114] R. Skomski, H. Zeng, M. Zheng, and D. J. Sellmyer, Phys. Rev. B **62**, 3900 (2000).
[115] R. Skomski, J. Appl. Phys. **91**, 7053 (2002).
[116] R. Coehoorn, D. B. de Mooij, J. P. W. B. Duchateau, and K. H. J. Buschow, J. de Physique **49**, C-8 669 (1988).
[117] E. F. Kneller and R. Hawig, IEEE Trans. Magn. **27**, 3588 (1991).
[118] I. A. Al-Omari and D. J. Sellmyer, Phys. Rev. B **52**, 3441 (1995).
[119] J. P. Liu, C. P. Luo, Y.Liu, and D. J. Sellmyer, Appl. Phys. Lett. **72**, 483 (1998).
[120] E. E. Fullerton, J. S. Jiang, C. H. Sowers, J. E. Pearson, and S. D. Bader, Appl. Phys. Lett. **72**, 380 (1998).
[121] J. Ding, Y. Liu, P. G. Mccormick, and R. Street, J. Appl. Phys. **75**, 1032 (1994).
[122] E. Callen, Y. J. Liu, and J. R. Cullen, Phys. Rev. B **16**, 263 (1977).
[123] K.-H. Müller, J. Schneider, A. Handstein, D. Eckert, P. Nothnagel, and H. R. Kirchmayr, Mat. Sci. Eng. **A133**, 151 (1991).
[124] A. Manaf, P. A. Buckley, and H. A. Davies, J. Magn. Magn. Mater. **128**, 302 (1993).
[125] H.-W. Zhang, B.-H. Li, J. Wang, J. Zhang, Sh.-Y. Zhang, and B.-G. Shen, J. Phys. D: Appl. Phys. **33**, 3022 (2000).
[126] E. E. Fullerton, S. J. Jiang, and S. D. Bader, J. Magn. Magn. Mater. **200**, 392 (1999).
[127] M. Sawicki, G. J. Bowden, P. A. J. de Groot, B. D. Rainford, J. M. L. Beaujour, R. C. C. Ward, and M. R. Wells, Phys. Rev. B **62**, 5817 (2000).
[128] R. J. Astalos and R. E. Camley, Phys. Rev. B **58**, 8646 (1998).
[129] G. C. Hadjipanayis, J. Magn. Magn. Mater. **200**, 373 (1999).
[130] G. J. Bowden, J. M. L. Beaujour, S. Gordeev, P. A. J. de Groot, B. D. Rainford, and M. Sawicki, J. Phys.: Condens. Matter **12**, 9335 (2000).
[131] Y. Yoshizawa, S. Oguma, and K. Yamauchi, J. Appl. Phys. **64**, 6044 (1988).
[132] R. Coehoorn, D. B. de Mooij, and C. de Waard, J. Magn. Magn. Mater. **80**, 101 (1989).
[133] R. Street and J. C. Wooley, Proc. Phys. Soc. **A62**, 562 (1949).

[134] E. Kneller, in: "Handbuch der Physik XIII/2: Ferromagnetismus", Ed.: H. P. J. Wijn, Springer, Berlin 1966, p. 438.
[135] L. Néel, Ann. Géophys. **5**, 99 (1949).
[136] E. Kneller, "Ferromagnetismus", Springer, Berlin 1962.
[137] W. F. Brown, Phys. Rev. **130**, 1677 (1963).
[138] R. Zwanzig, Phys. Rev. **124**, 983 (1961).
[139] H. Mori, Prog. Theor. Phys. 33, 423 (1965).
[140] R. Skomski, R. D. Kirby, and D. J. Sellmyer, J. Appl. Phys. **85**, 5069 (1999).
[141] W. F. Brown, J. Appl. Phys. **30**, 625 (1959).
[142] S. V. Vonsovskii, "Ferromagnetic Resonance", Pergamon Press, Oxford 1966.
[143] S. V. Vonsovskii, "Magnetism", John Wiley, New York 1974.
[144] Multidimensional relations are obtained as straightforward vector generalization in spin space, $s \rightarrow \underline{s}$ [5].
[145] H. A. Kramers, Physica **7**, 284 (1940).
[146] P. Hänggi, P. Talkner, and M. Borkovec, Rev. Mod. Phys. **62**, 251 (1990).
[147] F. Bloch, Z. Phys. **61**, 206 (1930).
[148] F. J. Dyson, Phys. Rev. **102**, 1217 (1956).
[149] J. R. Eshbach and R. W. Damon, Phys. Rev. **118**, 1208 (1960).
[150] J. Shen, R. Skomski, M. Klaua, H. Jenniches, S. S. Manoharan, and J. Kirschner, Phys. Rev. B **56**, 2340 (1997).
[151] R. Arias, and D. L. Mills, Phys. Rev. B **63**, 134439 (2001).
[152] J. Jorzick, S. O. Demokritov, B. Hillebrands, M. Bailleul, C. Fermon, K. Y. Guslienko, A. N. Slavin, D. V. Berkov, and N. L. Gorn, Phys. Rev. Lett. **88**, 047204 (2002).
[153] M. I. Chipara, R. Skomski, and D. J. Sellmyer, J. Magn. Magn. Mater. **249**, 246 (2002).
[154] Z. K. Wang, M. H. Kuok, S. C. Ng, D. J. Lockwood, M. G. Cottam, K. Nielsch, R. B. Wehrspohn, and U. Gösele, Phys. Rev. Lett. **89**, 027201 (2002).
[155] K. Y. Guslienko, B. A. Ivanov, V. Novosad, Y. Otani, H. Shima, and K. Fukamichi, J. Appl. Phys. **91**, 8037 (2002).
[156] U. Ebels, J. -L. Duvail, P. E. Wigen, L. Piraux, L. D. Buda, and K. Ounadjela, Phys. Rev. B **64**, 144421 (2001).
[157] P. W. Anderson, Phys. Rev. B **109**, 1492 (1958).
[158] H. Zeng, R. Skomski, L. Menon, Y. Liu, S. Bandyopadhyay, and D. J. Sellmyer Phys. Rev. B **65**, 134426 (2002).
[159] D. J. Sellmyer, M. Yu, R. A. Thomas, Y. Liu, and R. D. Kirby, Phys. Low-Dim. Struct. **1-2**, 155 (1998).
[160] C. P. Bean and J. D. Livingston, J. Appl. Phys. **30**, 120S (1959).
[161] E. P. Wohlfarth, J. Phys. F: Met. Phys. **14**, L155 (1984).
[162] S. W. Charles, in: "Studies of Magnetic Properties of Fine Particles and their Relevance to Materials Science", Eds.: J. L. Dormann and D. Fiorani, Elsevier, Amsterdam 1992, p. 267.
[163] S. F. Edwards and P. W. Anderson, J. Phys. F, **5**, 965 (1975).
[164] The magnetization dynamics of ferrofluids is characterized by the distinction between Brownian and Néel relaxations. Brownian relaxation refers to the mechanical rotation

of the particles in a magnetic field, whereas Néel relaxation involves jumps over magnetic energy barriers; the latter dominates in small nanoparticles.
[165] L. Néel, J. de Phys. Rad. **12**, 339 (1951).
[166] P. Gaunt, Phil. Mag. B **48**, 261 (1983).
[167] P. Gaunt, J. Appl. Phys. **59**, 4129 (1986).
[168] R. Skomski and V. Christoph, phys. stat. sol. (b) **156**, K149 (1989).
[169] D. Givord and M. F. Rossignol, in "Rare-earth Iron Permanent Magnets", Ed.: J. M. D. Coey, University Press, Oxford 1996. p. 218.
[170] D. J. Sellmyer, M. Yu, and R. D. Kirby, "Nanostructured Mater". **12**, 1021 (1999).
[171] D. Givord, A. Lienard, P. Tenaud, and T. Viadieu, J. Magn. Magn. Mater. **67**, L281 (1987).
[172] D. Givord, Q. Lu, M. F. Rossignol, P. Tenaud, and T. Viadieu, J. Magn. Magn. Mater. **83**, 183 (1990).
[173] M. P. Sharrock, J. Appl. Phys. **76**, 6413 (1994).
[174] L. Néel, J. de Phys. Rad. **11**, 49 (1950).
[175] R. H. Victora, Phys. Rev. Lett. **63**, 457 (1989); **65**, 1171 (1990).
[176] D. J. Sellmyer and R. Skomski, Sripta Materialia **47**, 531 (2002).
[177] J. Moritz, B. Dieny, J. P. Nozières, Y. Pennec, J. Camarero, and S. Pizzini, Phys. Rev. B 71, 100402R (2005).
[178] J. D. Livingston and C. P. Bean, J. Appl. Phys. **32**, 1964 (1961).
[179] E. F. Kneller and F. E. Luborsky, J. Appl. Phys. **34**, 656 (1959).
[180] R. Skomski and D. J. Sellmyer, J. Appl. Phys. **89**, 7263 (2001).
[181] J. P. Liu, R. Skomski, Y. Liu, and D. J. Sellmyer, J. Appl. Phys. **87**, 6740 (2000).
[182] F. Preisach, Z. Phys. **94**, 277 (1935).
[183] G. Bertotti and V. Basso, J. Appl. Phys. **73**, 5827 (1993).
[184] P. Wohlfarth, J. Appl. Phys. **29**, 595 (1958).
[185] D. Henkel, Phys. Stat. Sol. **7**, 919 (1964).
[186] X.-D. Che and N. H. Bertram, J. Magn. Magn. Mater. **116**, 121 (1992).
[187] V. Basso, M. LoBue, and G. Bertotti, J. Appl. Phys. **75**, 5677 (1994).
[188] R. J. Veitch, IEEE Trans. Magn. **26**, 1876 (1990).
[189] G. Herzer, Scripta Metal. **33**, 1741 (1995).
[190] H. Fukunaga and H. Inoue, Jpn. J. Appl. Phys. **31**, 1347 (1992).
[191] J. J. Versluijs, M. A. Bari, and J. M. D. Coey, Phys. Rev. Lett. **87**, 026601 (2001).
[192] E. Goto, N. Hayashi, T. Miyashita, and K. Nakagawa, J. Appl. Phys. **36**, 2951 (1965).
[193] Y. Imry and S.-K. Ma, Phys. Rev. Lett. **35**, 1399 (1975).
[194] R. Harris, M. Plischke, and M. J. Zuckermann, Phys. Rev. Lett. **31**, 160 (1973).
[195] E. M. Chudnovsky, W. M. Saslow, and R. A. Serota, Phys. Rev. B **33**, 251 (1986).
[196] D. J. Sellmyer, M. J. O'Shea, in: "Recent Progress in Random Magnets", Ed. D. H. Ryan, World Scientific, Singapore 1992, p. 71.
[197] G. Herzer, J. Magn. Magn. Mater. **112**, 258 (1992).
[198] C. H. Bennet and D. P. DiVincenzo, Nature (London) **404**, 247 (2000).
[199] A. Ekert and R Josza, Rev. Mod. Phys. **68**, 733 (1996).
[200] L. K. Grover, Phys. Rev. Lett. **79**, 325 (1997).
[201] G. Lagmago Kamta and A. F. Starace, Phys. Rev. Lett. **88**, 107901 (2002)
[202] F. Meier, J. Levy, and D. Loss, Phys. Rev. Lett. **90**, 047901 (2003).

[203] J. Tejada, E. M. Chunovsky, E. del Barco, J. M. Hernandez, and T. P. Spiller, Nanotechnology **12**, 181 (2001).

[204] R. Skomski, A. Y. Istomin, A. F. Starace, and D. J. Sellmyer, Phys. Rev. A **70**, 062307 (2004).

[205] R. Skomski, A. Kashyap, Y. Qiang, and D. J. Sellmyer, J. Appl. Phys. **93**, 6477 (2003).

[206] S. Hill and W. K. Wootters, Phys. Rev. Lett. **78**, 5022 (1997).

[207] R. Skomski, A. Y. Istomin, J. Zhou, A. F. Starace, and D. J. Sellmyer, J. Appl. Phys. **97**, 10R511 (2005).

[208] M. S. S. Brooks and B. Johansson, in: "Handbook of Magnetic Materials", Vol. 7, Ed.: K. H. J. Buschow, Elsevier, Amsterdam 1993, p. 139.

Chapter 4

NANOMAGNETIC SIMULATIONS

T. Schrefl
Department of Engineering Materials
University of Sheffield
Sheffield, S1 3JD, UK

D. Suess, G. Hrkac, M. Kirschner, O. Ertl, R. Dittrich, and J. Fidler
Solid State Physics
Vienna University of Technology
A-1040 Vienna, Austria

Abstract The theoretical and numerical background for the simulation of hysteresis properties and dynamic magnetization processes is reviewed. Nanomagnetic particle arrays may be well approximated by a modified Stoner-Wohlfarth theory which takes into account interactions owing to particle agglomeration. Energy minimization techniques are applied to calculate remanence enhancement in exchange spring permanent magnets. The influence of damping on the magnetization reversal time is discussed and examples of spin wave excitations during the switching of NiFe nano-elements are given. Finally, methods for the simulation of thermal magnetization noise and the thermal stability of magnetic states are presented.

1. INTRODUCTION

The development of nanomagnetic materials and nanomagnetic devices such as magnetic sensors and recording heads requires a precise understanding of the magnetic behavior. These applications require a reproducible magnetic domain structure and a well-defined switching field of the individual magnetic elements [1]. As the size of the magnetic components approach the nanometer regime, detailed predictions of the magnetic properties become possible using micromagnetic simulations. Micromagnetism is a continuum theory for the treatment of magnetization processes in ferromagnetic materials. The micromagnetic equations describe the relation between the magnetic properties and the physical and chemical microstructure of the material. Micromagnetic simulation and magnetic imaging [2] are complementary tools to investigate domain formation and magnetization reversal. The comparison of simulations and experiments can provide useful hints for the artificial structuring of the material, in order to tailor the magnetic properties according to their specific applications.

Microstructural features such as particle shape and grain structure influence both static and dynamic properties. The static hysteresis properties comprise the domain structure of the remanent state, the coercive field, and the coercive or loop squareness. Dynamic properties include the waiting time required to initiate the formation of vortices, the transient magnetic states during irreversible switching and the total reversal time. Both, static and dynamic micromagnetic simulations start from the total Gibbs or magnetic free energy of the system. A local minimum of this free energy corresponds to an equilibrium configuration of the magnetization (Ch. 3). The equilibrium state depends on the local magnetic field that is created by the magnet's microstructure and the applied magnetic field. The subsequent minimization of the Gibbs free energy for different applied fields provides the hysteresis loop of the magnet [3].

The coercive field of a magnet is a dynamic property. The measured coercivity significantly depends on the rate of change of the external field. This is seen at the sweep-rate dependence of the coercivity (Ch. 3), and several experiments show an enhancement of the coercive field with decreasing pulse width of the external field [4]. The dynamic coercivity becomes important in ultra-high density and high data rate magnetic storage [5]. In addition to thermal effects, the gyromagnetic precession causes the increase of the coercive field at short times [6]. The magnetization dynamics at zero temperature is described by the Gilbert equation of motion. It combines the torque equation for gyromagnetic motion with a viscous damping term [7]. The solution of the Gilbert equation gives the path the system follows towards equilibrium, whereas static hysteresis calculations give the final equilibrium configuration as a function of the external field.

Simplifying somewhat, micromagnetism can be regarded as an extension of the classical Stoner-Wohlfarth theory [8] to irregularly shaped, interacting particles. Starting from the change in energy of a magnetic particle upon rotation of the magnetization, Stoner and Wohlfarth derived the hysteresis loops of a single particle under an oblique applied external field. The magnetization within the particle is assumed to remain uniform during magnetization reversal. The integration over different particle orientations gives the hysteresis loop of non-interacting, single-domain particles. Micromagnetism replaces the atomic magnetic moments by a continuous function of position, the magnetization vector. It describes magnetization processes on a length scale which is large enough to replace the atomic magnetic moments by a magnetization vector and small enough to resolve the transition of the magnetization within a domain wall. It also describes the non-uniform magnetic state of irregularly shaped magnetic particles. Similar to the Stoner-Wohlfarth theory, micromagnetics starts from the total Gibbs free energy of a system which now contains interaction terms. In addition to the anisotropy energy and the Zeeman energy of the external field, the ferromagnetic exchange energy and the magnetostatic energy are included.

Numerical micromagnetics, which may be based either on the finite difference or finite element method, resolve the local arrangement of the magnetization which arises from the interaction between intrinsic magnetic properties such as the magnetocrystalline anisotropy and the physical and chemical microstructure of the material. The numerical solution of the equation of motion also provides information on how the magnetization evolves in time. The time and space resolution of numerical micromagnetic simulations is in the order of nanometers and nanoseconds, respectively.

In some modern applications of magnetic nanoparticles, both non-uniform magnetic states within a single particle and magnetostatic interactions between the particles can be neglected. Examples are arrays of FePt or CoPt particles, where the particle diameters are in the order of 4 nm or below [9]. The particle diameter is comparable to the domain wall width, so that the magnetization remains nearly uniform during magnetization reversal, irrespective of particle shape. The magnetostatic interaction field is only a few percent of the magneto-crystalline anisotropy field and thus magnetostatic interactions can be ignored. However, individual FePt particles may agglomerate during heat treatment forming multiple twinned particles [10]. The exchange interactions between the crystallites of the multiple twinned particles considerably influence the hysteretic properties. The exchange interactions between the crystallites were found to increase the remanence and decrease the coercive field with respect to the classical Stoner-Wohlfarth theory [11], in agreement with the experimental findings. A modified Stoner-Wohlfarth theory that takes into account exchange interactions between the crystallites of a particle is sufficient for the description of the hysteresis properties of high coercive particle arrays (Compare Ch. 13).

Classical miromagnetics neglects thermal fluctuations. The change of the intrinsic magnetic properties with temperature can be included by the use of temperature dependent constants in the expression for the total Gibbs free energy (Ch. 3), but, the spontaneous switching of the magnetization at the onset of superparamagnetism is not taken into account. These thermally activated magnetization processes become increasingly important, with decreasing size of the magnetic systems. Prominent examples are magnetization noise in magnetic sensor elements [12-14] and the thermal stability of magnetic MRAM (Magnetic Random Access Memory) cells [15] or magnetic storage media [16]. Magnetic sensors require a high sensitivity so that small magnetic fields can be detected. On the other hand, thermal fluctuations that will lead to thermal noise should be suppressed. The free layer of read head for high density magnetic storage may approach a size of $50 \times 40 \times 2$ nm^3. Thermally induced magnetization processes may cause local or global magnetization rotations which cause the magnetization noise.

With decreasing lateral extension of the elements thermal fluctuations become more pronounced. Magnetic storage elements have their magnetization oriented in two allowed directions. The shape or magneto-crystalline anisotropy should guarantee a life time of the stored bit of about 10 years. The energy barrier for thermally induced magnetization reversal decreases with increasing size of the storage elements. The corresponding time scales for thermal noise and thermally induced switching of storage elements differ by several orders of magnitude: Thermal noise arises on a time scale of a few nanoseconds; thermally induced switching of the magnetization over energy barriers extends over seconds to years. Random thermal fluctuations of the magnetization are the underlying physical process which causes both thermal noise and spontaneous switching. Depending on the relevant time scale different methods are used to treat thermal effects theoretically. For thermal processes on a time scale of several nano-seconds, a random fluctuation field or Langevin term may be added to the total effective field [17, 18], which changes the Gilbert equation to a stochastic system of ordinary differential equations [19]. For long term processes, the height of the energy barrier between the two stable energy minima is proportional to the relaxation time. Here path finding methods as originally introduced in physical chemistry [20] can be applied in micromagnetism. These methods calculate the minimum energy path and the saddle points for the transitions between two stable equilibrium configurations of a magnetic nano-element [21].

2. MODIFIED STONER-WOHLFARTH THEORY FOR HARD-MAGNETIC PARTICLE ARRAYS

2.1. Switching Field of Small Magnetic Particles

Traditional magnetization reversal has been discussed in the framework of the nucleation theory. The critical fields which cause an equilibrium magnetic state to change were calculated using stability analysis. In addition to the switching field these theories also revealed the reversal mode [22]. The total energy of a magnetic particle may be written in terms of the magnetization distribution. In equilibrium, the total free energy has a local minimum.

An example is an ellipsoidal particle magnetized parallel to its long axis. The total energy as a function of the magnetization angle φ between the magnetization vector and the anisotropy axis shows a local minimum at $\varphi = 0$. If an external field is applied at an angle $\pi - \psi$, the energy surface changes, and the position of the local minimum moves from $\varphi = 0$ to $\varphi > 0$. As a consequence the magnetization rotates towards its new equilibrium position. At a critical value of the external field, the local minimum vanishes. The second derivative of the energy vanishes, the magnetization configuration becomes unstable, and the magnetization switches into a reversed state which is again a local minimum of the energy. This process is illustrated in Fig. 1.

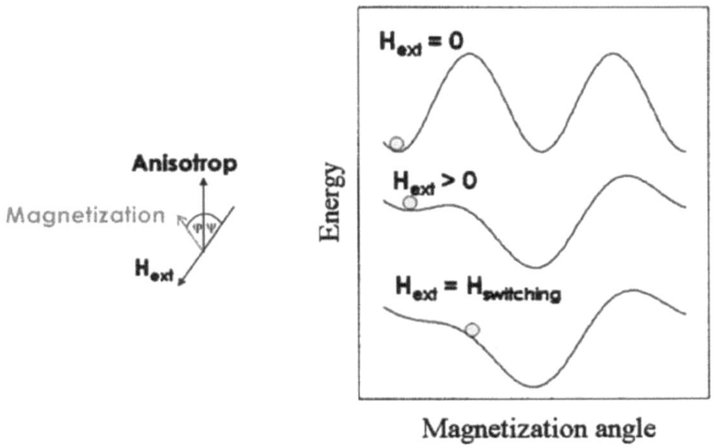

Figure 1. Total energy of an ellipsoidal magnetic particle as function of the magnetization angle φ.

The total energy density of the particle is the sum of the anisotropy energy density and the Zeeman energy density in the external field:

$$E(\varphi) = K_1 \sin^2 \varphi - J_s H_{ext} \cos(\pi - \psi - \varphi) \tag{1}$$

Here, K_1 is the uniaxial anisotropy constant, $J_s = \mu_0 M_s$ is the spontaneous magnetic polarization, and H_{ext} is the external field. In equilibrium the first derivative of Eq. (1) with respect to φ has to vanish. The equilibrium condition is

$$\frac{dE}{d\varphi} = K_1 \sin(2\varphi) - J_s H_{ext} \sin(\pi - \psi - \varphi) = 0. \tag{2}$$

The particle switches if the equilibrium condition and the instability condition

$$\frac{d^2E}{d\varphi^2} = 2K_1\cos(2\varphi) + J_sH_{ext}\cos(\pi - \psi - \varphi) = 0 \tag{3}$$

are fulfilled. Equations (2) and (3) give the maximum deviation angle φ_N of the magnetization and the switching field, $H_N = H_{switching}$.

The equilibrium angle of the magnetization for an external field smaller than the switching field can be easily calculated from Eq. (2) for small ψ [23]

$$\varphi = \frac{J_sH_{ext}}{2K_1 - J_sH_{ext}}\psi \tag{4}$$

under the influence of an oblique applied field. The magnetization rotates reversibly until φ reaches the critical angle φ_N at $H_{ext} = H_N$. Then the magnetization switches direction and the angle between the magnetization and the negative anisotropy direction is

$$\varphi' = \frac{J_sH_{ext}}{2K_1 + J_sH_{ext}}\psi \tag{5}$$

The critical angle and the switching field was given by Stoner-Wohlfarth [8]

$$\tan(\varphi_N) = \tan^{1/3}(\psi) \tag{6}$$

$$H_N = \frac{2K_1}{J_s\cos(\psi)}\frac{1}{(1+\tan^{2/3}\psi)^{3/2}} \tag{7}$$

2.2. Switching Modes and Critical Particle Size

The interplay between the ferromagnetic exchange energy and magnetostatic energy determines the magnetization reversal mode. A uniform magnetization throughout the particle keeps the ferromagnetic exchange energy low but causes magnetic surface charges which in turn increase the magnetostatic energy. In order to avoid magnetic surface charges the magnetization has to lie parallel to the surface. This leads to a nonuniform magnetic state which increases the exchange energy. In small magnetic particles, the ferromagnetic exchange interactions keep the magnetic moments parallel to each other and the particles reverse uniformly. During uniform rotation, the magnetization becomes aligned normal to the long axis of the particle, thereby increasing the magnetostatic energy. In larger particles the magnetostatic energy becomes important. When the magnetization may become arranged inhomogeneously without a significant expense of exchange energy, nonuniform magnetization reversal is energetically more favorable. Therefore, magnetization curling is the preferred reversal mode, if the diameter of the particle exceeds a critical length, d_{crit}. The curling mode avoids the creation of additional surface charges during magnetization reversal. Uniform rotation occurs below the critical diameter, d_{crit}, which is typically in the range of 20 nm to 30 nm. This value is significantly smaller than the single domain size, D. If a particle is smaller than D, the single domain state has a lower energy than a multidomain state for zero external field.

For a spherical particle with uniaxial anisotropy constant K_1 and a spontaneous magnetic polarization J_s the critical diameter for uniform rotation is given by [22, 24]

$$d_{crit} = 3.68\sqrt{\frac{6\mu_0 A}{J_s^2}} \tag{8}$$

and Kittel's single domain size [25]

$$D = \frac{72\mu_0}{J_s^2}\sqrt{AK_1} \tag{9}$$

For application in sensor and storage elements magnetic particles should be single domain. The particle is in one of two possible metastable states and thus can store information. The preferred reversal mode is not necessarily uniform rotation and depends on the application. Uniform rotation is favorable in magnetic sensors [26], as the rotation is reversible and thus the magnetization angle is a measure of the strength or direction of the applied field. Fast switching of the magnetization between the two metastable states is one of the prerequisites of magnetic storage elements [27]. Whether uniform reversal or inhomogeneous switching has a lower switching time depends on various parameters like the shape of the particle and the intrinsic damping mechanisms.

2.3. Example: Hard Magnetic Particle Arrays

Small magnetic particles are the basic structural units of magnetic recording media. To achieve a high storage density and obtain a high thermal stability a small particle size and high magneto-crystalline anisotropy are required.

Recently, Sun and coworkers [9, 28] fabricated monodisperse FePt nanoparticles. The particles were chemically synthesized, with various compositions and sizes. The particles self-assemble into a three-dimensional superlattice when deposited onto a suitably prepared substrate. The particles can be made with sizes varying from 3 to 10 nm with very narrow size range distributions. The magneto-crystalline anisotropy of the particles after annealing was found to be of the order of 10^6 J/m^3. The large anisotropy, the small particle size, and the high packing fraction makes self assembled FePt superlattices an ideal candidate for future high-density storage media with an areal density in the Tb/in^2 regime.

As synthesized FePt nanoparticles possess disordered fcc structure and are superparamagnetic at room temperature. At 5 K the FePt nanoparticles show ferromagnetic behavior. By raising the temperature the coercivity drops drastically. This is consistent with the small magnetocrystalline anisotropy of the disordered fcc phase. Heat treatment induces the Fe and Pt atoms to rearrange in long range ordered fct structure that has good hard magnetic properties. Experiments with $Fe_{56}Pt_{44}$ nanoparticles annealing at 500 °C, 550 °C, and 580 °C showed a continuously increase of the coercivity with increasing annealing temperature [9]. Transmission electron microscopy studies show that the phase transformation occurs at 530 °C [10]. The particles are randomly oriented. With increasing annealing temperature, the monodisperse particles coalesce during annealing and form multiple twined particles. The measured coercivity of optimally annealed at 580°C $Fe_{55}Pt_{45}$ nanoparticles reaches 716 kA/m. The reduced remanence (J_r/J_s) exceeds 0.5 which indicates either some crystallographic texture or strong interactions between the particles. In summary, the hysteresis properties of annealed FePt nanoparticles show two characteristic features which are clearly different from the Stoner Wohlfarth behavior for randomly oriented, non-interacting particles: (1) A coercive field which is only about 16% of the Stoner Wohlfarth value, and (2) a remanence ratio of about 0.6. This behavior can be explained by a modified Stoner-Wohlfarth model. Each multiple twined nanoparticle is represented by a triplet of three exchange coupled Stoner-Wohlfarth particles with mutually orthogonal easy axes. Magnetostatic interactions are neglected. Thus there is no interaction between particles of different triplets.

2.3.1. Model system

For the above mentioned FePt particles, the particle diameter is clearly smaller than the critical particle size given by Eq. (8) for coherent rotation. Furthermore the strength of the magnetostatic interaction field acting on nearest neighbor particles is only about 2% of the anisotropy field for a particle distance of 2 nm. Thus the Stoner-Wohlfarth theory can be applied.

However, the classical Stoner-Wohlfarth theory is modified, in order to take into account the agglomeration of particles.

The hysteresis loop of the classical Stoner-Wohlfarth theory follow from the subsequent minimization of the micromagnetic energy

$$E_{\mathrm{SW}} = \sum_i V_i \left(-K_1 \left(\frac{\mathbf{u}_i \mathbf{J}_i}{J_{\mathrm{s}}} \right)^2 - \mathbf{H}_{\mathrm{ext}} \mathbf{J}_i \right) \tag{10}$$

In Eq. (10) the sum is over all crystallites in the particle ensemble. V_i is volume of particle *i*, K_1 is the uniaxial anisotropy constant, $\mathbf{u}_i$ is the anisotropy direction of particle *i*, and $\mathbf{J}_i$ is the magnetic polarization vector of the particle. In addition to the classical energy terms of the Stoner Wohlfarth theory we have to consider the exchange energy between the crystallites that agglomerate to a particle. We assume that in average three crystallites agglomerate as schematically shown in Fig. 2.

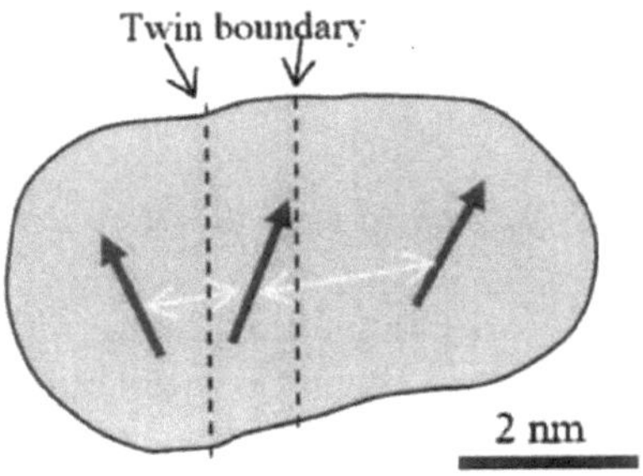

Figure 2. A multiple twined FePt particle consisting of three crystallites. A uniform magnetization is assumed within each crystallite. The crystallites within the particle are mutually exchange coupled. For the simulations the total number of particles was 500.

The radii of these three crystallites are assigned randomly under the constraint that the volume of a particle is $3(\pi r^3/3)$ with r = 2.5 nm. According to transmission electron microscopy studies the crystallites within a particle have mutually orthogonal anisotropy axes. Within a particle the crystallites are exchange coupled to each other. The exchange energy between crystallite *n* and crystallite *m* is approximated by

$$E_{\mathrm{ex}} = -\frac{A}{a} F_{nm} \frac{\mathbf{J}_n \cdot \mathbf{J}_m}{J_{\mathrm{s}}^2} \tag{11}$$

if crystallite n and crystallite m are within the same particle and zero otherwise. Here A is the exchange constant, $a = 0.68$ nm the lattice constant of FePt, and F_{nm} is the contact area between two crystallites given by

$$F_{nm} = r_{nm}^2 \pi = \left(\frac{r_n + r_m}{2} \right)^2 \pi , \tag{12}$$

where r_i is the radius of a crystallite. (For corrections to this 'hard interface' approximation, see Ch. 3.)

A total of 1500 crystallites and 500 particles was assumed for the calculations. The particles were randomly oriented. The anisotropy constant of the crystallites was $K_1 = 5.6\ 10^6$ J/m³, $J_s = 1.31$ [28]. The exchange constant was varied in the range of $A = 0.1\ 10^{-11}$ J/m to $A = 2.2\ 10^{-11}$J/m. Five percent of all crystallites are assumed to remain in the disordered fcc phase. To represent the fcc phase we simply set $K_1 = 0$. For the chosen parameters the critical size for magnetization reversal by uniform rotation, d_{crit}, is in the range from 7.8 to 36 nm which is greater than the assumed particle diameter of 5 nm. For comparison, the critical single domain diameter, D, is in the range from 130 to 590 nm.

2.3.2. *Remanence and coercivity of FePt particle arrays*

Figure 3 shows two hysteresis loops calculated with the modified Stoner-Wohlfarth model with high exchange interactions and low exchange interactions between the crystallites of a particle. Loop shape, remanence and coercive field are in good agreement with experimental results reported in [9].

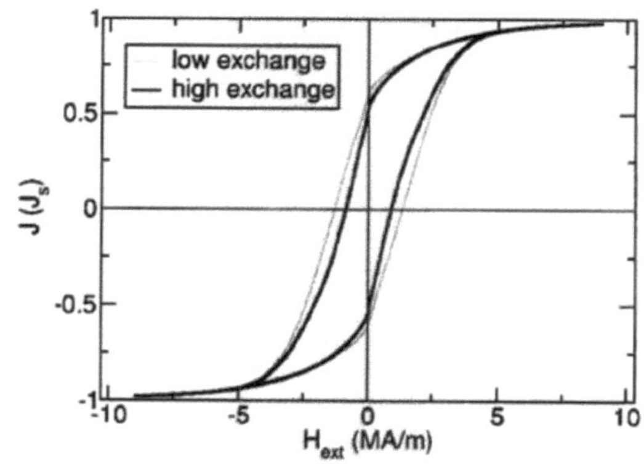

Figure 3. Hysteresis loops for FePt particles calculated using the modified Stoner-Wohlfarth model for high exchange ($A = 1.4\ 10^{-11}$ J/m) and low exchange ($A = 0.4\ 10^{-11}$ J/m).

The results clearly show that with increasing exchange interactions the coercivity drops to about 0.9 MA/m for an exchange of $A = 1.4\ 10^{-11}$ J/m. This is about 1/10 of the anisotropy field. The decrease of the coercive field can be explained in analogy to the random anisotropy model [29]. With increasing exchange between the crystallites, the effective magnetocrystalline anisotropy decreases and the coercivity drops. In the limit of high exchange interactions between the three crystallites of a particle, the particle will be uniformly magnetized. Each multiple twined particle behaves like a uniaxial particle with weak uniaxial anisotropy.

Figure 4 shows the calculated effective anisotropy energy density of a particle consisting of three crystallites with volume fractions 0.1, 0.2, and 0.7. The three crystallites have perpendicular easy axis and are strongly exchange coupled. Note if the three volumes are the same, no anisotropy occurs. For comparison, Fig. 4. gives the energy density surface for cubic anisotropy.

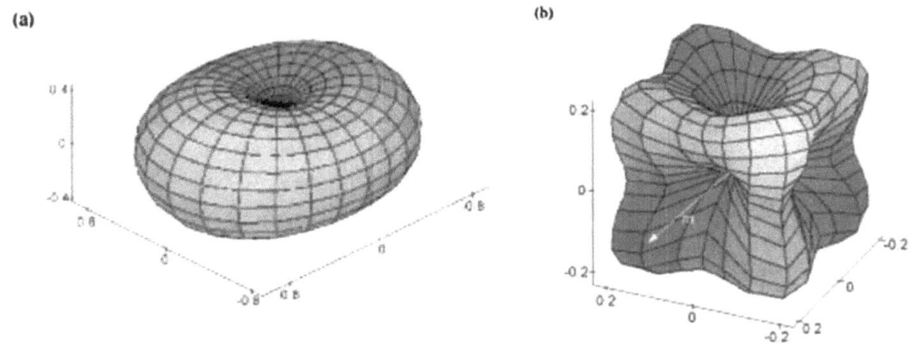

Figure 4. Energy density plots of the crystalline anisotropy in arbitrary units. For every direction of the magnetization, the corresponding crystalline energy density is the distance from the origin to the point of the surface lying along the direction of the magnetization. (a) Magnetocrystalline anisotropy energy density for a multiple twined particle with three orthogonal easy axes; (b) Energy density for cubic anisotropy.

For high exchange, the particles behave like randomly oriented, uniaxial particles. Since there are no interactions between the multiple twined particles the expected remanence ratio is in accordance with the predictions of the classical Stoner-Wohlfarth theory. Indeed, the calculated remanence ratio is 0.5. As the strength of the exchange interactions between the crystallites of a multiple twined particle is reduced, the magnetization may become non-uniformly arranged within a particle. The magnetization vectors of the crystallites within a particle are not parallel but point in slightly different directions. This gives rise to remanence enhancement. Figure 5 shows that the remanence increases as the exchange constant is reduced from $A = 2.2\ 10^{-11}$ J/m to $A = 0.25\ 10^{-11}$ J/m. The maximum remanence is reached when the deviation of the magnetization towards direction in which the sample was saturated becomes largest. When the exchange constant is decreased below $A = 0.2\ 10^{-11}$ J/m, the local anisotropy field of each crystallite becomes larger than the exchange field. The magnetization vector rotates more towards the local anisotropy axis than towards the saturation direction. With further decrease of the exchange constant the remanence decreases. For zero exchange coupling between the crystallites all crystallites behave like an assembly of non-interacting single domain particles and the remanence ratio becomes 0.5. This situation is similar to the remanence enhancement observed in nanocrystalline single-phase permanent magnets [30]. The coercive field decreases with increasing exchange interactions between the crystallites of the multiple twinned particles. The higher the exchange between the crystallites, the more the local anisotropy directions become averaged. This behavior is similar to nanocrystalline soft magnets and nanocrystalline hard magnets, where the effective anisotropy is low for strong coupling between the grains [29, 30].

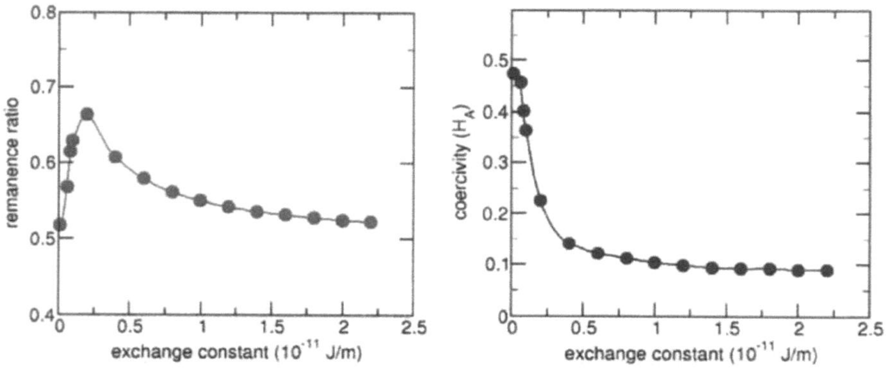

Figure 5. Remanence and coercivity as a function of the strength of the exchange interactions between the crystallites of a multiple twined particle.

Experimental observations of FePt nanoparticles arrays show a remanence enhancement reminiscent of materials with significant exchange coupling between magnetic grains. Remanence enhancement was produced within the model only when the magnetization within the FePt nanoparticles was inhomogeneous. To achieve inhomogeneous magnetization within the FePt nanoparticles require both multiple c-axes within a nanoparticle and reduced exchange stiffness in comparison to literature values for FePt bulk alloys [31]. With this two assumptions, remanence, coercivity, and loop shape calculated by the modified Stoner-Wohlfarth theory are close to the experimental observed data. Agglomeration that leads to multiple twined particles is observed experimentally by high resolution transmission electron microscopy [10]. It has been suggested that a reduction in the exchange interactions in ultra fine particles can be caused by the lower coordination of near surface atoms [32].

3. STATIC MICROMAGNETICS

3.1. Basic Principles

The basic concept of micromagnetism is to replace the atomic magnetic moments by a continuous function of position. In a continuum theory the local direction of the magnetic moments may be described by the magnetic polarization vector

$$\mathbf{J}(\mathbf{r}) = \mu_0 \mathbf{M}(\mathbf{r}) = \mu_0 \mathbf{m} / V \tag{13}$$

The magnetic polarization, **J**, is proportional to the magnetization, **M**, which is given by the magnetic moment, **m**, per unit volume, V. μ_o = $4\pi \times 10^{-7}$ Vs/Am is the magnetic permeability of vacuum. The second principle of micromagnetism treats the magnitude of the magnetization as a function of temperature only. The modulus of the magnetic polarization

$$|\mathbf{J}(\mathbf{r},T)| = J_s(\mathbf{r},T) \tag{14}$$

is assumed to be a function of temperature and to be independent of the local magnetic field. Of course, J_s may depend on space as in a two-phase nano-structured magnet [33]. Thus the magnetic state of the system can be uniquely described by the directions cosines, $a_i(\mathbf{r})$ of the magnetic polarization, $\mathbf{J} = \mathbf{a}J_s$. In a metastable equilibrium state, $a_i(\mathbf{r})$ minimizes the total Gibbs free energy of the system.

3.2. Micromagnetic Energy Contributions

The contributions to the total magnetic Gibbs free energy are derived from classical electrodynamics, condensed matter physics, and quantum mechanics so that the continuous expressions for the energy describe the interactions of the spins with the external field, the crystal lattice, and the interactions of the spins with one another. The latter consists of long-range magnetostatic interactions and short-range quantum-mechanical exchange interactions.

The competitive effects of the micromagnetic energy contributions upon minimization determine the equilibrium distribution of the magnetization.

The minimization of the ferromagnetic exchange energy aligns the magnetic moments parallel to each other, whereas the minimization of the magnetostatic energy favors the existence of magnetic domains. The magnetocrystalline anisotropy energy describes the interaction of the magnetization with the crystal lattice. Its minimization orients the magnetization preferably along certain crystallographic directions. The minimization of the Zeeman energy of the magnetization in an external field rotates the magnetization parallel to the applied field.

3.2.1. *Exchange energy*

In the approach of micromagnetism the quantum mechanical spin operators are substituted by classical vectors, **S**. Thus, the exchange energy in the Heisenberg model [34]

$$E_{exch} = -\sum_{i=1}^{N} \sum_{j \neq i}^{N} J^{ij} \mathbf{S}^i \cdot \mathbf{S}^j \tag{15}$$

where J^{ij} represents the exchange integral. In case of ferromagnetism the exchange integral is positive and leads to the observed strong coupling and parallel alignment of spins. Considering only nearest neighbors (n.n.) angular moments and assuming a constant exchange integral $J^{ij} \cong J$ leads to

$$E_{\text{exch}} = -J\sum_{i=1}^{N}\sum_{j \neq i}^{n.n.} \mathbf{S}^i \cdot \mathbf{S}^j \tag{16}$$

The assumption of a constant exchange integral is justified, because experiments show only small differences when measuring along different crystallographic directions. When the angle between $\mathbf{S}^i$ and $\mathbf{S}^j$ is small Eq. (16) can be written

$$E_{\text{exch}} = -J\sum_{i=1}^{N}\sum_{j \neq i}^{n.n.} JS^2 \cos\varphi_{ij} \approx \frac{1}{2}J\sum_{i=1}^{N}\sum_{j \neq i}^{n.n.} JS^2\varphi_{ij}^2 + \text{const.} \tag{17}$$

where S is the magnitude of the spin. The angle φ_{ij} can now be expanded in terms of the direction cosines of the magnetization [35]. Replacing the sum over pairs of spins by an integral over the magnet's volume the exchange energy finally can be given by

$$E_{\text{exch}} = A\int_V \left((\nabla a_1)^2 + (\nabla a_2)^2 + (\nabla a_3)^2 \right) dV \, . \tag{18}$$

The micromagnetic exchange constant (exchange stiffness) A [J/m] is proportional to the exchange integral J. For a cubic lattice having one nonequivalent atom, A is given by [36]

$$A = \frac{nJS^2}{a}, \tag{19}$$

where $n = 1$ for a simple cubic lattice, $n = 2$ for a body-centered cubic lattice, and $n = 4$ for a face centered cubic lattice. Since the exchange integral J is proportional to the Curie temperature, T_C [37], the exchange constant A is also linearly proportional to T_C. For monatomic cubic lattices, the following relations hold [36]:

$J = 0.54\ k_B\ T_C$ for a simple cubic lattice ($S = 1/2$),
$J = 0.34\ k_B\ T_C$ for a body-centered cubic lattice ($S = 1/2$), and
$J = 0.15\ k_B\ T_C$ for a body-centered cubic lattice ($S = 1$),

where Boltzmann's constant $k_B = 1.380658\ 10^{-23}$ J/K.

Equation (18) is the form of the exchange energy which is used in micromagnetics. Mostly, the value of A is taken from experiments. For example, Smith and co-workers [38] determined the exchange constant of NiFe thin films by magnetoresistive measurements ($A_{NiFe} = 1.05\ 10^{-11}$ J/m) and Livingston [39] calculated the exchange constant of Nd-Fe-B sintered magnets from measurements of the domain width ($A_{NdFeB} = 1.7\ 10^{-11}$ J/m).

3.2.2. Magnetostatic energy

The divergence of the magnetization creates a magnetic field, the so-called stray field or demagnetizating field. The magnetostatic energy of the magnetic polarization in this field is the stray field energy

$$E_S = -\frac{1}{2}\int_V \mathbf{J} \cdot \mathbf{H}_S\, dV \tag{20}$$

The factor ½ takes into account that the stray field is caused by the magnetization itself. The stray field can be determined by calculating the magnetic scalar potential Ψ from the divergence of the magnetic polarization and its intersection with the magnet's surface as will be outlined in the following.

The strayfield does not arise from any current, thus from Ampere's law it follows that

$$\operatorname{curl} \mathbf{H}_S = 0 \tag{21}$$

So $\mathbf{H}_S$ is a irrotational field and can therefore be expressed as gradient of a scalar field Ψ, called magnetic potential.

$$\mathbf{H}_s = -\operatorname{grad} \Psi \tag{22}$$

The magnetic potential of a volume V with a given magnetic polarization $\mathbf{J}$ can be calculated by solving the Poisson equation

$$\Delta\Psi = -\frac{\rho_m}{\mu_0} \tag{23}$$

with the jump condition

$$\left. \frac{\partial \Psi^{out}}{\partial \mathbf{n}} - \frac{\partial \Psi^{in}}{\partial \mathbf{n}} \right|_{\partial V} = -\frac{\sigma_m}{\mu_0} \tag{24}$$

at the boundary ∂V. Here ρ_m and σ_m are magnetic charge densities in the volume V and at its boundary ∂V. These charges are only virtual, because there are no magnetic monopoles in reality. These virtual charges are useful to calculate the magnetic potential of a polarization distribution. They can be obtained by taking the divergence and the face divergence of the magnetic polarization $\mathbf{J}$.

$$\rho_m = -\operatorname{div} \mathbf{J}, \ \sigma_m = -\operatorname{Div} \mathbf{J} \tag{25}$$

The face divergence is defined as difference of the normal components of outer and inner field.

$$\operatorname{Div} \mathbf{J} = \left(\mathbf{J}^{out} - \mathbf{J}^{in} \right) \cdot \mathbf{n} = J_n^{out} - J_n^{in} \tag{26}$$

Assuming that the magnetic potential is regular at infinity,

$$\Psi \to \frac{1}{r} \quad \text{for } r \to \infty, \tag{27}$$

the solution of (23) and (24) is

$$\Psi = \frac{1}{4\pi\mu_0}\left(\int_V \frac{\rho_m}{r}\,dV + \oint_{aV} \frac{\sigma_m}{r}\,dA\right) \tag{28}$$

3.2.3. Magnetocrystalline anisotropy energy

In general a ferromagnetic material does not show isotropic magnetization behavior (Ch. 3). Experimentally certain directions can be found, which favor magnetization (easy axes) and others which do not (hard axes). This leads to a new energy term.

$$E_{\text{ani}} = \int_V f_{\text{ani}}(\mathbf{J})\,dV \tag{29}$$

Here f_{ani} denotes a direction dependent energy density. In case of hexagonal systems (e.g. Co) a uniaxial anisotropy can be assumed. Therefore f_{ani} can be expressed as series expansion

$$f_{\text{ani}}(\theta) = K_0 + K_1 \sin^2\theta + K_2 \sin^4\theta + \ldots. \tag{30}$$

θ denotes the angle between **J** and the c-axis of the hexagonal system and K_0, K_1, K_2, etc. are the so called anisotropy constants. In practice, all higher terms are neglected.

In case of cubic systems, taking cubic symmetry into account f_{ani} can be written as

$$f_{\text{ani}}(a_1, a_2, a_3) = K_0 + K_1\left(a_1^2 a_2^2 + a_2^2 a_3^2 + a_3^2 a_1^2\right) + K_2\left(a_1^2 a_2^2 a_3^2\right) + \ldots \tag{31}$$

where a_1, a_2, and a_3 denote the direction cosines of the cubic lattice vectors to the direction of the magnetic polarization **J**. In both cases the first term K_0 can be neglected, because it does not depend on angle and is therefore a constant term.

3.2.4. Zeeman energy

The energy associated with a magnetized body in the presence of an external field $\mathbf{H}_{ext}$ is called Zeeman energy and can be determined by

$$E_H = -\int_V \mathbf{J} \cdot \mathbf{H}_{ext}\, dV \tag{32}$$

This energy will cause the magnetization to align parallel to the external field in order to minimize the energy.

3.3. Energy Minimization and the Finite Element Method

Many micromagnetic investigations focus on the optimization of the magnet's nanostructure, in order to tailor its magnetic properties for specific applications. For these studies dynamic effects are less important and static energy minimization techniques can be applied.

The microstructural effects such as grain size, particle shape, phase distribution, or grain boundary segregation result in a specific local arrangement of the magnetic polarization which in turn determine the macroscopic properties. Finite-element modeling treats magnetization processes within the framework of micromagnetic theory, taking into account the complex microstructure of a magnet. Realistic models of the magnet or device geometry are taken from transmission electron microscopy or from grain growth simulation. For example, the microstructure of a two-phase nanocrystalline magnet, which is shown in Fig. 6, is obtained from the construction of a modified Voronoi diagram [40]. Once a finite element model has been built which represents the real geometry of the device or a magnet's nanostructure, the basic geometric units are subdivided into tetrahedral finite elements. Then the intrinsic material properties such as the exchange constant, the spontaneous magnetic polarization, and the anisotropy constant are assigned to the individual elements, depending on the spatial variation of these properties. Now the total Gibbs free energy,

$$E_{\mathrm{total}} = E_{\mathrm{exch}} + E_{\mathrm{S}} + E_{\mathrm{ani}} + E_{\mathrm{H}}\,, \tag{33}$$

which is an integral over the total volume of the magnetic structure, breaks up into a sum over all tetrahedrons. Each tetrahedron contributes to the total energy.

Within each tetrahedron the direction cosines of the magnetic polarization vector, $a_i(\mathbf{r})$, are interpolated by linear functions. Thus all energy terms but the stray field energy can be easily evaluated using equations (18), (29), and (32). The evaluation of (20) requires the calculation of the stray field beforehand which can be done by applying standard finite element techniques for the solution of (22), (23), and (24) together with a effective method to treat the boundary conditions (27), at infinity [41, 42]. This is much faster then the direct integration of (28) which scales with n^2 in storage and CPU time, where n is the number of grid points in the finite element mesh.

The subsequent minimization of (33) with respect to the direction cosines for different external fields, subject to the constraint

$$a_1^2(\mathbf{r})+a_2^2(\mathbf{r})+a_3^2(\mathbf{r})=1, \tag{34}$$

provides a stable magnetization distribution. In order to transform the minimization problem into an unconstraint minimization problem, polar coordinates can be introduced:

$$\begin{aligned} a_1(\mathbf{r}_k) &= a_{k1} = \sin(\theta_k)\cos(\phi_k), \\ a_2(\mathbf{r}_k) &= a_{k2} = \sin(\theta_k)\sin(\phi_k), \\ a_3(\mathbf{r}_k) &= a_{k3} = \cos(\theta_k). \end{aligned} \tag{35}$$

Then the constraint (34) will be automatically fulfilled at each grid point k of the finite element mesh. Conjugate gradient based minimization techniques [43] require only the energy and the gradient of the energy to search the local minima. Using polar coordinates, the gradient of the total energy may be written

$$\begin{aligned} \frac{\partial E_{\text{total}}}{\partial \theta_k} &= -\sum_{i=1}^{3} \mu_0 m_k H_{\text{eff},ki} \frac{\partial a_{ki}}{\partial \theta_k} \\ \frac{\partial E_{\text{total}}}{\partial \phi_k} &= -\sum_{i=1}^{3} \mu_0 m_k H_{\text{eff},ki} \frac{\partial a_{ki}}{\partial \phi_k} \end{aligned} \tag{36}$$

where m_k is the total magnetic moment associated with grid point k of the finite element mesh and $H_{\text{eff},i}$ is the i-th Cartesian component of the effective field

$$H_{\text{eff},i}(\mathbf{r}_k) = H_{\text{eff},ki} = -\frac{1}{\mu_0 m_k}\frac{\partial E_{\text{total}}}{\partial a_{ki}} \tag{37}$$

Formally, the effective field is the negative variational derivative of the total Gibbs free energy with respect to the magnetic polarization. To evaluate the effective field numerically, we apply a technique called mass lumping on the finite element grid. Physically, we can think of a special type of space discretization where discrete magnetic moments sit on the grid points of an irregular finite element mesh.

3.4. Example: Remanence Enhancement in Exchange-Spring Systems

Nanocomposite permanent consists of a mixture of magnetically hard and soft phases. Nanocomposite magnets show a high remanence and a reasonable large coercive field, if both phases are sufficiently exchange coupled. Nanocomposite magnets with excellent hard magnetic properties were obtained for various different compositions. The soft magnetic phase is either α-Fe or Fe_3B [44, 45]. Possible application for nanocomposite permanent magnets are bonded magnets used in consumer electronic applications, where the miniaturization requires magnets that are easy to magnetize [46, 47].

Generally, isotropic permanent magnets with a remanence ration $J_r/J_s >$ 0.5 are referred as remanence enhanced materials. However, the theoretical limit for the remanence of noninteracting grains depends on the crystal symmetry, the crystallographic orientation of the easy directions, and the volume fraction and the saturation polarization of the phases. The remanence ratio a two-phase nanocomposite magnet is

$$J_r/J_s = (1/J_s)\{v_1 p_1 J_{s1} + (1 - v_1) p_2 J_{s2}\} \qquad (38)$$

where p_1 and p_2 denote the remanence ratio of phase 1 and phase 2, and v_1 is the volume fraction of phase 1. Without any interactions, the remanence ratio of $Nd_2Fe_{14}B$ (uniaxial) and Fe_3B (easy-plane) is $p_1 = 0.5$ and $p_2 = 0.79$, respectively.

Figure 6 shows the calculated magnetization distribution of a 50 vol% $Nd_2Fe_{14}B$ / 50 vol% Fe_3B magnet with a mean grain size of 15 nm for zero applied field. In order to calculate the magnetic state of Fig. 6 the total energy was subsequently minimized for decreasing external field, starting with a high saturation field. The calculated remanence ratio $J_r/J_s = 0.71$ exceeds the theoretical limit for noninteracting particles ($J_r/J_s = 0.65$). This result clearly indicates that intergrain exchange interactions enhance the remanence of nanocomposite magnets, provided that the grain size is sufficiently small. The magnetic polarization **J** is nearly parallel to the anisotropy direction within the hard magnetic grains. Within the Fe_3B grains, **J** is parallel to the average magnetization direction of all neighboring

$Nd_2Fe_{14}B$ grains. This average direction corresponds to the direction of the saturation field.

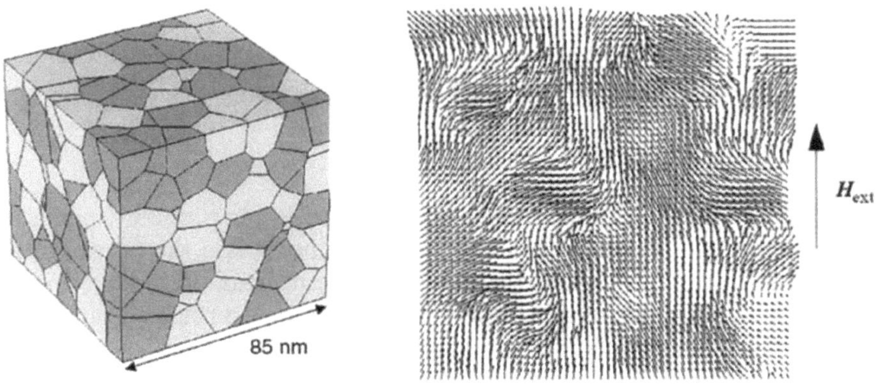

Figure 6. Remanence enhancement in a two-phase $Nd_2Fe_{14}B/Fe_3B$ magnet containing 343 grains. Left: Finite element model of the grain structure. Right: Magnetization distribution in a slice plane for zero applied field. The arrows denote the magnetization direction projected on a slice plane.

4. DYNAMIC MICROMAGNETICS

4.1. Equation of Motion

The dynamic response of a magnetic particle subject to an applied field follows from the Gilbert equation of motion [7]

$$\frac{\partial \mathbf{J}}{\partial t} = -|\gamma| \mathbf{J} \times \mathbf{H}_{\text{eff}} + \frac{\alpha}{J_s} \mathbf{J} \times \frac{\partial \mathbf{J}}{\partial t} \tag{39}$$

This equation is mathematically equivalent to the Landau-Lifshitz-Gilbert equation (LLG)

$$\frac{\partial \mathbf{J}}{\partial t} = -\frac{|\gamma|}{(1+\alpha^2)} \mathbf{J} \times \mathbf{H}_{\text{eff}} - \frac{|\gamma|}{J_s} \frac{\alpha}{(1+\alpha^2)} \mathbf{J} \times (\mathbf{J} \times \mathbf{H}_{\text{eff}}) \tag{40}$$

Here γ is the gyromagnetic ratio $|\gamma| = 2.210175 \times 10^5 \,\text{m/As}$. The first terms in the equations describe the precessional movement of the magnetic polarization vector $\mathbf{J}$ around the effective field $\mathbf{H}_{\text{eff}}$. The second phenomenological term stands for the energy dissipation during the movement and causes a movement into the direction of the effective field.

The strength of the damping is characterized by the dimensionless Gilbert damping constant α. Large α means strong damping. There are many processes which contribute to the damping in magnetic solids, as for example magnon-magnon, magnon-phonon interactions or eddy currents. So the damping constant is an intrinsic property of matter and can also be varied in practice, for instance by doping [48].

4.2. The Role of Damping

The influence of the damping constant on the switching speed depends on the particle shape. The system is critically damped when the reversal time reaches its minimum. Classically, critical damping was investigated under the assumption that the magnetization remains uniform during magnetization reversal. Whereas the critical damping constant for a spherical particle is $\alpha = 1$, the critical damping reaches $\alpha = 0.01$ in a thin film [49]. Here, magnetization reversal is accelerated by the precession around the self-demagnetizing field perpendicular to the film plane once an initial out of plane component of the magnetization is built up.

For more complex geometries, the magnetization may not remain uniform during reversal. Generally, the switching becomes faster when the magnetization stays more coherent during reversal. An important example is the magnetization process in write heads for magnetic recording. The write head consists of the large soft magnetic yoke with extensions up to micro-meters, whereas the pole tip extensions are in the range of 100 nanometers or below. The head is driven be the non-uniform magnetic field produced by the coil. The huge span of length scales and the inhomogeneous driving field give rise to non-uniform magnetization reversal. Clearly, the head field rise time and the maximum data rate depend on the Gilbert damping constant.

The fastest field rise time is found for intermediate values of the Gilbert damping constant [50, 51]. Small damping gives rise to phase decoherence of the magnetization in different parts of the yoke which increase the time required to saturate the pole tips. Large damping simply slows down magnetization motion too much. As a consequence the shortest field rise time are obtained for a Gilbert damping constant of about $\alpha = 0.5$. Figure 7 compares the field rise time for different values of the Gilbert damping constant.

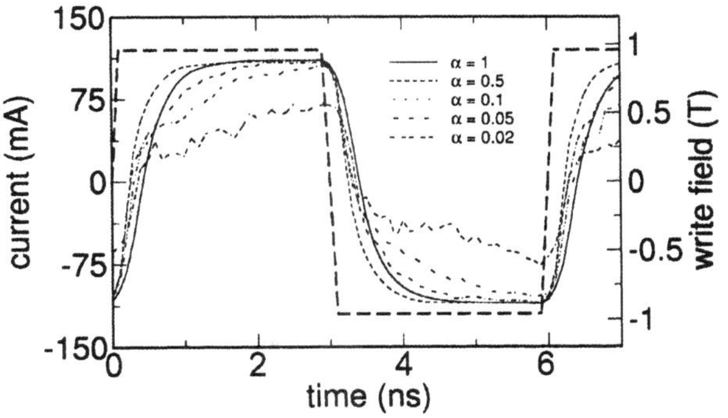

Figure 7. Head field strength as a function of time for different Gilbert damping constants within the head. The dashed line gives the current profile The solid lines give the head field for $\alpha = 1$, 0.5, 0.1, and 0.02. The fastest field rise time is achieved with intermediate damping ($\alpha = 0.5$).

4.3. Numerical Time Integration Methods

After the evaluation of the magnetic moments and the effective fields at the nodes of the finite element grid a system of coupled ordinary differential equations has to be solved. Due to exchange interactions, the system may become stiff, which suggest the use of higher order backward differentiation formula (BDF methods) for the solution of the coupled system of ordinary differential equations [52]. BDF methods are implicit and thus need to solve a nonlinear system of equations at each time step. However, only a few Newton steps are required to obtain convergence. At each Newton step a linear system of equations has to solved. Due to the long-range magnetostatic interactions, which couples all the nodes of the computational grid, the system matrix of this equation is fully populated. The use a generalized minimum residual method (GMRES) for its solution avoids the storage of the system matrix. A considerable speed up for the solution of the linear system can be obtained with proper preconditioning of the linear system. For preconditioning an approximate Jacobian matrix calculated which includes the exchange and anisotropy term but omits the magnetostatic contributions. Thus the approximate Jacobian remains sparse and can be calculated on the fly [53].

4.4. Example: Magnetization Ringing in Magnetic Nano-Elements

Dynamic micromagnetics simulations show that shape, size, and damping constant significantly influence the switching behavior of thin film elements. The magnetization reversal of submicron NiFe elements was found to be a two stage process. At first, the magnetization rotates almost in plane towards the applied field direction. Then, the precessional motion of the magnetization around the effective field leads to the excitation of spin waves. The time required for the initial rotation of the magnetization decreases with decreasing damping constant and is independent of the element shape. However, the element shape influences the decay rate of the oscillations. A more rapid decay is observed in elements with slanted ends. Here surface charges cause a transverse demagnetizing field that suppresses the oscillations. Figure 8 clearly shows that non-uniform spin wave modes are more pronounced in the rectangular element.

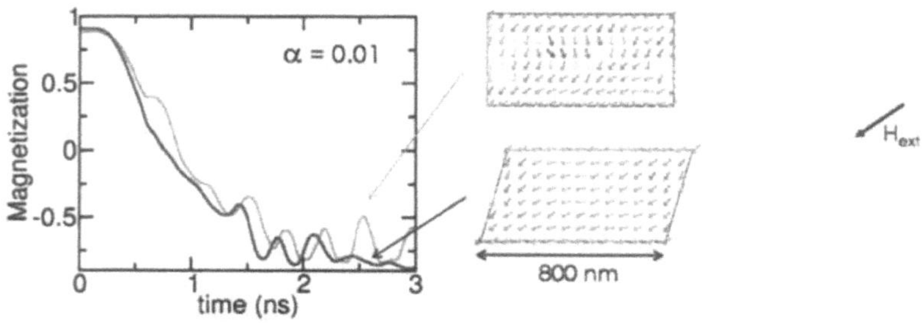

Figure 8. Dynamic magnetization reversal of small NiFe elements. The magnetization component parallel to the long axis is given as function of time. The rise time of the external field is 0.5 ns, then the field remains constant for 2 ns and finally decays to zero within 0.5 ns.

5. TEMPERATURE EFFECTS

5.1. The Stochastic LLG Equation

In order to treat thermal effects on small time scales, a random thermal field, $\mathbf{H}_{\text{th}}$, is added to the effective field in the Landau-Lifshitz Gilbert equation. The thermal field is a Gaussian random process with the following statistical properties

$$\left\langle H_{\text{th},ki} \right\rangle = 0 \tag{41}$$

$$\left\langle H_{\text{th},ki} H_{\text{th},lj} \right\rangle = D\delta_{ij}\delta_{kl}\delta(t-t') \tag{42}$$

Here k and l are position indices and denote the node number of the finite element grid, i and j denote the Cartesian space direction. Equation (41) states the average of the thermal field take over different realization is zero in each space direction. Equation (42) shows that the thermal field is neither correlated in space nor in time. The strength of the thermal field at node k of the computational grid is given by the fluctuation-dissipation theorem

$$D_k = \frac{2\alpha k_B T}{|\gamma| \mu_0 m_k} \tag{43}$$

Here m_k is the magnetic moment associated with node k of the computational grid. Again the interpretation of the finite element discretization as interacting discrete magnetic moments is useful. The thermal field is added independently for each magnetic moment and the method may be regarded as an extension of Brown's work [17] for single magnetic particles.

In a micromagnetic simulation, each computational cell is supposed to be magnetized to the bulk value of the magnetic polarization $J_s(T)$, which corresponds to the experimentally measured temperature dependent value of the magnetic polarization. This common assumption is valid only if the computational cell contains enough atomistic spins to justify the use of statistics and the laws of thermodynamics. If the size of the computational cell becomes small, the thermal motion of the atomistic spins leads to a total magnetic polarization of the cell, which is different from $J_s(T)$. The proper value of J_s to be used in a simulation will depend on the size of the computational cell, h, and on the temperature, T. It can be calculated from Metropolis Monte Carlo simulations using atomistic spins. For materials with a simple cubic lattice the Monte Carlo simulations give the following scaling law [54]

$$J_{s,\mathrm{cell}}(h,T) = J_s(T) - \left(J_{s,0} - J_s(T)\right)\left(\frac{a}{h}\right)^{3/2} \tag{44}$$

where $J_{s,0}$ is the magnetic polarization at zero temperature, $J_{s,0} = J_s(T=0)$, and a is the lattice parameter. Similar scaling laws hold for the anisotropy constant and the exchange constant. The straight forward use of (41) to (43) without rescaling of the intrinsic magnetic parameters overestimates the temperature by about 4%.

5.2. Energy Barrier Calculation

The calculation of the thermal stability requires the estimation of transition rates between stable equilibrium states of the magnet. The calculation of transition rates needs a detailed characterization of the energy landscape along the most probable path which is taken by the system from its initial state to a final state. Henkelman and Jónsson [20] proposed the *nudged elastic band method* to calculate minimum energy paths. Starting from an initial guess for the path which connects two local minima of the system, a highly probable path is found by moving the points along the path according to an algorithm which resembles tensioning an elastic band across a mountain.

First we construct a sequence of magnetic states in such a way as to form a discrete representation of a path from the initial magnetization state. An optimization algorithm is then applied until at any point along the path the gradient of the energy is only pointing along the path. This path is called minimum energy path which means that the energy is stationary for any degree of freedom perpendicular to the path. From minimum energy path the energy barrier height between the initial state and the saddle point can be evaluated.

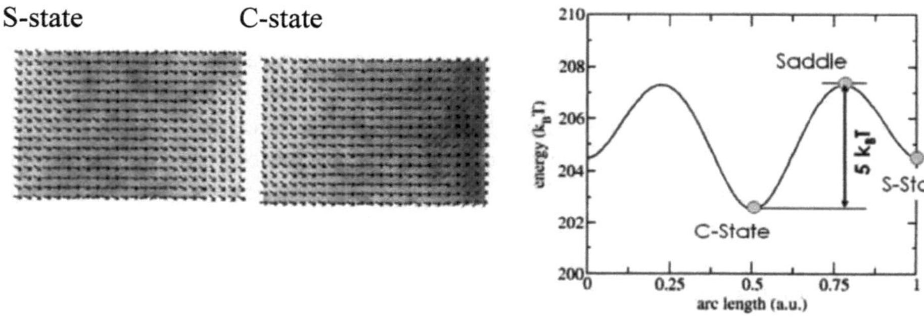

Figure 9. Left: Starting from the S-sate the C-state is reached after 4 ns at $T = 350$ K. Right: Magnetization configurations and energy along the minimum energy path.

As an example, we consider the switching of end domains in NiFe Elements. Small rectangular NiFe elements show two distinct ground states: The S-state and the C-state. Thermal fluctuations may induce a transition between the different states. The C-state has a slightly lower energy than the S-state. A transition time of about 4 ns at a temperature of $T = 350$ K was derived from the numerical solution of the stochastic Landau-Lifshitz Gilbert equation. The transition between the grounds states was also calculated using the path finding method. Figure 9 shows the energy along the final path. The magnetic states along the minimum energy path correspond to the magnetic states which are visited during the stochastic motion of the magnetization according to the stochastic Landau-Lifshitz-Gilbert equation.

Acknowledgement

The authors wish to acknowledge the valuable discussions with Robert C. Woodward. This work is supported by the Austrian Science Fund (Y132-N02).

References

[1] K. J. Kirk, J. N. Chapman, and C. D. W. Wilkinson, Appl. Phys. Lett. **71**, 539 (1997).
[2] E. D. Dahlberg and J. G. Zhu, Physics Today **48**, 34 (1995).
[3] D. Kinderlehrer and L. Ma, IEEE Trans. Magn. **30**, 4380 (1994).
[4] L. He, W. D. Doyle, L. Varga, H. Fujiwara, and P. J. Flanders, J. Magn. Magn. Mater **155**, 6 (1996).
[5] A. Moser, K. Takano, D. T. Margulies, M. Albrecht, Y. Sonobe, Y. Ikeda, S. Sun, and E. E. Fullerton, J. Phys. D: Appl. Phys. **35**, R157 (2002).
[7] T. L. Gilbert, Phys. Rev. **100**, 1243 (1955).
[6] R. W. Harrell, IEEE Trans. Magn. **37**, 533 (2001).
[8] E. C. Stoner and E. P. Wohlfarth, Phil. Trans. R. Soc. **240**, 599 (1948).
[9] S. Sun, E. E. Fullerton, D. Weller, and C. B. Murray, IEEE Trans. Magn. **37**, 1239 (2001).
[10] Z. R. Dai, S. Sun, and Z. L. Wang, Nanoletters **1**, 443 (2001).
[11] T. Schrefl, G. Hrkac, D. Suess, W. Scholz, and J. Fidler, J. Appl. Phys. **93**, 7041 (2003).
[12] N. Smith and P. Arnett, Appl. Phys. Lett. **78**, 1448 (2001).
[13] H. N. Bertram, V. L. Safonov, and Z. Jin, IEEE Trans. Magn. **38**, 2514 (2002).
[14] J.-G. Zhu, J. Appl. Phys. **91**, 7273 (2002).
[15] N. D. Rizzo, M. DeHerrera, J. Janesky, B. Engel, J. Slaughter, and S. Tehrani, Appl. Phys. Lett. **80**, 2335 (2002).
[16] D. Weller and A. Moser, IEEE Trans. Magn. **35**, 4423 (1999).
[17] W. F. Brown, IEEE Trans. Magn. Magn. **15**, 1196 (1979).
[18] A. Lyberatos and R. W. Chantrell, J. Appl. Phys. **73**, 6501 (1993).
[19] J. L. Garcia-Palacios and F. J. Lazaro, Phys. Rev. B **58**, 14937 (1998).
[20] G. Henkelman and H. Jónsson, J. Chem. Phys. **113**, 9978 (2000).
[21] R. Dittrich, T. Schrefl, D. Suess, W. Scholz, H. Forster, and J. Fidler, J. Magn. Magn. Mater. **250**, 12 (2002)
[22] A. Aharoni, "Introduction to the Theory of Ferromagnetism", Oxford University Press, New York 1996.
[23] H. Kronmüller, K.-D. Durst, and G. Martinek, J. Magn. Magn. Mater. **69**, 149 (1987).
[24] M. Fähnle and H. Kronmüller, "Micromagnetism and the Microstructure of Ferromagnetic Solids", Cambridge University Press, Cambridge 2003.
[25] C. Kittel, Phys. Rev. **70**, 965 (1946).
[26] J. N. Chapman, P. R. Aitchison, K. J. Kirk, S. McVitie, J. C. S. Kools, and M. F. Gillies, J. Appl. Phys. **83**, 5321 (1998).
[27] J. C. Mallinson, IEEE Trans. Magn. **36**, 1976 (2000).

[28] S. Sun, C. B. Murray, D. Weller, L. Folks, and A. Moser, Science **287**, 1989 (2000).
[29] G. Herzer, IEEE Trans.Magn. **26**, 1397 (1990).
[30] T. Schrefl, H. Kronmüller, and J. Fidler, Phys. Rev. B. **49**, 6100 (1994).
[31] A = 10.8 pJ/m estimated from the Curie temperature and published values of A in CoPt in "Permanent Magnetism" R. Skomski and J. M. D. Coey, IOP Publishing, Bristol 1999.
[32] F. T. Parker, M. W. Foster, D. T. Margulies, and A. E. Berkowitz, Phys. Rev. B **47**, 7885 (1993).
[33] R. Skomski and J. M. D. Coey, Phys. Rev. B **48**, 15812 (1993).
[34] P. Mohn, "Magnetism in the Solid State, an Introductory Course", Springer, Vienna 2002.
[35] C. Kittel, Rev. Mod. Phys. **21**, 541 (1949).
[36] S. Chikazumi, "Physics of Ferromagnetism", Oxford University Press, New York 1997.
[37] P. R. Weiss, Phys. Rev. **74**, 1493 (1948).
[38] N. Smith, D. Markham, and D. LaTourette, J. Appl. Phys. **65**, 4362 (1989).
[39] J. D. Livingston, J. Appl. Phys. **57**, 4137 (1985).
[40] T. Schrefl and J. Fidler, J. Magn. Magn. Mater. **111**, 105 (1992).
[41] D. R. Fredkin and T. R. Koehler, IEEE Trans. Magn. **26**, 415 (1990).
[42] H. Forster, T. Schrefl, R. Dittrich, W. Scholz, and J. Fidler, IEEE Trans. Magn. **39**, 2513 (2003).
[43] P. E. Gill, W. Murray, and M. H. Wright, "Practical Optimization", Academic Press, London 1993.
[44] R. Coehoorn, D. B. De Mooij, and C. De Waard, J. Magn. Magn. Mater. **80**, 101 (1989).
[45] E. F. Kneller, IEEE Trans. Magn. **27**, 3588 (1991).
[46] V. Panchanathan, IEEE Trans. Magn. **31**, 3605 (1995).
[47] J. J. Croat, J. Appl. Phys. **81**, 4804 (1997).
[48] W. Bailey, P. Kabos, F. Mancoff, and S. Russek, IEEE Trans. Magn. **37**, 1749 (2001).
[49] R. Kikuchi, J. Appl. Phys. **27**, 1352 (1956).
[50] K. Takano, IEEE Trans. Magn. **40**, 257 (2004).
[51] O. Ertl, T. Schrefl, D. Suess, M. Schabes, J. Magn. Magn. Mater. **290-291**, 518 (2005).
[52] A. C. Hindmarsh and L. R. Petzold, Computers in Physics **9**, 148 (1995).
[53] D. Suess, V. Tsiantos, T. Schrefl, J. Fidler, W. Scholz, H. Forster, R. Dittrich, and J. J. Miles, J. Magn. Magn. Mater. **248**, 298 (2002).
[54] M. Kirschner, T. Schrefl, F. Dorfbauer, G. Hrkac, D. Suess, and J. Fidler, J. Appl. Phys. **97**, 10E301 (2005).

Chapter 5

NANOSCALE STRUCTURAL AND MAGNETIC CHARACTERIZATION USING ELECTRON MICROSCOPY

David J. Smith and Martha R. McCartney
Department of Physics and Astronomy and Center for Solid State Science
Arizona State University
Tempe AZ 85287-1504, USA

Rafal E. Dunin-Borkowski
Department of Materials Science and Metallurgy
University of Cambridge
Cambridge, CB2 3QZ, United Kingdom

Abstract The transmission electron microscope (TEM) is a powerful instrument for structural, chemical and magnetic characterization at the nanoscale. Imaging, diffraction and microanalytical information can be combined with complementary micromagnetic information to provide a more thorough understanding of magnetic behavior. The first part of this chapter provides a brief overview of TEM operating modes that are suitable for examination of magnetic materials. The latter part provides examples that serve to illustrate the diverse range of materials that can be usefully studied.

1. INTRODUCTION

Novel physical properties are demonstrated by magnetic materials as their dimensions are reduced towards the nanoscale. Important examples include the giant magnetoresistance (GMR) effect in magnetic/non-magnetic multilayers [1], and oscillatory exchange coupling between magnetic thin films, which depends on the thickness of the non-magnetic spacer layer [2]. Such unique properties of magnetic thin films and nanostructures hold great promise for the development of novel nanoscale storage devices and sensors. However, for successful utilization, it is essential to correlate the magnetic behavior of the materials with their microstructure. In particular, such characteristics as interface roughness, film texture, crystallographic defects, and chemical mixing can have adverse effects on magnetic response. An understanding of the interdependence between growth and synthesis conditions, microstructure and magnetic properties is essential.

The transmission electron microscope (TEM) enables the local microstructure and magnetic properties of a material to be visualized directly with high spatial resolution, unlike bulk techniques such as X-ray diffraction and Rutherford backscattering. Technical advances in the high-resolution electron microscope (HREM) have resulted in performance improvements that enable atomic-scale detail to be resolved on a routine basis [3]. A range of imaging, diffraction and microanalytical modes is readily available in modern TEMs, and can be used to obtain this essential nanoscale information. Imaging techniques include transmission and high-resolution electron microscopy and Z-contrast annular-dark-field imaging, while nanospectroscopy approaches include electron-energy-loss and energy-dispersive X-ray spectroscopy as well as energy-filtered imaging. The TEM can also be used to provide micromagnetic information about materials over a wide range of length scales with a spatial resolution that cannot be approached by most other instruments. TEM methods that are specifically used to highlight micromagnetic features include Lorentz microscopy [4] and electron holography [5]. We begin this chapter with a brief introduction to these various imaging, nanospectroscopy and micromagnetic operating modes. We then describe in more detail some representative applications of these techniques to the characterization of nanostructured magnetic materials.

2. ELECTRON MICROSCOPY METHODS

In the transmission electron microscope, the finely focused electron beam, with a wavelength typically of much less than 0.1 nm, is used to form a highly magnified image of the specimen region of interest. The conventional transmission electron microscope (CTEM) uses a broad circular beam with a diameter of about 0.5-1.0 microns, which is transmitted through a suitably thinned specimen. In the scanning transmission electron microscope (STEM), a focused nanoprobe is rastered across the sample. The transmitted electrons are then used to form a modulated image on a display monitor that is synchronized with the incident scan. CTEM images have historically been recorded on photographic film, but digital recording with charge-coupled-device (CCD) cameras enables quantitative analysis, and has become widespread. Unavoidable electron lens aberrations normally restrict image resolutions to the 0.25-0.1 nm range for electron energies of 200-1250 keV. Nevertheless, atomic arrangements can be resolved in many different inorganic materials when they are imaged in low-index crystallographic orientations. When the nanoprobe of a STEM is stopped at specific sample locations, the detection of characteristic X-rays emitted by the sample or the examination of the energy-loss spectrum of the transmitted electrons enables element-specific information to be obtained. These microanalytical signals can be used to provide highly useful complementary information about local defects or inhomogeneities in magnetic materials.

2.1. Transmission Electron Microscopy

In the standard operating mode of the TEM, usually referred as amplitude or diffraction contrast imaging, electrons that have been transmitted by the sample are used to form a highly magnified image of the specimen feature of interest on a final viewing screen or recording medium. A small objective aperture is typically used to prevent most of the scattered (or diffracted) electrons from reaching the final image plane, in order to determine the image contrast. Crystalline samples are usually aligned by reference to their electron diffraction pattern (EDP) to ensure that the specimen orientation relative to the incident electron beam direction satisfies a strongly diffracting condition. Under such conditions, many common types of defects then have a highly characteristic appearance, potentially allowing identification of any second-phase materials. Crystal lattice spacings and the angles between lattice planes can also be determined by reference to the EDP.

The availability of a crystalline substrate or supporting medium often provides a useful method for orientation of the sample in the electron microscope. By using the substrate EDP as a reference, it becomes possible to align an internal interface of interest perpendicular to the electron beam direction. Changes in the microstructure of thin films and multilayers, either laterally or as a function of distance from the substrate can then be determined. As an example, Fig. 1 shows diffraction contrast images of thin CoFe films grown directly on the native oxide of a Si substrate at room temperature and at 100 °C, and also grown on a CoO buffer layer [6]. Considerable changes in the morphologies of the CoFe thin films, including the average widths of the columnar grains, result from the different growth conditions. These changes can be correlated with marked changes in the measured coercivities of the films. Finally, it should be appreciated here that the examination of complex multilayered samples can represent a serious challenge to the electron microscopist because it can be difficult to prepare regions that are electron-transparent across the entire region of interest simultaneously. Different approaches for preparing electron-transparent specimens are described below.

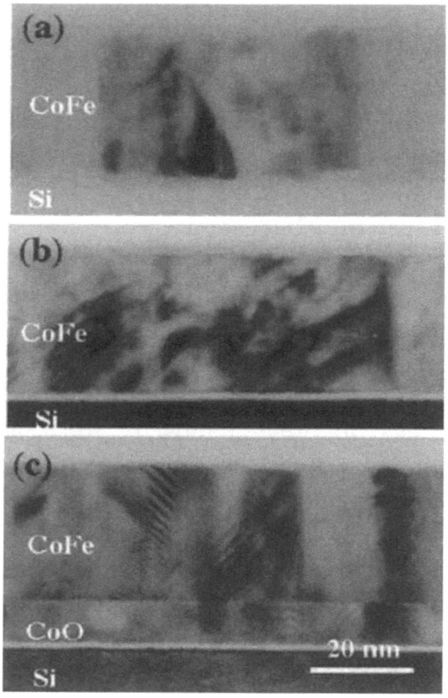

Figure 1. Cross-section diffraction contrast electron micrographs showing changes in morphology of 30-nm CoFe thin films depending on specific growth conditions: (a) grown at room temperature on Si substrate; (b) grown at 100 °C on Si substrate; (c) grown at 100 °C on CoO (10 nm) intermediary layer [6].

2.2. High-Resolution Electron Microscopy

In high-resolution or phase-contrast imaging, a large objective aperture (or sometimes no aperture at all) is used. One or more diffracted beams interfere with the directly transmitted beam to form the image, and the image contrast then depends on the relative phases of the various beams. This mode of operation can thus be termed phase-contrast imaging. When the imaging conditions (correct lens defocus, no image astigmatism, accurate incident beam alignment) are properly optimized, phase-contrast images can be directly interpreted in terms of the projected crystal potential. In recent generations of TEMs, the image resolution is usually better than 0.2 nm, so that individual atomic columns can be resolved in many crystalline inorganic materials. For direct image interpretation with the highest possible spatial resolution, the specimen thickness must be less than about 10 nm. High electron doses, typically 500 to 2000 electrons per square Ångstrom, are required to record such images, which means that specimens for high-resolution studies must be relatively resistant to electron irradiation effects. It is impossible to examine most organic materials and polymers directly under such intense imaging conditions. By using a specimen heating holder, and by adding a TV-rate image-pickup system to the base of the electron microscope lens column, dynamic events can be followed in real time without significant loss of spatial resolution.

Over the past 40 years, the technique of high-resolution electron microscopy has been used to characterize a wide range of inorganic materials. Important applications include determining the detailed microstructure of defects, interfaces and grain boundaries, investigating nanocrystalline features in amorphous films, and studying small particles in heterogeneous catalysts. The characterization of magnetic thin films and multilayers continues to be very important, since layer continuity and defect microstructure are crucial to the viability of recording media. High-resolution images are able to provide site-specific details that are usually unavailable using other techniques. As an example, Figs. 2(a) and (b) show two high-resolution electron micrographs that illustrate the amorphous or polycrystalline nature of the barrier layers in simple magnetic tunnel junctions (MTJs) grown by dc reactive sputtering [7]. The layer sequences in the images are: (a) Co (50 nm)/ HfO_2 (10 nm)/Fe (50 nm), and (b) Co (50 nm)/ CoO (10 nm)/Fe (50 nm).

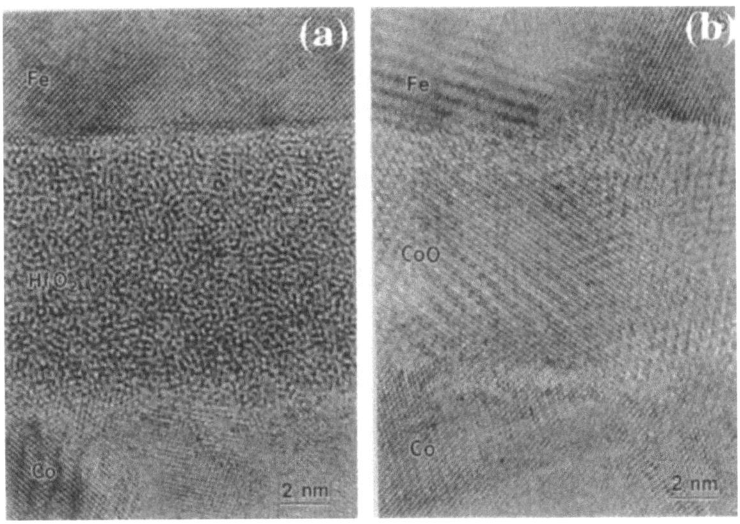

Figure 2. (a) High-resolution electron micrograph of Co(50 nm) / HfO_2 (10 nm) / Fe(50 nm) tunnel junction showing amorphous nature of barrier layer; (b) Cross-sectional electron micrograph of Co/CoO/Fe tunnel junction showing polycrystallinity prevailing in CoO barrier layer [7].

2.3. Nanospectroscopy

Many modern electron microscopes are equipped with high-brightness field-emission-gun electron sources which are able to provide a small, high intensity electron probe at the level of the specimen. Such nanoprobes allow complementary chemical information to be obtained from sub-nanometer-sized regions. Specimens that are intended for microanalysis must be thin (< 50 - 100 nm) in order to avoid multiple scattering effects that would lead to loss of spectral resolution and sensitivity. Their surfaces must be carefully cleaned if quantitative analysis is to be performed since substantial carbon contamination may otherwise occur during small-probe observations. It is thus becoming common-place to clean both the TEM sample, and the sample holder, in a plasma-cleaning device immediately prior to insertion in the electron microscope. Moreover, it must be appreciated that an intense stationary electron probe can itself alter the local microstructure and composition of a sample significantly.

The technique of electron-energy-loss spectroscopy (EELS) is used to measure the energy-loss spectrum of electrons that have been transmitted through a specimen. The EELS spectrum consists of a monotonically decreasing background and a series of peaks, which typically have characteristic energies in the range of 100 to several thousand eV, which are directly related to inelastic scattering processes that are specific to the elements in the specimen. Information can thus be obtained from an EELS spectrum about elemental composition, as well as about chemical bonding and electronic structure. The spatial resolution of this information is determined primarily by the focused probe diameter, provided that the specimen is not too thick. Quantification of elemental composition can attain sensitivity levels approaching better than one atomic percent. EELS can also be used to measure changes in near-edge fine structure, enabling oxidation states to be determined. EELS can be applied to magnetic materials when chemical information about interfacial abruptness is needed, for example after multilayer deposition or after extended annealing cycles. Diffusion profiles across interfaces can then be extracted from EELS data, provided that the effects of the finite probe size on the elemental profiles are properly taken into account. An interesting example of an $[Fe(2nm)/oxide]_{x50}$ multilayer structure is shown in Fig. 3. In this study [8], convolution of the expected composition (solid line) with an appropriate probe function gave a convoluted profile (dotted line) that was in close agreement with the experimental points (filled circles), indicating unexpectedly that the Fe:O stoichiometry of the native oxide was close to unity.

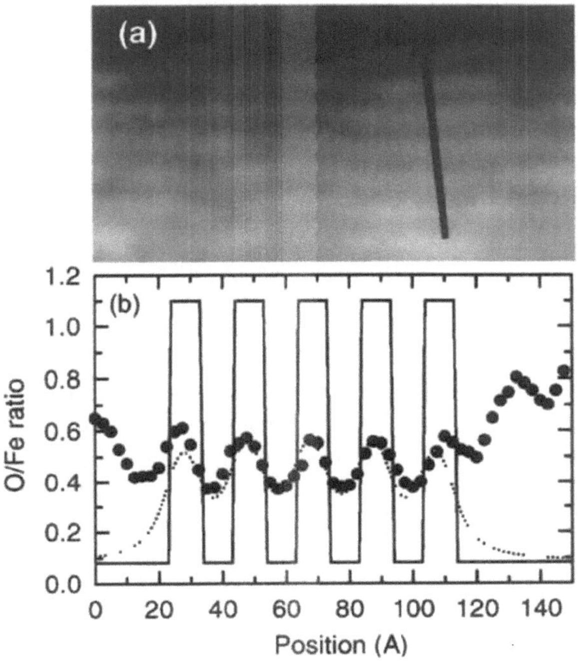

Figure 3. (a) Annular-dark-field electron micrograph of [Fe (2 nm)/ oxide]$_{x50}$ showing path of EELS probe; (b) O/Fe ratio along line scan for model profile (solid curve), convolved profile (dotted curve) and experimental points (filled circles) [8].

The technique of energy-dispersive X-ray spectroscopy (EDXS) utilizes the characteristic spectrum of X-rays that is emitted by a sample following initial excitation of an inner-shell electron by the high-energy electron beam to a vacant higher-energy level. Highly localized information about the elemental composition of the sample can again be obtained, with the spatial resolution determined primarily by the probe size and by any beam broadening that has occurred within the specimen. EDXS is a relatively straightforward technique experimentally, and it can be used to provide rapid qualitative microanalysis. Quantitative elemental analysis can also be achieved, with an accuracy that can approach a few atomic percent. However, thin specimens are required for quantitative EDXS analysis because of the need to avoid spurious signals arising from any backscattered electrons. Moreover, (pre-)calibration of the specific EDXS analysis system using standards of known composition is necessary. EDXS is better suited than EELS for detecting elements of high atomic number (Z). However unlike EELS, it is unable at present to provide any information about local bonding. Low-Z elements ($Z < 11$) are not detectable by some EDXS systems due to absorption by the detector window, but this shortcoming can be overcome by using ultrathin or even window-less detectors. Care is required when performing EDXS analysis because of the possibility of detecting X-rays originating from the microscope column or from regions of the specimen away from the area of interest.

2.4. Energy-Filtered Imaging

Energy-filtered imaging is a TEM technique based on EELS that allows information about the spatial distribution of elements in a sample to be obtained. By using an energy-selecting imaging spectrometer, located either in the post-specimen lens column or below the final viewing screen, only those electrons that have lost energies within a specific energy range are used to form the final image. Since electrons in the low-loss energy region of the spectrum (plasmon or valence-loss electrons) are not very sensitive to variations in composition, those originating from characteristic inner-shell ionization edges are instead normally used for image formation.

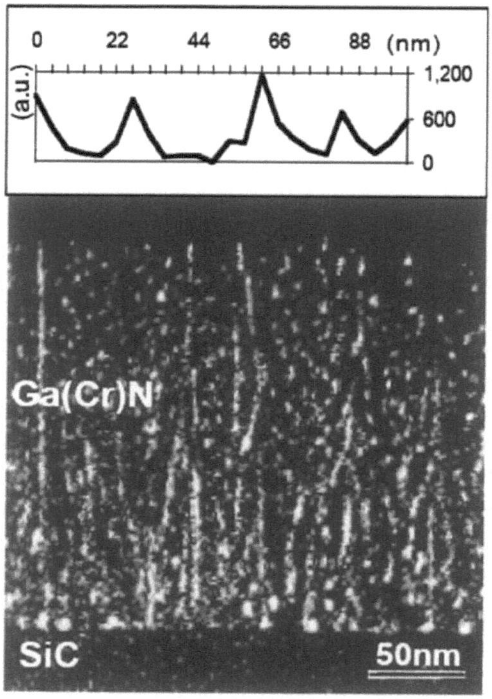

Figure 4. Energy-filtered image of Cr-doped GaN showing considerable Cr clustering. Inset above shows Cr EELS line profile [9].

The selection of electrons for imaging is accomplished by using an energy-selecting slit of variable width, and some sort of background subtraction routine is required in order to separate the desired element-specific contributions from the background signal. Although the quality and spatial resolution of energy-filtered images suffer from poor signal-to-noise due to low inelastic scattering cross-sections, elemental mapping provides a highly useful technique for extracting valuable compositional information about magnetic materials. As an example, Fig. 4 shows an energy-filtered image of a cross-sectioned Cr-doped GaN epilayer, as grown by reactive molecular-beam epitaxy at 825 °C, under investigation for possible 'spintronic' applications [9]. Unlike companion samples grown at lower temperatures which had relatively uniform Cr doping, this specimen showed considerable Cr clustering, which was believed to account for its lower coercivity when compared with the other samples.

2.5. Lorentz Microscopy

The magnetic microstructure of a TEM specimen is not normally visible under conventional operating conditions. In particular, the strong magnetic field (about 2 T) at the specimen level generated by the objective lens modifies or even completely destroys any intrinsic magnetic domain structure inside a TEM specimen. Such severe restrictions for studying magnetic materials using the TEM have led to the development of several alternative imaging modes. These are collectively known as Lorentz microscopy, since they are based on the sideways deflection of the charged high-energy electron inside a magnetic specimen by the Lorentz force. Since the objective lens is usually switched off for Lorentz imaging, the spatial resolution of the magnetic detail revealed in Lorentz micrographs is substantially degraded when compared with conventional imaging. However, this loss of resolution is not normally considered as a serious obstacle since magnetic fields only rarely change abruptly on a sub-nanometer scale.

The simplest and most common technique for imaging magnetic domain structures is the so-called Fresnel or defocus mode of Lorentz microscopy. Electrons traversing domains with different magnetization orientations are deflected in different directions, creating regions that have either an excess or a deficit of electrons immediately below the sample. By imaging in an under- or over-focus condition, lines of black or white contrast appear at the positions of the domains walls. This configuration is illustrated schematically in Fig. 5(a), and an example of a Lorentz micrograph taken in the plan-view geometry from a CoPtCr /alumina /Co MTJ is shown in Fig. 5(b) [10].

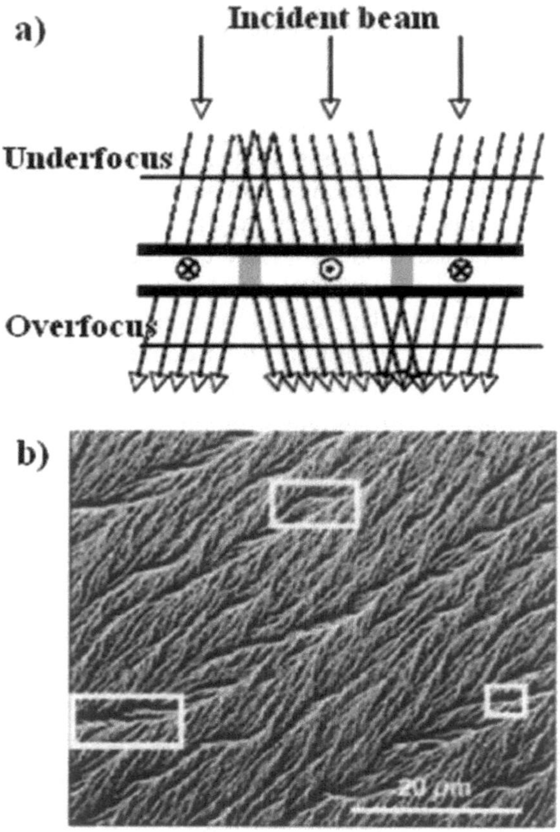

Figure 5. (a) Lorentz imaging (schematic) and (b) Lorentz micrograph from a CoPtCr (15 nm)/ alumina /Co (10 nm) magnetic tunnel junction during field-cycling. Domain walls and ripple contrast due primarily to the soft Co layer [10].

Measuring the apparent domain-wall width as a function of (pre-calibrated) defocus yields a quantitative estimate of the wall width. The alternative Foucault mode of Lorentz microscopy involves the use of a small objective aperture to select for imaging only those electrons that have been deflected in a specific direction. The image contrast of Foucault images is highly sensitive to aperture position, and quantitative information is virtually unobtainable. However, domain configurations are easily recognized using this approach. Despite inherent spatial resolution limitations of Fresnel or Foucault imaging, either technique allows useful real-time viewing of dynamic processes.

2.6. Electron Holography

The technique of electron holography permits retrieval *and* quantification of both the amplitude and the phase shift of the electron beam that has traversed through a TEM specimen, unlike the situation for normal imaging where all phase information is lost [5]. Since the phase shift of the electron which has passed through the sample can be directly related to magnetic (and electric) fields, electron holography provides a powerful way to probe magnetic microstructure. There are many types of electron holography, but the mode known as off-axis or sideband electron holography, as illustrated in Fig. 6, is most commonly used.

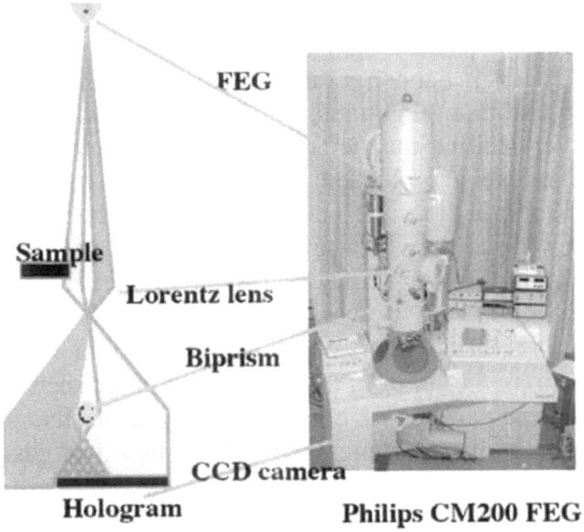

Figure 6. (Left) Schematic showing basic TEM configuration for off-axis electron holography; and (b) photograph showing electron microscope suitably equipped for recording holograms of magnetic materials.

A field-emission electron gun is used to provide a coherent source of electrons. A thin, positively charged wire (or 'biprism') is then used to provide interference between the reference wave that has passed through vacuum with the object wave that has passed through the specimen. The biprism is usually located in the selected-area aperture plane of the microscope below the sample. Historically, photographic film was used for recording holograms but it has recently become commonplace to record the final holographic interference pattern using a CCD camera. Digital recording greatly benefits the process of hologram reconstruction and phase quantification. Once the phase shift of the electron wave has been reconstructed, quantitative information with high spatial resolution about the magnetic (and electric) fields in a sample can be easily extracted. The convenient availability of the so-called 'Lorentz minilens' just below the sample enables moderate magnifications (about 50,000 x) to be achieved during observation of magnetic materials with the normal objective lens completely switched off and with the sample located in a magnetic-field-free condition. Alternatively, a weak vertical field can be achieved by slightly exciting the objective lens, thus enabling in-plane magnetic fields to be applied *in situ* by tilting the sample. Hysteresis loop response of FM nanostructures can be determined directly, as shown later in Section 3.5.

3. NANOSTRUCTURED MAGNETIC MATERIALS

Many different types of nanostructured magnetic specimens are of interest, with a host of possible applications. For example, a sample may consist of granular or polycrystalline magnetic thin films or multilayers, complicated structures such as exchange-biased magnetic tunnel junctions or magnetic tunneling transistors, or it may contain nanoscale patterned features. It can sometimes be a serious drawback of TEM methods that thin sections of a material are required for examination. Since the micromagnetic features usually extend into at least two, and more usually three, dimensions, specimens for observation must often be prepared in two or even three orthogonal directions, i.e. in both plan-view and cross-sectional geometries. Plan-view specimens can be straightforward to prepare using methods such as chemical etching or mechanical polishing. In other cases, mechanical polishing and dimple-grinding are required, followed by ion-beam milling at low beam energies. This latter approach will generally lead to larger regions of thinned material. The preparation of suitable cross-sections sometimes necessitates the use of costly ancilliary equipment such as a focused-ion-beam system, and it can be time-consuming and tedious to achieve thin specimen areas that are genuinely representative.

The cross-sectional viewing geometry is necessary when the features of interest are buried within the material or else distributed throughout the depth of the sample because of its layered structure. The planarity of buried interfaces can be determined, and the distribution, width and composition of interfacial defects and the presence of any secondary phases such as pinholes across tunnel junctions can be assessed. Cross-sectional specimens are typically prepared by glueing the material of interest face-to-face onto additional supporting material. Mechanical polishing of a thin slice of this stack, followed by dimple-grinding, is used to reduce the total overall thickness to about 5 to 10 microns. Final milling with an argon-ion-beam is used to achieve perforation of the cross-sectional film. Thin areas of the interface are created where the perforated hole intersects the glued faces. The damage to some magnetic materials that occurs during cross-sectional specimen preparation can be reduced by ion-milling with the specimen held at liquid nitrogen temperature.

Specimens that contain materials with very different ion-milling rates, such as metallic multilayers grown on silicon substrates, often tend to form bridges of material across the perforated area. Ion-milling at very low angles of incidence (~ 1-2°) in a direction parallel to the interface can sometimes be used to overcome or at least alleviate these bridging problems. Finally, it should be noted that the use of a crystalline substrate such as silicon provides a convenient reference material for specimen orientation purposes in the TEM. Examination of the substrate EDP can be used to ensure that the substrate normal is aligned exactly perpendicular to the electron beam direction. The thin-film microstructure can then be easily determined.

3.1. Thin Films and Multilayers

Magnetic thin films and multilayers are typically prepared by sputtering or molecular beam epitaxy (MBE), often using computer control of deposition parameters to ensure high reproducibility. Single crystal Si with various surface normals, and sometimes GaAs or MgO, are commonly used as substrates for growth. MBE facilitates epitaxial growth of high quality crystalline samples. Any orientation dependence of the magnetic behavior can then be investigated by choosing suitable substrates that induce the films to grow in specific orientations. In contrast, sputtering generally provides polycrystalline films and permits greater sample throughput, allowing trends in behavior with changes in deposition parameters to be correlated rapidly and systematically. Indeed, sputtering was used in early studies of polycrystalline Co/Cu multilayers that exhibited GMR values as high as 65% at room temperature and 115% at 4.2 K [11]. Later studies showed that highly-oriented (111), (110) and (100) Co/Cu superlattices of excellent crystallinity could be grown by sputtering onto substrates of sapphire, MgO (110) or MgO (100), respectively, and using thin (1.2 - 1.5 nm) intermediary seed layers of Pt or Pd [12].

Figure 7(a) shows part of a {111}-oriented [Co (1.0 nm)/Cu (0.9 nm)]$_{x24}$ superlattice. Strong, two-dimensional {111} lattice fringes are visible, revealing considerable structural disorder, including stacking faults and twins, although the individual layers cannot be seen. In order to highlight the anticipated compositional contrast from the individual Co and Cu layers, it was necessary to record defocused diffraction contrast images with a small objective aperture, as shown in Fig. 7(b).

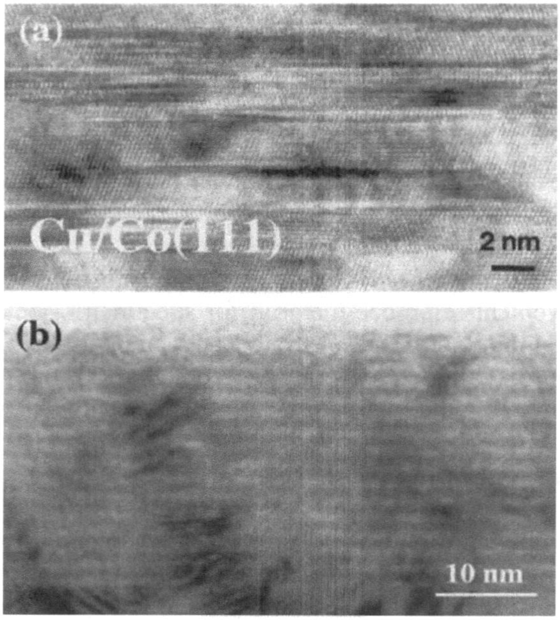

Figure 7. (a) High-resolution image showing part of {111}-oriented Cu(0.9 nm) /Co(1.0 nm) GMR multilayer. (b) Defocused low-magnification image of Cu/Co multilayer showing clear delineation of separate Cu and Co layers [12].

Crystalline FePt thin films are of much interest because of the possibility of forming the chemically ordered $L1_0$ tetragonal phase, which has very high magnetocrystalline anisotropy. Figure 8 shows an example of a high quality, epitaxial Fe_xPt_{1-x} ($x \approx 0.6$) film grown by MBE on an MgO substrate held at 300 °C during deposition [13]. The use of a thin (~ 1.5 nm) Pt seed layer in this case was found to be necessary in order to establish the desired single growth orientation of the alloy film. Doubling of the {200} lattice-fringe period was observed along the horizontal direction of this sample (surface normal vertical), indicative of the desired compositional modulation. Further TEM observations of films grown at 500 °C established the presence of more extensive ordered regions, consistent with the much larger short-range-order parameter (S = 0.81) relative to that obtained for growth at 300 °C (where S = 0.39).

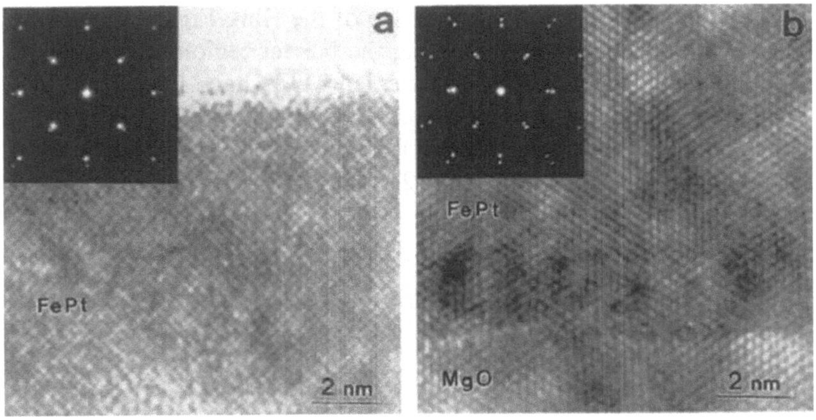

Figure 8. High-resolution electron micrographs showing cross-section of FePt/Pt/MgO sample: (a) [001] projection. Note surface flatness; (b) [110] projection showing FePt/Pt/MgO interface region. Insets show corresponding selected-area electron diffraction patterns [13].

3.2. Tunnel Junctions and Spin-Tunneling Transistors

Spin-dependent magnetic tunnel junctions (MTJs), which consist of two thin ferromagnets (FMs) separated by a narrow insulating barrier, display large conductance changes when the relative magnetization alignments of the FM layers are reversed. Devices based on this alignment effect, such as magnetic random access memory (MRAM), utilize the large MR difference between the parallel/anti-parallel configurations for sensing and recording purposes. Two basic types of MTJ have been developed. The simplest consists of two FM materials that have different switching fields, usually referred to as 'hard' and 'soft'. A more complex MTJ structure has an additional antiferromagnetic layer adjacent to one of the FM layers, which has the effect of exchange-biasing or 'pinning' the magnetic moment of this particular layer so that its hysteresis loop is offset sideways from zero field by the exchange field. In both of these MTJ geometries, successful device fabrication and operation depends on several factors, including the uniformity, stoichiometry, and thermodynamic stability of the insulating layer, and the quality (smoothness and compositional abruptness) of the FM-insulator interfaces.

Figure 9 shows an example of the simpler type of MTJ structure [14]. The micrograph in Fig. 9(a) displays the entire cross-section of an MTJ consisting of magnetically hard $Co_{75}Pt_{12}Cr_{13}$ (15 nm)/ alumina barrier (~ 2.5 nm)/ magnetically soft $Co_{88}Pt_{12}$ (15 nm), with an underlying layer of Cr-V which is used here as a seed layer to promote growth of the *hcp* Co-Pt-Cr alloy with its easy (or *c*) axis in the plane of the film. Figure 9(b) is a high-resolution electron micrograph showing the barrier region after annealing at 350 °C. Extensive observations of annealed MTJs have been used to show that the alumina layers remained continuous and relatively flat after annealing, although the mean barrier thickness decreases slightly from 2.7 nm for the as-grown samples to about 2.5 nm after annealing.

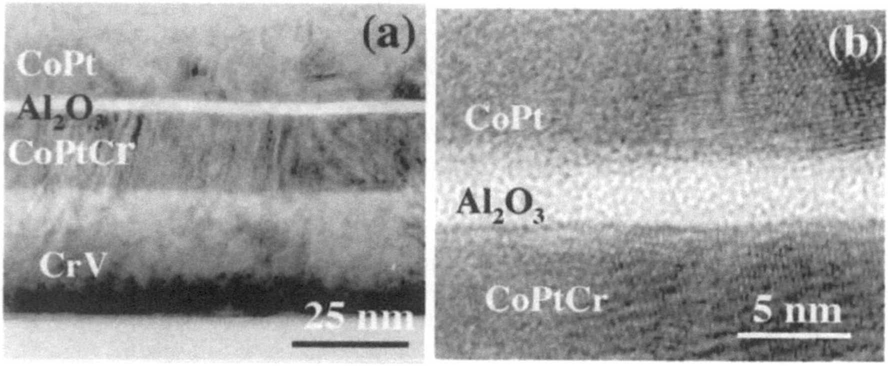

Figure 9. Magnetic tunnel junctions: (a) cross-section of simple MTJ before annealing and (b) high-resolution image of barrier region after annealing at 350 °C [14].

Much attention has recently been directed to 'spintronic' devices, which aim to utilize the spins of the conduction electrons as well as their charge for very high density information storage purposes. The magnetic tunneling transistor (MTT), which is based on an exchange-biased MTJ of the second type described above, has emerged as a possible candidate for providing highly efficient spin injection. The core of the MTT consists of an FM emitter, a tunnel barrier, an FM base, and a final semiconductor collector. Since spin-polarized electrons must overcome a Schottky barrier at the base/collector interface, the use of thin intermediary seed layers represents a possible mechanism for enhancing device sensitivity. Figure 10(a) shows a cross-sectional electron micrograph of a possible Ru-based MTT. The enlargement in Fig. 10(b) shows the region near the Ru seed layer in greater detail [15]. Some interdiffusion or reaction with the Si substrate, also confirmed by small-probe microanalysis, has taken place, and could account for the low junction breakdown voltages measured for the Ru-based MTT devices.

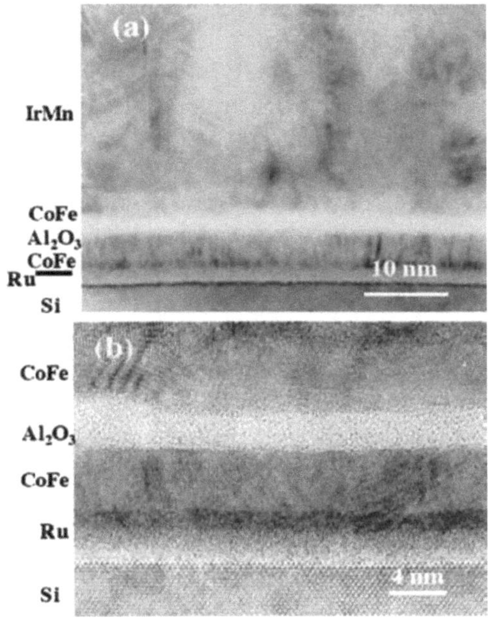

Figure 10. Cross-sectional electron micrographs of possible Ru-based magnetic tunneling transistor: (a) low magnification image, also showing antiferromagnetic IrMn layer; (b) high-magnification showing central barrier and Ru layer interaction with Si substrate [15].

3.3. Discontinuous Multilayers and Metal-Insulator Granular Films

Discontinuous multilayers, which consist of planes of metallic FM particles interspersed within an insulating matrix are easy to prepare, and they offer several distinct advantages over continuous magnetic layers and especially MTJs. The benefits include less susceptibility to electrical breakdown and greatly reduced likelihood of pinhole development across the insulating barrier which often serve to short-circuit the current in a regular MTJ structure. Figure 11(a) shows a plan-view image of a simple trilayer system consisting of SiO_2 (3 nm)/Co (2 nm)/SiO_2 (3 nm) [16]. In this case, the Co particles (dark contrast) form short chains of clusters of 3.5-nm approximate width. It is clear that the layer is indeed discontinuous.

Figure 11(b) shows a cross-section electron micrograph of a discontinuous magnetic tunnel junction (DMTJ), consisting of two planes of magnetic Co nanoparticles sandwiched between three insulating SiO_2 layers. After moderate annealing, this simple bilayer DMTJ showed an MR of ~ 4%

at room temperature. Annealing (300 – 350 °C) of similar multilayers consisting of 20 bilayers, led to increased numbers of superlattice (SL) reflections visible in electron diffraction patterns, suggesting some sharpening of the metal/insulator interfaces, whereas over-annealing (≥ 400 °C) led to a decrease in the number of SL reflections implying deterioration of the interfaces, most likely due to interdiffusion or possibly metal oxidation.

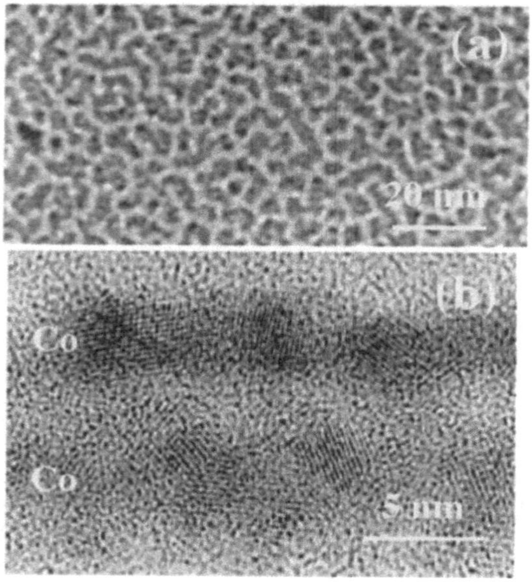

Figure 11. Discontinuous magnetic layers: a) plan-view of single layer - SiO_2(3 nm) /Co(2 nm) /SiO_2(3 nm); b) cross-section of Co/SiO_2/Co DMTJ [16].

Granular films, rather than multilayers, which consist of FM nanoparticles dispersed within a non-magnetic insulating matrix, display interesting spin-dependent transport behavior that is determined primarily by particle size. Electron micrographs from co-deposited Co-SiO_2 granular films, such as the example shown in Fig. 12, confirmed that the films basically consisted of small *hcp* Co particles distributed relatively uniformly throughout the SiO_2 matrix [17]. Further TEM observations as a function of increasing Co volume fraction up to percolation gave good agreement with estimates of particle size based on magnetic susceptibility measurements.

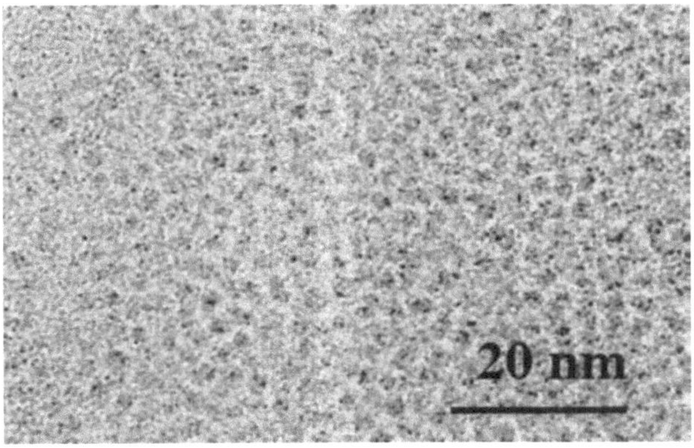

Figure 12. Electron micrograph showing co-deposited Co-SiO_2 granular film containing approximately 32 vol.% Co [17].

3.4. Nanostructures

Characterization of the micromagnetic states of FM nanostructures using off-axis electron holography becomes increasingly challenging when their dimensions are reduced below 100 nm, primarily as a result of a fundamental signal-to-noise limit on the detection of small phase gradients, rather than the available instrumental spatial resolution. Nevertheless, recent results demonstrate the feasibility of studying the magnetization behavior of closely-spaced nanoparticles and nanowires that have one or more dimension in the 5 - 100 nm range. As an example, Figs. 13 (a) and (b) show the magnetic remanent states of several $Fe_{0.56}Ni_{0.44}$ nanoparticle chains, as determined by electron holography [18]. Interpretation of the phase contours suggests the presence of magnetic vortex states surrounding a central flux channel, as drawn schematically in Figs. 13 (c) and (d). The unexpected complexity of these observations, which was confirmed by extensive micromagnetic simulations, reinforces the need to understand and control the effects of size, shape and proximity of magnetic nanoparticles intended for potential storage applications.

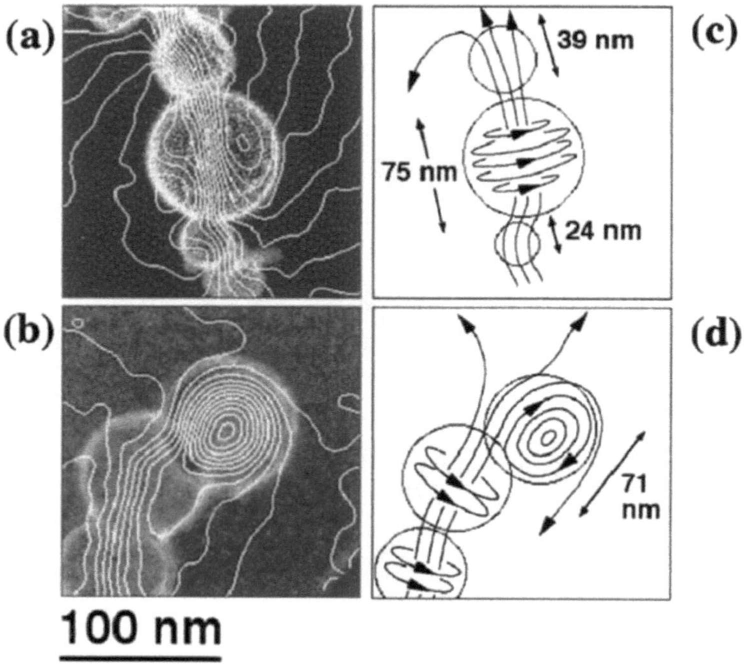

Figure 13. (a) and (b) Experimental magnetic phase contours showing the strength of the local induction (integrated in the electron beam direction) in two different chains of $Fe_{0.56}Ni_{0.44}$ particles. (c) and (d) show schematic representations of the magnetic microstructure in the chains. Magnetic vortices spinning about the chain axis are visible in (c) and (d). A vortex spinning perpendicular to the chain axis is also visible in (d). [18].

Another intriguing example of magnetostatic interactions between closely-coupled nanoparticles is illustrated in Fig. 14 [19]. This study involved Co nanoparticles, with diameters in the range of 20 to 30 nm, which had spontaneously formed rings, chains and other close-packed aggregates. Examination by electron holography established the existence of flux closure remanent states associated with each cyclic ring ensemble, having either clockwise or counterclockwise configurations, as shown in Figs. 14 (b) and (c).

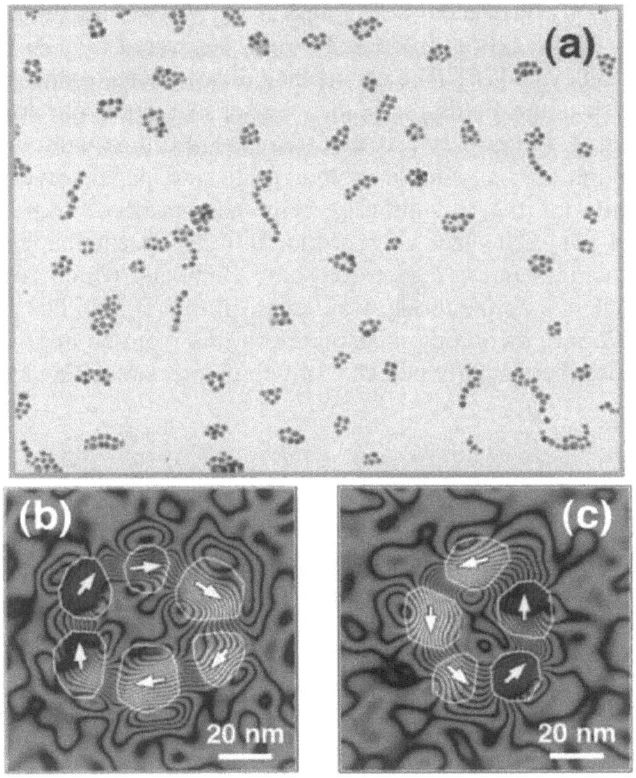

Figure 14. Co nanoparticle rings and chains: (a) low magnification bright-field image of self-assembled Co nanoparticle rings and chains deposited onto an amorphous carbon support film, where each Co particle has a diameter of between 20 and 30 nm, (b) and (c) magnetic phase contours (128 × amplification; 0.049 radian spacing), formed from the magnetic contribution to the measured phase shift, in two different nanoparticle rings. The outlines of the nanoparticles are marked in white, while the arrows indicate direction of the measured magnetic induction [19].

3.5. Patterned Nanostructures

For a fuller understanding of nanoparticle interactions and magnetic response, it is desirable to have better control over particle placement and geometry. Moreover, stable and reproducible magnetization states are essential for potential device applications. Patterning of FM nanostructures enables systematic electron-holographic studies of remanent states and magnetization reversal processes to be made as a function of element size, thickness and shape. Magnetostatic interactions between closely separated particles can also be investigated. The left part of Fig. 15 shows some representative results of reconstructed holograms from two 30-nm-thick rectangular Co elements (275 x 220 and 275 x 300 nm, separated by 170 nm) over an entire hysteresis cycle [20]. The occurrence of solenoidal (or vortex) states at close to zero in-plane fields for both directions of the cycle coincides with minimal fringing fields between the elements. Further work using micromagnetic simulations, as shown in the right hand panel, revealed that the close proximity of the neighboring cells had assisted in trapping these domain states [21]. The lack of reproducibility of particular domain structures during magnetization cycles for some elements, which is obvious importance for device applications, was noted in this study. The formation of vortex states during switching is also undesirable from an applications point of view because they usually persist to much greater switching fields.

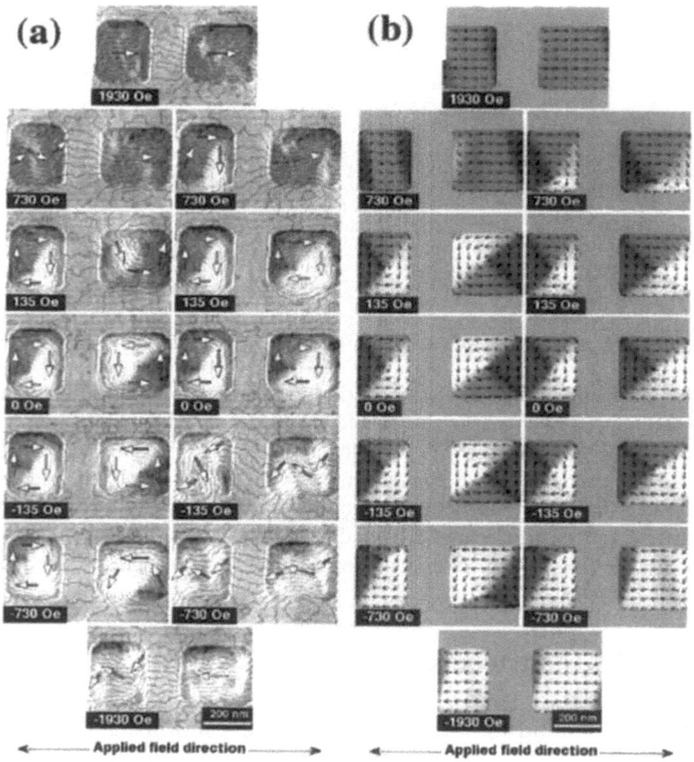

Figure 15. Rectangular Co elements: (a) magnetic contributions to measured phase shifts of holograms of 30-nm-thick elements for entire magnetization reversal cycle, for applied in-plane fields indicated, and (b) best-fitting micromagnetic simulations of the hysteresis cycle shown in (a). Initial states in the top image are an S-state for the smaller rectangle and a C-state for the larger rectangle [21]. In (a), the contour spacing is 0.21 π radians, and the images should be followed in a counter-clockwise sense [20].

Recent attention has been directed towards circular magnetization modes (i.e., clockwise and counterclockwise), rather than linear modes as an alternative basis for storage logic [22].

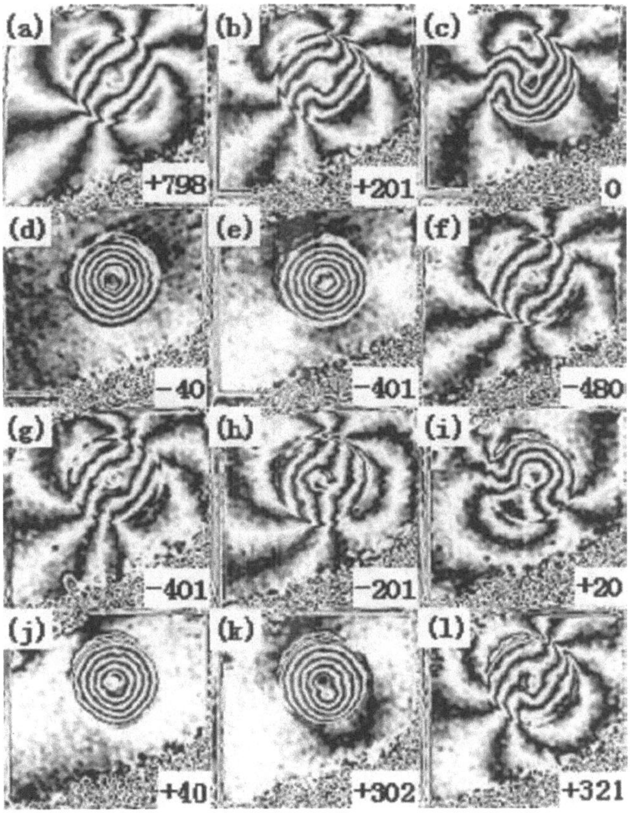

Figure 16. Amplified (8 ×) phase images of 400 nm/ 50 nm Co ring during entire hysteresis cycle. In-plane filed magnitude indicated below each image (units of Oe). Note vortex states visible in (d), (e) and (j), (k) [23].

However, variability and reproducibility of reversal processes for nano-scale disk-shaped elements are likely to be a serious issue, prompting consideration of modified shapes such as rings, as well as slotted disks and slotted rings. Ring-shaped elements have switching behavior that varies significantly depending on the ratio of the outer-ring/ inner-ring diameters [23]. Reversal proceeds either via a coherent domain-wall rotation or via the formation of an intermediate vortex state, as shown by the example in Fig. 16. Interestingly, these vortex states were not observed in any remanent states of the rings.

3.6. Minerals and Biogenic Crystals

Closely spaced magnetic nanocrystals that are found in nature are often more perfect in their sizes, shapes and arrangements than their synthetic counterparts. As a result, they can be chosen as model systems to study the effect of particle size, morphology, crystallography and spacing on magnetic microstructure.

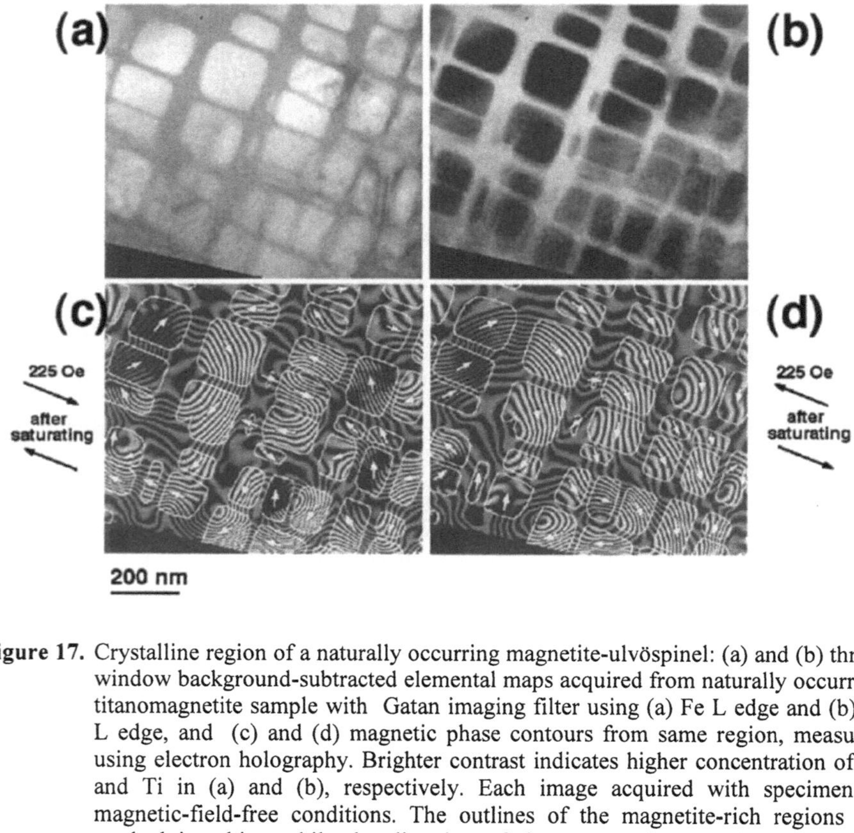

Figure 17. Crystalline region of a naturally occurring magnetite-ulvöspinel: (a) and (b) three-window background-subtracted elemental maps acquired from naturally occurring titanomagnetite sample with Gatan imaging filter using (a) Fe L edge and (b) Ti L edge, and (c) and (d) magnetic phase contours from same region, measured using electron holography. Brighter contrast indicates higher concentration of Fe and Ti in (a) and (b), respectively. Each image acquired with specimen in magnetic-field-free conditions. The outlines of the magnetite-rich regions are marked in white, while the direction of the measured magnetic induction is indicated using arrows. (c) was acquired after applying a large field of about 10 kOe towards top left, then the indicated field towards bottom right, after which the external magnetic field was removed for hologram acquisition. (d) was acquired after applying identical fields in the opposite directions [24].

Figures 17(a) and (b) show chemical maps of a crystalline region of a naturally occurring magnetite-ulvöspinel (Fe_3O_4-Fe_2TiO_4) mineral, which has exsolved during slow cooling and yields a three-dimensional intergrowth of ferrimagnetic magnetite-rich blocks that are separated by non-magnetic ulvöspinel-rich lamellae [24]. Figures 17(c) and (d) show two remanent magnetic states recorded from the same region using off-axis electron holography. The domain structure in the images is extremely complex, and a saturated domain structure was never observed at remanance. The abundance of single domain states implies that they have a lower energy than vortex states in the presence of strong interactions with neighboring magnetic crystals. Similar magnetic interactions have been observed in linear chains of ferrimagnetic magnetite crystals in magnetotactic bacteria, which are oriented passively in the direction of the earth's magnetic field in aquatic environments [25]. A magnetic induction map recorded from a single helical bacterial cell containing a chain of equidimensional magnetite crystals is shown in Fig. 18. The magnetic phase contours are highly constrained to be parallel to each other within the crystals and to follow the chain axis. Unlike the more complicated magnetic arrangement seen in Fig. 17, the magnetic moment in the linear bacterial chain is maximized, and its remanent magnetic state is almost equivalent to its saturated state.

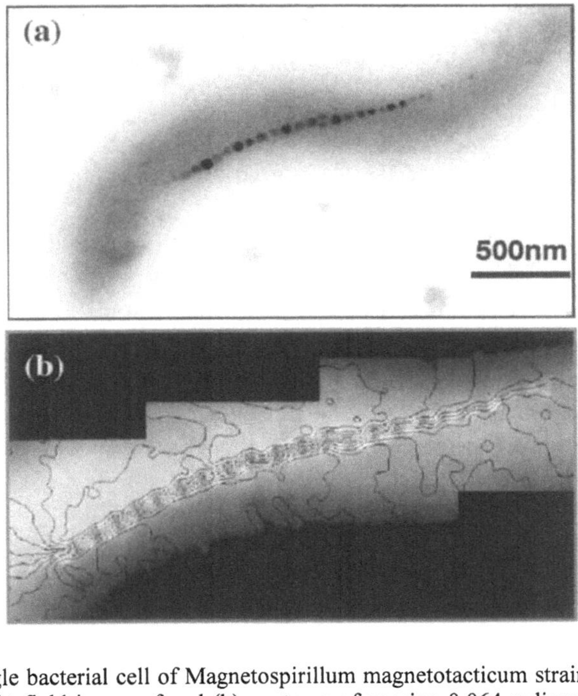

Figure 18. Single bacterial cell of Magnetospirillum magnetotacticum strain MS-1: (a) TEM bright-field image of and (b) contours of spacing 0.064 radians formed from the magnetic contribution to the holographic phase shift. The contours, which spread out at the chain ends, have been overlaid onto the mean potential contribution to the phase to allow the positions of the crystals to be correlated with the magnetic contours [25].

4. CONCLUDING REMARKS

The examples described in this chapter have been carefully chosen to demonstrate the broad diversity of information about magnetic thin films and nanostructures that becomes available from examination in the electron microscope. The powerful combination of microstructural, microanalytical and micromagnetic information at the nanoscale facilitates a more detailed understanding of basic magnetic properties. Current resolution levels for imaging and microanalysis are more than adequate, but future micromagnetic studies would benefit from enhancement in the strength of the available magnetic signals. The combination of electron holography with electron tomography could also enable the determination of three-dimensional magnetic vector fields in and surrounding nanoscale objects.

Acknowledgements

We are pleased to acknowledge friends and colleagues for their involvement in the work described here. Many of these studies involved use of the facilities in the John M. Cowley Center for High Resolution Electron Microscopy at Arizona State University.

References

[1] S. S. P. Parkin, Ann. Revs. Mater. Sci. **25**, 357 (1995).

[2] S. S. P. Parkin, N. More, and K.P. Roche, Phys. Rev. Lett. **64**, 2304 (1990).

[3] D. J. Smith, Rep. Prog. Phys. **60**, 1513 (1997).

[4] J. N. Chapman and M. R. Scheinfein, J. Magn. Magn. Mater. **200**, 729 (1999).

[5] R. E. Dunin-Borkowski, M. R. McCartney, and D. J. Smith, in "Encyclopedia of Nanoscience and Nanotechnology", Ed. H.S. Nalwa, American Scientific, Stevenson Ranch 2003, Volume X, Pp. 1-59.

[6] C. L. Platt, A. E. Berkowitz, D. J. Smith and M. R. McCartney, J. Appl. Phys. **88**, 2058 (2000).

[7] D. J. Smith, M. R. McCartney. C. L. Platt, and A. E. Berkowitz, J. Appl. Phys, **83**, 5154 (1998).

[8] G. S. D. Beach, F. T. Parker, D. J. Smith, P. A. Crozier, and A. E. Berkowitz, Phys. Rev. Lett. **91**, 267201 (2003).

[9] L. Gu, S. Wu, H.-X. Liu, R. Singh, N. Newman, and D. J. Smith, J. Magn. Magn. Mater. **291**, 1395 (2005).

[10] M. R. McCartney, R. E. Dunin-Borkowski, M. R. Scheinfein, D. J. Smith, S. Gider, and S. S. P. Parkin, Science, **286**, 1337 (1999).

[11] S. S. P. Parkin, Z. G. Li, and D. J. Smith, Appl. Phys. Lett. **58**, 2910 (1991).

[12] D. J. Smith, A. R. Modak, T. Rabedeau, and S. S. P. Parkin, Appl. Phys. Lett. **71**, 1480 (1997).

[13] R. F. C. Farrow, D. Weller, R. F. Marks, M. F. Toney, D. J. Smith, and M. R. McCartney, J. Appl. Phys. **84**, 934 (1998).

[14] S. S. P. Parkin, K.-S. Moon, K. E. Pettit, D. J. Smith, R. E. Dunin-Borkowski, and M. R. McCartney, Appl. Phys. Lett. **75**, 543 (1999).

[15] H. Wang, M. R. McCartney, D. J. Smith, X. Jiang, R. Wang, S. van Dijken, and S. S. P. Parkin, J. Appl. Phys. **97**, 104514 (2005).

[16] S. Sankar, A. E. Berkowitz, and D. J. Smith, Appl. Phys. Lett. **73**, 535 (1998).

[17] S. Sankar, A. E. Berkowitz, and D. J. Smith, Phys. Rev. B **62**, 14273 (2000).

[18] M. J. Hÿtch, R. E. Dunin-Borkowski, M. R. Scheinfein, J. Moulin, C. Duhamel, F. Mazaleyrat, and Y. Champion, Phys. Rev. Lett. **91**, 257207 (2003).

[19] S.L. Tripp, R.E. Dunin-Borkowski, and A. Wei, Angew. Chem. Int. Ed. **42**, 5591 (2003).

[20] R.E. Dunin-Borkowski, M.R. McCartney, B. Kardynal, and D.J. Smith, J. Appl. Phys. **84**, 374 (1998).

[21] R.E. Dunin-Borkowski, M.R. McCartney, B. Kardynal, D.J. Smith, and M.R. Scheinfein, App;l. Phys. Lett. **75**, 2641 (1999).

[22] J.G. Zhu, Y. Zheng, and G.A. Prinz, J. Appl. Phys. **87**, 6668 (2000).

[23] H. Hu, H. Wang, M.R. McCartney, and D.J. Smith, J. Appl. Phys. **97**, 054305 (2005).

[24] R.J. Harrison, R.E. Dunin-Borkowski, and A. Putnis, Proc. Nat. Acad. Sci (US) **99**, 16556 (2002).

[25] R.E. Dunin-Borkowski, M.R. McCartney, R.B. Frankel, D.A. Bazylinski, M. Posfai, and P.R. Buseck, Science, **282**, 1868 (1998).

Chapter 6

MOLECULAR NANOMAGNETS

W. Wernsdorfer
Laboratoire Louis Néel, CNRS
BP166
38042 Grenoble Cedex 9, France

Abstract The interface between classical and quantum physics has always been an interesting area, but its importance has grown with the current explosive thrusts in nanoscience. Taking devices to the limit of miniaturization where quantum effects become important makes it essential to understand the interplay between classical macroscopic properties and microscopic quantum properties. This is particularly true in nanomagnetism, where many potential applications require monodisperse, magnetic nanoparticles. One source of such species are single-molecule magnets (SMMs), individual molecules that function as single-domain magnetic particles. Below their blocking temperature, they exhibit magnetization hysteresis, the classical macroscale property of a magnet, as well as quantum tunneling of magnetization (QTM) and quantum phase interference, as properties of a micro-scale entity. Quantum tunneling is advantageous for some potential applications of single-molecule magnets, for example in providing the quantum superposition of states for quantum computing, but is a disadvantage in others such as information storage. This chapter introduces the basic concepts that are needed to understand the quantum phenomena observed in molecular nanomagnets.

1. INTRODUCTION

The perspective of this review is outlined in Fig. 1, which shows a ranging from macroscopic down to nanoscopic sizes. The unit of this scale is the number of magnetic moments in a magnetic system. At macroscopic sizes, a magnetic system is described by magnetic domains (Weiss 1907) that are separated by domain walls [1]. Magnetization reversal occurs via the nucleation, propagation, and annihilation of domain walls. One example is the hysteresis loop on the left in Fig. 1, which was measured on an individual elliptic CoZr particle of area $1\,\mu m \times 0.8\,\mu m$ and thickness 50 nm [2]). Shape and width of domain walls depend on the material of the magnetic system, on its size, shape and surface, and on its temperature [3, 4].

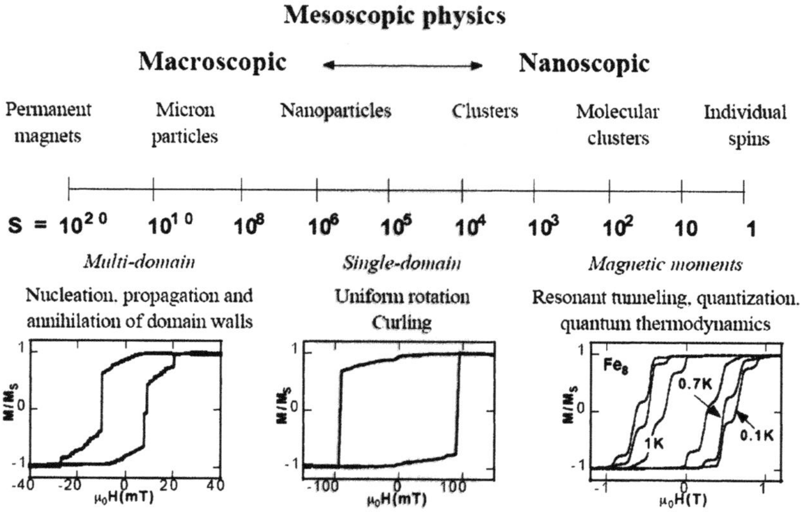

Figure 1. Size scale going from macroscopic down to nanoscopic sizes. The unit of this scale is the number of magnetic moments in a magnetic system, roughly corresponding to the number of atoms. The hysteresis loops are typical examples of magnetization reversal via nucleation, propagation, and annihilation of domain walls (left), via uniform rotation (middle), and via quantum tunneling (right).

When the system size is of the order of magnitude of the domain wall width or the exchange length, the formation of domain walls requires too much energy. Therefore, the magnetization remains in the so-called single-domain state, and the magnetization might reverse by uniform rotation, curling or other nonuniform modes (see the hysteresis loop in the middle of Fig. 1). For system sizes well below the domain-wall width or the exchange length, one must take into account explicitly the magnetic moments (spins) and their couplings. The theoretical description is complicated by the particles' boundaries.

Magnetic molecular clusters (also called molecular nanomagnets or single-molecule magnets SMM) are the final point in the series of smaller and smaller units from bulk matter to atoms (Fig. 2). Up to now, they have been the most promising candidates for observing quantum phenomena because they have a well-defined structure with well-characterized spin ground state and magnetic anisotropy. These molecules can be regularly assembled in large crystals where all molecules often have the same orientation. Hence, macroscopic measurements can give direct access to single molecule properties.

This chapter introduces the basic concepts that are needed to understand the quantum phenomena observed in molecular nanomagnets. All tunneling studies presented here were performed by magnetization measurements on single crystals using an array of micro-SQUIDs [4].

2. OVERVIEW OF MOLECULAR NANOMAGNETS

Molecular nanomagnets or single-molecule magnets (SMMs) are mainly organic molecules that have one or several metal centers with unpaired electrons. These polynuclear metal complexes are surrounded by bulky ligands (often organic carboxylate ligands). The most prominent examples are a dodecanuclear mixed-valence manganese-oxo cluster with acetate ligands, short Mn_{12} acetate [5], and an octanuclear iron (III) oxo-hydroxo cluster of formula $[Fe_8O_2(OH)_{12}(tacn)_6]^{8+}$ where tacn is a macrocyclic ligand, short Fe_8 [6]. Both systems have a spin ground state of $S = 10$ and an Ising-type magnetic anisotropy, which stabilizes the spin states with $m = \pm 10$ and generates an energy barrier for the reversal of the magnetization of about 67 K for Mn_{12} acetate [7-9] and 25 K for Fe_8 [10].

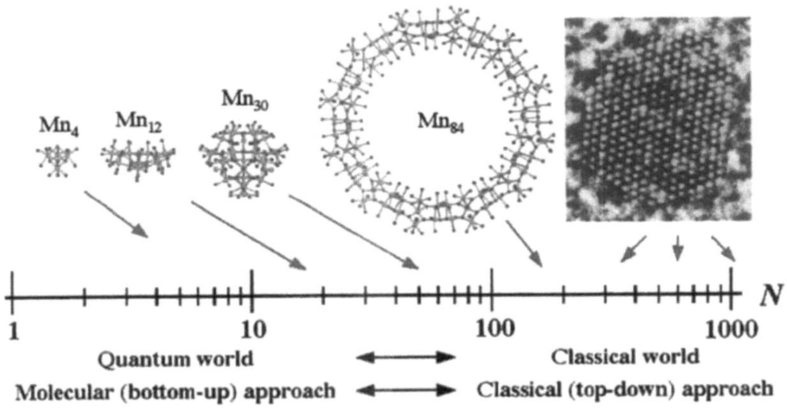

Figure 2. Size scale spanning atomic to nanoscale dimensions. On the far right is shown a high-resolution transmission electron microscopy view along a [110] direction of a typical 3 nm diameter cobalt nanoparticle exhibiting a face-centered cubic structure and containing about 1000 Co atoms [22]. The Mn_{84} molecule is a 4.2 nm diameter particle. Also shown for comparison are small Mn nanomagnets, which are drawn to scale. An alternative means of comparison is the Néel vector (N, the sum of the individual spins), which is the scale shown. The green arrows indicate the magnitude of the Néel vectors for the indicated SMMs, which are 7.5, 22, 61, and 168 for Mn_4, Mn_{12}, $Mn_{30}$and Mn_{84}, respectively.

Thermally activated quantum tunneling of the magnetization has first been evidenced in both systems [11-15]. Theoretical discussion of this assumes that thermal processes (principally phonons) promote the molecules up to high levels with small quantum numbers |m|, not far below the top of the energy barrier, and the molecules then tunnel inelastically to the other side. Thus the transition is almost entirely accomplished via thermal transitions and the characteristic relaxation time is strongly temperature-dependent. For Fe_8, however, the relaxation time becomes temperature-independent below 0.36 K [15, 16] showing that a pure tunneling mechanism between the only populated ground states $m = \pm S \pm 10$ is responsible for the relaxation of the magnetization. During the last years, many new molecular nanomagnets were presented (see, for instance, Refs. [17-20]) which show also tunneling at low temperatures. The largest molecular nanomagnets is currently a Mn_{84} molecule [21] that as a size of a magnetic nanoparticle (Fig. 2).

3. GIANT SPIN MODEL FOR NANOMAGNETS

A magnetic molecule, that behaves like a small nanomagnet, must have a well-defined magnetic anisotropy and a large ground state spin. A typical example is the above-mentioned octanuclear iron (III) oxo-hydroxo cluster (Fig. 3).

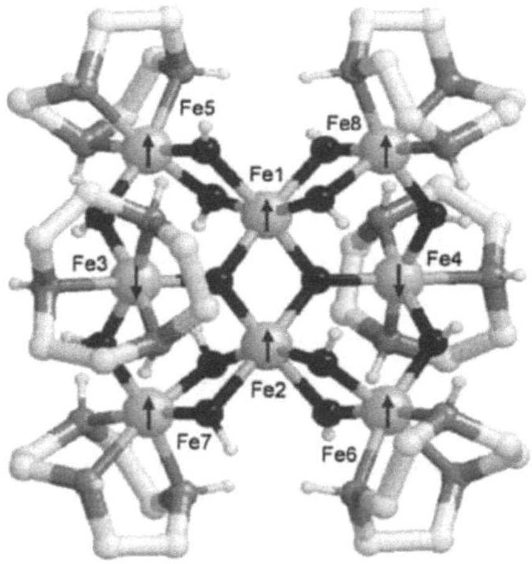

Figure 3. Schematic view of the magnetic core of the Fe_8 cluster. The oxygen atoms are black, the nitrogen atoms are gray, and carbon atoms are white. The arrows represent the spin structure of the ground state $S = 10$.

The internal iron (III) ions are octahedrally coordinated by the two oxides and four hydroxo bridges. The outer iron(III) ions coordinate three nitrogens and three hydroxyls. Spin polarized neutron scattering showed that all Fe ions have a spin 5/2, six spins up and two down [23]. This rationalizes the $S = 10$ spin ground state that is in agreement with magnetization measurements.

In principle, a multi-spin Hamiltonian can be derived taking into account of all exchange interactions and the single-ion magnetic anisotropies. However, the Hilbert space is very large ($6^8 \approx 10^6$) and the exchange coupling constants are not well known. A giant spin model is therefore often used that describes in an effective way the ground spin state multiplet. For a nanomagnet like the Fe_8 molecular cluster it has the following Hamiltonian

$$\mathsf{H} = -DS_z^2 + E\left(S_x^2 - S_y^2\right) + g\mu_B\mu_0\vec{S}\cdot\vec{H} \tag{1}$$

S_x, S_y, and S_z are the three components of the spin operator, D and E are the anisotropy constants which were determined via high-frequency electron paramagnetic resonance ($D/k_B \approx 0.275$ K and $E/k_B \approx 0.046$ K [10]), and the last term of the Hamiltonian describes the Zeeman energy associated with an applied field $\vec{H}$.

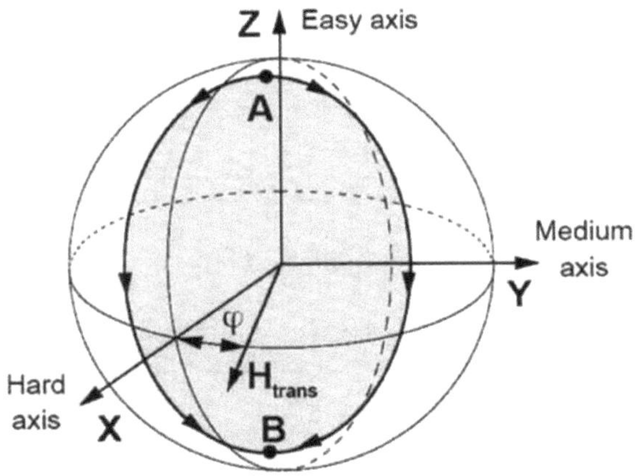

Figure 4. Unit sphere showing degenerate minima **A** and **B** joined by two tunnel paths (heavy lines). The hard, medium, and easy axes are taken in x-, y-, and z-direction, respectively. The constant transverse field H_{trans} for tunnel-splitting measurements is applied in the xy-plane, at an azimuthal angle φ. At zero applied field, $\vec{H} = 0$, the giant spin reversal results from the interference of two quantum spin paths of opposite direction in the easy anisotropy yz-plane. For transverse fields in direction of the hard axis, the two quantum spin paths are in a plane which is parallel to the yz-plane, as indicated in the figure. By using Stokes' theorem it has been shown [24] that the path integrals can be converted in an area integral, yielding that destructive interference—that is a quench of the tunneling rate—occurs whenever the shaded area is $k\pi/S$, where k is an odd integer. The interference effects disappear quickly when the transverse field has a component in the y-direction because the tunneling is then dominated by only one quantum spin path.

This Hamiltonian defines hard, medium, and easy axes of magnetization in x, y, and z directions, respectively (Fig. 4). It has an energy level spectrum with $(2S+1)=21$ values which, to a first approximation, can be labeled by the quantum numbers m = -10, -9,..., 10, choosing the z-axis as quantization axis. The energy spectrum, shown in Fig. 5, can be obtained by using standard digitalization techniques of the $[21\times 21]$ matrix describing the spin Hamiltonian S = 10. At $\vec{H}=0$, the levels $m=\pm 10$ have the lowest energy. When a field H_z is applied, the energy levels with m < - 1 increase, while those with m > 1 decrease (Fig. 5).

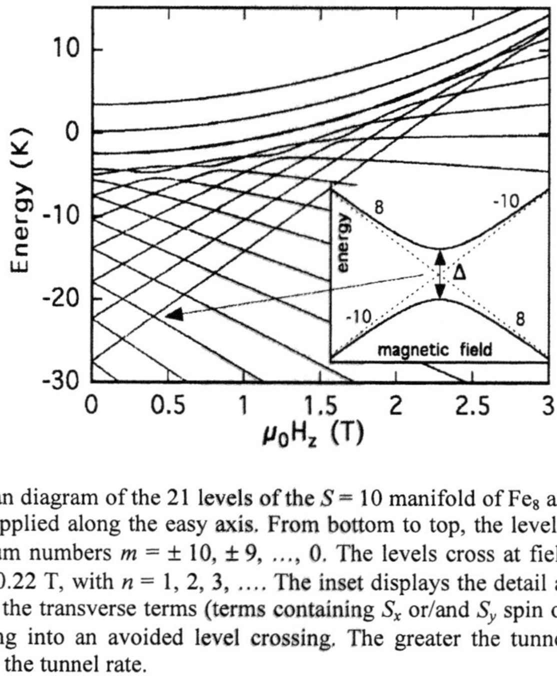

Figure 5. Zeeman diagram of the 21 levels of the S = 10 manifold of Fe_8 as a function of the field applied along the easy axis. From bottom to top, the levels are labeled with quantum numbers m = ± 10, ± 9, ..., 0. The levels cross at fields given by $\mu_0 H_z \approx n \times 0.22$ T, with n = 1, 2, 3, ... The inset displays the detail at a level crossing where the transverse terms (terms containing S_x or/and S_y spin operators) turn the crossing into an avoided level crossing. The greater the tunnel splitting Δ, the higher the tunnel rate.

Therefore, energy levels of positive and negative quantum numbers cross at certain fields H_z. It turns out that for Fe_8 the levels cross at fields given by $\mu_0 H_z \approx n\times$ 0.22 T, with $n=1,2,3,...$. The inset of Fig. 5 displays the details at a level crossing where transverse terms containing S_x or S_y spin operators turn the crossing into an "avoided level crossing". The spin S is "in resonance" between two states when the local longitudinal field is close to an avoided level crossing. The energy gap, the so-called "tunnel splitting" Δ, can be tuned by an applied field in the xy-plane (Fig. 4) via the $S_x H_x$ and $S_y H_y$ Zeeman terms (Section 3.2).

Figure 6 shows the effect of these avoided level crossings can be seen in hysteresis loop measurements. When the applied field is near an avoided level crossing, the magnetization relaxes faster, yielding steps separated by plateaus. As the temperature is lowered, there is a decrease in the transition rate due to reduced thermal-assisted tunneling.

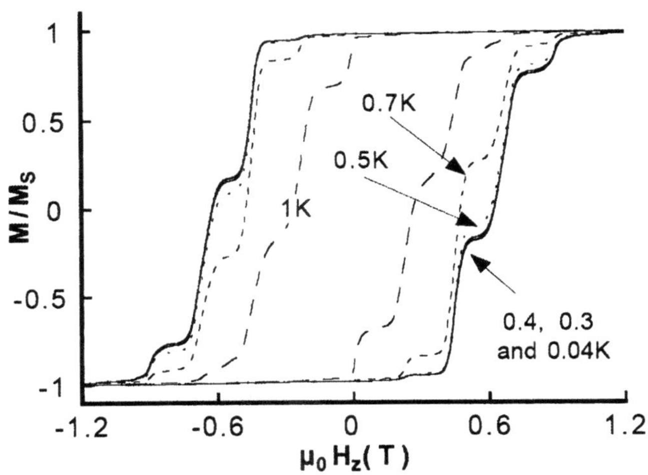

Figure 6. Hysteresis loops of a single crystal of Fe_8 molecular clusters at different temperatures. The longitudinal field (*z*-direction) was swept at a constant sweeping rate of 0.014 T/s. The loops display a series of steps, separated by plateaux. As the temperature is lowered, there is a decrease in the transition rate due to reduced thermal assisted tunneling. The hysteresis loops become temperature independent below 0.35 K, demonstrating quantum tunneling at the lowest energy levels.

3.1. Landau-Zener Tunneling in Fe_8

The nonadiabatic transition between the two states in a two-level system has first been discussed by Landau, Zener, and Stückelberg [25-27]. The original work by Zener concentrates on the electronic states of a bi-atomic molecule, while Landau and Stückelberg considered two atoms that undergo a scattering process. Their solution of the time-dependent Schrödinger equation of a two-level system could be applied to many physical systems and it became an important tool for studying tunneling transitions. The Landau-Zener model has also been applied to spin tunneling in nanoparticles and clusters [28-32].

The tunneling probability P when sweeping the longitudinal field H_z at a constant rate over an avoided energy level crossing (Fig. 7) is given by

$$P_{m,m'} = 1 - \exp\left(- \frac{\pi \Delta^2_{m,m'}}{2\hbar g \mu_B \mid m - m' \mid \mu_0 \mathrm{d}H_z / \mathrm{d}t} \right) \tag{2}$$

Here, m and m' are the quantum numbers of the avoided level crossing, $\Delta_{m,m'}$ denote the tunnel splitting between the levels (Fig. 7), $\mathrm{d}H_z/\mathrm{d}t$ is the constant field sweep rate, $g \approx 2$, μ_B is the Bohr magneton, and $\hbar$ is Planck's constant.

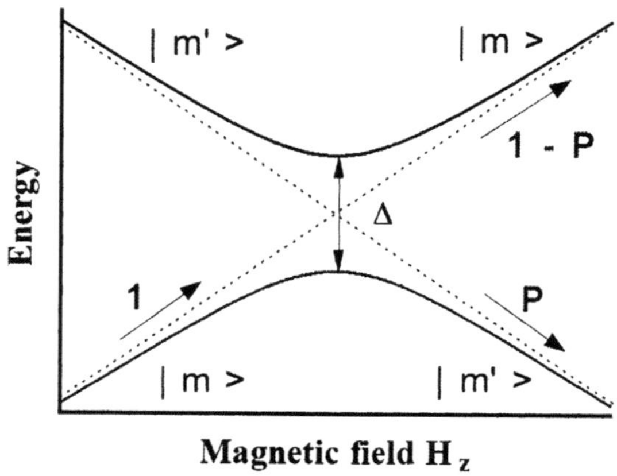

Figure 7. Detail of the energy level diagram near an avoided level crossing. m and m' are the quantum numbers of the energy level. $P_{m,m'}$ is the Landau-Zener tunnel probability when sweeping the applied field from the left to the right over the anticrossing. The greater the gap Δ and the slower the sweeping rate, the higher is the tunnel rate, Eq. (2).

With the Landau-Zener model in mind, we can understand qualitatively the hysteresis loops (Fig. 6). Let us start at a large negative magnetic field H_z. At very low temperature, all molecules are in the m = -10 ground state (Fig. 5). When the applied field H_z is ramped down to zero, all molecules will stay in the m = -10 ground state. When ramping the field over the $\Delta_{-10,10}$-region at $H_z \approx 0$, there is a Landau-Zener tunnel probability $P_{-10,10}$ to tunnel from the m = -10 to the m = 10 state. $P_{-10,10}$ depends on the sweeping rate, Eq. (2); that is, the slower the sweeping rate, the larger the value of $P_{-10,10}$. This is clearly demonstrated in the hysteresis loop measurements showing larger steps for slower sweeping rates [33]. When the field H_z is further increased, there remains a fraction of molecules in the $m = -10$ state which became a metastable state. The next chance to escape from this state is when the field reaches the $\Delta_{-10,9}$ region. There is a Landau-Zener tunnel probability $P_{-10,9}$ to tunnel from the m = -10 to the $m = 9$ state. As $m = 9$ is an excited state, the molecules in this state desexcite quickly to the m = 10 state by emitting a phonon. An analogous procedure happens when the applied field reaches the $\Delta_{-10,10-n}$ regions (n = 2, 3,) until all molecules are in the m = 10 ground state; that is, the magnetization of all molecules is reversed. As phonon emission can only change the molecule state by Δm = 1 or 2, there is a phonon cascade for higher applied fields.

In order to apply quantitatively the Landau-Zener formula Eq. (2), we first saturated the crystal of Fe_8 clusters in a field of H_z = -1.4 T, yielding an initial magnetization M_{in} = $-M_s$. Then, we swept the applied field at a constant rate over one of the resonance transitions and measured the fraction of molecules which reversed their spin. This procedure yields the tunneling rate $P_{-10,10-n}$ and thus the tunnel splitting $\Delta_{-10,10-n}$, Eq. (2), with n = 0, 1, 2,.

We first checked the predicted Landau-Zener sweeping field dependence of the tunneling rate. We found a good agreement for sweeping rates between 10 and 0.001 T/s [33]. The deviations at lower sweeping rates are mainly due to the *hole-digging* mechanism [34] which slows down the relaxation (see below). Our measurements showed for the first time that the Landau-Zener method is particularly adapted for molecular clusters because it works even in the presence of dipolar fields which spread the resonance transition provided that the field sweeping rate is not too small.

3.2. Oscillations of Tunnel Splitting

An applied field in the *xy*-plane can tune the tunnel splittings $\Delta_{m,m'}$ via the S_x and S_y spin operators of the Zeeman terms that do not commute with the spin Hamiltonian. This effect can be demonstrated by using the Landau--Zener method (Section 3.1). Fig. 8 presents a detailed study of the tunnel splitting $\Delta_{\pm 10}$ at the tunnel transition between $m = \pm 10$, as a function of transverse fields applied at different angles φ, defined as the azimuth angle between the anisotropy hard axis and the transverse field (Fig. 4).

For small angles φ the tunneling rate oscillates with a period of ~ 0.4 T, whereas no oscillations showed up for large angles φ [33]. In the latter case, a much stronger increase of $\Delta_{\pm 10}$ with transverse field is observed. The transverse field dependence of the tunneling rate for different resonance conditions between the state m = -10 and $(10-n)$ can be observed by sweeping the longitudinal field around $\mu_0 H_z = n \times 0.22$ T with n = 0, 1, 2,. The corresponding tunnel splittings $\Delta_{-10,10-n}$ oscillate with almost the same period of ~ 0.4 T (Fig. 8). In addition, comparing quantum transitions between $m = -10$ and $(10-n)$, with n even or odd, revealed a parity (or symmetry) effect that is analogous to the Kramers' suppression of tunneling predicted for half-integer spins [35, 36]. A similar strong dependence on the azimuth angle φ was observed for all studied resonances.

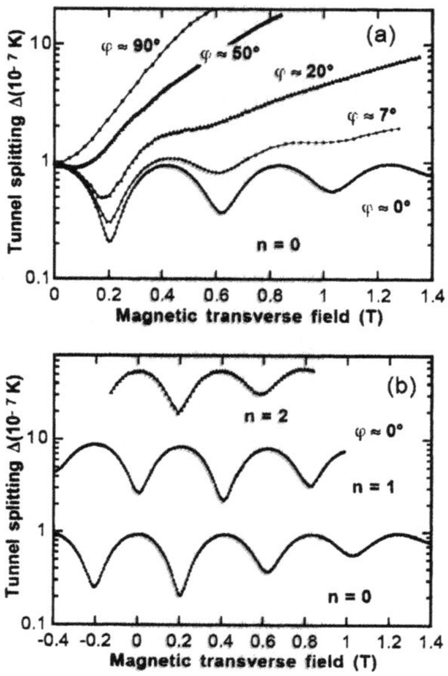

Figure 8. Measured tunnel splitting Δ as a function of transverse field for (a) several azimuth angles φ at $m = \pm 10$ and (b) $\varphi \approx 0°$, as well as for quantum transition between $m = -10$ and (10 - n). Note the parity effect that is analogous to the suppression of tunneling predicted for half-integer spins. It should also be mentioned that internal dipolar and hyperfine fields hinder a quench of Δ which is predicted for an isolated spin.

3.2.1. Semiclassical descriptions

Before showing that the above results can be derived by an exact numerical calculation using the quantum operator formalism, it is useful to discuss semiclassical models. The original prediction of oscillation of the tunnel splitting was done by using the path integral formalism [37]. Here [24], the oscillations are explained by constructive or destructive interference of quantum spin phases (Berry phases) of two tunnel paths (instanton trajectories, Fig. 4). Since our experiments were reported, the Wentzel-Kramers-Brillouin theory has been used independently by Garg [38] and Villain and Fort [39]. The surprise is that although these models [24, 38, 39] are derived semiclassically, and should have higher-order corrections in $1/S$, they appear to be exact as written! This has first been noted in Refs. [38] and [39] and then proven in Ref. [40] Some extensions or alternative explications of Garg's result can be found in Refs. [41-44].

The period of oscillation is given by [24]

$$\Delta H = \frac{2k_B}{g\mu_B}\sqrt{2E(E+D)} \tag{3}$$

where D and E are defined in Eq. (1). We find a period of oscillation of $\Delta H = 0.26$ T for $D = 0.275$ K and $E = 0.046$ K as in Ref. [10]. This is somewhat smaller than the experimental value of ~ 0.4 T. We believe that this is due to higher-order terms of the spin Hamiltonian which are neglected in Garg's calculation. These terms can easily be included in the operator formalism as shown in the next subsection.

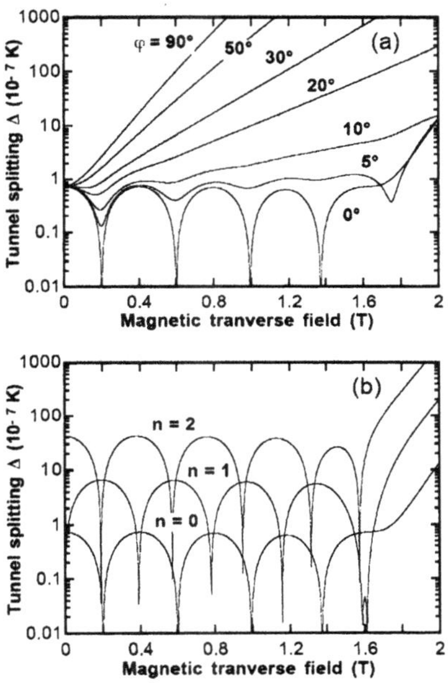

Figure 9. Calculated tunnel splitting Δ as a function of transverse field for (a) quantum transition between $m = \pm 10$ at several azimuth angles φ and (b) quantum transition between m = -10 and (10 - n) at $\varphi \approx 0°$. The fourth-order terms suppress the oscillations of Δ for large transverse fields $|H_x|$.

3.2.2. *Exact numerical diagonalization*

In order to quantitatively reproduce the observed periodicity we included fourth-order terms in the spin Hamiltonian, Eq. (1), as employed in the simulation of inelastic neutron scattering measurements [45, 46] and performed a diagonalization of the $[21 \times 21]$ matrix describing the $S = 10$ system. For the calculation of the tunnel splitting we used $D = 0.289$ K, $E = 0.055$ K [Eq. 1] and the fourth-order terms as defined in [45] with

$B_4^0 = 0.72 \times 10^{-6}$ K, $B_4^2 = 1.01 \times 10^{-5}$ K, $B_4^4 = -0.43 \times 10^{-4}$ K, which are close to the values obtained by EPR measurements [47] and neutron scattering measurements [46].

The calculated tunnel splittings for the states involved in the tunneling process at the resonances $n = 0$, 1, and 2 are reported in Fig. 9, showing the oscillations as well as the parity effect for odd resonances.

4. QUANTUM DYNAMICS OF A DIMER OF NANOMAGNETS

Here we present a new family of dimers of nanomagnets [48, 49] in which antiferromagnetic coupling between two single-molecule magnets (SMMs) results in quantum behaviour different from that of the individual SMMs. Each SMM acts as a bias on its neighbor, shifting the quantum tunneling resonances of the individual SMMs. Hysteresis loop measurements on a single crystal of SMM-dimers established quantum tunneling of the magnetization via entangled states of the dimer. This shows that the dimer really does behave as a quantum-mechanically coupled dimer, and also allows the measurement of the longitudinal and transverse superexchange coupling constants.

The compound $[Mn_4]_2 \cdot 2C_6H_{14}$ crystallizes in the hexagonal space group $R3$(bar) with two Mn_4 molecules per unit cell lying head-to-head on a crystallographic S_6 symmetry axis [48] (Fig. 10). Each Mn_4 monomer has a ground state spin of $S = 9/2$, well separated from the first excited state $S = 7/2$ by a gap of about 300K [50]. The Mn-Mn distances and the Mn-O-Mn angles are similar and the uniaxial anisotropy constant is expected to be the same for the two dimer systems. These dimers are held together via six C-H···Cl hydrogen bonds between the pyridine (py) rings on one molecule and the Cl ions on the other, and one Cl···Cl Van der Waals interaction. These interactions lead to an antiferromagnetic superexchange interaction between the two Mn_4 units of the $[Mn_4]_2$ dimer [48]. Dipolar couplings between Mn_4 molecules can be easily calculated and are more than one order of magnitude smaller than the exchange interaction.

Before presenting the measurements, we summarize a simplified spin Hamiltonian describing the $[Mn_4]_2$ dimer [48]. Each Mn_4 SMM can be modeled as a *giant spin* of $S = 9/2$ with Ising-like anisotropy [Eq. (1)]. The corresponding Hamiltonian is given by

$$\mathrm{H}_i = -D S_{z,i}^2 + \mathrm{H}_{\mathrm{trans},i} + g\mu_B\mu_0\vec{S}_i \cdot \vec{H} \quad (4)$$

where $i = 1$ or 2 (referring to the two Mn_4 SMMs of the dimer), D is the uniaxial anisotropy constant, and the other symbols have their usual meaning.

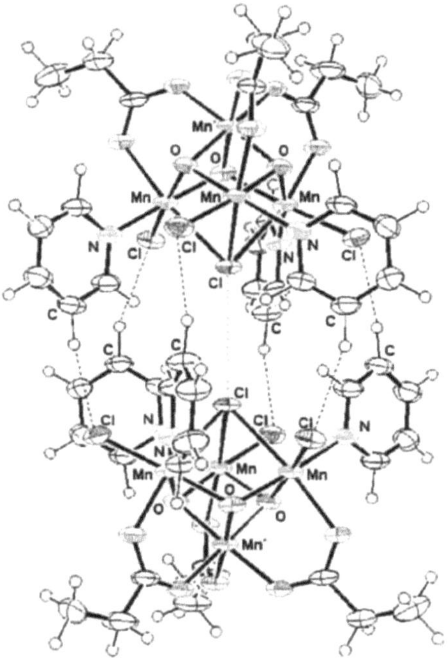

Figure 10. The structure of the $[Mn_4]_2$ dimer of $[Mn_4O_3Cl_4(O_2CEt)_3(py)_3]$. The small circles are hydrogen atoms. The dashed lines are C-H···Cl hydrogen bonds and the dotted line is the close Cl···Cl approach. The labels Mn and Mn' refer to Mn^{III} and Mn^{IV} ions, respectively.

Tunneling is allowed in these half-integer ($S = 9/2$) spin systems because of a small transverse anisotropy $\mathsf{H}_{trans,i}$ containing $S_{x,i}$ and $S_{y,i}$ spin operators and transverse fields (H_x and H_y). The exact form of $\mathsf{H}_{trans,i}$ is not important in this discussion. The last term in Eq. 4 is the Zeeman energy associated with an applied field. The Mn_4 units within the $[Mn_4]_2$ dimer are coupled by a weak superexchange interaction via both the six C-H···Cl pathways and the Cl···Cl approach. Thus, the Hamiltonian (H) for $[Mn_4]_2$ is

$$\mathsf{H} = \mathsf{H}_1 + \mathsf{H}_2 + J_z S_{z,1} S_{z,2} + J_{xy}(S_{x,1}S_{x,2} + S_{y,1}S_{y,2}) \qquad (5)$$

where J_z and J_{xy} are respectively the longitudinal and transverse superexchange interactions. $J_z = J_{xy}$ is the case of isotropic superexchange. The $(2S + 1)^2 = 100$ energy states of the dimer can be calculated by exact numerical diagonalization and are plotted in Fig. 11 as a function of the field applied along the easy axis.

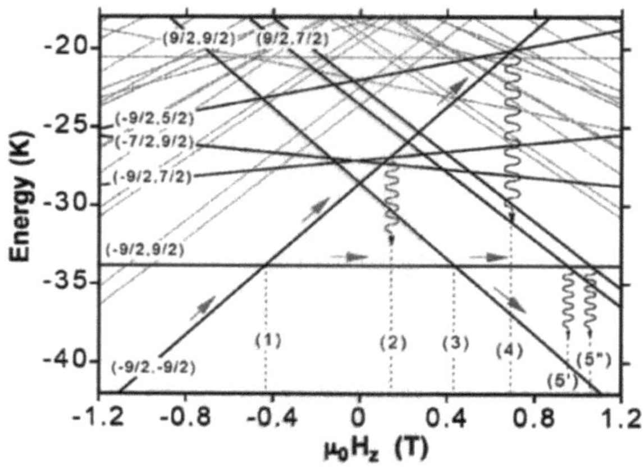

Figure 11. Low lying spin-state energies of the $[Mn_4]_2$ dimer, calculated by exact numerical diagonalization using Eq. 2 with D = 0.77 K and J = 0.13 K, as a function of applied magnetic field H_z (Zeeman diagram). The bold energy levels are labeled with two quantum numbers (M_1, M_2). Dotted lines, labeled **1** to **5**, indicate the strongest tunnel resonances: **1**: (-9/2,- 9/2) to (-9/2,9/2); **2**: (-9/2,-9/2) to (-9/2,7/2), followed by relaxation to (-9/2,9/2); **3**: (-9/2,9/2) to (9/2,9/2); **4**: (-9/2,-9/2) to (-9/2,5/2), followed by relaxation to (-9/2,9/2); **5**: (-9/2,9/2) to (7/2,9/2), followed by relaxation to (9/2,9/2). For clarity, degenerate states such as (M, M') and (M', M) and lifted degenerate states such as (M, $M \pm 1$) and (M, $M \pm 2$) are not both listed. For example, the (9/2, 7/2) and (7/2, 9/2) states are strongly split into a symmetric (labeled 5") and antisymmetric (labeled 5') combination of (9/2, 7/2) and (7/2, 9/2) states. This splitting is used to measure the transverse super-exchange interaction constant J_{xy}. Co-tunneling and other two-body tunnel transitions have a lower probability of occurrence and are neglected [51].

Each state of $[Mn_4]_2$ can be labeled by two quantum numbers (M_1, M_2) for the two Mn_4 SMMs, with M_1 = -9/2, -7/2,..., 9/2 and M_2 = -9/2, -7/2,..., 9/2. The degeneracy of some of the (M_1, M_2) states is lifted by transverse anisotropy terms. For the sake of simplicity, we will discuss mainly the effect of the transverse superexchange interaction $J_{trans} = J_{xy} (S_{x1}S_{x2} + S_{y1}S_{y2}) = J_{xy} (S_{+1}S_{-2} + S_{-1}S_{+2})/2$, where $S_{+,i}$ and $S_{-,i}$ are the usual spin raising and lowering operators. Because J_{trans} acts on $(M, M \pm 1)$ states to first order of perturbation theory, the degeneracy of those states is strongly lifted. For example, the (9/2, 7/2) and (7/2, 9/2) states are strongly split into a symmetric (labeled 5") and antisymmetric (labeled 5') combination of (9/2, 7/2) and (7/2, 9/2) states. Similarly for the (-9/2, -7/2) and (-7/2, -9/2) states. Measuring this energy splitting allows us to determine the transverse superexchange interaction constant J_{xy} because the latter is proportional to the former.

Figure 12 shows typical hysteresis loops (magnetization versus magnetic field scans) with the field applied along the easy axis of magnetization of $[Mn_4]_2$, that is, parallel to the S_6 axis. These loops display step-like features separated by plateaus. The step heights are temperature-independent below about ~ 0.35 K [48]. The steps are due to resonant quantum tunneling of the magnetization (QTM) between the energy states of the $[Mn_4]_2$ dimer (see figure caption 11 and 12 for a discussion of 5 tunnel transitions). QTM has been previously observed for most SMMs, but the novelty for $[Mn_4]_2$ dimers is that the QTM is now the collective behavior of the complete $S = 0$ dimer of exchange-coupled $S = 9/2$ Mn_4 quantum systems. This coupling is manifested as an exchange bias of all tunneling transitions, and the resulting hysteresis loop consequently displays unique features, such as the absence for the first time in a SMM of a QTM step at zero field [48].

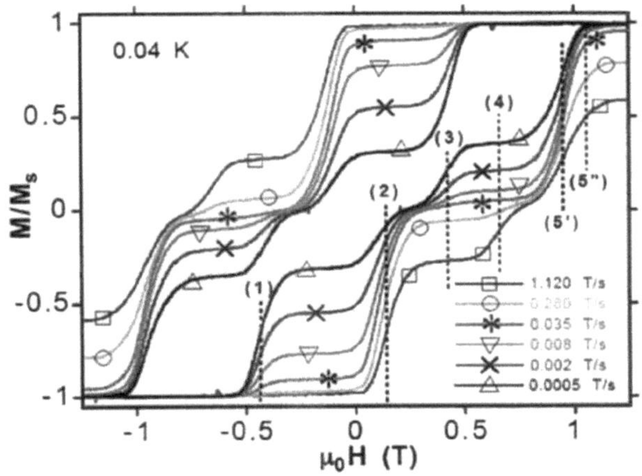

Figure 12. Hysteresis loops for the $[Mn_4]_2$ dimer at several field sweep rates and 40 mK. The tunnel transitions (manifested by steps) are labeled from **1** to **5**, see Fig. 1.

Even though the five strongest tunneling transitions are observed in Fig. 12, fine structure was not observed. For example, the hysteresis loops do not show the splitting of the $(9/2, 7/2)$ states (labeled 5' and 5"), which we suspected might be due to line broadening. Usually, line broadening in SMMs is caused by dipolar and hyperfine interactions [52], and distributions of anisotropy and exchange parameters. In most SMMs, the zero-field resonance is mainly broadened by dipolar and hyperfine interactions because distributions of anisotropy parameters do not affect the zero-field resonance.

For an antiferromagnetically coupled dimer, however, this resonance is shifted to negative fields. Therefore, a distribution of the exchange coupling parameter J_z can further broaden this resonance. In fact, we show in the following that the latter is the dominant source of broadening. We then use the 'quantum hole-digging' method [34, 52-55] to provide direct experimental evidence for the transitions 5' and 5", which establishes tunneling involving entangled dimer states and allows us to determine J_{xy}.

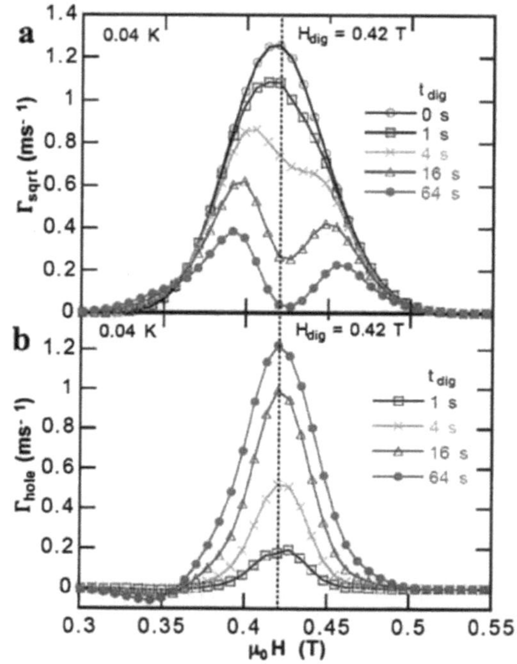

Figure 13. Short-time relaxation in dimmers: (a) field dependence of the short-time square-root relaxation rates Γ_{sqrt} are presented on a logarithmic scale showing the depletion of the molecular spin states by quantum tunneling at $H_{dig} = 0.42$ T for various waiting times t_{dig} and (b) difference between the relaxation rate in the absence and in the presence of digging, $\Gamma_{hole} = \Gamma_{sqrt}(0) - \Gamma_{sqrt}(t_{dig})$.

The quantum hole-digging method is a relatively new method that can, among other things [56], study line broadening and its evolution during relaxation [34, 52-55]. It amounts to a reshuffling of dipolar states, as contrasted to the hole burning in optics, which eliminates a certain subgroup of states. The idea is that the short-time magnetization relaxation after a rapid field change is directly related to the number of molecules in resonance at the applied field. Prokof'ev and Stamp proposed [52] that this short time relaxation should follow a square-root time dependence. Thus, the magnetization of the $[Mn_4]_2$ dimers in the crystal was first saturated with a large positive field, and then a 'digging field' H_{dig} was applied at 0.04 K for a chosen 'digging time' t_{dig}. Then, the fraction (and only that fraction) of the molecules that is in resonance at H_{dig} can undergo magnetization tunneling. After t_{dig}, a field H_{probe} is applied and the magnetization relaxation rate is measured for short time periods; from this is calculated the short-time relaxation rate Γ_{sqrt}, which is related to the number of $[Mn_4]_2$ dimers still available for QTM [4]. The entire procedure is then repeated at other H_{probe} fields.

Figure 13 shows the resulting plot of Γ_{sqrt} versus H_{probe}, Fig. 13(a) reflects the distribution of spins still available for tunneling after t_{dig}. In the limit of very short digging times, the difference between the relaxation rates in the absence and in presence of digging, $\Gamma_{hole} = \Gamma_{sqrt}(0) - \Gamma_{sqrt}(t_{dig})$ is approximately proportional to the number of molecules which reversed their magnetization during the time t_{dig} (Fig. 13(b)). Γ_{hole} is characterized by a width that can be called the 'tunnel window'.

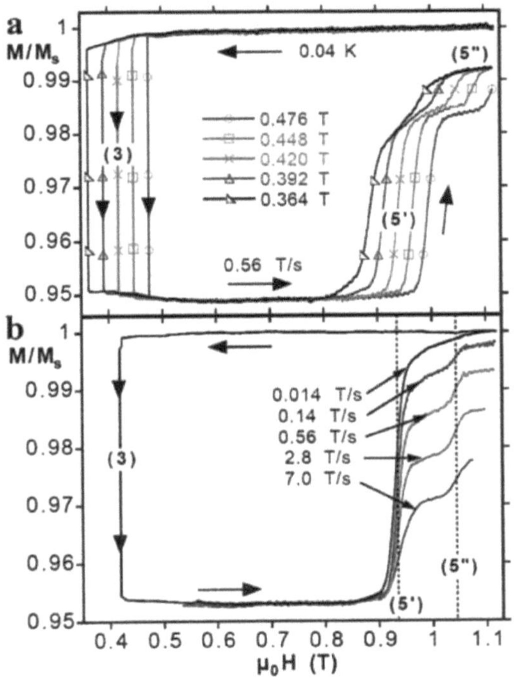

Figure 14. Minor hysteresis loops: (a) various digging fields and (b) various field sweep rates. After positive saturation, a digging field H_{dig} was applied to reverse ≈ 2.5% of the molecules that are in resonance at H_{dig} (transition **3** in Fig. 11). Then, the applied field is swept back to a large positive field. 5' and 5" are the first tunnel transitions allowing the reversed molecules to tunnel back to positive saturation.

The width of the distribution in the absence of digging, about 80 mT (a), is too large to be exclusively due dipolar (~ 20 mT) and hyperfine-coupling (~ 10 mT). The following result suggests that it is due to a distribution of the exchange-coupling parameter J_z.

First, the magnetization of the $[Mn_4]_2$ dimers was saturated with a large positive field, and then a digging field H_{dig} was applied to reverse a fraction of the molecules that are in resonance at H_{dig} (transition **3** in Fig. 11). After the reversal of 2.5% of the molecules, the applied field is swept back to a large positive field. 5' and 5" are the first tunnel transitions that can allow the reversed molecules to tunnel back to positive saturation. Figures 14(a) and (b) show the respectively minor hysteresis loops for several 'digging fields' and field sweep rates. Both figures clearly show the expected tunnel transitions 5' and 5", that were not resolved in the major hysteresis loops (Fig. 12). This suggests that the broadening of tunneling transition **3** –the distribution in the absence of digging in Fig. 13(a)– is dominated by a distribution of the exchange-coupling parameter J_z. During the application of the digging field, a subgroup of molecules is selected with an exchange coupling constant $J_z \approx g\mu_B\mu_0 H_{dig} / S$ that can tunnel back at the fields of transitions 5' and 5".

This interpretation is supported by the field values of 5' and 5" as a function of digging field, that is J_z, exhibiting a nearly linear variation, Fig. 15(a). The field difference between transition 5' and 5" can be used to find the J_{xy}, as shown in Fig. 15(b). This shows that the superexchange interaction of the dimers is nearly isotropic ($J_{xy} \approx J_z$). This is due to the relativistic character of the exchange anisotropy, as outlined in Ch. 3. It is important to mention that the transitions 5' and 5" are well separated, suggesting long coherence times compared to the time scale of the energy splitting.

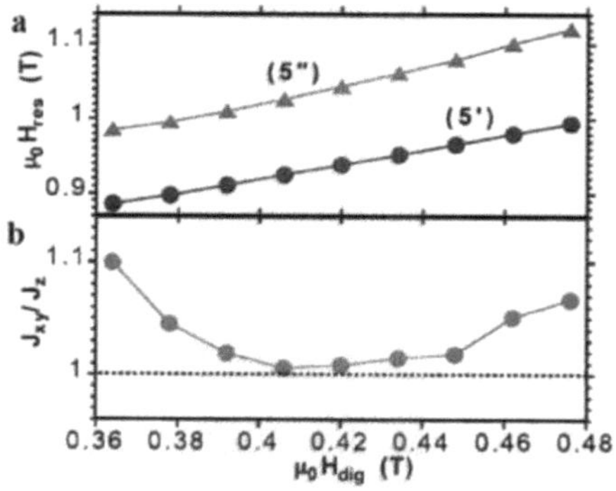

Figure 15. (a) Resonance field positions H_{res} of 5' and 5" and (b) normalized transverse superexchange interaction J_{xy}/J_z as a function of digging field H_{dig}.

The above results demonstrate for the first time tunneling via entangled states of a dimer of exchange coupled SMMs, showing that the dimer really does behave as a quantum mechanically coupled system. This result is of great importance if such systems are to be used for quantum computing.

5. RESONANT PHOTON ABSORPTION IN THE LOW-SPIN MOLECULE V_{15}

It is acknowledged that single-molecule magnets (SMMs) have a great potential for quantum computation [57], in particular because they are extremely small and almost identical, allowing to obtain, in a single measurement, statistical averages of a larger number of qubits. Each spin being carried by a big molecule of ~ 1 nm^3, these magnets are naturally diluted (with 10^{19}-10^{20} molecules/cm^3), which is highly desirable to minimize dipolar interactions and thus preserve quantum coherence. Despite such natural dilutions, dipolar interactions remain rather important (0.1 – 0.5 K) in large spin SMMs such as Mn_{12} or Fe_8 (S = 10). Reductions of dipolar interactions can easily be achieved by decreasing the value of the spin S. At a given concentration, the dipolar energy scales with the square of S. Thus, decreasing the spin from $S = 10$ to $1/2$ leads to a reduction of dipolar energy by a factor of $(1/20)^2 \sim 10^{-3}$ [58]. At first glance, low-spin systems seem more suitable for quantum computation than large-spin systems.

In this section we report the first study of the micro-SQUID response of a low-spin molecular system, V_{15}, to electromagnetic radiation. The advantages of our micro-SQUID technique in respect to pulsed electron paramagnetic resonance (EPR) techniques consist in the possibility to perform time-resolved experiments (below 1 ns) [59] on submicrometer sizes samples (about 1000 spins) [22] at low temperature (below 100 mK). Our first results on V_{15} open the way for time-resolved observations of quantum superposition of spin-up and spin-down states in SMMs. Other results obtained in similar systems but with large spins concern for example EPR measurements [10], resonant photon-assisted tunneling in a Fe_8 SMM [60].

The molecular complex V_{15} forms a lattice with space group R_{3c} containing two molecules per unit cell [61, 62]. The third-order symmetry axis of the unit cell is also the symmetry axis of the two V_{15} clusters. In each molecule the fifteen V^{IV} ions (S = 1/2), are placed on a quasi-spherical layered structure formed of a triangle sandwiched between two non-planar hexagons. Each hexagon contains three pairs of strongly coupled spins and each spin at a corner of the inner triangle is coupled to two of those pairs – one belonging to the upper hexagon and one belonging to the lower hexagon [62]. In principle, this molecule should be discussed in terms of the entanglement of the 15 spins 1/2, with a Hilbert space dimension of 2^{15}.

However, considerable simplifications occur at low temperature where the molecule can be described by a $S = 1/2$ spin ground state and a $S = 3/2$ excited state. The energy separation of both spin states of $\approx 3.8K$ was accurately determined by susceptibility and high-field magnetization measurements [63] and by inelastic neutron scattering experiments [64]. Magnetization measurements performed down to 30 mK showed that the $S = 1/2$ spin ground state is split in zero field by $\approx$ 80 mK [58, 65]. The origin of this splitting for a half-integer spin is interpreted by the interplay between intra-molecular, hyperfine and Dzyaloshinskii-Moriya interactions [58, 65]. Inter-molecular interactions – dipolar or/and residual exchange interactions – can be evaluated from the low temperature Curie-Weiss law associated with the spin ground state, giving the Curie constant $C = 0.686\mu_B$ K/T and the paramagnetic temperature $\theta_p \approx 12$ mK, corresponding to a mean internal field of $\approx$ 12 mT. This value, one order of magnitude larger than calculated dipolar interactions is nevertheless much smaller than usual super-exchange interactions; it is attributed to non-trivial exchange paths.

The measurements were made at a cryostat temperature of 40 mK, using a 50 μm sized single crystal of V_{15}. The magnetic probe was a micro-SQUID array [4] equipped with three coils allowing to apply a field in any direction and with sweep rates up to 10 T/s. The electromagnetic radiation was generated by a frequency synthesizer (Anritsu MG3694A) triggered with a nanosecond pulse generator. This setup allows a continuous variation of the frequency from 0.1 Hz to 20 GHz, with pulse lengths form ~ 1 ns to 1 s [66]. The AC radiation field B_{ac} was directed in a plane perpendicular to the applied static field B_0.

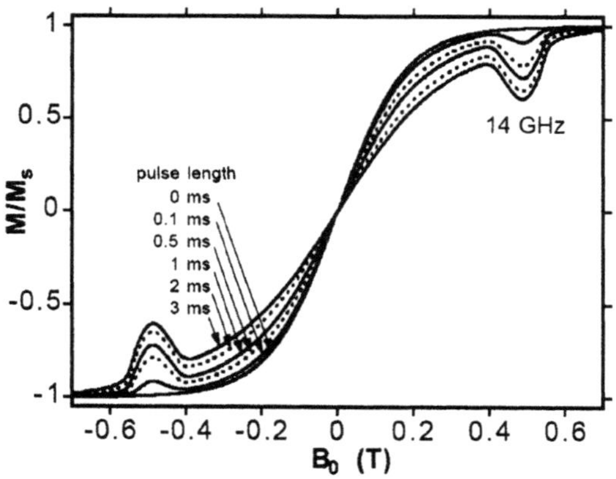

Figure 16. Equilibrium magnetization curves measured with or without irradiation. The cryostat temperature was 40 mK. The electromagnetic radiation was pulsed with a period of 10 ms and a pulse length going from 0 (no radiation) to 3 ms.

The field dependence of the magnetization was measured in the quasi-static regime with a field sweep rate slow enough to keep the system at equilibrium (1 mT/s). The phonon-bottleneck regime has a characteristic spin-phonon relaxation time to the cryostat τ_s of a few seconds [58]. AC radiation pulses of 0.1 to 0.3 ms were applied every 10 ms. Due to the large value of τ_s, the relaxation in the intervals between pulses is negligible. As a consequence, the effects of each pulse are additive, leading to an equilibrium magnetization after a time being larger than τ_s.

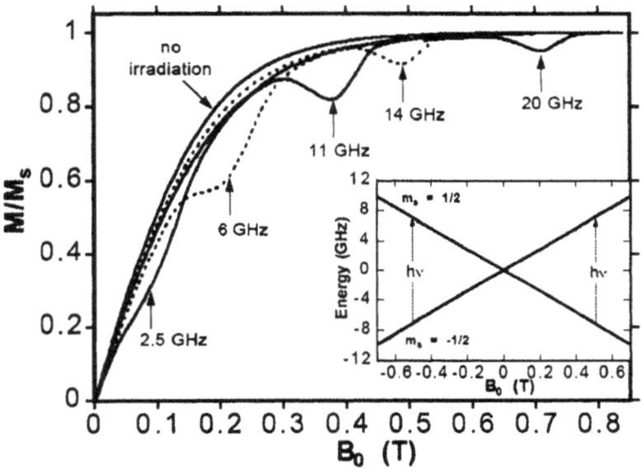

Figure 17. Equilibrium magnetization curves similar to Fig. 16 but at several frequencies (pulse length 1 ms). Inset: calculated energy levels for a spin 1/2 with and g = 2.02.

Figure 16 shows $M(B_0)$ curves for several pulse lengths, measured at ν = 14 GHz. Two symmetrical dips are clearly visible at $B_\nu = \pm$ 0.491 T. They result from resonant absorptions of photons associated with $m_s = 1/2$ to $-1/2$ spin transitions, as indicated in the inset of Fig. 17. Typical measurements at other frequencies are also shown in Fig. 17.

Apart from the resonant absorptions, there is also a small absorption at all applied fields. To elucidate the origin of this absorption, Fig. 18 presents the same data as in Fig. 16, but the magnetization M is now converted into a paramagnetic spin temperature T_s using the equation [67]

$$M(T_s)/M_s = \tanh(g\mu_B SB_0/k_B T_s) \tag{6}$$

with $S = 1/2$ and $g = 2.02$ (see below).

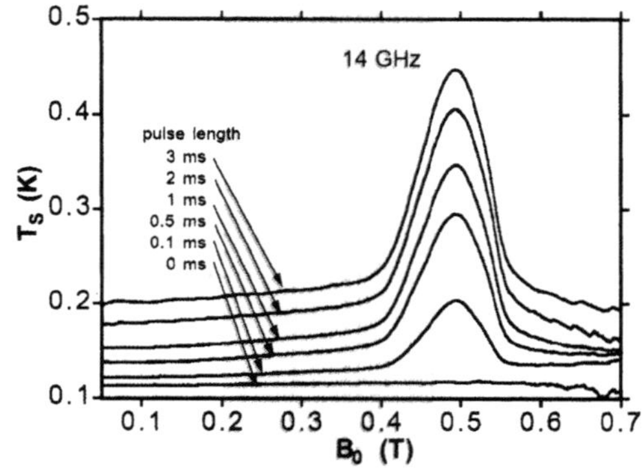

Figure 18. Spin temperature T_s obtained by inversion of (6) where $M(T_s)$ is given by the magnetization curves of Fig. 16. The resonant absorption near 0.491 T is superimposed to a non-resonant background absorption corresponding to a nearly constant spin-temperature.

Figure 18 contains the increase of the spin temperature T_s due to resonant absorptions and a nearly-field independent increase of T_s. Before going into a deeper discussion, we show in Fig. 19 the experimental confirmation of the linear evolution of the resonant field with frequency $\nu = \gamma B_\nu$ as predicted from the inset of Fig. 17. The measured slope $d\nu / dB_\nu = \gamma \approx$ 28.3 GHz/T yields a gyromagnetic ratio $g = 2h\gamma / \mu_B \approx$ 2.02, a value close to the one obtained by Ajiro *et al.* [68] in single-crystal EPR measurements at 2.4 K ($g \approx 1.98$).

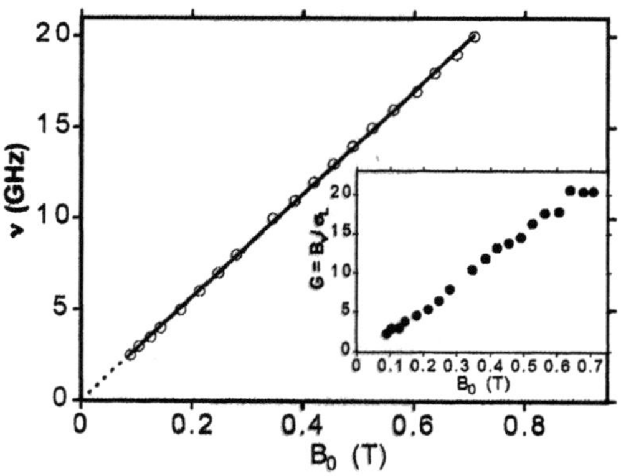

Figure 19. Measured linear variation of the resonance frequency, corresponding to absorption maxima, on the applied magnetic field. The slope gives $g = 2.02$. Inset: similar variation obtained for the quality factor $G = B_\nu/\sigma_L$ representing a lower bound of the coherence time of molecular spin precession.

In order to analyze the resonance lines we will consider the well known Rabi model for a spin $1/2$, which stipulates that the probability P of finding the system in the state $m_s = -1/2$ at time t, if it was in the state $m_s = 1/2$ at $t = 0$, is given by [67, 69, 70]:

$$P = \frac{(\gamma B_{\mathrm{ac}})^2}{(\gamma B - \nu)^2 + (\gamma B_{\mathrm{ac}})^2} \sin^2(\omega_{\mathrm{Rabi}} t) \tag{7}$$

where $\omega_{\mathrm{Rabi}} = 1/2\sqrt{(\gamma B - \nu)^2 + (\gamma B_{\mathrm{ac}})^2}$, B is total field seen by the spin, and B_{ac} the amplitude of the electromagnetic field. In the limit of long times, Eq. 7 becomes proportional to the time, $P = \pi(\gamma B_{\mathrm{ac}}/2)^2 t\delta(\gamma B - \nu)$, and the transition rate $\Gamma = dP/dt = \pi(\gamma B_{\mathrm{ac}}/2)^2 \delta(\gamma B - \nu)$ turns out to be a constant. Integrating Γ over a distribution of local fields $\mathsf{F}(B_{\mathrm{L}})$ and considering that $B = B_0 + B_{\mathrm{L}}$, where B_0 is the applied field, we obtain the field-dependent spin-photon transition rate:

$$\Gamma_{\mathrm{L}} = \frac{\pi}{4} \gamma B_{\mathrm{ac}}^2 \mathsf{F}(B_{\mathrm{L}} = B_\nu - B_0) \tag{8}$$

This expression shows that the field-dependence of the transition rate is given by the distribution function of local fields. Below, we analyze the results in terms of two limiting distributions, a Gaussian distribution and a Lorentzian one.

The measured magnetization dips corresponding to resonant power absorption (Figs. 16-18), result from the balance between the induced transition rate $\Gamma = \Gamma_{\mathrm{L}} + \Gamma_0$ and τ_{s}. Γ_0 is taking into account a small heating effect of the environment. $\Gamma\tau_{\mathrm{s}}$ can be determined using the measured equilibrium magnetization with and without irradiation, $M_{\mathrm{RF}}(B_0)$ and $M_0(B_0)$ respectively, and is given by [67]:

$$\Gamma\tau_{\mathrm{s}} = \frac{1}{2}\left(\frac{M_{\mathrm{RF}}(B_0)}{M_0(B_0)} - 1\right) \tag{9}$$

Figure 20 shows $\Gamma_{\mathrm{L}}\tau_{\mathrm{s}}$ obtained from the data in Figs. 16 and 17. Taking the example of the resonance at ν = 14 GHz, B_0 = 0.491 T, we obtain $\Gamma_{\mathrm{L}}\tau_{\mathrm{s}} \approx$ 0.2 (Fig. 20). As mentioned above, $\tau_{\mathrm{s}} \approx$ 1 s is rather large due to the phonon bottleneck effect. This allows to obtain sufficiently large values of $\Gamma_{\mathrm{L}}\tau_{\mathrm{s}}$ to observe significant radiation absorption even if the rate $\Gamma_{\mathrm{L}} \approx$ 0.2 s^{-1} is relatively small. Using expression (3) yield $B_{\mathrm{ac}} \sim$ 1 μT, and similar values were obtained for all measured frequencies.

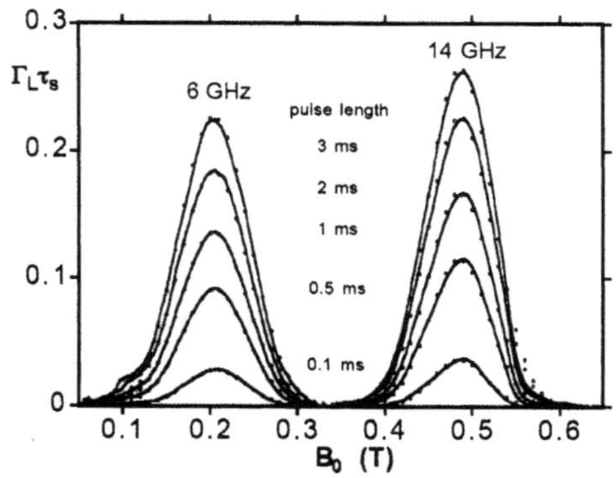

Figure 20. Examples of the field dependence of $\Gamma_L\tau_s$. The curves are obtained from the measured magnetization curves and Eq. 9, at 6 and 14 GHz, after subtraction of the nearly field independent non-resonant absorption background. The dots represent the fit of these curves using Eqs. 8 and 10 (Gaussian distributions) yielding $\sigma_L \approx 33$ - 38 mT.

The resonant absorption lines of Fig. 20 are well described by a Gaussian distribution

$$F(B_L) = \frac{1}{\sigma_L\sqrt{2\pi}} e^{-\frac{(B_0 - B_\nu)^2}{2\sigma_L^2}} \tag{10}$$

yielding $\sigma_L \approx 35$ mT for frequencies between 2 and 20 GHz. The Gaussian and Lorentzian distributions essentially differ when $|B_0 - B_\nu| > \sigma_L$, the Gaussian distribution being more localized (non correlated fields) than the Lorentzian (correlated fields). Our result suggests therefore that the line-broadening is due to fluctuations of non-correlated intra-molecular nuclear spins and not to correlated inter-molecular interactions. This is confirmed by several features:

(i) $\sigma_L \approx 35$ mT (Fig. 20) is much larger than the mean dipolar field seen by the molecules (< 1 mT);

(ii) σ_L is also larger than intermolecular interactions (12 mT);

(iii) σ_L is close to the hyperfine coupling in V_{15} ($\approx$ 40 mT) [58, 63, 65]. This value is derived from the hyperfine field of 11.2 T/μ_B obtained in different systems of V^{IV} [71] and giving a field of about 20 mT for a single V^{IV} spin;

(iv) finally, σ_L is very close to the level width (40 mK) obtained from magnetic relaxation experiments near the $m_s = \pm 1/2$, avoided level crossing. Its origin was mainly attributed to hyperfine and weak Dzyaloshinskii-Moriya interactions [58, 63, 65].

This hyperfine broadening of resonant absorption is consistent with theoretical investigations according to which the main dephasing mechanism of magnetic molecules is connected to nuclear spins [52, 72-75].

The inset of Fig. 19 shows the field dependence of the quality factor $G = B_\nu / \sigma_L$, which gives a lower bound of the precessional coherence of V_{15} molecule spins. A linear increase of G is observed, reaching $G = 20$ at 0.7 T, that is rather far from values of 10^4 needed for quantum computation.

Finally, we discuss the possibility of observing Rabi oscillations with the present set-up. Due to inhomogeneous broadening, only a lower bound of the coherence time can be estimated from the resonance lines: $\tau_c^{-1} = \gamma\sigma_L$. The corresponding number of coherent flips of the spin system is given by $N = \tau_c / \tau_{Rabi} = B_{ac} /(2\pi\sigma_L)$ (Rabi coherence). Using the values of B_{ac} and σ_L obtained from the fit of the resonance lines, we get $\tau_c \approx 2$ ns – comparable to recent EPR measurements [76] – and $N \approx 10^{-6}$, showing that there is no hope to see Rabi oscillations in the present conditions. In order to get $N >> 1$, it will be necessary to increase the radiation field B_{ac} by orders of magnitude. Note that $\Gamma_{L,Max} \approx \gamma B_{ac}^2 / 4\sigma_L$ increases with electromagnetic power and decreases with increasing width of the distribution of local fields. An important consequence of intra-molecular inhomogeneous broadening is the possibility to significantly longer coherence times accessible by spin-echo techniques. This should lead to a transverse relaxation time $\tau_2 >> \tau_c$ which will be necessary for quantum computation.

To summarize this section, we have presented a new technique for studying radiation absorption in the molecular system V_{15} constituting a first step towards the observation of Rabi oscillations in molecular nanomagnets. The main results are the observation of relatively narrow resonant absorption lines that are dominated by hyperfine interaction. In order to observe Rabi oscillations in a magnetic system, an important requirement is a large AC field amplitude.

6. ENVIRONMENTAL DECOHERENCE EFFECTS IN NANOMAGNETS

At temperatures below 0.36 K, Fe_8 molecular clusters display a clear crossover from thermally activated relaxation to a temperature-independent quantum regime, with a pronounced resonance structure of the relaxation time as a function of the external field (Section 3). It was surprising, however, that the observed relaxation of the magnetization in the quantum regime was found to be nonexponential and the resonance width orders of magnitude too large [15, 16]. The understanding of this seemingly anomalous behavior involves the hyperfine fields as well as the evolving distribution of the weak dipole fields of the nanomagnets themselves [52].

Both effects were shown to be the main source of decoherence at very low temperature. At higher temperatures, phonons are another source of decoherence.

In the following sections, we focus on the low temperature and low field limits, where phonon-mediated relaxation is astronomically long and can be neglected. In this limit, the $m = \pm S$ spin states are coupled due to the tunneling splitting $\Delta_{\pm S}$, which is about 10^{-7} K for Fe_8 having $S = 10$. In order to tunnel between these states, the longitudinal magnetic energy bias $\xi = g\mu_B S H_{\mathrm{local}}$ due to the local magnetic field H_{local} on a molecule must be smaller than $\Delta_{\pm S}$, implying a local field smaller than 10^{-8} T for Fe_8 clusters. Since the typical intermolecular dipole fields are of the order of 0.05 T, it seems at first that almost all molecules should be blocked from tunneling by a very large energy bias. Prokof'ev and Stamp have proposed a solution to this dilemma by proposing that fast dynamic nuclear fluctuations broaden the resonance, and the gradual adjustment of the dipole fields in the sample caused by the tunneling brings other molecules into resonance and allows continuous relaxation [52]. Some interesting predictions are briefly reviewed in the following section.

6.1. Prokof'ev-Stamp Theory

Prokof'ev and Stamp were the first who realized that there are localized couplings of environmental modes with mesoscopic systems which cannot be modeled with an "oscillator bath" model [77] describing delocalized environmental modes such as electrons, phonons, photons, and so on. They found that these localized modes such as nuclear and paramagnetic spins are often strong and described them as a spin bath model [73]. We do not review this theory [78] but focus on one particular application which is interesting for molecular clusters [52]. Prokof'ev and Stamp showed that at a given longitudinal applied field H_z, the magnetization of a crystal of molecular clusters should relax at short times with a square-root time dependence which is due to a gradual modification of the dipole fields in the sample caused by the tunneling

$$M(H_z,t) = M_{\mathrm{in}} + (M_{\mathrm{eq}}(H_z) - M_{\mathrm{in}})\sqrt{\Gamma_{\mathrm{sqrt}}(H_z)t} \tag{11}$$

Here M_{in} is the initial magnetization at time $t = 0$ (after a rapid field change), and $M_{\mathrm{eq}}(H_z)$ is the equilibrium magnetization at H_z. The rate function $\Gamma_{\mathrm{sqrt}}(H_z)$ is proportional to the normalized distribution $P(H_z)$ of molecules which are in resonance at H_z

$$\Gamma_{\mathrm{sqrt}}(H_z) = c\frac{\xi_0}{E_D}\frac{\Delta_{\pm S}^2}{4\hbar}P(H_z) \tag{12}$$

where ξ_0 is the line width coming from the nuclear spins, E_D is the Gaussian half-width of $P(H_z)$, and c is a constant of the order of unity which depends on the sample shape. If these simple relations are exact, then measurements of the short time relaxation as a function of the applied field H_z give directly the distribution $P(H_z)$, and they allows one to measure the tunnel splitting $\Delta_{\pm S}$ which is described in the next section.

6.2. Hole Digging Method to Study Dipolar Distributions and Hyperfine Couplings

Motivated by the Prokof'ev-Stamp theory [52], we developed a new technique—which we call the *hole digging method*—that can be used to observe the time evolution of molecular states in crystals of molecular clusters. It allowed us to measure the statistical distribution of magnetic bias fields in the Fe_8 system that arise from the weak dipole fields of the clusters themselves. A hole can be "dug" into the distribution by depleting the available spins at a given applied field. Our method is based on the simple idea that after a rapid field change, the resulting short time relaxation of the magnetization is directly related to the number of molecules which are in resonance at the given applied field. Prokof'ev and Stamp have suggested that the short time relaxation should follow a square-root t relaxation law [Eq. 11]. However, the hole digging method should work with any short time relaxation law—for example, a power law

$$M(H_z,t) = M_{\mathrm{in}} + (M_{\mathrm{eq}}(H_z) - M_{\mathrm{in}})(\Gamma_{\mathrm{short}}(H_z)t)^{\alpha} \tag{13}$$

where Γ_{short} is a characteristic short time relaxation rate that is directly related to the number of molecules which are in resonance at the applied field H_z, and $0 < \alpha < 1$ in most cases. $\alpha = 0.5$ in the Prokof'ev-Stamp theory [Eq. 11] and Γ_{sqrt} is directly proportional to $P(H_z)$ [Eq. 12]. The *hole digging method* can be divided into three steps (Fig. 21):

(i) *Preparing the initial state.* A well-defined initial magnetization state of the crystal of molecular clusters can be achieved by rapidly cooling the sample from high down to low temperatures in a constant applied field H_z^0. For zero applied field (H_z = 0) or rather large applied fields (H_z > 1 T), one yields the demagnetized or saturated magnetization state of the entire crystal, respectively. One can also quench the sample in a small field of few milliteslas yielding any possible initial magnetization $M_{\mathtt{in}}$. When the quench is fast (< 1 s), the sample's magnetization does not have time to relax, either by thermal or by quantum transitions. This procedure yields a frozen thermal equilibrium distribution, whereas for slow cooling rates the molecule spin states in the crystal might tend to certain dipolar ordered ground state.

(ii) *Modifying the initial state—hole digging.* After preparing the initial state, a field $H_{\mathtt{dig}}$ is applied during a time $t_{\mathtt{dig}}$, called "digging field and digging time", respectively. During the digging time and depending on $H_{\mathtt{dig}}$, a fraction of the molecular spins tunnel (back and/or fourth); that is, they reverse the direction of magnetization. (The field sweeping rate to apply $H_{\mathtt{dig}}$ should be fast enough to minimize the change of the initial state during the field sweep.)

(iii) *Probing the final state.* Finally, a field $H_z^{\mathtt{probe}}$ is applied (Fig. 21) to measure the short time relaxation from which one obtains $\Gamma_{\mathtt{short}}$, Eq. 13. This is related to the number of spins which are still free for tunneling after step (ii).

The entire procedure is then repeated many times but at other fields $H_z^{\mathtt{probe}}$ yielding $\Gamma_{\mathtt{short}}(H_z, H_{\mathtt{dig}}, t_{\mathtt{dig}})$ which is related to the distribution of spins $P(H_z, H_{\mathtt{dig}}, t_{\mathtt{dig}})$ that are still free for tunneling after the hole digging. For $t_{\mathtt{dig}} = 0$, this method maps out the initial distribution.

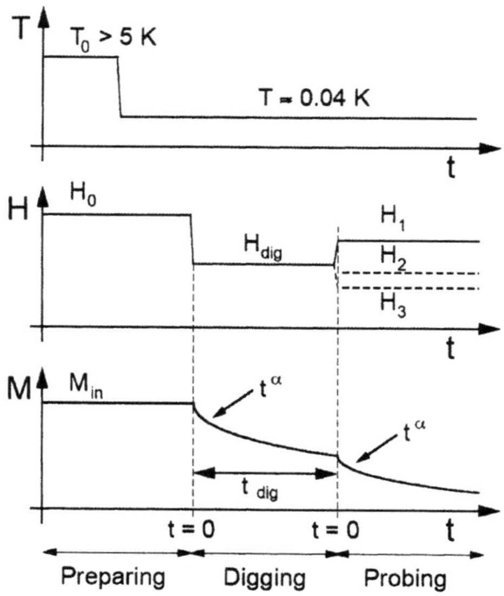

Figure 21. Schema of the hole digging method presenting the time dependence of temperature, applied field, and magnetization of the sample.

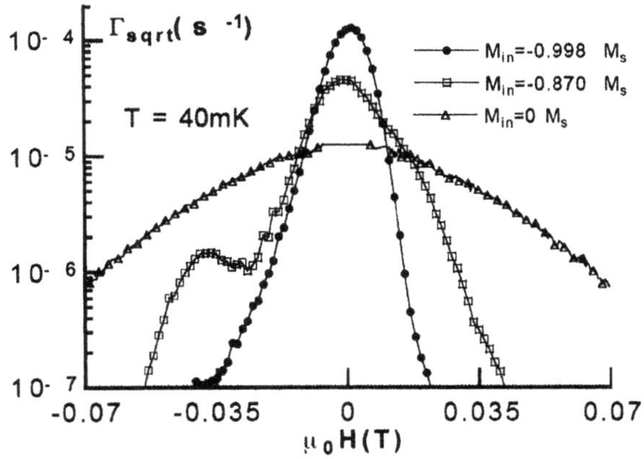

Figure 22. Field dependence of the short time square-root relaxation rates $\Gamma_{sqrt}(H_z)$ for three different values of the initial magnetization M_{in}. According to Eq. (12), the curves are proportional to the distribution $P(H_z)$ of magnetic energy bias due to local dipole field distributions in the sample. Note the logarithmic scale for Γ_{sqrt}. The peaked distribution labeled M_{in} = -0.998 M_s was obtained by saturating the sample, whereas the other distributions were obtained by thermal annealing. M_{in} = -0.870 M_s is distorted by nearest neighbor lattice effects. The peak at 0.04 T as well as the shoulder at 0.02 T and 0.04 T are originated by the clusters which have one nearest-neighbor cluster with reversed magnetization: The peak at 0.04 T corresponds to the reversal of the neighboring cluster along the crystallographic a-axis, which almost coincides with the easy axis of magnetization, while the shoulder at 0.02 T and 0.04 T are due to the clusters along the b and c axes.

6.3. Intermolecular Dipole Interaction in Fe_8

We applied the hole digging method to several samples of molecular clusters and quantum spin glasses. The most detailed study has been done on the Fe_8 system. We found the predicted square-root t relaxation, Eq. (11), in experiments on fully saturated Fe_8 crystals [16, 79] and on nonsaturated samples [34]. Figure 22 analyzes the dipolar distributions, revealing a remarkable structure due to next-nearest-neighbor effects [34]. These results are in good agreement with simulations [80, 55].

For a saturated initial state, the Prokof'ev-Stamp theory allows us to estimate the tunnel splitting $\Delta_{\pm S}$. Using Eqs. (3), (9), and (12) of Ref. [52], along with integration, we find $\int \Gamma_{sqrt} d\xi = \xi_0 D^2_{\pm S}/4E_D \hbar$ where c is a constant of the order of unity which depends on the sample shape. With E_D = 15 mT, ξ_0 = 0.8 mT, c = 1, and Γ_{sqrt} [34, 53], we find $\Delta_{\pm 10} = 1.2\times 10^{-7}$ K which is close to the result of $\Delta_{\pm 10} = 1.0\times 10^{-7}$ K obtained by using a Landau--Zener method (Section 3.1) [33]. Note that the hole-digging method probes the longitudinal dipolar distribution (H_z direction), whereas the

Landau-Zener method can be used to probe the transverse dipolar distribution by measuring the tunnel splittings Δ around a topological quench.

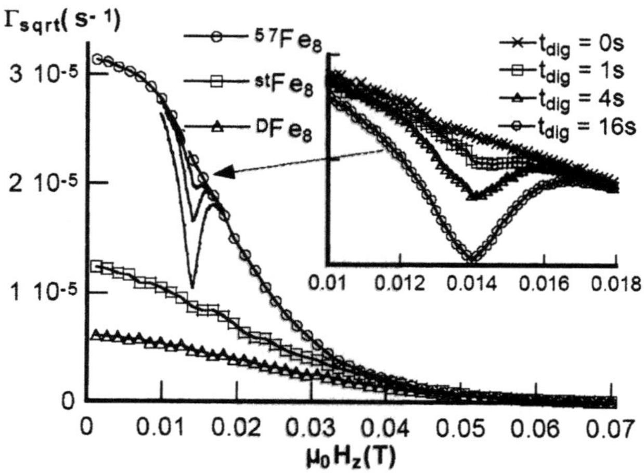

Figure 23. Comparison of the short time relaxation rates of three different Fe_8 samples at T = 40 mK with H_{trans} = 0 and M_{init} = 0. The inset displays an typical example of a hole which was dug into the distribution by allowing the sample to relax for the time t_{dig} at $\mu_0 H_{dig}$ = 14 mT.

6.4. Hyperfine Interaction in Fe_8

The strong influence of nuclear spins on resonant quantum tunneling in the molecular cluster Fe_8 was demonstrated by comparing the relaxation rate of the standard Fe_8 sample with two isotopic modified samples: (i) ^{56}Fe is replaced by ^{57}Fe, and (ii) a fraction of ^{1}H is replaced by ^{2}H [53]. By using the hole digging method, we measured an intrinsic broadening which is driven by the hyperfine fields (Fig. 23), and our measurements are in good agreement with numerical hyperfine calculations [53, 55]. For $T >$ 1.5 K, the influence of nuclear spins on the relaxation rate is less important, suggesting that spin-phonon coupling dominates the relaxation rate.

7. CONCLUSION

In summary, we have reviewed some basic features of molecular nanomagnets and discussed various aspects of our recent research in the feld. Molecular nanomagnets offer a unique opportunity to explore the quantum dynamics of a large but finite spin. We focused our discussion on the Fe_8 molecular nanomagnets, because it is the first system where studies in the pure quantum regime were possible. The tunneling in this system is remarkable because it does not show up at the lowest order of perturbation theory.

A new family of supramolecular, antiferromagnetically exchange-coupled dimers of single-molecule magnets (SMMs) has recently been reported [48]. Each SMM acts as a bias on its neighbor, shifting the quantum tunneling resonances of the individual SMMs. Hysteresis loop measurements on a single crystal of SMM-dimers have established quantum tunneling of the magnetization via entangled states of the dimer. This showed that the dimer really does behave as a quantum-mechanically coupled dimer. The transitions are well separated, suggesting long coherence times compared to the time scale of the energy splitting [49]. This result is of great importance if such systems are to be used for quantum computing.

Molecules with small spin have also been discussed. For example, time-resolved magnetization measurements were performed on a spin 1/2 molecular complex, so-called V_{15} [81]. Despite the absence of a barrier, magnetic hysteresis is observed over a time scale of several seconds. A detailed analysis in terms of a dissipative two-level model has been given, in which fluctuations and splittings are of the same energy. Spin-phonon coupling leads to long relaxation times and to a particular "butterfly" hysteresis loop [58, 82].

What remains still debated is the possibility of observing quantum coherence between states of opposite magnetization. Dipole--dipole and hyperfine interactions are source of decoherence. In other words, when a spin has tunneled through the barrier, it experiences a huge modification of its environment (hyperfine and dipolar) which prohibits the back tunneling. Prokof'ev and Stamp suggested three possible strategies to suppress the decoherence [83]. (i) Choose a system where the NMR frequencies far exceed the tunnel frequencies making any coupling impossible, (ii) Isotopically purify the sample to remove all nuclear spins, and (iii) apply a transverse field to increase the tunnel rate to frequencies much larger than hyperfine field fluctuations. Several groups are currently working on such proposals.

Concerning the perspectives of the field of single-molecule magnets, we expect that chemistry is going to play a major role through the synthesis of novel larger spin clusters with strong anisotropy. We want to stress that there are already many other molecular nanomagnets (the largest is currently a Mn_{84}) which are possible model systems. We believe that more sophisticated theories are needed which describe the dephasing effects of the environment onto the quantum system. These investigations are important for studying the quantum character of molecular clusters for applications like quantum computers. The first implementation of Grover's algorithm with molecular nanomagnets has been proposed [57]. However, many experimantal difficulties are still waiting for solutions.

References

[1] A. Hubert and R. Schäfer, "Magnetic Domains: The Analysis of Magnetic Microstructures", Springer-Verlag, Berlin Heidelberg New York 1998.

[2] W. Wernsdorfer, K. Hasselbach, D. Mailly, B. Barbara, A. Benoit, L. Thomas, and G. Suran, J. Magn. Magn. Mat. **145**, 33 (1995).

[3] A. Aharoni, "An Introduction to the Theory of Ferromagnetism", Oxford University Press, London 1996.

[4] W. Wernsdorfer, Adv. Chem. Phys. **118**, 99 (2001).

[5] T. Lis, Acta Cryst. B **36**, 2042 (1980).

[6] K. Wieghardt, K. Pohl, I. Jibril, and G. Huttner, Angew. Chem. Int. Ed. Engl. **23**, 77 (1984).

[7] A. Caneschi, D. Gatteschi, J. Laugier, P. Rey, R. Sessoli, and C. Zanchini, J. Am. Chem. Soc. **113**, 5873 (1991).

[8] R. Sessoli, H.-L. Tsai, A. R. Schake, S. Wang, J. B. Vincent, K. Folting, D. Gatteschi, G. Christou, and D. N. Hendrickson, J. Am. Chem. Soc. **115**, 1804 (1993).

[9] R. Sessoli, D. Gatteschi, A. Caneschi, and M. A. Novak, Nature **365**, 141 (1993).

[10] A.-L. Barra, P. Debrunner, D. Gatteschi, Ch. E. Schulz, and R. Sessoli, EuroPhys. Lett. **35**, 133 (1996).

[11] M. A. Novak and R. Sessoli, in "Quantum Tunneling of Magnetization-QTM'94", Vol. 301 of NATO ASI Series E: Applied Sciences, Eds. L. Gunther and B. Barbara, Kluwer Academic Publishers, London 1995, pp. 171-188.

[12] C. Paulsen and J.-G. Park, in "Quantum Tunneling of Magnetization-QTM'94", Vol. 301 of NATO ASI Series E: Applied Sciences, Eds. L. Gunther and B. Barbara, Kluwer Academic Publishers, London 1995, pp. 189--205.

[13] J. R. Friedman, M. P. Sarachik, J. Tejada, and R. Ziolo, Phys. Rev. Lett. **76**, 3830 (1996).

[14] L. Thomas, F. Lionti, R. Ballou, D. Gatteschi, R. Sessoli, and B. Barbara, Nature **383**, 145 (1996).

[15] C. Sangregorio, T. Ohm, C. Paulsen, R. Sessoli, and D. Gatteschi, Phys. Rev. Lett. **78**, 4645 (1997).

[16] T. Ohm, C. Sangregorio, and C. Paulsen, Euro. Phys. J. B **6**, 195 (1998).

[17] A. Caneschi, D. Gatteschi, C. Sangregorio, R. Sessoli, L. Sorace, A. Cornia, M. A. Novak, C. Paulsen, and W. Wernsdorfer, J. Magn. Magn. Mat. **200**, 182 (1999).

[18] S. M. J. Aubin, N. R. Dilley, M. B. Wemple, G. Christou, and D. N. Hendrickson, J. Am. Chem. Soc. **120**, 839 (1998).

[19] D. J. Price, F. Lionti, R. Ballou, P. T. Wood, and A. K. Powell, Phil. Trans. R. Soc. Lond. A **357**, 3099 (1999).

[20] J. Yoo, E. K. Brechin, A. Yamaguchi, M. Nakano, J. C. Huffman, A. L. Maniero, L.-C. Brunel, K. Awaga, H. Ishimoto, G. Christou, and D. N. Hendrickson, Inorg. Chem. **39**, 3615 (2000).

[21] A. J. Tasiopoulos, A. Vinslava, W. Wernsdorfer, K.A. Abboud, and G. Christou, Angew. Chem. Int. Ed. Engl. **43**, 2117 (2004).

[22] M. Jamet, W. Wernsdorfer, C. Thirion, D. Mailly, V. Dupuis, P. Melinon, and A.Perez, Phys. Rev. Lett. **86**, 4676 (2001).

[23] Y. Pontillon, A. Caneschi, D. Gatteschi, R. Sessoli, E. Ressouche, J. Schweizer, and E. Lelievte-Berna, J. Am. Chem. Soc. **121**, 5342 (1999).

[24] A. Garg, EuroPhys. Lett. **22**, 205 (1993).

[25] L. Landau, Phys. Z. Sowjetunion **2**, 46 (1932).

[26] C. Zener, Proc. R. Soc. London, Ser. A **137**, 696 (1932).

[27] E. C. G. Stückelberg, Helv. Phys. Acta **5**, 369 (1932).

[28] S. Miyashita, J. Phys. Soc. Jpn. **64**, 3207 (1995).

[29] S. Miyashita, J. Phys. Soc. Jpn. **65**, 2734 (1996).

[30] G. Rose and P. C. E. Stamp, Low Temp. Phys. **113**, 1153 (1998).

[31] M. Thorwart, M. Grifoni, and P. Hänggi, Phys. Rev. Lett. **85**, 860 (2000).

[32] M. N. Leuenberger and D. Loss, Phys. Rev. B **61**, 12200 (2000).

[33] W. Wernsdorfer and R. Sessoli, Science **284**, 133 (1999).

[34] W. Wernsdorfer, T. Ohm, C. Sangregorio, R. Sessoli, D. Mailly, and C. Paulsen, Phys. Rev. Lett. **82**, 3903 (1999).

[35] D. Loss, D. P. DiVincenzo, and G. Grinstein, Phys. Rev. Lett. **69**, 3232 (1992).

[36] J. von Delft and C. L. Henley, Phys. Rev. Lett. **69**, 3236 (1992).

[37] R. P. Feynman, R. B. Leighton, and M. Sand, "The Feynman Lectures on Physics", Addison-Wesley Publishing Company, London 1970, Vol. 3.

[38] A. Garg, Phys. Rev. Lett. **83**, 4385 (1999).

[39] J. Villain and A. Fort, Euro. Phys. J. B **17**, 69 (2000).

[40] E. Kececioglu and A. Garg, Phys. Rev. B **63**, 064422 (2001).

[41] S. E. Barnes, cond-mat/9907257 **0**, 0 (1999).

[42] J.-Q. Liang, H. J. W. Mueller-Kirsten, D. K. Park, and F.-C. Pu, Phys. Rev. B **61**, 8856 (2000).

[43] Sahng-Kyoon Yoo and Soo-Young Lee, Phys. Rev. B **62**, 3014 (2000).

[44] Hui Hu, Jia-Lin Zhu, Rong Lu, and Jia-Jiong Xiong, cond-mat/0005527 **0**, 0 (2000).

[45] R. Caciuffo, G. Amoretti, A. Murani, R. Sessoli, A. Caneschi, and D. Gatteschi, Phys. Rev. Lett. **81**, 4744 (1998).

[46] G. Amoretti, R. Caciuffo, J. Combet, A. Murani, and A. Caneschi, Phys. Rev. B **62**, 3022 (2000).

[47] A. L. Barra, D. Gatteschi, and R. Sessoli, Chem. Eur. J. **6**, 1608 (2000).

[48] W. Wernsdorfer, N. Aliaga-Alcalde, D.N. Hendrickson, and G. Christou, Nature **416**, 406 (2002).

[49] R. Tiron, W. Wernsdorfer, D. Foguet-Albiol, N. Aliaga-Alcalde, and G. Christou, Phys. Rev. Lett. **91**, 227203 (2003).

[50] D. N. Hendrickson *et al.*, J. Am. Chem. Soc. **114**, 2455 (1992).

[51] W. Wernsdorfer, S. Bhaduri, R. Tiron, D. N. Hendrickson, and G. Christou, Phys. Rev. Lett. **89**, 197201 (2002).

[52] N. V. Prokof'ev and P. C. E. Stamp, Phys. Rev. Lett. **80**, 5794 (1998).
[53] W. Wernsdorfer, A. Caneschi, R. Sessoli, D. Gatteschi, A. Cornia, V. Villar, and C. Paulsen, Phys. Rev. Lett. **84**, 2965 (2000).
[54] J. J. Alonso and J. F. Fernandez, Phys. Rev. Lett. **87**, 097205 (2001).
[55] I. Tupitsyn and P. C. E. Stamp, cond-mat/0305371 **5**, 5371 (2003).
[56] W. Wernsdorfer, M. Soler, D. N. Hendrickson, and G. Christou, cond-mat/0306303 **6**, 303 (2003).
[57] M. N. Leuenberger and D. Loss, Nature **410**, 789 (2001).
[58] I. Chiorescu, W. Wernsdorfer, A. Müller, H. Bögge, and B. Barbara, Phys. Rev. Lett. **84**, 3454 (2000).
[59] I. Chiorescu, Y. Nakamura, C. J. P. M. Harmans, and J. E. Mooij, Science **299**, 1869 (2003).
[60] L. Sorace, W. Wernsdorfer, C. Thirion, A.-L. Barra, M. Pacchioni, D. Mailly, and B. Barbara, Phys. Rev. Lett. **68**, 220407 (2003).
[61] A. Müller and J. Döring, Angew. Chem. Intl. Ed. Enl. **27**, 1721 (1988).
[62] D. Gatteschi *et al.*, Molecular Engineering **3**, 157 (1993).
[63] B. Barbara, I. Chiorescu, W. Wernsdorfer, H. Bögge, and A. Müller, Prog. Theor. Phys. Suppl. **145**, 357 (2002).
[64] G. Chaboussant, S.T. Oschenbein, A. Sieber, H. U. Güdel, H. Mukta, A. Müller, and B. Barbara, EuroPhys. Lett. **1**, 1 (2004).
[65] I. Chiorescu, W. Wernsdorfer, A. Müller, S. Miyashita, and B. Barbara, Phys. Rev. B **67**, 020402 (2003).
[66] C. Thirion, W. Wernsdorfer, and D. Mailly, Nature Mat. **2**, 524 (2003).
[67] A. Abragam and B. Bleaney, "Electron Paramagnetic Resonance of Transition Ions", Clarendon Press, Oxford 1970.
[68] Y. Ajiro, M. Itoh, Y. Inagaki, T. Asano, Y. Narumi, K. Kindo, T. Sakon, M. Motokawa, A. Cornia, D. Gatteschi, A. Müller, and B. Barbara, in "Proc. on the French-Japanese Symposium on Quantum Properties of Low-Dimensional Antiferromagnets", Eds. Y. Ajiro and J. P. Boucher, Kyushu University Press, 7-1-146, Hakozaki Higashi-ku, Fukuoka-shi 812-0053, Japan, 2002, p. 80.
[69] I. I. Rabi, Phys. Rev. **21**, 652 (1937).
[70] M. Grifoni and P. Hanggi, Physics Reports **304**, 229 (1998).
[71] G. C. Carter, H. H. Bennett, and D. J. Kahan, in "Metallic Shifts in NMR", Vol. 20 of "Progress in Material Science", Eds. B. Chalmers, J. W. Christian, and T. B. Massalki, Pergamon Press, London 1977.
[72] A. Garg, Phys. Rev. Lett. **74**, 1458 (1995).
[73] N. V. Prokof'ev and P. C. E. Stamp, J. Low Temp. Phys. **104**, 143 (1996).
[74] D. A. Garanin, E. M. Chudnovsky, and R. Schilling, Phys. Rev. B **61**, 12204 (2000).
[75] P. C. E. Stamp and I. S. Tupitsyn, Phys. Rev. B **69**, 014401 (2004).
[76] S. Hill, R. S. Edwards, N. Aliaga-Alcalde, and G. Christou, Science **302**, 1015 (2003).
[77] R. P. Feynman and F. L. Vernon, Ann. Phys. **24**, 118 (1963).
[78] N. V. Prokof'ev and P. C. E. Stamp, Rep. Prog. Phys. **63**, 669 (2000).
[79] T. Ohm, C. Sangregorio, and C. Paulsen, J. Low Temp. Phys. **113**, 1141 (1998).
[80] A. Cuccoli, A. Fort, A. Rettori, E. Adam, and J. Villain, Euro. Phys. J. B **12**, 39 (1999).

[81] I. Chiorescu, W. Wernsdorfer, B. Barbara, A. Müller, and H. Bögge, J. Appl. Phys. **87**, 5496 (2000).

[82] V. V. Dobrovitski, M. I. Katsnelson, and B. N. Harmon, Phys. Rev. Lett. **84**, 3458 (2000).

[83] N. V. Prokof'ev and P. C. E. Stamp, in "Quantum Tunneling of Magnetization-QTM'94, Vol. 301 of NATO ASI Series E: Applied Sciences, Eds. L. Gunther and B. Barbara, Kluwer Academic Publishers, London 1995, p. 369.

Chapter 7

MAGNETIC NANOPARTICLES

Michael J. Bonder, Yunhe Huang, and
George C. Hadjipanayis
Department of Physics and Astronomy
University of Delaware
Newark, DE 19716, USA

Abstract As the size of a magnetic particle is reduced down to the nanoscale there is a convergence of the physical length scales with those that mediate the properties of the material. This chapter introduces the physics of magnetic nanoparticles with attention paid to size effects, including anisotropy and superparamagnetism. We then review the research involved with the formation and study of magnetic nanoparticles in a nonmagnetic matrix resulting from the post synthesis annealing of multilayer precursor systems. The effects of phase formation in the case of the FePt/C system and the more general effects of inter-particle spacing on the magnetic properties will be discussed. In the third section the work involving epitaxial growth of FePt nanoparticle islands on single crystal MgO substrates is discussed.

1. FUNDAMENTALS

1.1. Particle Magnetism

Bulk magnets consist of a large number of magnetic domains that form in an effort to minimize the magnetostatic energy of the material. A schematic of a simplified domain configuration is shown in Fig. 1(a). For this reason the microstructure of bulk magnetic materials plays a large role in determining the magnetic properties. As the particle size is reduced down to the nano-scale the particle size and the exchange length converge allowing for single domain states to stabilize as seen in Fig. 1(b). The magnetism and hence the magnetic properties of nanoparticles are predominantly dictated by the intrinsic properties of a material such as the anisotropy and saturation magnetization.

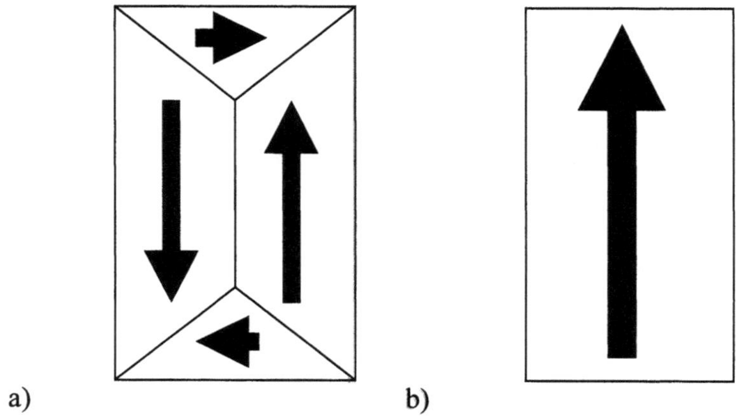

Figure 1. Schematic of magnetic domain configuration for a) multi-domain and b) single domain ferromagnets.

The transition from multi-domain to single domain becomes very apparent when one considers the coercivity as a function of particle size as shown in Fig. 2 below.

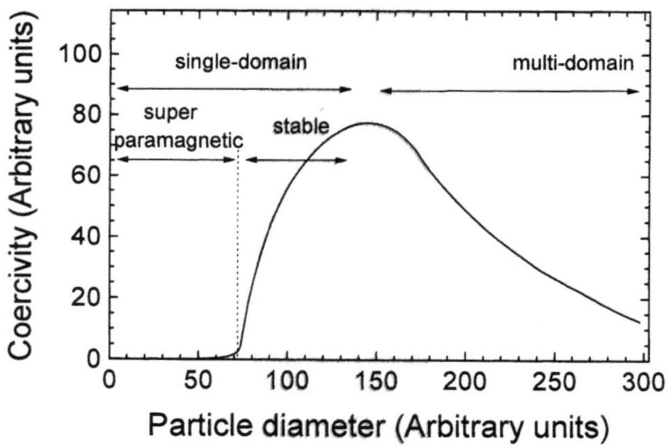

Figure 2. Coercivity as a function of particle size.

For large sizes the particles are multi-domain becoming more bulk-like with increasing size. The mechanism of magnetization reversal is domain wall nucleation and motion for the multi-domain case [1]. As the particle size is reduced the increase of energy due to domain wall formation dominates over the decrease in energy attributed to the formation of domains. Thus, below a critical particle size domain walls will no longer form due to energy considerations and single domain particles are stable. This critical size corresponds to the peak in the coercivity in Fig. 2. The peak position on the particle size axis is dependent on the values of anisotropy and magnetization as will be seen later in the chapter. Below the coercivity maximum the particles remain stable until the effects of temperature destroy the ferromagnetic order. The particles are then superparamagnetic. The superparamagnetic size strictly depends on the magnetocrystalline anisotropy of the material.

1.2. Superparamagnetism

Superparamagnetism refers to the influence of thermal energy on a ferromagnetic nanoparticle. In the superparamagnetic size regime the moments of the nanoparticle fluctuate due to thermal energy. This fluctuation tends to randomize the moments of the nanoparticle unless a magnetic field is applied. The basis of superparamagnetism can be summarized by a double potential well as shown in Fig. 3.

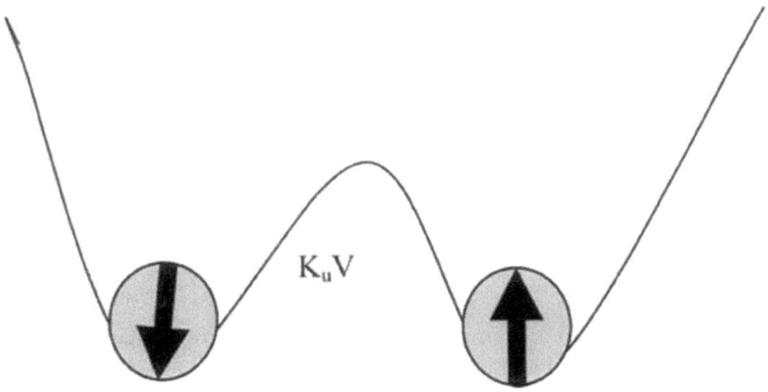

Figure 3. Double potential well (schematic) as a model of superparamagnetism.

As seen in the schematic figure, the two states of magnetization of a uniaxial magnetic particle are separated by an energy barrier, K_uV, where K_u is the anisotropy energy density and V is the particle volume (Chs. 3 and 4). If the thermal energy, kT, becomes comparable to the barrier height there is an increased probability of the magnetization reversing. This thermally activated switching follows the Arrhenius-Néel equation

$$f = f_0 e^{-\frac{KV}{kT}t} \qquad (1)$$

where f is the frequency of switching between magnetization states and f_0 is the proportionality constant equal to 10^9 s^{-1} (Ch. 3). Thus, the minimum requirement for thermal stability for magnetic recording purposes is $K_u V > 25\ kT$ [2].

Superparamagnetism is identified by three characteristics of magnetization curves. A superparamagnet has no remanence or coercivity; there is a temperature below which ferromagnetic order reappears, known as the blocking temperature (T_B) and the magnetization curves superimpose when plotting M as a function of H/T above T_B. Figure 4(a) and (b) depict the first two requirements respectively, where the blocking temperature is determined by measuring the magnetization as a function of temperature in the zero field cooled (ZFC) state followed by the field cooled (FC) curve. The peak in the ZFC curve in Fig. 4(b) indicates the blocking temperature. Below the blocking temperature, the effects of thermal energy are diminished and ferromagnetic order is regained.

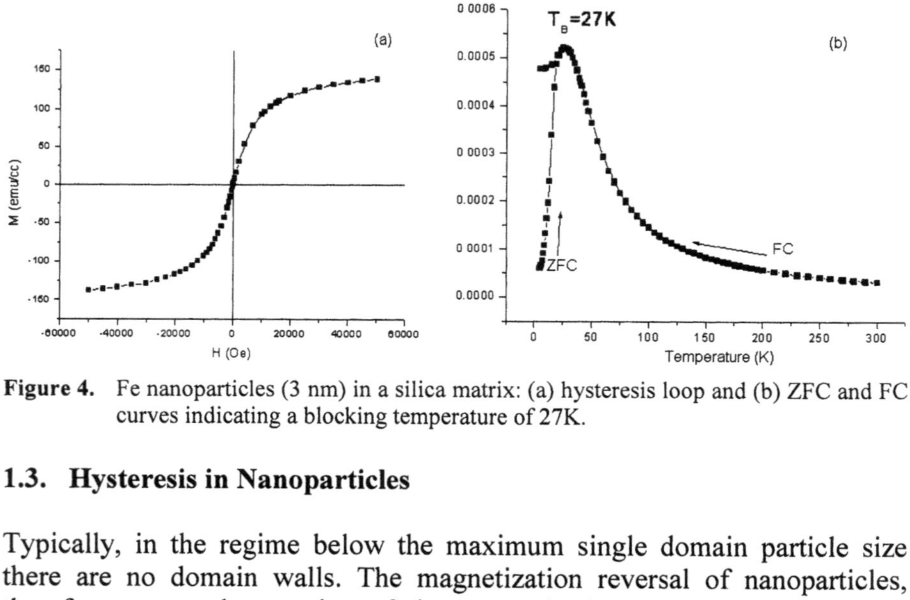

Figure 4. Fe nanoparticles (3 nm) in a silica matrix: (a) hysteresis loop and (b) ZFC and FC curves indicating a blocking temperature of 27K.

1.3. Hysteresis in Nanoparticles

Typically, in the regime below the maximum single domain particle size there are no domain walls. The magnetization reversal of nanoparticles, therefore, occurs by rotation of the magnetization upon application of a magnetic field. By considering a random distribution an of non-interacting, single domain particles with uniaxial anisotropy that reverse by rotation Stoner and Wohlfarth in 1948 theoretically determined that the coercivity, (H_c), of these particles along the anisotropy axis should follow the relationship [3]

$$H_c = \frac{2K_u}{M_s} \tag{2}$$

where K_u is the anisotropy constant and M_s is the saturation magnetization. Considering the effect of thermal fluctuations, the coercivity as a function of particle size, D, approximately follows the relation [2]

$$H_c(D) = 0.5H_k\left[1 - \left(\frac{D_p}{D}\right)^{\frac{3}{2}}\right] \tag{3}$$

where H_k is the anisotropy field of the material in question, D is the measured particle size and D_p is the superparamagnetic particle size. From this relationship the superparamagnetic particle size can be extracted from hysteresis loops of systems with varied particle size.

These models, however, make the erroneous assumption that the nanoparticles are non-interacting. This is certainly not the case for most realizable systems of magnetic nanoparticles. The proximity of particles has a large affect on the hysteresis as they become increasingly exchange coupled with decreasing distance between particles [4].

In order to describe this phenomenon, remanence curves provide insight into magnetic interactions through the use of the delta M (δM) formalism which classifies the nature of interactions present via the relation [5]

$$\delta M = \overline{M_d}(H) - [\ 1 - \overline{M_r}(H)\ ] \tag{4}$$

where M_r is the isothermal remanent magnetization and M_d is the DC demagnetization curve and the bar refers to reduced values normalized by $M_r(\infty)$. The values of δM indicate the extent of magnetostatic or exchange interactions between magnetic particles. Values greater than zero correspond to exchange coupling while values below zero correspond to magnetostatic interactions. A δM of zero indicates a non-interacting system of particles (an idealized Stoner-Wohlfarth system).

2. NANOPARTICELS FROM MULTILAYER PRECURSORS

This section looks at high anisotropy magnetic nanoparticles embedded in a nonmagnetic matrix. First we discuss the study of FePt/C samples. As discussed in chapter 9, self assemblies of FePt nanoparticles agglomerate at the temperatures needed to form the $L1_0$ high anisotropy phase. One solution is the use of the MnAl τ phase which we discuss after FePt. The advantage of these high anisotropy nanoparticles is that they remain thermally stable as the particle is reduced down to a few nanometers. As technologies continue to require smaller magnetic particles there is a subsequent need for higher anisotropy materials. The highest anisotropy material known to date is $SmCo_5$ and we look at this system at the end of the section in looking to continue to reduce particle sizes. Alloying effects are inherent in these multilayer precursor structures due to the need for annealing. This is detrimental to the magnetic properties as it reduces the overall anisotropy. The last part of this section discusses the use of the gas condensation technique of chapter 8 in formulating as deposited high anisotropy FePt and $SmCo_5$ nanoparticles.

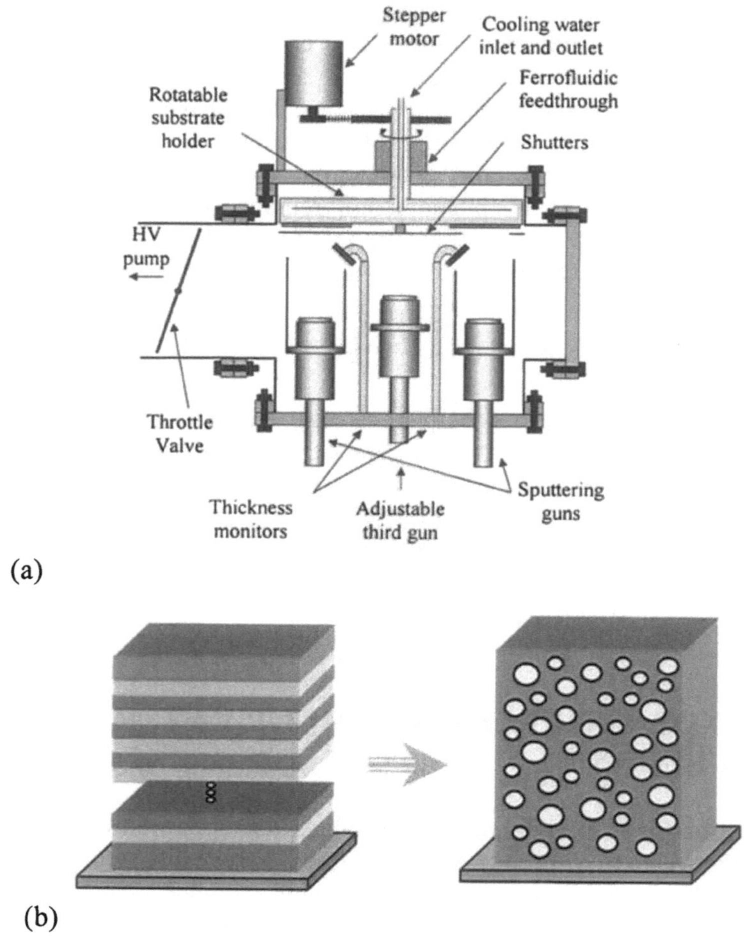

Figure 5. Schematic representation of (a) sputtering chamber showing three magnetrons and water cooled rotating substrate platter. (b) nanoparticle formation from multilayer precursors.

2.1. FePt Nanoparticles Embedded in a Matrix

By annealing multilayers of metallic species and carbon deposited using a high vacuum system like that depict in Fig. 5(a) it is possible to form magnetic nanoparticles embedded in a matrix as schematically shown in Fig. 5(b).

With the application of heat to the multilayers the metallic layers will break up forming particles as they coalesce with the second material encapsulating the nanoparticles as they form. In general the matrix material needs to be immiscible with the nanoparticles to avoid the effects of alloying on the properties of the nanoparticles. By varying the annealing conditions the particle size and inter-particle distances can be tailored. Figure 6 shows the effect of the matrix on particle size for four different initial volume fractions of matrix material for FePt/C multilayers. For each volume % of carbon the FePt particles increase in size with annealing time. The particle growth is slower with increased amounts of carbon allowing for a way to tailor the particle size [6].

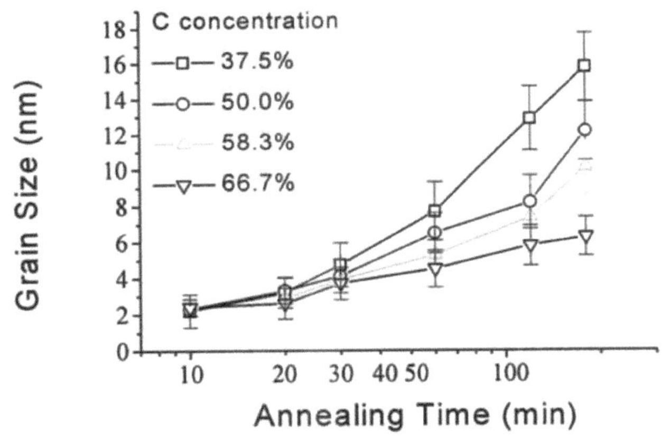

Figure 6. Variation of grain size with annealing time annealed at 600 °C for the vol% C listed.

2.2. Inter-Particle Spacing

Along with the increase in particle size there is an increase in the distance between particles in part due to an increased amount of matrix and continual aggregation as they grow in size. The TEM images in Fig. 7 show the development of particle size and spacing as a function of annealing time for samples annealed at 600 °C. In the as-made state the material is very fine grained with grain sizes less than 5 nm. In the initial stages of growth there is a nucleation of a large number of particles which subsequently grow with increasing spacing between particles.

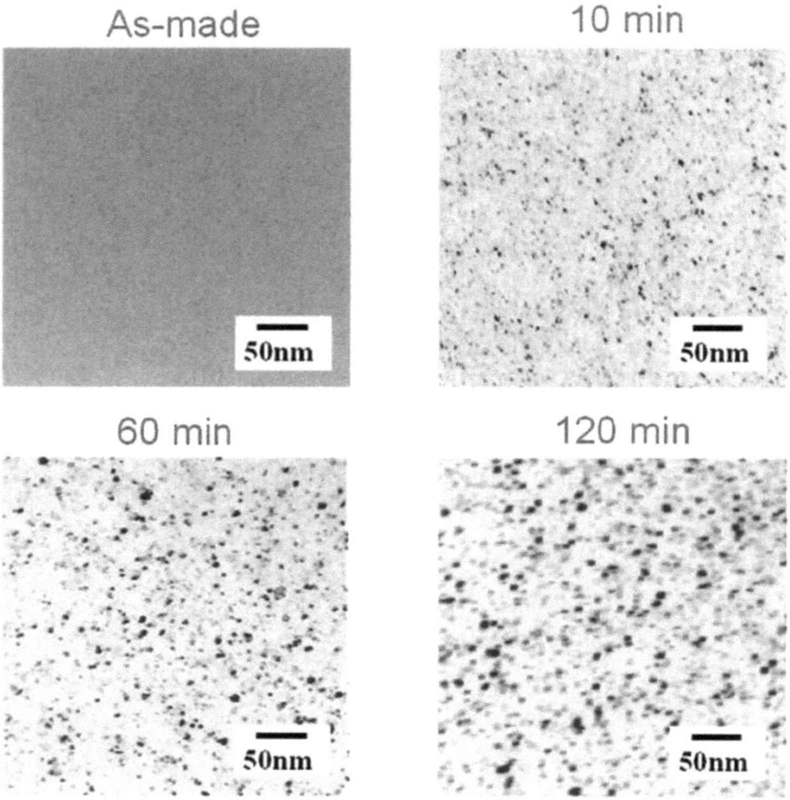

Figure 7. TEM Bright Field images for a FePt/C sample with 58.3 vol% C annealed at 600°C for a) as made, b) 10, c) 60 and d) 120 minutes.

With control of particle size and spacing as per annealing conditions, if one considers Eq. 3 it is possible to extract the critical superparamagnetic particle size. Figure 8 shows a plot of H_c as a function of particle size for FePt/C samples. As expected the coercivity asymptotically approaches a maximum value as the particle size increases. A fit to the data with Eq. 3 reveals that the superparamagnetic particle size for FePt is between 2 and 3 nm [6]. The value of the critical superparamagnetic size can be seen in the figure to be the same, within experimental error, for each of the volume fractions of carbon in the samples.

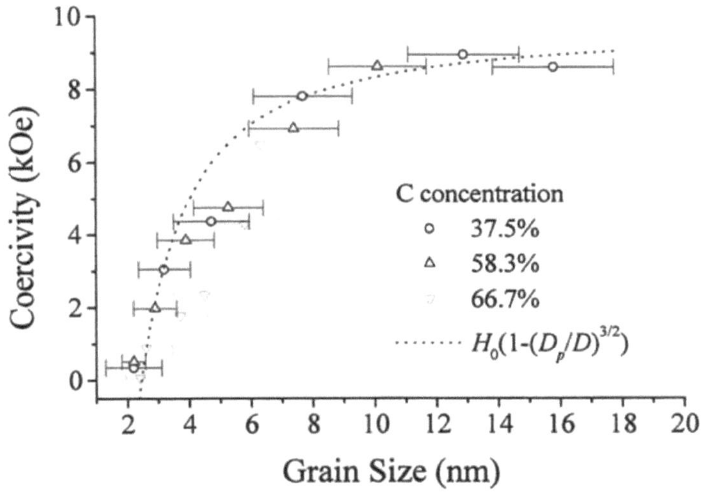

Figure 8. Grain size as a function of coercivity for 37.5, 58.3 and 66.7 vol% C along with the theoretical curve for particle size dependence of coercivity.

2.3. Inter-Particle Interactions

As the volume fraction of carbon increases, the particle size and spacing is varied. By varying these two parameters the magnetic interactions are subsequently affected [4]. Nanoparticles in close proximity will be exchange coupled while those that are at larger distances will be magnetostatically coupled. For exchange coupled systems the hysteresis loop is square. With respect to the hysteresis loop itself this can be quantitatively extracted if we consider the ratio of the remanence (M_r) to the saturation magnetization (M_S). This remanence ratio provides a gauge of the magnetic interactions such that a ratio greater than 0.5 indicates exchange coupling while a ratio less than 0.5 indicates magnetostatic coupling. Assuming the particles are uniaxial, a ratio of 0.5 indicates the nanoparticles are non-interacting. Figure 9 shows the hysteresis loops for samples having the same particle size and coercivity, while having different carbon content. As the volume fraction of carbon decreases the loops become squarer with the remanence ratio approaching 1.0. Thus, as the carbon acts to separate and restrict the growth of the particles the magnetic interactions are subsequently affected, providing for a means to tailor the properties of these nanoparticle systems.

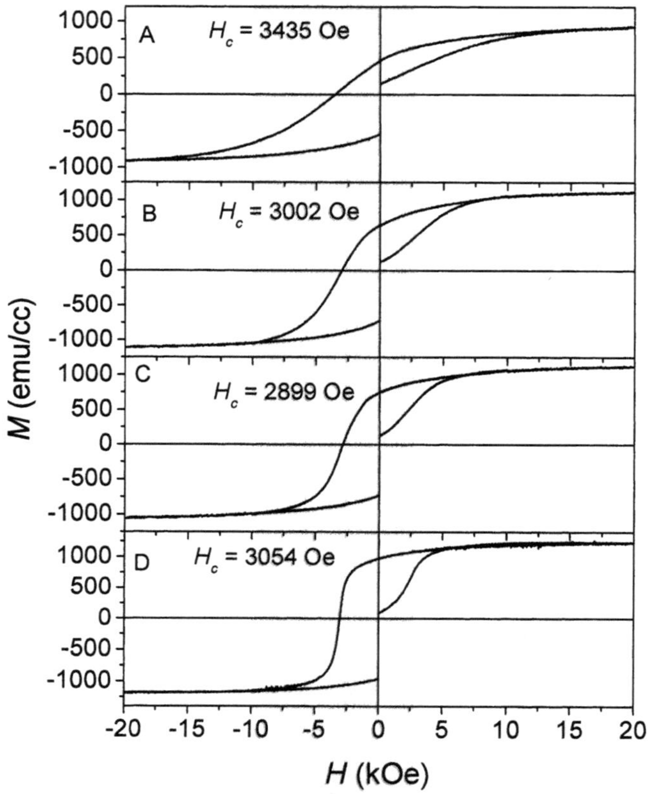

Figure 9. Hysteresis loops of FePt/C samples with particle size of 3 nm and carbon vol% of A) 66.7 B) 58.3 C) 50.0 and D) 37.5 percent.

When the δM formalism of Eq. 4 is applied to samples with varied carbon vol%, the plot in Fig. 10 results. As with the remanence ratio considerations the lowest carbon volume fraction corresponds to a δM that is positive. This is indicative of exchange interactions as discussed above with regard to Eq. 4. With increased carbon content the δM curves become increasingly negative indicating magnetostatic interactions in agreement with the remanence ratio analysis.

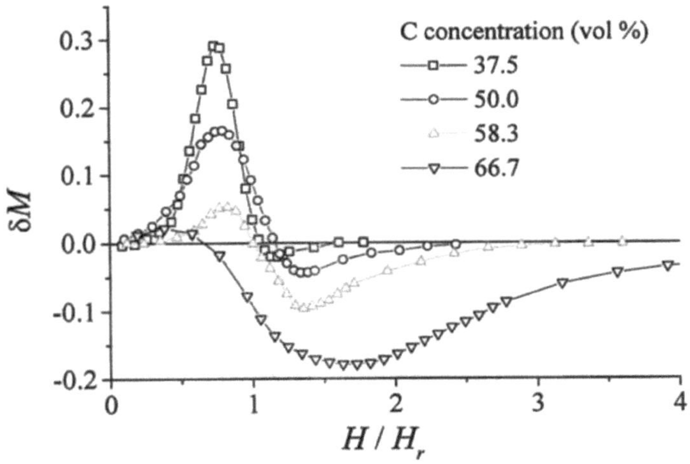

Figure 10. δM plots for FePt/C samples with different carbon contents.

Further evidence of this can be extracted from time dependent measurement of the magnetization. By determining the viscosity from time dependent magnetization measurements the effective switching volume of a sample is derived [7]. When this is compared to the particle sizes determined from TEM images as a function of carbon volume fraction the plot in Fig. 11 results. As the separation between particles increases there is a convergence of the magnetic volume that coherently reverses with the actual particle size. Thus the particles are less affected by inter-particle interaction and approaching the idealized Stoner-Wohlfarth condition.

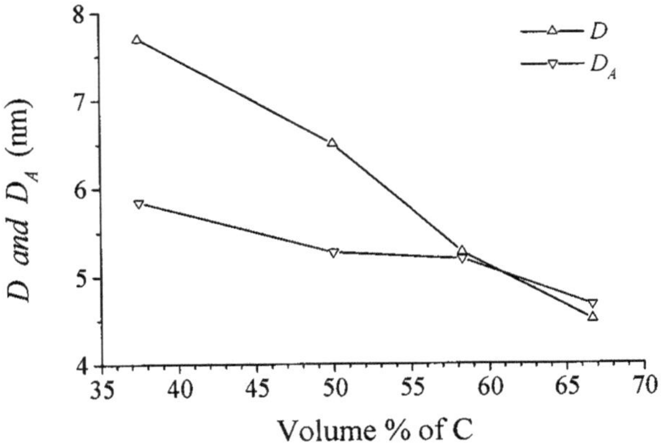

Figure 11. Plot of the activation volume and actual particle size as determined by TEM.

2.4. FePt Phase Transformation

The FePt alloy undergoes a phase transformation to the high anisotropy FCT (face-center-tetragonal) upon annealing above 600 °C. The transformation occurs with the nucleation of FCT and growth at the expense of the FCC (face-center-cubic) material. Mossbauer spectra taken at the early stages of growth confirm this for the nanoparticles. The spectra in Fig. 12 correspond to measurements taken at room and Liquid nitrogen temperatures for the same sample annealed at 700 °C for two minutes which showed a 12% volume fraction of FCT. The room temperature measurement contains a large paramagnetic component that disappears at low temperature which is consistent with superparamagnetism. In addition, the FCT component increases by the same relative amount as the decrease in the superparamagnetic component. Hence the FCT is nucleating at very small sizes [6].

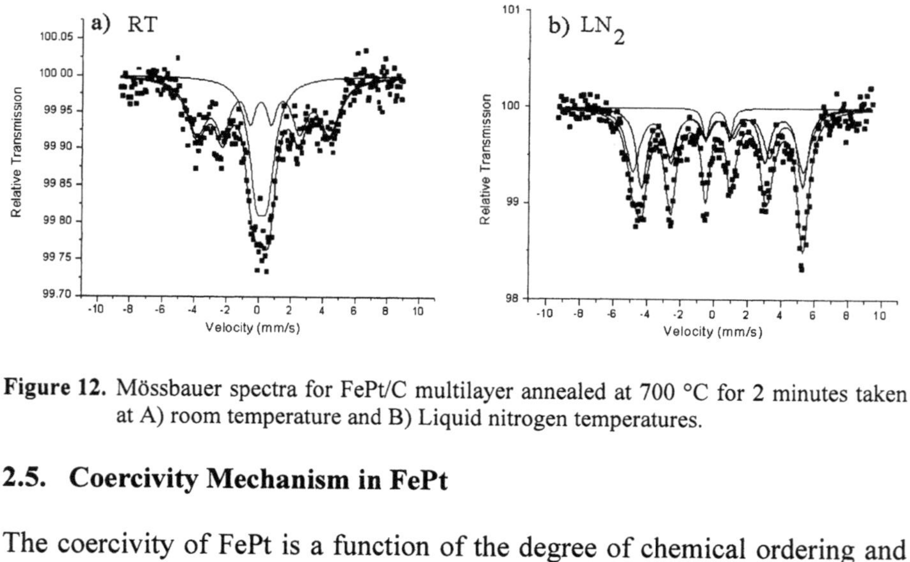

Figure 12. Mössbauer spectra for FePt/C multilayer annealed at 700 °C for 2 minutes taken at A) room temperature and B) Liquid nitrogen temperatures.

2.5. Coercivity Mechanism in FePt

The coercivity of FePt is a function of the degree of chemical ordering and the particle size. The chemical ordering is a measure of the tetragonality of the FePt lattice and related to the anisotropy. The grain size effect on the coercivity was phenomenologically depicted in Fig. 2 and more specifically shown for FePt in Fig. 8. When the degree of ordering (S_0) is related to the inverse grain size as in Fig. 13 the size dependence becomes very obvious due to the linear relationship shown. In light of the progression of the $L1_0$ phase transformation Fig. 13 says that small particles have a lower degree of order than larger particles. This is consistent with the Mössbauer results of Fig. 12 since the volume fraction of the FCT phase increases at the expense of the FCC material with increased annealing and hence grain growth.

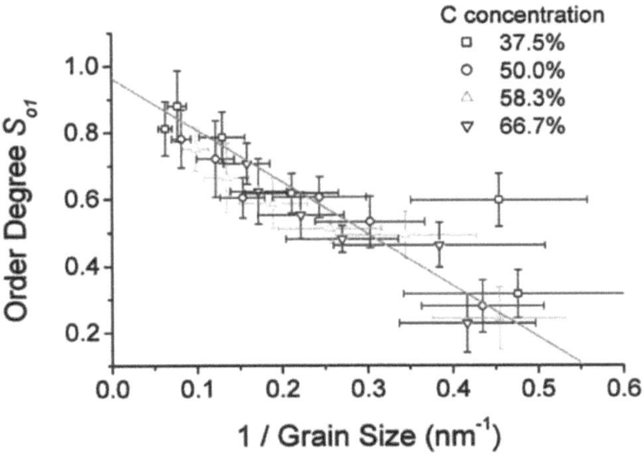

Figure 13. Degree of atomic ordering as a function of the inverse grain size for FePt/C samples with different C content.

2.5.1. MnAl nanoparticles

MnAl also goes through an $L1_0$ transformation with an anisotropy on the order of 10^7ergs/cm^3 when in the metastable τ phase. By depositing MnAl/Ag multilayers at relatively low temperatures the formation of ferromagnetic τ phase is observed as seen in the x-ray diffraction (XRD) patterns in Fig. 14.

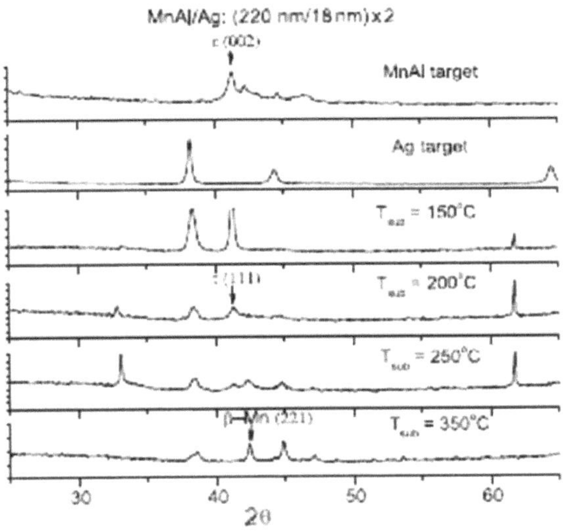

Figure 14. XRD patterns of MnAl/Ag multilayers deposited at different temperatures.

The multilayers were deposited at temperatures ranging from 150-350°C. The τ phase appears at 200 °C and ceases to form when deposited at 350°C [8]. The τ phase is the only ferromagnetic phase of MnAl and hence for the two deposition temperatures for which the phase is formed there is a development of coercivity. Reports in the literature indicate a coercivity approaching 2 kOe for thin films [9]. As seen in Fig. 15 the coercivity increases up to 3.7 kOe. The formation of MnAl nanoparticles would provide an answer to the problem of agglomeration of FePt nanoparticles due to the lower annealing temperatures required to transform the particles to the FCT.

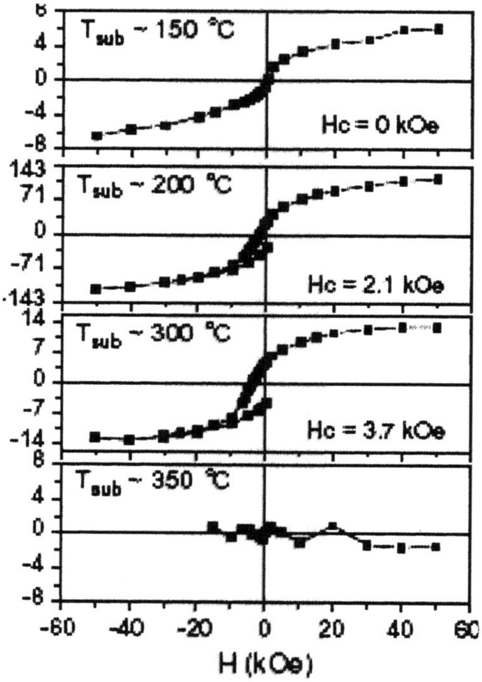

Figure 15. Hysteresis loops of MnAl/Ag films deposited at different temperatures.

2.5.2. *$SmCo_5$ nanoparticles*

With an anisotropy constant in excess of $1.4 \cdot 10^8$ ergs/cm^3 (14 MJ/m^3), $SmCo_5$ has the highest anisotropy constant of any material currently known. By utilizing the multilayer precursor technique it is possible to embed $SmCo_5$ nanoparticles in a matrix such as Cr and W [10]. "By varying the thickness of the $SmCo_5$ layer the coercivity increases up to 10 kOe as the layer thickness reduces to 5 nm as seen in Fig. 16.

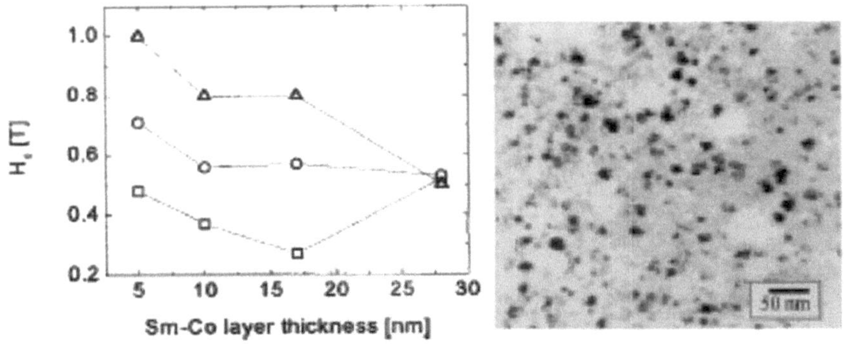

Figure 16. Coercivity as a function of Sm-Co layer thickness for three different annealing conditions, 600 °C for 60 min (squares); 700 °C for 20 min (circles); 800 °C for 5 min (triangles). The W layer thickness for each was 28 nm. The TEM image shows the $SmCo_5$ particles sample annealed at 700 °C for 5 min.

The TEM image in the figure shows particles less than 10 nm in size corresponding to a coercivity of 7 kOe. The values of coercivity here do not approach those expected for the anisotropy of pure $SmCo_5$ phase due to alloying with the tungsten. When multilayered with Cr the coercivity reaches 30 kOe for similar particle sizes [11]. These high anisotropy materials like FePt and $SmCo_5$ are very sensitive to composition as will be discussed in the next section.

2.6. Nanoparticles by the Cluster Gun

The multilayer precursor technique is susceptible to alloying effects in the magnetic nanoparticles. Depending on the miscibility of the matrix material and the magnetic nanoparticles and the fact that the formation of the nanoparticles requires post-synthesis annealing during which intermixing occurs. To avoid this it is important to form the nanoparticles prior to embedding them in the matrix. By utilizing an inert gas condensation technique similar to that described in chapter 8 high anisotropy nanoparticles have been embedded in a carbon matrix.

In the case of the FePt nanoparticle system the nanoparticles must be subjected to elevated temperature in order to undergo the $L1_0$ phase transformation. This is achieved by passing the particles through a furnace region as seen in Fig. 17, which consists of an assembly of halogen lamps surrounding a quartz tube situated above the particle formation region.

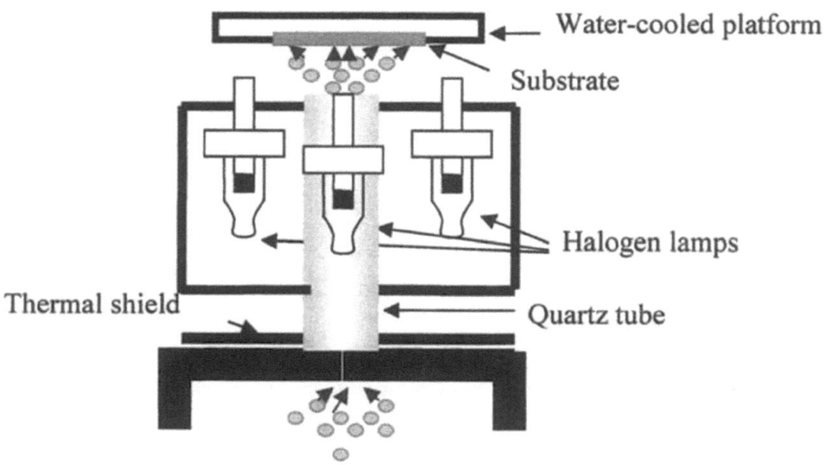

Figure 17. Schematic of a halogen lamp heater stage for an inert gas condensation particle deposition system.

Under the right conditions the particles pass through the quartz tube being transformed by the heat caused by the radiated light from the bulbs. Figure 18 shows TEM bright field images, SAD pattern and particle size distributions for samples with and without the lights on [12]. The diffraction pattern with the lights on includes the reflection from the superlattice while the unirradiated samples show an FCC pattern typical of the disordered FePt. Thus the heater stage has transformed the particles and no subsequent annealing is necessary. The bright field images of both samples are very similar indicative of the fact that the heater stage is not causing morphological changes beyond transformation. This is also evidence from the particle size distributions of the two samples given that the two distributions are centered on the same particle size within experimental limits designated by the standard deviation.

For the Sm-Co system the necessity of post synthesis annealing is not attributed to a phase transformation but rather the requirement of crystallinity. Hence $SmCo_5$ nanoparticles must also be heat treated prior to depositing them into a matrix material. With the correct composition and increased crystallinity the magnetic properties depend on the inter-particle spacing as mentioned above. By varying the amount of matrix the magnetic properties in these systems are tailored for potential applications as seen for the coercivity of Sm-Co particles embedded in a carbon matrix. Figure 19(a) and (b) show the coercivity as a function of matrix volume and the bright field image of the particles.

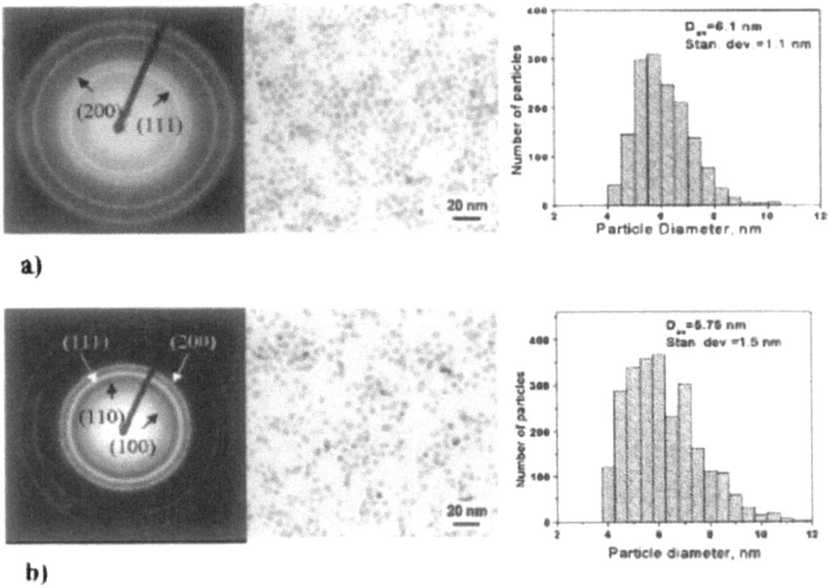

Figure 18. SAD pattern, Bright field image and particle size distribution of FePt nanoparticles with a) lights off and b) lights on.

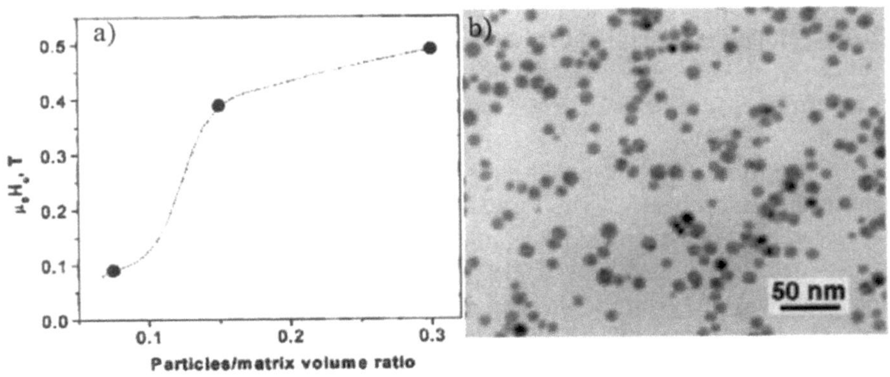

Figure 19. Plot of the coercivity *vs.* matrix volume percentage for Sm-Co nanoparticles (a) and bright field image of the nanoparticles (b).

As the distance between nanoparticles decrease there is an increase in the coercivity of the nanoparticles. As particles get closer the exchange interaction become stronger. The particles in the bright field image are less than 10 nm and well below the single domain particle size making them susceptible to thermal effects. The increase in coercivity with decreasing spacing appears to be improving the thermal stability as has been shown elsewhere [13].

3. FORMATION AND SUPERSTRUCTURAL DEVELOPMENT OF EPITAXIALLY GROWN FePt NANOPARTICLES

Epitaxial growth occurs when there is a slight mismatch in the lattice spacing between the substrate and a deposited material. This produces stress in the deposited material that leads to crystallographic alignment at the interface. As the thickness of the film increases there is a relaxation, to the bulk lattice parameter, of the deposited material as it continues to grow. This phenomenon has allowed for the development of semiconductor technologies and when applied to a magnetic material produces artificially textured films. In this section epitaxially grown FePt nanoparticles with perpendicular magnetic anisotropy embedded in nonmagnetic matrices M (M= Ag, C) have been fabricated by sputtering FePt/M multilayer films onto single crystal MgO [001] at temperatures above 300 °C. In the case of FePt there is a slight mismatch between the FePt (001) and the MgO (001). By depositing FePt onto single crystal substrates, at elevated temperatures, highly oriented $L1_0$ FePt thin films result [14]. By varying the thickness from a few angstroms to several nanometers an island structure is produced with island size and spacing dependent upon substrate temperature and deposition rate.

3.1. Fabrication of FePt/Ag(C) on MgO

(FePt/Ag) and (FePt/C) multilayers were fabricated by DC magnetron sputtering onto heated MgO [001] substrates using $Fe_{50}Pt_{50}$, Ag and C targets. The base pressure of the chamber was 2 x 10^{-7} Torr and a high purity Argon gas flow with a pressure of 5 mTorr was used during sputtering. The substrate was attached to a ceramic heater with adjustable temperature up to 650 °C. FePt /Ag thin films with different bilayer thickness (t_{FePt}=1 – 10 nm) were sputtered in order to study the effect of film thickness as well as the effect of Ag addition on the magnetic and structural properties of the films. A 20 nm carbon capping layer was sputtered on all the samples in order to protect samples from oxidation.

3.2. FePt/MgO: Effect of Film Thickness

On a heated substrate the deposited material forms islands as the increased surface mobility and surface tension cause the deposit to coalesce on the surface. With increased deposition the islands grow laterally until the islands become interconnected to form a continuous layer. For FePt the substrate temperature also transforms the sample into the FCT. The island sizes vary with the nominal thicknesses of the deposited layer. Figure 20 is a plot of the coercivity *vs.* layer thickness for FePt deposited on MgO (001) single crystal substrates at the two temperatures listed on the plot. The coercivity follows a functional form very similar to that predicted for particles in Fig. 2.

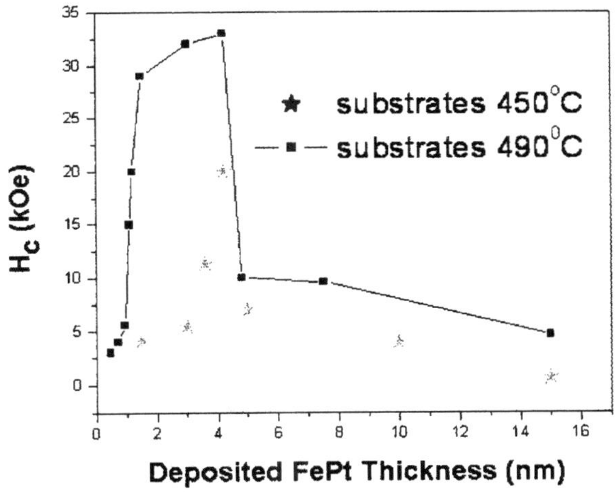

Figure 20. Coercivity vs. Layer thickness for FePt deposited on heated (450 and 490 °C) single crystal MgO (001) substrates.

Hence, the islands are behaving like particles. The increase in coercivity for each thickness due to temperature is attributed to a higher degree of chemical ordering.

Looking at the microstructure for samples above and below the maximum in coercivity, Fig. 21 shows that the FePt islands become interconnected above the coercivity maximum while below the maximum the islands are well separated. In the insets of Fig. 21(a) and (b) are the selected area diffraction (SAD) patterns for the samples. These indicate a single crystal FCT pattern with (001) orientation. Adjacent to the FePt diffraction spots are the (001) MgO single crystal spots indicating a slight mismatch in the lattice spacing of the two materials and a good epitaxial relationship between the two.

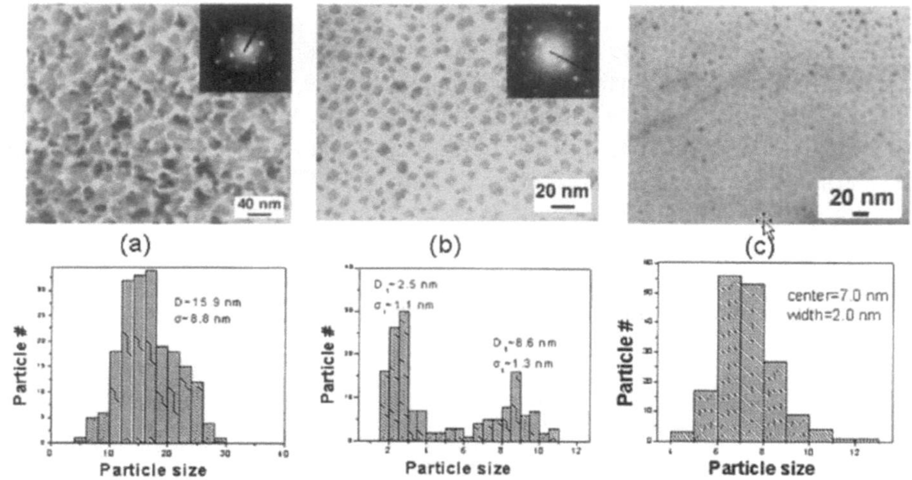

Figure 21. TEM images, SAD patterns, and particle size distribution for FePt/MgO samples with different thicknesses and deposition rates, deposited at 490 °C: (a) 7 nm at 0.15 nm/sec, (b) 1.5 nm at 0.15 nm/sec, (c) 0.96 nm at 0.12 nm/sec.

The histograms of the size distribution of the islands are below the respective bright field images. As the thickness of the deposited layer is decreased the island size distribution becomes bimodal with much smaller islands situated between larger ones. This bimodal distribution is detrimental to recording technologies since the two sizes have different coercivity. By reducing the deposition rate of the FePt the bimodal distribution disappears as seen in Fig. 21(c). Reducing the deposition rate reduces the rate of nucleation of new islands allowing for a more uniform growth.

The hysteresis loops of selected samples are shown in Fig. 22. For each of the thicknesses shown the loops are square with high coercivity when measured with the applied field perpendicular to the film plane. When magnetization curves are measured parallel to the plane of the film a straight line results which is characteristic of a hard axis loop. The SAD patterns in Fig. 21 showed a (001) texture which corresponds to the c axis of FePt along the direction perpendicular to the plane. Since the c axis is the easy axis of the FCT the loops in Fig. 22 confirm the perpendicular orientation seen in the electron diffraction pattern.

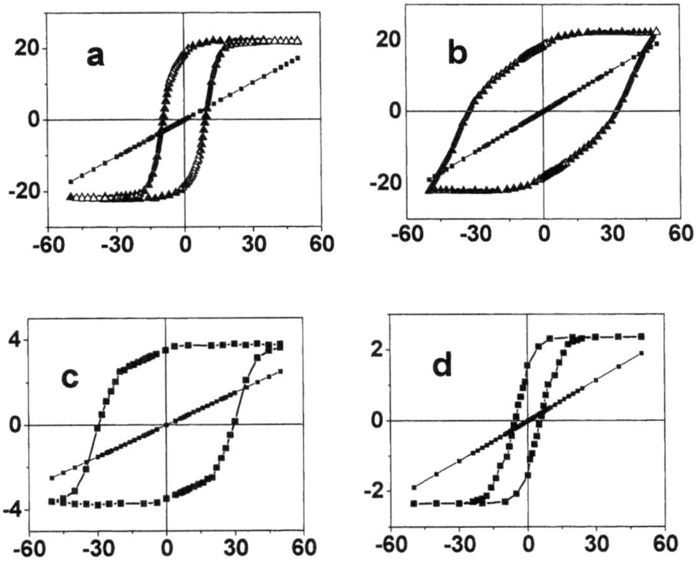

Figure 22. Magnetization curves for FePt films made at a substrate temperature of 490 °C with different film thickness: a) 7 nm, b) 4.2 nm, c) 1.5 nm and d) 1 nm. The figures show the magnetization in arbitrary units as a function of the field in kOe.

3.3. FePt/Ag Multilayer Thin Films on MgO

Silver has been reported to reduce the transformation temperature of FePt nanoparticles made by both sputtering and chemical synthesis [15, 16]. Furthermore, Ag, MgO and FePt have very similar lattice parameters (a_{Ag} = 0.40 nm) and therefore allows for a study of the transformation temperature without destroying the perpendicular texture. Here we describe the study of FePt/Ag multilayers deposited on MgO single crystal substrates at temperatures up to 400 °C. The X-ray diffraction patterns of the (FePt 2 nm/Ag 16 nm) pseudo-multilayers deposited on MgO [001] substrates at different temperatures are shown in Fig. 23. The superlattice (001) peak is present in all samples at temperatures above 280 °C, indicating the formation of $L1_0$ phase.

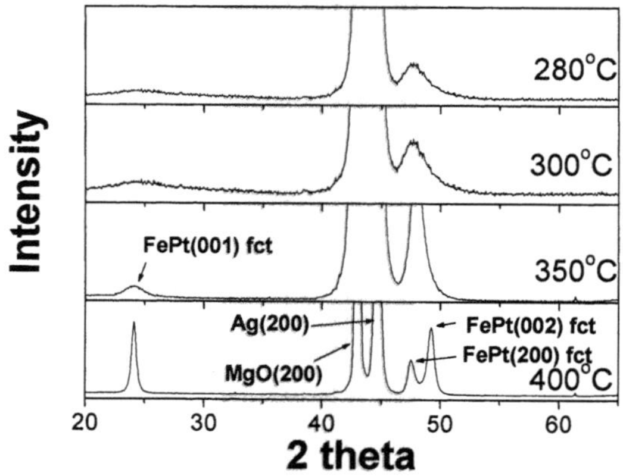

Figure 23. X-ray diffraction patterns of (FePt 2nm/Ag 16nm) ×5 films deposited at different temperatures.

The strong (001) and (002) peaks and the absence of other peaks such as (111) and (110) indicate the presence of a strong c-axis texture. X-ray data of similar pseudo-multilayers with carbon matrix, (FePt 2 nm/C 16 nm) ×5 deposited at different substrate temperatures show no evidence of the formation of the $L1_0$ phase when the deposition temperature is below 350 °C. This XRD data suggests that Ag helps to lower the transformation temperature. More interestingly, FePt/Ag films deposited at 400 °C showed both the (200) and (002) peaks, which has not been observed in films without Ag, indicating a possible competition between in-plane and out-of-plane texture [17, 18]. By increasing the deposition temperature, the (002) peak shifts to a higher angle indicating the decrease of lattice constant along the c direction and the increase of the degree of ordering. The FePt particle size calculated from the XRD data using Scherrer's formula showed that the particles grow rapidly with increased deposition temperature (the estimated particle size change from 2 to 13 nm as the deposition temperature increased from 280 to 400 °C). The grain growth mechanism is similar to that in post-synthesis annealed multilayer precursors [19] but it occurs at much lower temperatures.

Figure 24 shows hysteresis loops of (FePt 2 nm/Ag 16 nm)×5 pseudo-multilayers deposited at different temperatures, measured perpendicular to the film plane, and those of (FePt 2 nm/ Ag t nm, t = 1, 2, 4, 8)×5 deposited at 400 °C. The coercivity of (FePt 2 nm/Ag 16 nm) ×5 films is increased with increasing deposition temperature. At 300 °C the coercivity has reached the value of 1.5 kOe, which is consistent with the onset of chemical ordering as shown by the XRD data. Also, the coercivity of (FePt 2 nm/ Ag t nm, t =

1, 2, 4, 8) ×5 films is decreased with decreasing silver thickness due to a weaker perpendicular texture and a smaller particle size.

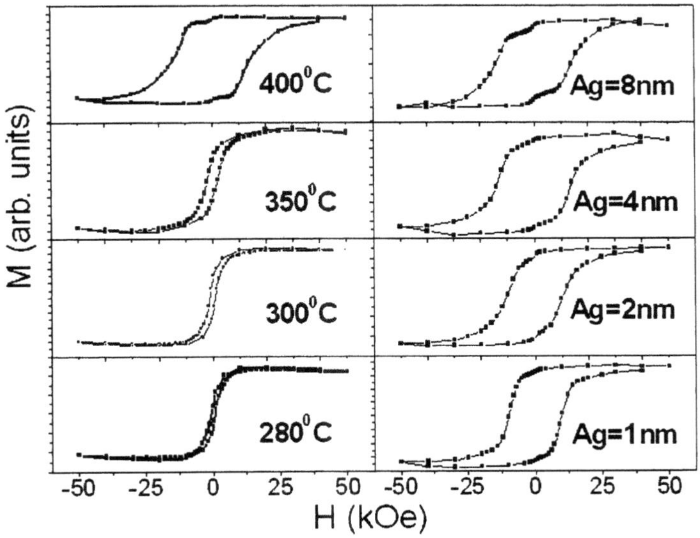

Figure 24. Hysteresis loops of (FePt 2 nm/Ag 16 nm) × 5 films deposited at different temperatures (left column) and (FePt 2 nm/Ag) × 5 films deposited at 400 °C (right column).

4. CONCLUSION

The phenomena associated with the field of magnetic nanoparticles provide the experimentalist a vast and rich field of research. The properties of nanoparticles are very sensitive to the composition, structure, size and immediate surroundings as described above. The multilayer precursor and epitaxial growth described above provide the archetypical structures and the inert gas condensation technique is expected to garner the next generation of novel nanoparticle systems. As technologies require smaller and smaller magnetic features, the field of magnetic nanoparticles is well ahead of the game, with stable and highly anisotropic ferromagnetic alloys. Down to a few nanometers, they provide the building blocks for artificially engineered structures. Utilizing the fabrication techniques discussed in this chapter provides not only a way to improve and create technologies but also serves as a platform for probing the physics of magnetic nanoparticles.

Acknowledgements

The authors would like to thank the National Science Foundation for its support of the research presented here under grant DMR-9972035.

References

[1] S. Chikazumi, "Physics of Ferromagnetism", Oxford University Press, (1999) or in "Magnetic Properties of Fine Particles", J.L. Dorman and D. Fiorani, North-Holland 1992.

[2] B. D. Cullity, "Introduction to Magnetic Materials", Addison-Wesley Publishing 1972.

[3] E. C Stoner and E. P. Wohlfarth, Trans. Roy. Soc. (London) A **240**, 599 (1948).

[4] C. Verdes, S. M. Thompson, R. W. Chantrell, and Al. Stancu, Phys. Rev. B **65** (17), 174417 (2002).

[5] P. E. Kelly, K. O'Grady, P. I. Mayo, and R. W. Chantrell, IEEE Trans. Mag. **25**, 449 (1989).

[6] J. A Christodoulides, M. J. Bonder, Y. Huang, Y. Zhang, S. Stoyanev, G. C. Hadjipanayis, A. Simopoulos, and D. Weller, Phys. Rev. B **68**, 054428 (2003).

[7] R. Cammarano, G. McCormick, and R. Street, J. Appl. Phys. **75** (10), 5483 (1994).

[8] Z. C. Yan, Y. Huang, Y. Zhang, G. C. Hadjipanayis, W. Soffa, and D. Weller, Scripta Mater. (accepted, 2005).

[9] see for instance, G. A. Fischer, M. L. Rudee, J. Magn. Magn. Mater. **213** (3), 335 (200).

[10] P. Farber, H. Okumura, Y. Zhang, J. Christodoulides, and G. C. Hadjipanayis, J. Appl. Phys. **90** (12), 6397 (2001).

[11] Z. C. Yan, PhD. Thesis

[12] S. Stoyanov, V. Skumryev, Y. Zhang, Y. Huang, G. Hadjipanayis, and D. Weller, J. Appl. Phys. **93** (10), 7592 (2003).

[13] S. A. Majetich, K. M. Chowdry, and E. M. Kirkpatrick, IEEE Trans. Mag. **34**, 985 (1998).

[14] Y. Zhang, J. Wan, V. Skumryev, S. Stoyanov, Y. Huang, G. C. Hadjipanayis, and D. Weller, Appl. Phys. Lett. **85** (22), 5343 (2004).

[15] S. Kang, J. W. Harrell, and D. E. Nikles, NanoLetters **2** (10), 1033-1036 (2002)

[16] Z. L. Zhao, J. S. Chen, J. Ding, K. Inaba, and J. P. Wang, J. Magn. Magn. Mater. **282**, 105 (2004).

[17] Z. Zhang, K. Kang, and T. Suzuki, J. Appl. Phys. **93**, 7163 (2002).

[18] J. S. Chen, B. C. Lim, and J.P. Wang, Appl. Phys. Lett. **81**, 1848 (2003).

[19] J. A. Christodoulides, Y. Huang, Y. Zhang, G. C. Hadjipanayis, I. Panagiotopoulos, and D. Niarchos, J. Appl. Phys. **87**, 6938 (2000).

Chapter 8

CLUSTER-ASSEMBLED NANOCOMPOSITES

Y. F. Xu, M. L. Yan and D. J. Sellmyer
Center for Materials Research and Analysis and
Department of Physics and Astronomy
University of Nebraska
Lincoln, NE 68588, USA

Abstract This chapter focuses on a gas-aggregation technique to prepare magnetic nanoclusters with controllable cluster sizes and size distributions. The review includes current research on nanoclusters, such as uncoated and oxide-coated Fe, Co and Fe clusters and clusters made from alloys, but special emphasis is on highly anisotropic $L1_0$-ordered FePt clusters, which are of potential interest for magnetic recording with ultrahigh areal densities of more than 1 Tera bit/in^2. In particular, we discuss magnetic and structural properties of FePt nanoclusters and thin films. Another approach discussed in the chapter is to create cluster nanocomposites by multilayering with post-deposition annealing. The advantage of this method is that the clusters can be oriented along a desired easy axis. One example is $L1_0$-(001) oriented FePt nanocomposite films with a FePt cluster size of about 5 nm.

1. INTRODUCTION

Nanostructured materials, assembled using nanoparticles or nanoclusters as building blocks, can be synthesized to have a wide variety of controlled optical, electronic, magnetic, mechanical and chemical properties with potential or realized technological applications. Simple free-metal clusters have been thoroughly investigated as reviewed by Haberland [1], W. Ekardt [2], and Meiwes-Broer [3]. In this chapter we focus on magnetic clusters and magnetic cluster-assembled nanocomposites.

In the last decade, magnetic nanoclusters with diameters of 1-10 nm and containing tens-to-thousands of atoms have been of great interest because their properties are critically dependent on size [4-10]. The total magnetic moment in free Fe, Co, and Ni clusters was determined as a function of size by measuring their Stern-Gerlach deflections [4, 5, 9]. For example, at a low temperature of 120 K, for small Fe clusters ($25 \leq N \leq 130$ atoms), the moment was found to be 3 μ_B per atom, which is considerably higher than the bulk value (2.2 μ_B); for larger Fe clusters, the moment decreased with increasing cluster size and approached the bulk value at cluster sizes of about 500 atoms [5]. This size dependence may enable cluster-based materials to be produced with novel magnetic or electronic properties that can be used in technologies such as extremely high-density data storage, spin electronics, memory devices, and high-performance magnetic materials.

Recently there has been considerable interest in $L1_0$ ordered CoPt, FePt nanoparticles with the high magnetocrystalline anisotropy for extremely high density magnetic recording (EHDR) media [11-15]. The drive for the EHDR, i.e. from 100 Gb/in^2 to 1 Tb/in^2, imposes the need for grain sizes below 8 nm with extremely uniform grain size to maintain an optimum signal-to-noise ratio [16]. However, such small particles will cause the thermal instability of magnetic recording because of the superparamagnetic limitation. The equiatomic $L1_0$ ordered phase CoPt and FePt alloys have high anisotropy constant of ~5 x 10^7 erg/cm^3 and ~ 7 x 10^7 respectively [17-19], which are of crucial importance for EHDR media with a small grain size below 8 nm, because high magnetocrystalline anisotropy is needed to create a barrier to thermally activated switching of the magnetization. Most CoPt and FePt nanoparticles were prepared with normal magnetron sputtering or electron beam evaporation techniques [11-15]. However, Sun *et al.* prepared FePt nanoparticles with very uniform particle size and narrow size distribution by using the simultaneous reduction of platinum acetylacetonate and thermal decomposition of iron pentacarbonyl [20, 21], as described in chapter 9 by S. Sun.

In this chapter, we will introduce a gas-aggregation technique, in which magnetron sputtering is employed in the source [22], for the preparation of magnetic nanoclusters and cluster-assembled nanocomposites. This cluster-deposition technique can produce a large range of mean cluster sizes, from 3 nm to 12 nm, with high deposition rate (can be as high as 5 Å/s), and with the cluster size and size distribution controllable. Thus it is very suitable for preparation of magnetic clusters as building blocks for nanostructered materials. A major advantage of this type of system is that the clusters have much smaller size dispersion than grains in a typical vapor-deposition system. We will first introduce the cluster sources and the gas aggregation technique, then will review elemental and alloy clusters prepared by this technique, with special emphasis on the FePt, CoPt clusters and FePt cluster-based composites. Finally, multilayers that are processed by post-deposition annealing will be described as a method for preparing oriented FePt nanocomposites.

2. EXPERIMENT FOR CLUSTER PREPARATION

2.1. Cluster Source and Gas-Aggregation Technique

Formation of metal clusters by gas aggregation, in which metal atoms are evaporated or sputtered into a cooled inert gas flow at relatively high pressure, has been well established in last decade. By repeated collisions with the carrier gas, the supersaturated metal vapor nucleates and forms clusters. The mechanism of cluster formation can be explained with homogeneous and heterogeneous nucleation theories. The gas aggregation methods have been applied extensively to produce small clusters of metals such as zinc, copper, silver etc. [23-26]. In some cases this method was used in combination with a mass filter such as a quadruple or a time-of-flight spectrometer [27, 28]. The metal vapor for cluster source can be produced by either thermal evaporation [23-28] or sputter discharge [22, 29].

The advantages of a sputtering-based cluster source are: (1) it can produce a very large range of mean cluster sizes from 200 to 15000 atoms per cluster, (2) it has a high degree of ionization from 20% to 50%, depending on the target materials, and (3) a wide variety of elements and alloys can be used as source materials. Usually the deposition rate is about 3 Å/s for the total flux of the cluster beam. For charged clusters the deposition rate is about 1 Å/s. Currently sputtering-based gas aggregation sources are widely used in many laboratories and are commercially available from Oxford Applied Research or Mantis Deposition [30].

Figure 1 shows a schematic drawing of a cluster-deposition system with a sputtering-based gas-aggregation source [29]. It consists of four chamber: (1) the chamber for cluster formation containing a LN_2 cooled aggregation tube with a magnetron sputter discharge, (2) the deposition chamber with a substrate holder, (3) a sample load-lock chamber (not shown) for rapid substrate change, and (4) a chamber (not shown) containing the time-of-flight (TOF) mass spectrometer. Briefly, the clusters are generated in the aggregation tube, they go through two apertures B1 and B2, then are deposited onto the substrate S.

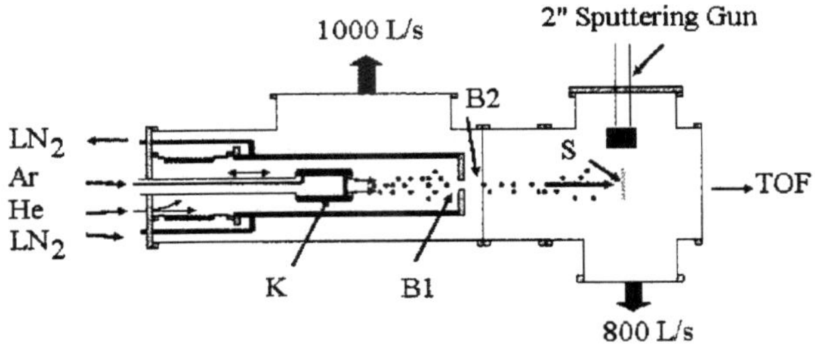

Figure 1. Schematic drawing of a cluster deposition system.

The cluster size can be adjusted by varying the source parameters including gas-aggregation tube temperature, gas pressure, gases Ar to He flux ratio, magnetron power and the condensation length in which the clusters aggregate. The most critical variables determining the cluster size are the magnetron power and the residence time of the clusters in the aggregation region, i.e. the condensation length. Figure 2 shows a photograph of a sputtering based gas-aggregation cluster-deposition system at the University of Nebraska. The system includes a sputtering gas-aggregation source in which clusters are formed in a 0.5-1 Torr inert gas, with diameters ranging from 3 to 12 nm. The clusters impinge upon a substrate alone or in combination with atoms from one or two other sputtering guns. The e-beam evaporation chamber can be used to provide a thin coating (e.g. 1-3 monolayers) of oxide or other molecules. The other two sputtering guns are used to form a matrix for the magnetic clusters and/or underlayer or coating layers.

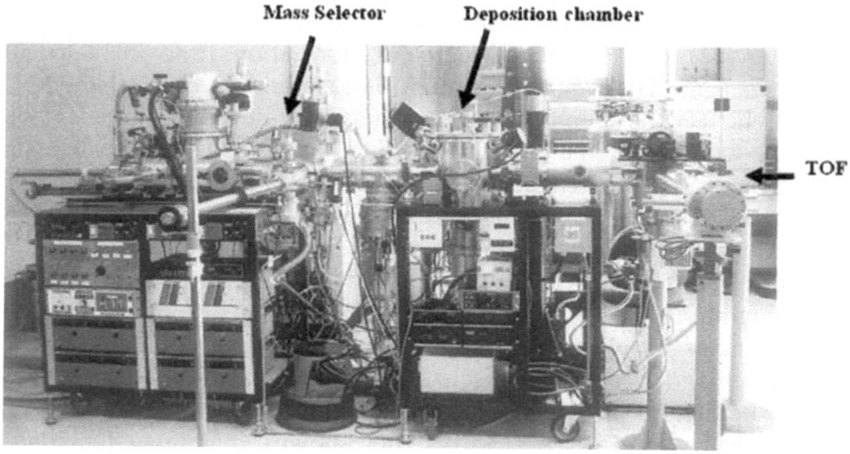

Figure 2. Cluster-deposition system at the University of Nebraska.

2.2. Multilayer and Post-Annealing

The non-epitaxial growth, orientated $L1_0$ nanocomposite films were prepared by multiplayer deposition plus post annealing. Figure 3 shows the schematic drawing of the system for preparing FePt:X(= Ag, C, B_2O_3, etc.) oriented $L1_0$ films. Fe/Pt/X layers were deposited individually on Si wafer or glass substrates by magnetron sputtering at room temperature and then annealed in vacuum for 5 to 30 minutes at various temperatures. The as-deposited multilayers were converted by phase separation into nanocomposite films with $L1_0$-phase FePt grains embedded in a nonmagnetic matrix.

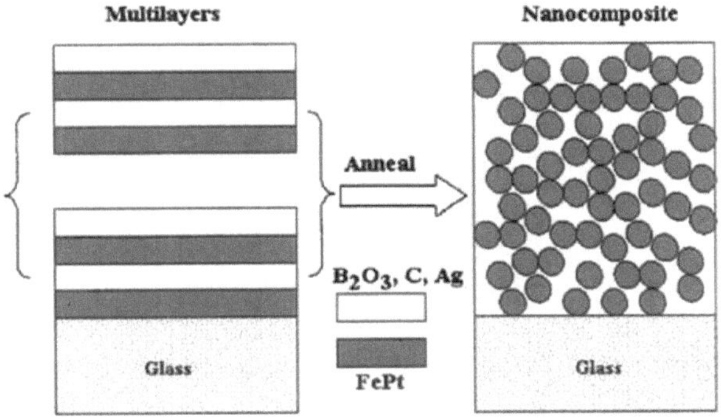

Figure 3. Preparation method of non-epitaxial growth $L1_0$-phase FePt:X nanocomposite films.

3. ELEMENTAL AND ALLOY CLUSTERS

3.1. Fe, Co and Ni Clusters

Fe clusters were prepared by a gas-aggregation technique, in which metal was evaporated from a high temperature crucible to form the clusters [31]. Figure 4 is a typical STM image of Fe clusters on the Si(111) surface. In the STM images the clusters appear as isolated topographic features that are randomly distributed across the surface. The Fe cluster size fitted with a lognormal distribution, agreed with theoretical prediction for vapor growth process [32]. The effect of exposure of the Fe clusters to the atmosphere results in a significant increase in cluster size, which is consistent with total oxidation of the clusters. Magnetic circular dichroism study shows that Fe clusters with size ranging from 1 to 5 nm at low coverage have an orbital magnetic moment about twice that of bulk Fe [33].

The detailed magnetic behavior of Fe clusters has been studied for the films with Fe clusters embedded in Ag [34]. In UHV conditions, preformed Fe clusters with a mean diameter of 3 nm from a gas-aggregation source were deposited in conjunction with atomic Ag vapor. In such films clusters can come into direct contact and interact via exchange. Films containing Fe cluster volume fraction from < 1 % (isolated clusters) to 100% (pure clusters with no matrix) have been studied at temperatures ranging from 2 to 300 K by magnetometry and field-cooled (FC)/zero-field-cooled (ZFC) measurements.

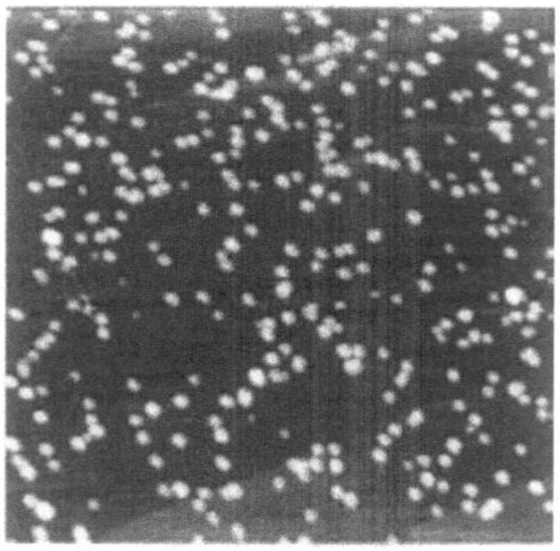

Figure 4. STM image of Fe clusters on the Si (111) surface. The image size is 200×200 nm^2, after [31].

The results are summarized in the magnetic phase diagram shown in Fig. 5. At very low cluster volume fractions (~1 %) and well above the cluster blocking temperature (~ 5 K), the films fulfill all the conditions for ideal superparamagnetism including an *H//T* scaling of the magnetization curves. At 2 K, the films demonstrate a single-particle blocking and a remanence close to 0.5 M_s indicating that the particles have a uniaxial anisotropy with a random distribution of easy axis. The anisotropy constant is 2.4×10^5 J/m^3. At volume fractions below the percolation threshold, the high-temperature behavior can be described by interacting superparamagnetic aggregates of clusters with their moments aligned by exchange and a dipolar interaction between the aggregates. At low temperature below a collective freezing of the moments the aggregates each have a uniaxial anisotropy axis in a random direction. At high volume fraction (> 50%), above 50 K the magnetic configuration is a correlated superspin glass.

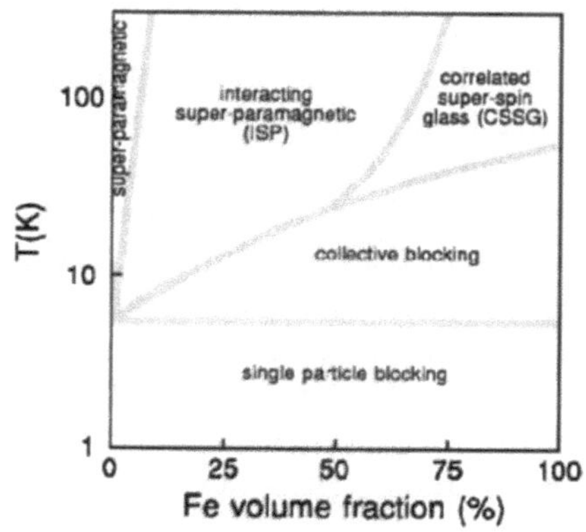

Figure 5. Magnetic phase diagram of Ag-capped Fe clusters (3 nm) in Ag matrix, after [34].

Qiang *et al.* studied Co clusters of a variable mean size of about 300-9000 atoms. These were produced by combining a magnetron sputtering source with a gas-aggregation tube at liquid nitrogen temperature [35]. The clusters land softly on the surface of a Si (100) substrate at room temperature simultaneously along with an atomic beam of Cu or SiO_2, which served as matrix, from a normal magnetron-sputtering gun. Figure 6(a) shows a high-resolution TEM image of a Co cluster (9000 atoms, d=6.2 nm) at 10 % concentration in Cu. It can be seen that there is some interface diffusion around the Co cluster. Figure 6(b) is a nanodiffraction from the Co cluster that shows a typical *fcc* structure. Magnetization per unit volume of Co (M_s) increases as one goes from (300-1000)-atom clusters to (4500-9000)-atom cluster. The 9000-atom clusters have higher M_s in an insulating matrix (SiO_2) than in a metallic matrix (Cu). Also, the M_s decreases with increasing Co concentration for all cluster size, and is always lower than bulk Co value.

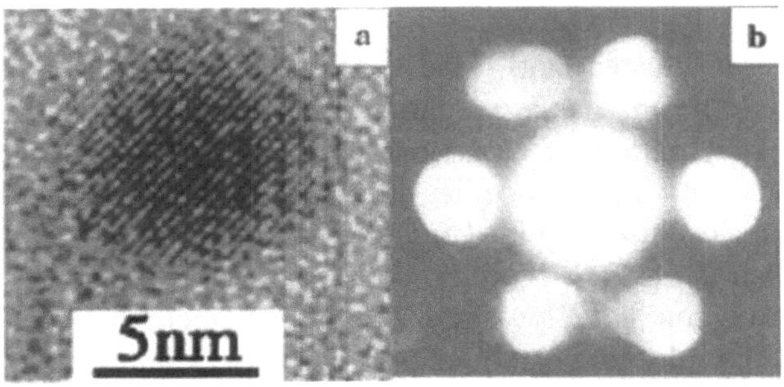

Figure 6. Co cluster having a diameter of 6.2 nm: (a) HRTEM image and (b) nanodiffraction pattern with a [110] zone axis.

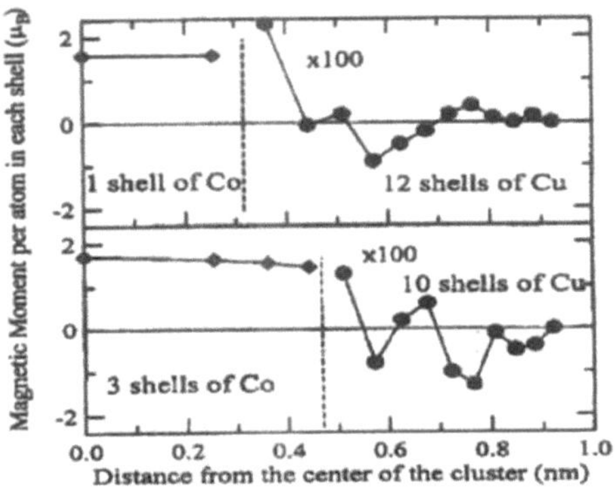

Figure 7. Calculated magnetic moments per Co (diamond) and Cu atoms (circle) as a function of the distance for one- and three-shell Co clusters.

First-principle self-consistent electronic-structure calculations have been carried out to analyze the observed magnetization of Co clusters as a function of their size and concentration. These calculations along with a plausible cluster-matrix interface model give magnetization in good agreement with the data for low Co concentration. The moment per atom as a function of the distance from the center of the cluster is shown in Fig. 7 for clusters with 1 and 3 Co shells surrounding the central Co atom. All finite-size clusters have Co magnetic moment comparable to the bulk Co value. The neighboring Cu shells develop very small oscillatory magnetic moments making negligible contribution to the overall magnetization [35].

Formation of Ni clusters by gas aggregation technique was reported by Hihara *et al.* [36]. The cluster growth models were discussed but there are no magnetic data for Ni clusters yet.

3.2. Oxide Coated Fe, Co Clusters

CoO-coated Co cluster and oxide-coated Fe cluster assemblies were prepared by a plasma-gas-aggregation cluster-beam-deposition technique [37-39]. For preparation of CoO-coated Co cluster assembly, oxygen gas was introduced through a nozzle near the skimmer into the deposition chamber. The Co clusters with CoO shells were formed before deposition onto the substrate [37]. Figure 8 shows a TEM image of the clusters produced at oxygen gas flow rate $R(O_2)$ = 1 sccm. Clusters are almost monodispersed, with the mean diameter of about 13 nm. Electron diffraction pattern indicated the coexistence of Co and CoO phases. The cluster assemblies were formed on a polyimide film with a thickness of about 100 nm.

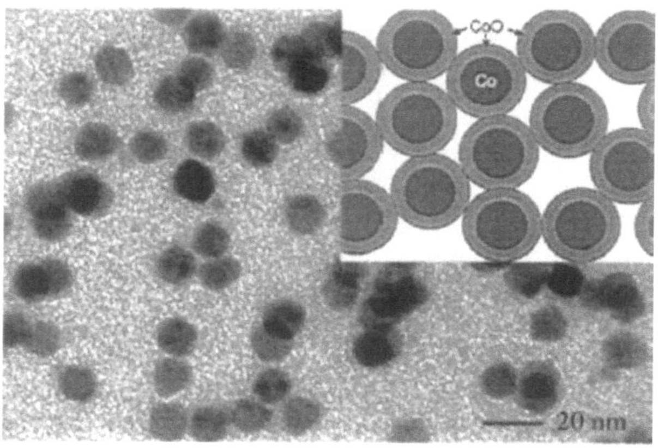

Figure 8. TEM image of CoO-coated Co clusters. The inset is a schematic drawing of the assembly of Co cores with CoO shells, after [37].

In order to examine the exchange coupling arising from the presence of the CoO phase, the hysteresis loops of CoO-coated Co cluster assemblies were measured at a temperature of 5K for both zero-field-cooled (ZFC) and field-cooled (FC) samples. The large loop shift against the direction of the cooling field is detected in the FC sample, in contrast to the symmetric feature of the ZFC sample [37]. This conforms the presence of the unidirectional exchange anisotropy due to the strong exchange coupling between the ferromagnetic Co core and the antiferromagnetic CoO shell.

The electrical conductivity σ, and tunnel-type magnetoresistance (TMR) were measured using a conventional four-probe method. For oxygen gas flow rate $R(O_2) < 0.24$ sccm, the resistivity revealed a minimum and showed a "ln *T*" dependence at low temperature, probably due to the weak localization of conduction electrons owing to the presence of thin oxide shell covering Co cores. A small negative MR was observed in this regime. For the $R(O_2) > 0.3$ sccm, tunnel-type temperature dependence of σ in the form of "ln σ *vs.* $1/T$" was observed between the temperatures 7 and 80 K. This differs from the well-known temperature dependence of "ln σ *vs.* $1/T^{1/2}$" for disordered granular materials. The MR ratio is negative and its absolute value increases sharply with decreasing temperature below 25 K: from 3.5 % at 25 K to 20.5 % at 4.2 K. This increase, by a factor of 6, is much larger than those observed for conventional metal-insulator granular systems. These results are ascribed to a prominent co-tunneling effect in the Coulomb blockade regime, arising from the uniform Co core size and CoO shell thickness in the present monodispersed cluster assemblies.

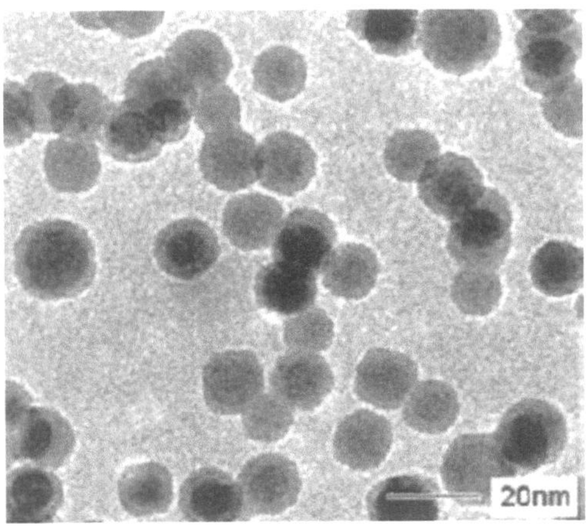

Figure 9. TEM image of the oxide-coated Fe clusters with mean diameter of 13 nm, after [39].

The oxide-coated Fe cluster assembly was prepared in a similar way as the above CoO-coated Co clusters [39]. Figure 9 shows a TEM micrograph of the oxide-coated Fe clusters with a mean diameter of 13 nm, prepared at oxygen gas flow rate of 3 sccm. Almost all clusters are characterized with a strong contrast in their "core" with a uniform grey contrast in their "shell". Electron diffraction pattern revealed that α-Fe phase, Fe_3O_4 or γ-Fe_2O_3 phase are coexisting, and confirmed the core-shell structure. Here the diffraction ring for Fe_3O_4 and γ-Fe_2O_3 overlapped, so it is not possible to distinguish them by the electron diffraction.

The coercivity and exchange bias of the oxide-coated Fe clusters strongly depend on the cluster size, temperature and the oxygen gas flow rate, i.e., the surface oxidization degree of the clusters. For the cluster size of 9 nm, the coercivity at 5 K (in both ZFC and FC cases) and the loop shift increase with the increasing oxygen gas flow rate; at room temperature, however, the coercivity becomes smaller with introduced O_2 gas. The H_c versus $T^{1/2}$ fitting gives the effective anisotropy constant K_{eff} values of 3.09×10^6, 4.29×10^6 and 5.63×10^6 erg/cm^3 for oxygen gas flow rate of 0, 1 and 3 sccm, respectively. The onset of loop shift and bifurcation between H_c(ZFC) and H_c(FC) below about $T = 50$ K can be ascribed to an exchange interaction between the ferromagnetic Fe core and the freezing layers of disordered surface spins of the oxide shell crystallites, which is also the predominant origin of an enhancement of the magnetic anisotropy at low temperatures.

3.3. Alloy Cluster Formation in Co-Al, Co-Pt Systems

From the viewpoint of application, formation of stable alloy clusters with intermetallic compound structure is highly desired. So far the formation of ordered intermetallic alloy clusters is quite limited. Although the sputtering source can generate almost any kind of metal vapors which makes the gas-aggregation technique very promising, the synthesis of alloy clusters that possess an ordered structure is quite challenging. Konno *et al.* studied the alloy cluster formation in CoAl and CoPt systems [40, 41]. CoAl and CoPt were chosen because of the well-known stability of the ordered B2 (CsCl-type) structure for the former and $L1_0$ (AuCu-type) structure for the latter near the composition range of Co:M = 1:1 (M = Al or Pt). Furthermore, the free energy change, i.e. the heat of formation for CoAl phase (B2, -60 kJ/mol) is much larger than that of CoPt ($L1_0$, -14 kJ/mol). They found that B2-CoAl clusters were formed for a composition range wider than that predicted by the Co-Al phase diagram, and that high-temperature *fcc*-CoPt clusters were formed in the CoPt system.

Compared with the heat of formation of $L1_0$ CoPt, another important $L1_0$ phase FePt has a larger value (-25 kJ/mol) [42]. The ordering temperature for the latter is about 100 °C lower than that of the former. This result may provide a hint: large heat of formation may be favorable for the formation of ordered alloy clusters in the as-deposited state.

4. $L1_0$ FePt, CoPt NANOCLUSTERS

4.1. Size Distribution and Nanostructure of Clusters

For free-clusters, the cluster size distribution can be measured by the time-of-flight mass spectrometer; for cluster films deposited on substrate by the cluster beam, the measurement of size distribution and observation of nanostructure are mostly done using transmission electron microscopy (TEM). In this section we will focus on the latter and pay special attention to FePt, CoPt clusters which have high anisotropy $L1_0$ phase after annealing [43-45]. For the TEM observations, FePt, CoPt nanoclusters, produced in a gas-aggregation chamber, in which high pressure Ar gas (~ 0.5-1Torr) was applied and cooled by LN_2, were directly deposited onto carbon-coated films supported by Cu grids.

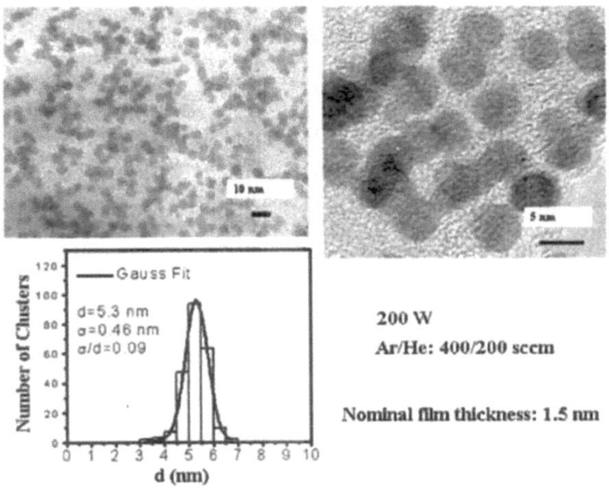

Figure 10. TEM images of FePt clusters with cluster size distribution. Right: larger magnification of the clusters.

The cluster size was controlled in a range from 3 nm to 6 nm by adjusting the preparation parameters such as sputtering power, Ar-He gases flow and gas pressure etc. Uniform cluster size distribution with standard deviation $\sigma/d < 0.1$ is obtained. The size distribution can be further improved if desired by applying a mass selector. Figure 10 shows the TEM image of as-deposited FePt nanoclusters. The corresponding cluster size distribution counted by more than 200 clusters from lower magnification TEM pictures is also included. Spherical clusters with an average size of 5.3 nm fitted with Gaussian distribution (with standard deviation σ = 0.46 nm, σ/d = 0.09) were obtained with sputtering power of 200 W and Ar/He gas flow of 400/200 sccm.

High resolution TEM image reveals some faceting of the FePt clusters as shown in Fig. 11. An enlarged FePt cluster image shows the lattice fringes, indicating the single-crystal structure.

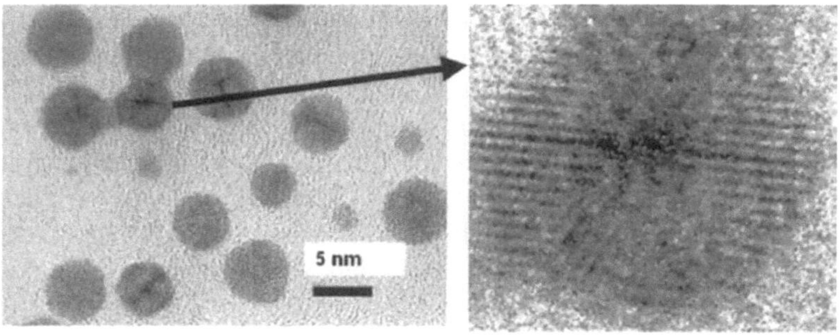

Figure 11. HRTEM images of FePt clusters with facet feature. Right: an enlarged cluster showing lattice fringe.

Similar results were also obtained for CoPt nanoclusters and shown in Fig. 12 is the high resolution TEM image for CoPt clusters with average size of 4.9 nm (σ = 0.46 nm, σ/d = 0.09), prepared with sputtering power of 160 W.

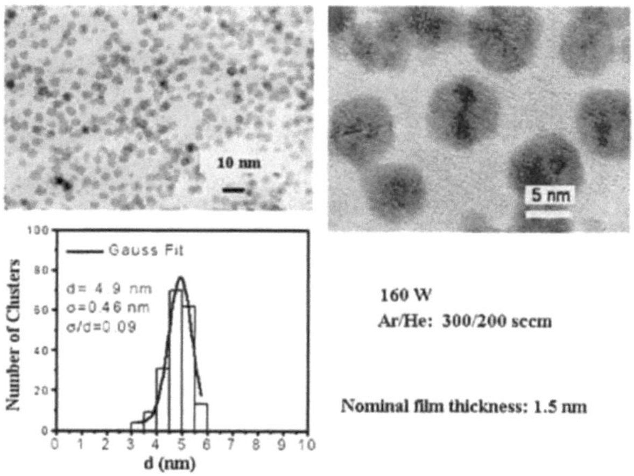

Figure 12. TEM images of CoPt clusters with cluster size distribution. Right: larger magnification of the clusters.

Figure 13 shows the sputtering power dependence of cluster size. It can be seen that FePt, CoPt cluster size has a nearly linear relationship with the sputtering power of the cluster source. It is noted that the cluster size follows a Gaussian distribution rather than a lognormal one as observed in Fe clusters [31]. This may be related to the cluster formation mechanism involved in the present method, which is different from the mechanisms for lognormal distribution. A similar Gaussian distribution was also reported in CoAl clusters [40].

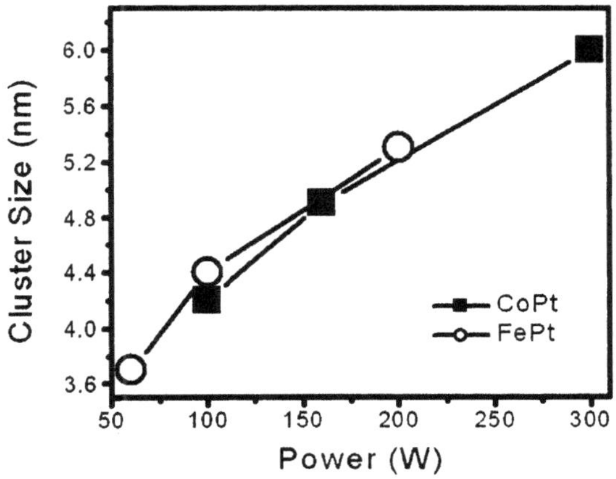

Figure 13. FePt, CoPt cluster size versus sputtering power.

In order to obtain $L1_0$ ordered FePt clusters, high temperature post-deposition annealing is usually required. However, post-deposition annealing leads to subsequent crystallite growth and concurrent agglomeration of the clusters on the substrate. Rellinghaus *et al.* employed a sintering oven between the cluster source and deposition chamber, allowing preparation and thermal sintering of FePt nanoclusters in the gas-phase prior to their deposition [46, 47]. The sintering temperature of the oven can be varied in the range from 293 to 1273 K; the residence time of the clusters within the oven is between 0.05 s and 0.9 s, depending on the gas flow rate in the oven. High resolution TEM observation reveals that the spherical shaped clusters are of icosahedral symmetry when prepared at gas pressure of 0.4 Torr and sintering temperature up to 1273 K, and the relative amount of icosahedral clusters increases with increasing sintering temperature as $T_s \geq 800$ K. Figure 14 shows a HRTEM image of an FePt icosahedron, prepared at gas pressure of about 0.4 Torr and sintering temperature of 1073 K, with its 3-fold symmetry axis oriented parallel to the electron beam. For FePt clusters prepared at a pressure > 0.4 Torr and high sintering temperature (1273 K), single-crystalline clusters with the $L1_0$-ordered phase were observed.

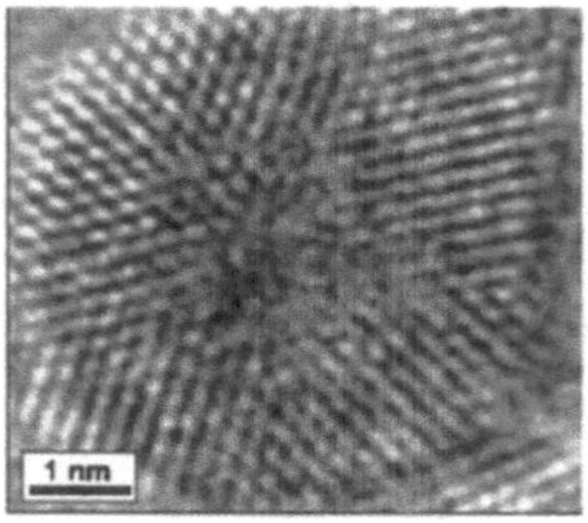

Figure 14. HRTEM image of an FePt cluster with icosahedral symmetry, after [47].

4.2. Magnetic Properties of FePt Clusters

Depending on the gas-pressure during preparation, the structure of the unannealed FePt clusters is icosahedral (~ 0.4 Torr) or polycrystalline *fcc* phase (>0.75 Torr) [47]. Temperature-dependent magnetization measurements revealed that these clusters are superparamagnetic at room temperature. Clusters of icosahedral structure withstand even the highest annealing or sintering temperature (1273 K) and are still superparamagnetic at room temperature. Polycrystalline fcc FePt nanoclusters, on the other hand, are transformed to the chemically ordered $L1_0$ phase by annealing at elevated temperature. Both the blocking temperature T_b and the coercivity increased with the increase of T_s. Figure 15 shows the hysteresis loops of polycrystalline FePt nanoclusters prepared at gas pressure of 1.1 Torr and sintering temperature of 1273 K. The magnetization loops measured at temperatures between 5 and 293 K. The coercivity increases with the decrease of temperature; a coercivity of 2.3 kOe at temperature of 5 K indicated the high anisotropy $L1_0$ structure, although the degree of $L1_0$ ordering is not very high.

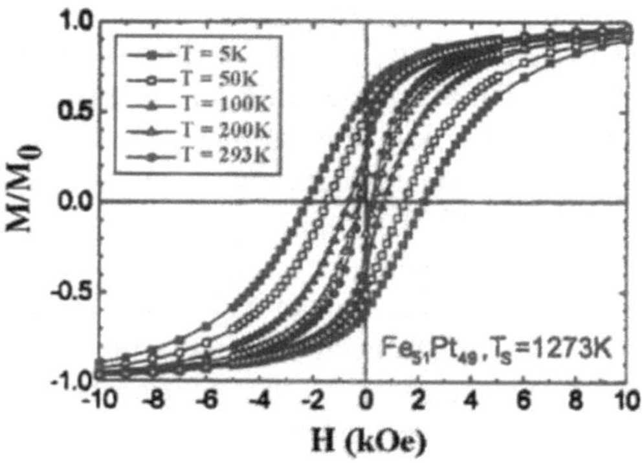

Figure 15. Hysteresis loops of polycrystalline FePt clusters sintered at 1273 K, after [47].

5. FePt:X (X=Ag, C) CLUSTER FILMS

For most technological applications, the properties of clusters embedded in magnetic or non-magnetic, metallic or non-metallic matrices are more important than those of free clusters. Figure 16 shows a schematic drawing of cluster film deposition: a cluster beam combined with two atomic beams from normal magnetron sputtering guns, which are used for matrix materials deposition, such as Ag, C, etc.

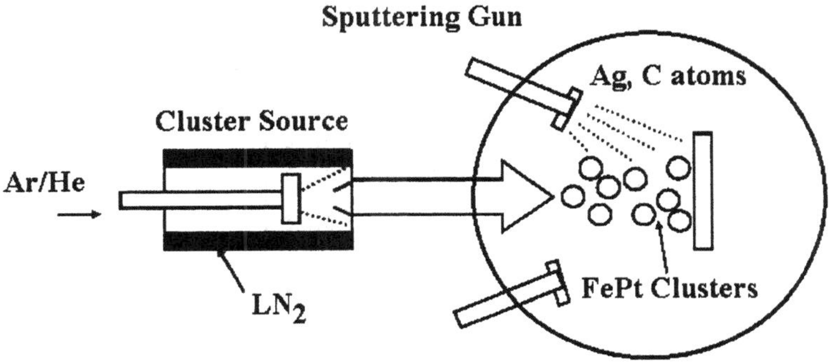

Figure 16. Schematic drawing of cluster deposition with matrix atoms beam.

Figure 17 gives two examples of cluster film structure: (a) clusters are embedded in matrix by co-deposition or, (b) isolated by a matrix via multilayering. The ability of the deposition technique to independently control these parameters makes it ideal for systematic studies of magnetic clusters. In this section, some recent work on FePt:Ag and FePt:C cluster films from our laboratory are reviewed.

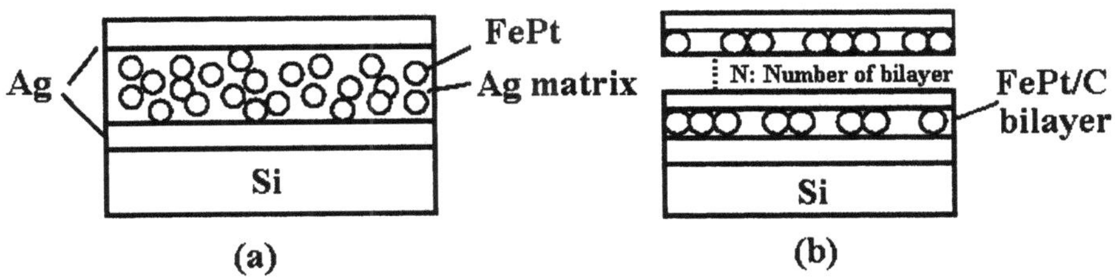

Figure 17. Cluster film structures: (a) Co-deposition and (b) multilayers.

5.1. FePt:Ag Cluster Films

FePt:Ag cluster films are deposited with a three-layer structure [Ag(2 nm)/FePt (12 nm)/Ag(2nm)] or co-sputtering FePt clusters and Ag atoms on a Si substrate. The FePt clusters are produced with a power of 100 W. The mean diameter of the clusters is about 4.4 nm. For the three-layer films, the magnetic layer thickness is about 12 nm, which is nearly 3 layers of FePt clusters [43]. X-ray diffraction (XRD) measurement confirmed the ordering of $L1_0$ phase by post-deposition annealing. Figure 18 shows the XRD (θ-2θ) scans of FePt:Ag cluster film, prepared by co-sputtering, annealed at 500, 600 and 700 °C for 10 min, respectively. The as-deposited film showed a broad FePt fcc (111) peak and Ag fcc (111) and (200) peaks. After annealing at higher temperature, the fct-(200) and (002) peaks are well-separated, and the intensity of fct-(001) peak increased with the increase of annealing temperature, indicating that the degree of $L1_0$ ordering increased with the annealing temperature. In all of these films, the orientation of the fct FePt clusters is random. Magnetic properties of the FePt:Ag cluster films were measured by SQUID.

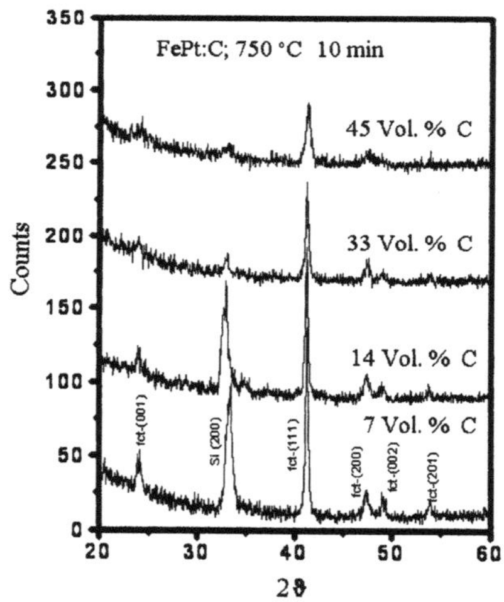

Figure 18. XRD spectra of FePt:Ag cluster films annealed at 500, 600 and 700 °C for 10 min.

Figure 19 shows the room temperature hysteresis loops of the FePt:Ag films annealed at 550 and 700 °C for 10 min. Both in-plane (H // film plane) and perpendicular (H ⊥ film plane) loops are presented. The perpendicular coercivity is almost the same as the in-plane coercivity, indicating that the easy axis of clusters is distributed randomly. This result is consistent with that obtained by XRD measurement (Fig. 18).

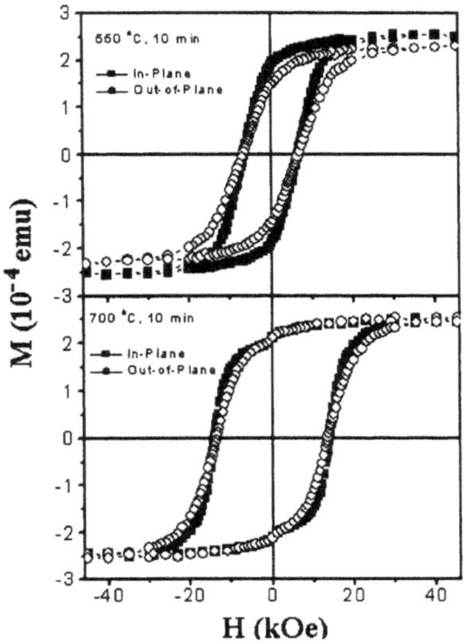

Figure 19. Hysteresis loops of FePt:Ag cluster films annealed at 550 and 700 °C for 10 min.

Figure 20 shows the dependence of the coercivity on annealing temperature for co-sputtered FePt:Ag films. The coercivity increases almost linearly with increase of annealing temperature. This can be understood through the increase of the degree of $L1_0$ ordering. The degree of ordering depends on the annealing temperatures; large coercivity of the cluster films originated from the large K_u of the *fct* phase. Note that a coercivity more than 4 kOe can be achieved just by annealing the film at 450 °C for 10 min, which otherwise could be obtained by annealing samples at 550 °C or higher for FePt cluster films without Ag addition. This result suggests that addition of Ag in FePt cluster films can lower the $L1_0$ ordering temperature and increase the coercivity of the cluster films significantly.

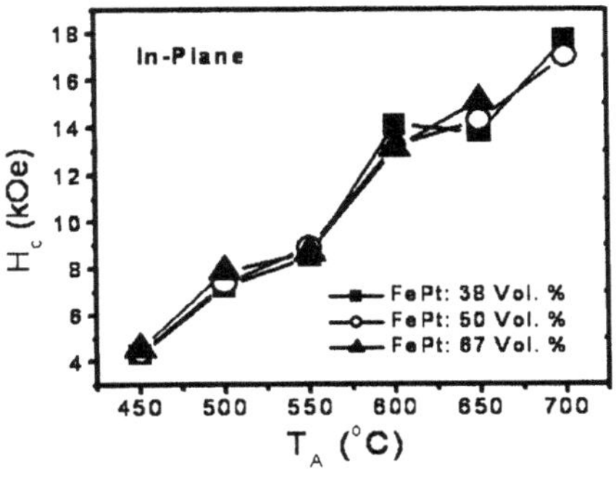

Figure 20. Effect of annealing temperature on coercivities for FePt:Ag cluster films

5.2. FePt:C Cluster Films

FePt:C nanocluster films were prepared with a multilayer method in which FePt cluster layers and C layers were alternately deposited onto a Si substrate. C served to isolate the clusters and to avoid cluster-aggregation during thermal annealing. The average FePt cluster size is about 4.5 nm. The nominal thickness of each cluster layer is about 1 nm, so that the FePt clusters can be well separated with C layer. The thickness of C layer was adjusted with C volume fraction changing from 7 to 45 % [48].

Figure 21 shows the TEM plan-view images for the FePt:C film with 45 vol. % C annealed at 650 °C for 10 min. It can be seen that FePt clusters are embedded in C matrix. Shown in Fig. 21 (b) is the high resolution TEM image. It shows that single FePt clusters with diameter about 4.5 nm were well separated by amorphous C matrix; the cluster growth during high temperature annealing has been remarkably reduced with the isolation of higher volume fraction of C.

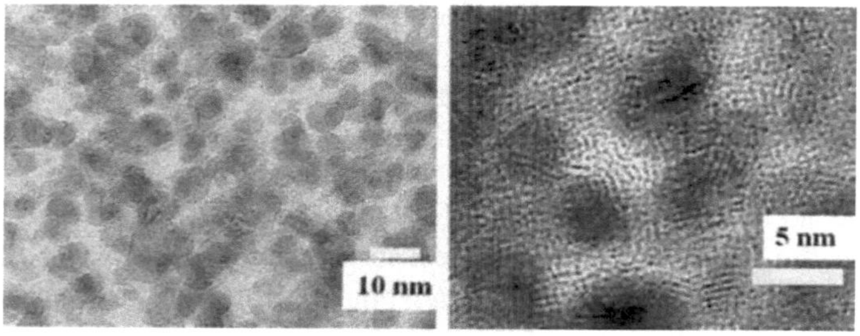

Figure 21. TEM images of FePt:C cluster film with 45 vol. % C, annealed at 650 °C fpr 10 min.

XRD measurements confirmed the ordering of the $L1_0$ structure [48]. The intensity of the $L1_0$-peaks increases with annealing temperature, indicating the increase of the degree of $L1_0$ ordering upon annealing. Figure 22 shows the effect of annealing temperature T_A on the in-plane and perpendicular coercivities of FePt:C films with 45 vol. % C. Both in-plane (H // film plane) and perpendicular (H ⊥ film plane) coercivities increased with increase of annealing temperature. The perpendicular coercivity larger than 13 kOe was achieved in the films after annealing at a temperature of 700 °C and above. In order to determine the thermal effect on coercivity, the hysteresis loops of FePt:C cluster films were measured at low temperature down to 10 K.

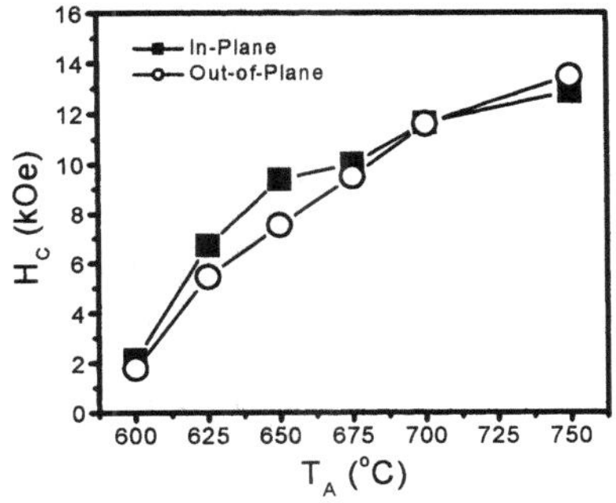

Figure 22. Effect of annealing temperature on coercivities of FePt:C cluster film with 45 vol. % C.

Figure 23 shows the coercivity dependence on temperature for FePt:C cluster films with 45 vol. % C from 10 to 300 K. The films were annealed at different temperature for 10 min. The coercivity decreases with increase of measuring temperature, for example, dropping from 19 kOe at 10 K to about 13 kOe at 300 K for a film annealed at 750 °C for 10 min. This may be caused by a contribution of intrinsic temperature dependence of the anisotropy and magnetization, and thermal activation effects [49].

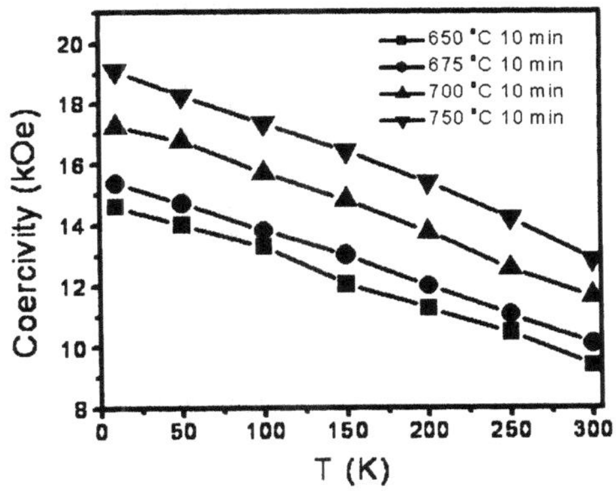

Figure 23. Temperature dependence of coercivity for FePt:C cluster films with 45 vol. % C, annealed at temperatures from 650 to 750 °C.

Figure 24 shows the Δm curve measured for the FePt:C film with 45 vol. % C annealed at 625 °C for 10 min [48]. The negative value of Δm peak (-0.17) suggests dipolar interactions.

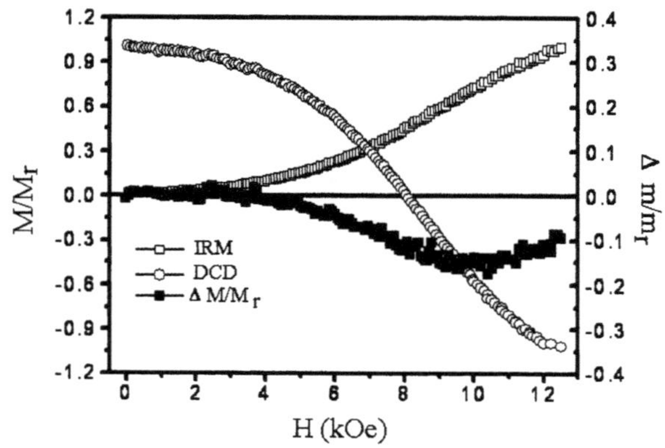

Figure 24. Δm curve for FePt:C cluster film with 45 vol. % C annealed at 625 °C for 10 min.

5.3. Dynamics of Magnetization Reversal in FePt:C Cluster Films

As grain size becomes smaller and smaller with increased recording areal density, the magnetic moment in a grain may reverse spontaneously via thermal activation. Therefore, understanding the magnetization reversal behavior of media is of essential importance to control and design new media with desired properties and good thermal stability. We carried out a set of moment decay measurements for FePt:C films with 45 vol. % C. The data were fitted with the Sharrock formula [50], and the thermal stability factor and anisotropy constant were extracted from the fitting parameters. The detailed experiment and fitting procedures can be found in Refs. [48, 51].

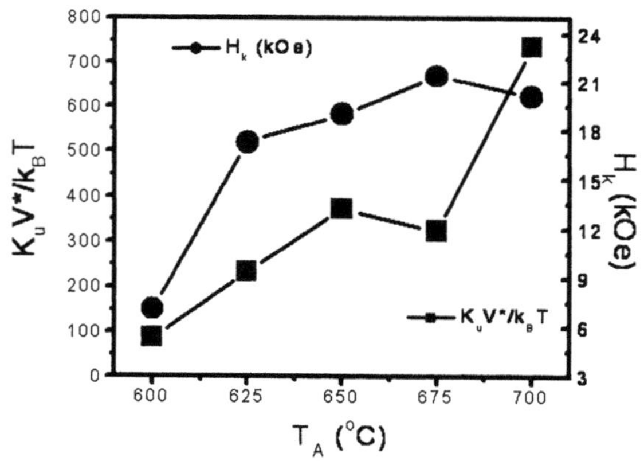

Figure 25. Effect of annealing temperature on thermal stability and magnetic anisotropy field of FePt:C cluster films with 45 vol. % C.

Figure 25 shows the annealing temperature effect on the thermal stability factor K_uV^*/k_BT [49] and anisotropy field H_k for the FePt:C films with 45 vol. %C. H_k was calculated from $H_0 = 0.48\ H_k$, H_0 was obtained from the Sharock fitting parameter. H_k increases rapidly with annealing temperature T_A between 600 and 625 °C, then increases slowly and to saturation for $T_A > 625$ °C. K_uV^*/k_BT increases linearly with T_A except for the point at 675 °C that might be caused by either experimental error or the activation volume V^* being unusually small. Since K_u would be constant after the completion of $L1_0$ ordering, the further increase of K_uV^*/k_BT with T_A is mainly due to the increase of V^*. As shown in Fig. 26, K_u is about 1.2×10^7 erg/cm^3 for $T_A > 625$ °C; V^* increases slowly initially and then increases more rapidly with T_A, which results in the quasi-linear increase of K_uV^*/k_BT.

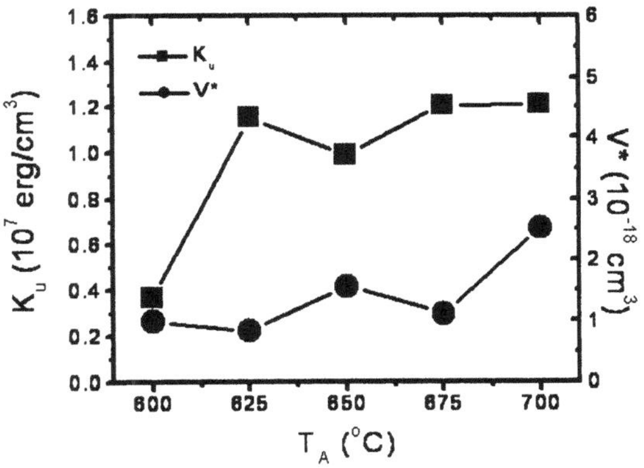

Figure 26. Effect of annealing temperature on magnetic anisotropy constant and activation volume of FePt:C cluster films with 45 vol. % C.

5.4. Dilute FePt:C Cluster Film: Stoner-Wohlfarth-Like Behavior

Understanding the magnetic properties of a collection of well-isolated clusters is of great interest for exploring FePt clusters as a media for EHDR beyond 1 Tera bit/in. Dilute FePt:C cluster films were prepared by the multilayer method as described earlier. Thus Stoner-Wohlfarth-like behavior has been observed in a FePt:C cluster film with FePt volume fraction of 5 %, [45]. Figure 27 shows the zero-field-cooled (ZFC) and field-cooled (FC) magnetization curves for as-deposited and annealed FePt:C cluster film with 5 vol. % FePt.

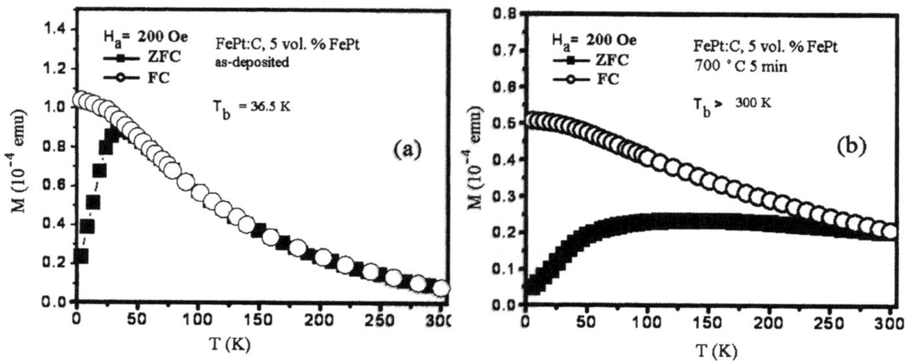

Figure 27. ZFC and FC magnetization curves of FePt:C cluster film with 5 vol. % FePt: (a) as-deposited; (b) annealed at 700 °C for 5 min.

An arrow in Fig. 27(a) indicates a block temperature T_b where ZFC and ZF curves are coincident. For as-deposited film, the T_b is about 36.5 K; after annealing at 700 °C for 5 min., the peak in ZFC curve was not observed, Fig. 27(b), suggesting a T_b higher than 300 K and the increase of anisotropy constant K_u due to order or partial order of the $L1_0$ phase. Figure 28 shows the temperature dependence of coercivity for FePt:C cluster films with 5 and 30 vol. % FePt annealed at 700 °C for 5 min. At room temperature, a few hundred Oe of coercivity was observed for both films; at low temperature (10 K), the coercivity reaches to 4 kOe for 5 vol. % FePt film and 5.6 kOe for 30 vol. % FePt. Increase of the FePt cluster volume fraction increases the coercivity. These results suggest the partial ordering of $L1_0$ phase at short-time annealing (5 min) and some very small clusters showing superpara-magnetic behavior at room temperature.

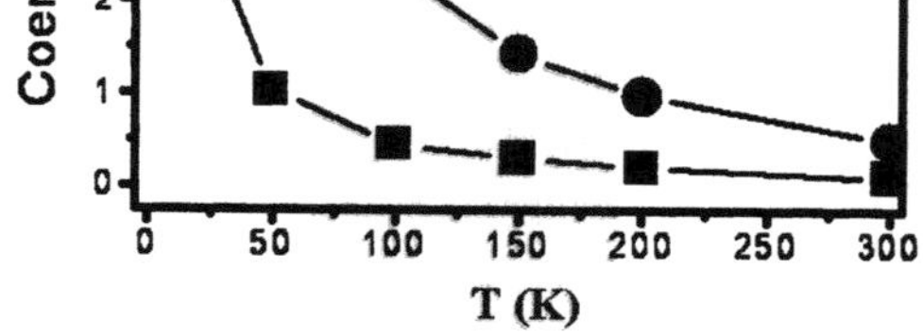

Figure 28. Temperature dependence of coercivity for dilute FePt:C cluster films annealed at 700 °C for 5 min.

Figure 29 shows a TEM image of FePt:C cluster film with 5 vol. % FePt annealed at 700 °C for 10 min. Well-isolated clusters with single crystal $L1_0$ structure are observed. The arrows indicate the clusters with lattice fringe observable, suggesting single crystalline clusters.

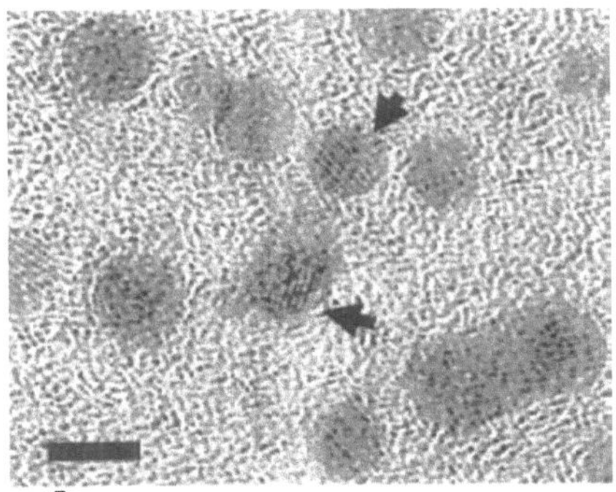

Figure 29. TEM image of FePt:C cluster film with 5 vol. % FePt annealed at 700 °C for 10 min.

Shown in Fig. 30 is the temperature dependence of coercivity for the FePt:C films annealed at 700 °C for 10, 30 and 60 min. respectively. High coercivities (> 21 kOe) at room temperature were observed and they reached about 40 kOe at low temperature. This result indicates the high degree of $L1_0$ ordering after annealing for longer time (> 10 min.).

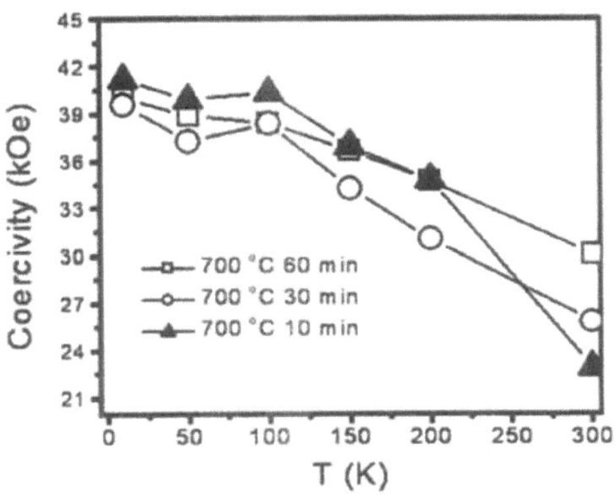

Figure 30. Temperature dependence of coercivity of dilute FePt:C cluster films with 5 vol. % FePt annealed at 700 °C for 10, 30 and 60 min.

Fig. 31 shows the in-plane and perpendicular plane hysteresis loops of the FePt:C film with 5 vol. % FePt annealed at 700 °C for 60 min, measured at 10 K with a SQUID magnetometer. A perpendicular coercivity of about 40 kOe was achieved. The loops are not saturated under an applied field of 70 kOe. The law-of-approach to saturation was applied and the ratio of M_r/M_s is estimated between 0.5 – 0.6. Assuming these clusters are non-interacting and the formula $H_c = 0.48\ H_k$ can be applied; an anisotropy field of 83 kOe is estimated.

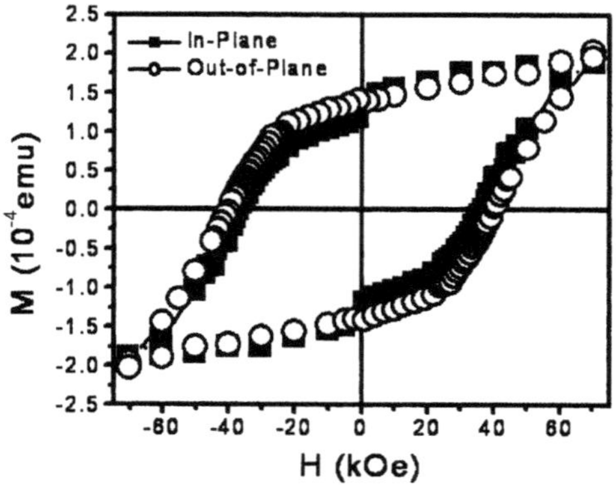

Figure 31. Hysteresis loops of FePt:C cluster film with 5 vol. % FePt annealed at 700 °C for 60 min, measured at 10 K.

In summary the FePt nanoclusters studied are thermally stable, have a narrow size distribution and controllable coercivities. Their properties are similar to those required for magnetic recording at extremely high densities, i.e. approaching an areal density of 1 Terabit/in^2. For practical application, the FePt cluster orientation and film roughness must be further investigated and controlled. From the viewpoint of fundamental studies, the possibility of fabricating nearly monodisperse clusters of a wide range of sizes and densities will enable us to investigate many phenomena involving the magnetic behavior of materials, especially for clusters with a few nano-meters in size as building blocks. Research on the magnetic properties of cluster-assembled materials is in its infancy and thus is fertile ground for both fundamental studies and applications.

6. NON-EPITAXIAL ORIENTED $L1_0$ FePt:X (X = Ag, C, B_2O_3) NANOCOMPOSITE FILMS

6.1. Non-Epitaxial Growth of (001) Textured $L1_0$ FePt:X Films

Under normal growth conditions, $L1_0$-phase FePt and FePt based films often possess (111) preferred or random orientations. In order to control crystal orientation, the epitaxial growth of FePt film is needed to obtain the oriented films. The most common methods to obtain the c axis normal to the film plane [(001) texture] are to use seed or buffer layers between a single-crystal substrate, such as MgO, and the FePt and FePt-based films. Recently, highly (001) textured $L1_0$ FePt [52,53] and FePt based films [54] have been obtained by a nonepitaxial growth method. Yan *et al.* obtained (001) textured FePt and FePt based films by using multilayer deposition plus post annealing. It was found that orientations of FePt grains are affected by many preparation parameters, such as initial as-deposited film structure, composition, annealing time, and so on [55,56].

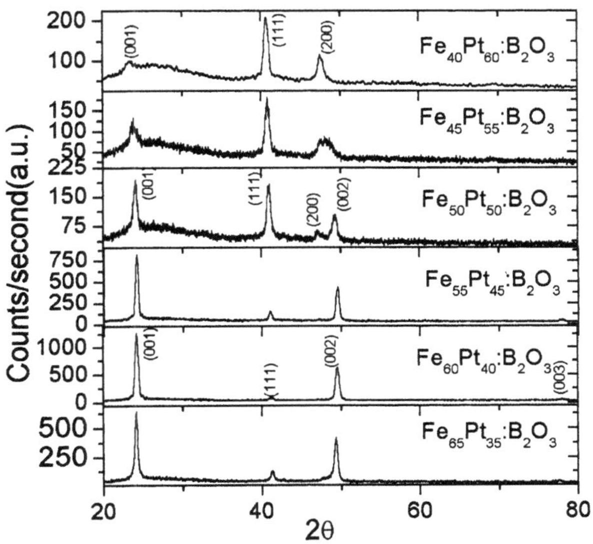

Figure 32. XRD patterns of the Fe_xPt_{1-x}:B_2O_3(11vol.%) nanocomposite films. Films deposited directly on glass substrate and annealed at 550 °C for 30 minutes. Film thickness is 15 nm.

Figure 32 shows the investigation of Fe concentration on orientation of $L1_0$ FePt:B_2O_3 nanocomposite films. In the Fe-40at.% film, XRD pattern shows (111), (200) peaks and relatively weak peak of (001). In the Fe-45at.% film, relative intensity of (00*l*) begins to increase and intensity of (111) and (200) decreases. The (200) and (002) peaks merge into one because of the close intensity and position of (200) and (002) peaks. This means the film is partially ordered. As Fe concentration further increases, the relative intensity of (111) peak decreases further. In Fe-60at.% film, the (00*l*) peak became predominant. The (111) peak is restrained and the (200) disappears. The high intensity ratio of $I_{(001)}/I_{(111)}$ indicates that (001) texture is favored and the c axis prefers to orient perpendicularly to the film plane in slightly Fe-rich films.

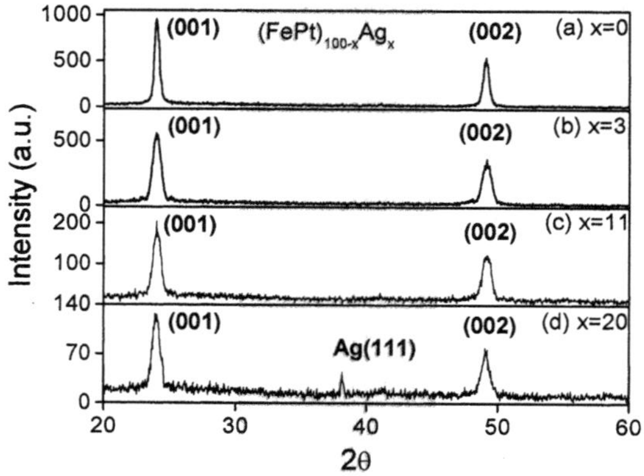

Figure 33. XRD patterns of FePt:Ag films with different Ag content. Films deposited directly on Si wafer and annealed at 600°C for 10 minutes. Film thickness is 10 nm.

The effect of matrix concentration on orientation, intergranular interactions and magnetic properties has been investigated for $L1_0$ FePt:Ag nanocomposite films. The Ag contents varied from 0 to 20 vol %. As shown in Fig. 33, all films clearly show the (001) superlattice peaks indicating that the Ag content (matrix) has relatively small affection on orientation.

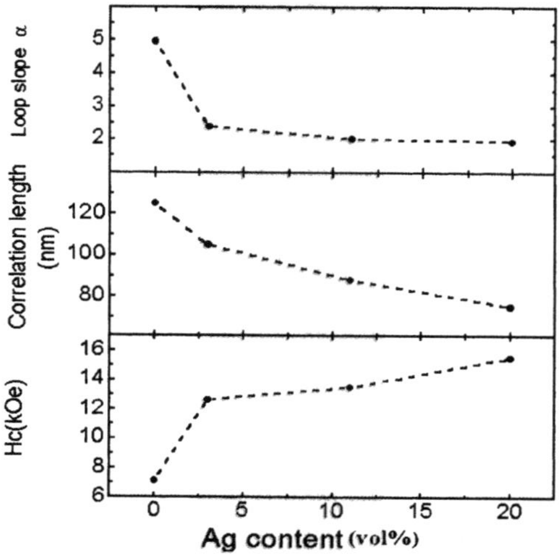

Figure 34. Hysteresis loop slope, correlation length and coercivity for various Ag contents (vol %).

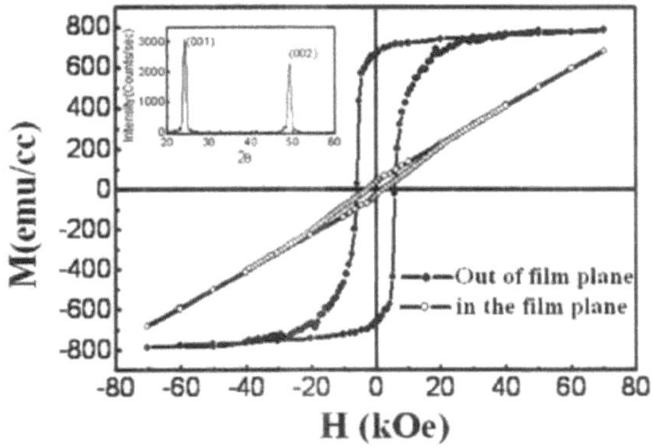

Figure 35. Hysteresis loop of FePt:C film. Film deposited directly on Si wafer and annealed at 550°C for 5 minutes. Film thickness is 16 nm. Inset is the XRD pattern of this sample.

A key aspect of the magnetism of nanocomposite films is the relation between intergranular interactions, the hysteresis-loop shape, and spatial correlations. Figure 34 shows the hysteresis loop slope, correlation length and coercivity of the FePt:Ag nanocomposite films with various Ag contents. Average interaction-domain sizes (correlation lengths) were deduced from MFM images. The coercivities and the hysteresis-loop slope parameter $\alpha = 4\pi dM/dH(H_c)$ were determined from the hysteresis loops. With increasing Ag content the hysteresis-loop slope and the correlation length decrease, whereas the coercivity increases. The Ag increases the separation between the particles, thereby effectively exchange decoupling the grains.

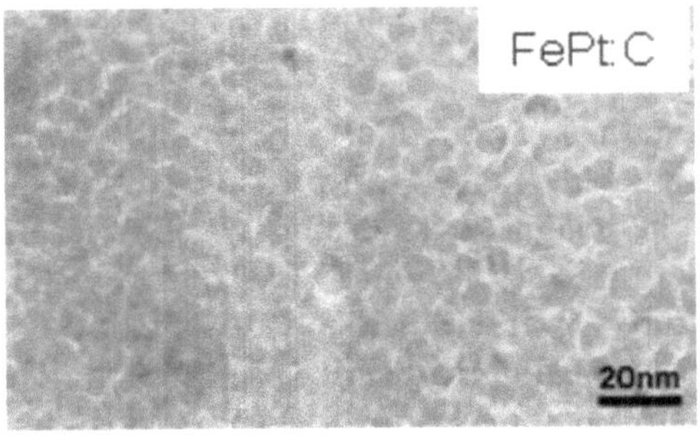

Figure 36. Bright-field TEM image of FePt:C film.

Hysteresis loops for $L1_0$ nanocomposite FePt:C film is shown in Fig. 35. Inset is the XRD patterns of the same FePt:C film, on which (001), (002) peaks appear indicating that FePt grains are (001) oriented. The full width at half maximum (FWHM), obtained from the rocking curve of (001) peak, is 1.68°, conforming a high degree of (001) texturing. The loop shows perpendicular anisotropy with square shape in perpendicular direction due to the enhanced preferential FePt $L1_0$ (001) texture. The perpendicular loop shows large coercivity (H_c = 6.2 kOe) and high remanence ratio (S = 0.9).

The nanostructure of nonepitaxially grown $L1_0$ FePt:C thin film was characterized by TEM. As shown in Fig. 36, the bright-field image reveals that FePt grains with uniform size less than 5nm are embedded in the carbon matrix and appear to be well isolated.

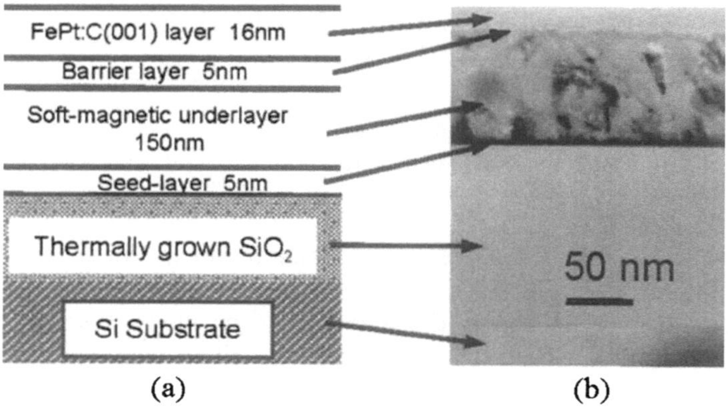

Figure 37. FePt:C double-layered perpendicular composite medium construction. (a) schematic sketch of the layers in the medium and (b) TEM cross-section image of this medium.

6.2. Double-Layered $L1_0$ FePt:C/FeCoNi Perpendicular Recording Media

For practical applications, a nonepitaxially grown, double-layered thin-film medium of nanocomposite FePt:C with a FeCoNi soft underlayer for high-density perpendicular magnetic recording was fabricated and investigated [56]. Figure 37(a) shows a schematic view of the cross section indicating substrate and each layer in the film. The FeCoNi soft underlayer (SUL) was first deposited on thermally oxidized Si substrates. The saturation magnetization flux density (B_s) for this layer is approximately 20 kG. The FePt:C layer was then deposited on the SUL at room temperature by a multilayer deposition technique and subsequent annealing. A 5 nm Ta barrier layer was introduced between SUL and FePt:C layer to reduce the diffusion between the SUL and FePt:C layer during the annealing process. A cross sectional TEM image, Fig. 37(b), verifies that this medium consists of nanocomposite

FePt:C layer with FePt nanoparticles embedded in a C matrix and soft FeCoNi underlayer.

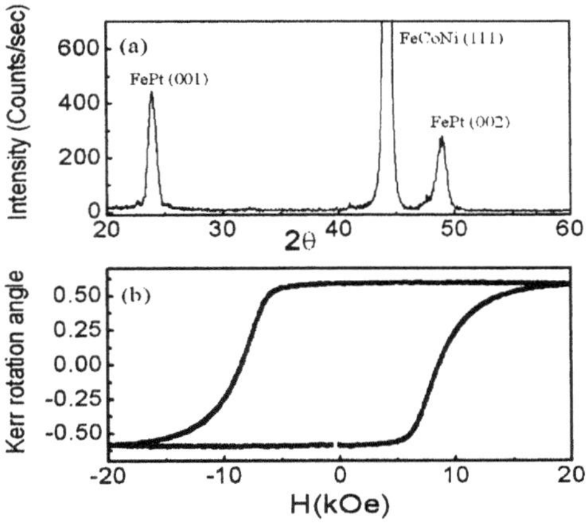

Figure 38. XRD pattern(a) and hysteresis loop(b) of FePt:C double-layered media. Loop measured by polar-Kerr Effect at room temperature.

Figure 38 shows the XRD pattern (a) and hysteresis loop (b) of FePt:C double-layered nanocomposite thin-film medium. The soft underlayer FeCoNi (111) peak and the $L1_0$ FePt (001) and (002) peaks are shown only in the XRD pattern. This means that the preferred crystal orientation of $L1_0$ FePt:C nanocomposite film is successfully obtained on this SUL by nonepitaxial growth. The polar-Kerr measurement shows a square loop that is only sensitive to the top layer; the Kerr effect data shown in this loop give the coercivity $H_c = 8.5$ kOe, nucleation field $H_n = 5.65$ kOe, remanence ratio $S = 1$, and loop slope (at H_c) $\alpha = 3.3$, respectively.

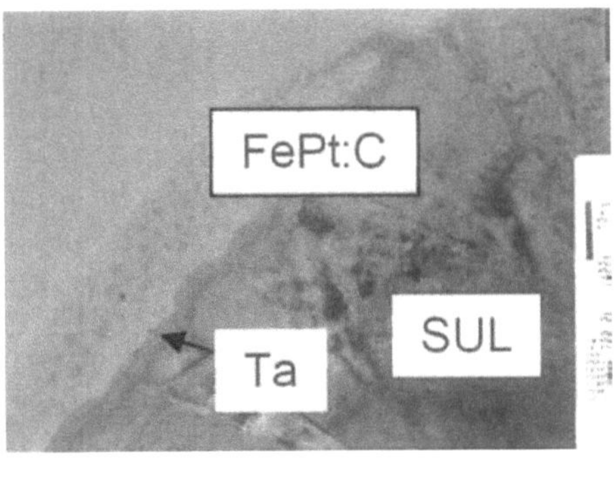

Figure 39. Cross-section TEM image of FePt:C double-layered media (scale bar: 20 nm).

The nanostructure of the double-layered nanocomposite FePt:C thin film medium was characterized by TEM. Electron diffraction shows that the crystallites are FePt with the $L1_0$ structure. Figure 39 shows the bright-field and cross-section TEM image, which verifies that FePt crystallites are imbedded in C matrix and well isolated from each other.

These results suggest that nonepitaxially grown, (001) textured double-layered nanocomposite $L1_0$ Fe(Co)Pt-based films with perpendicular anisotropy are a promising candidate to realize extremely high-density perpendicular recording.

Acknowledgements

The authors are grateful to many colleagues for assistance and helpful discussions, and to NSF-MRSEC, INSIC, DOE, ARO, and CMRA for financial support.

References

[1] H. Haberland (Ed.), "Clusters of Atoms and Molecules", Springer-Verlag, 1994.
[2] W. Ekardt (Ed.), "Metal Clusters", Whily 1999.
[3] K.-H. Meiwes-Broer (Ed.), "Metal Cluster at Surfaces", Springer 2000.
[4] J. P. Bucher, D. C. Douglass, and L.A. Bloomfield, Phys. Rev. Lett. **66**, 3052 (1991).
[5] I. M. L. Billas, J. A. Becker, A. Châtelain, and W. A. de Heer, Phys. Rev. Lett. **71**, 4067 (1993).
[6] I. M. L. Billas, A. Châtelain, and W. A. de Heer, Science **265**, 1682 (1994).
[7] I. M. L. Billas, W. A. de Heer, and A. Châtelain, J. Non-Cryst. Solids **179**, 316 (1994).
[8] A. J. Cox, J. G. Louderback, S. E. Apsel, and L.A. Bloomfield, Phys. Rev. B **49**, 12295 (1994).
[9] S. E. Apsel, J. W. Emmert, J. Deng, and L. A. Bloomfield, Phys. Rev. Lett. **76**, 1441 (1996).
[10] K. W. Edmonds, C. Binns, S. H. Baker, S.C. Thornton, and C. Norris, Phys. Rev. B **60**, 472 (1999).
[11] S. Stavroyiannis, I. Panagiotopoulos, D. Niarchos, J. A. Christodoulides, Y. Zhang, and G. C. Hadjipanayis, Appl. Phys. Lett. **73**, 3453 (1998).
[12] M. Yu, Y. Liu, A. Moser, D. Weller, and D. J. Sellmyer, Appl. Phys. Lett. **75**, 3992 (1999).
[13] B. Bian, K. Sato, and Y. Hirotsu, Appl. Phys. Lett. **75**, 3686 (1999).
[14] T. Suzuki, T. Kiya, N. Honda, and K. Ouchi, J. Magn. Magn. Mater. **235**, 312 (2001).
[15] T. Yang, E. Ahmad, and T. Suzuki, J. Appl. Phys. **91**, 6860 (2002).
[16] R. Wood, IEEE Trans. Magn. **36**, 36 (2000).
[17] R. A. McCurrie and P. Gaunt, Philos. Mag. **13**, 567 (1966).
[18] B. Zhang and W.A. Soffa, IEEE Trans. Magn. **26**, 1388 (1990).

[19] E. E. Fullerton, C. H. Sowers, J. E. Pearson, S. D. Bader, J. B. Patel, and X. Z. Wu, J. Appl. Phys. **81**, 5937 (1997).
[20] S. Sun, C. B. Murray, D. Weller, L. Folks, and A. Moser, Science **287**, 1989 (2000).
[21] H. Zeng, J. Li, J. P. Liu, Z. L. Wang, and S. Sun, Nature **420**, 395 (2002).
[22] H. Haberland, M. Karrais, M. Mall and Y. Thurner, J. Vac. Sci. Technol. A **10**, 3266 (1992).
[23] K. Sattler, J. Muhlbach, and E. Recknagel, Phys. Rev. Lett. **45**, 821 (1980).
[24] S. S. Riley, E. Parks, C. Mao, L. Pobo, and S. J. Wexler, Phys. Chem. **86**, 3911 (1982).
[25] F. Frank, W. Schulze, B. Tesche, J. Urban, and B. Winter, Surf. Sci. **156**, 90 (1985).
[26] H. Schaber, T. P. Martin, Surf. Sci. **156**, 64 (1985).
[27] S. H. Baker, S. C. Thornton, A. M. Keen, T. I. Preston, C. Norris, K. W. Edmonds, and C. Binns, Rev. Sci. Instrum. **68**, 1853 (1997).
[28] S. H. Baker, S. C. Thornton, K. W. Edmonds, M. J. Maher, C. Norris, and C. Binns, Rev. Sci. Instrum. **71**, 3178 (2000).
[29] H. Haberland, M. Mall, M. Moseler, Y. Qiang, T. Reiners, and Y. Thurner, J. Vac. Sci. Technol. A **12**, 2925 (1994).
[30] See: www.oareserch.co.uk and www.Mantisdeposition.com.
[31] M. D. Upward, B. N. Cotier, P. Moriarty, P. H. Beton, S. H. Baker, C. Binns and Edmonds, J. Vac. Sci. Technol. B **18**, 2646 (2000).
[32] J. Söderlund, L. B. Kiss, G. A. Niklasson, and C. G. Granqvist, Phys. Rev. Lett. **80**, 2386 (1998).
[33] K. W. Edmonds, C. Binns, S. H. Baker, S. C. Thornton, and C. Norris, Phys. Rev. B **60**, 472 (1999).
[34] C. Binns, M. J. Maher, Q. A. Pankhurst, D. Kechrakos, and K. N. Trohidou, Phys. Rev. B **66**, 184413 (2002).
[35] Y. Qiang, R. F. Sabiryanov, S. S. Jaswal, Y. Liu, H. Haberland and D. J. Sellmyer, Phys. Rev. B **66**, 064404 (2002).
[36] T. Hihara and K. Sumiyama, J. Appl. Phys. **84**, 5270 (1998).
[37] D. L. Peng, K. Sumiyama, S. Yamamuro, T. Hihara, and T. J. Konno, Appl. Phys. Lett. **74**, 76 (1999).
[38] D. L. Peng, K. Sumiyama, T. J. Konno, T. Hihara, S. Yamamuro, Phys. Rev. B **60**, 2093 (1999).
[39] D. L. Peng, T. Hihara, K. Sumiyama, and H. Morikawa, J. Appl. Phys. **92**, 3075 (2002).
[40] T. J. Konno, S. Yamamuro, and K. Sumiyama, J. Vac. Sci. Technol. B **20**, 834 (2002).
[41] T. J. Konno, S. Yamamuro, and K. Sumiyama, J. Appl. Phys. **90**, 834 (2001).
[42] F. R. de Boer, R. Boom, W. C. M. Mattens, A. R. Miedema, and A. K. Niessen, in "Cohesion in Metals Transition Metal Alloys", North-Holland, Amsterdam 1988.
[43] Y. Xu, Z.G. Sun, Y. Qiang, and D.J. Sellmyer, J. Appl. Phys. **93**, 8289 (2003).
[44] Y. Xu, Z.G. Sun, Y. Qiang, and D.J. Sellmyer, J. Magn. Magn. Mater. **266**, 164 (2003).
[45] Y. Xu, M. L. Yan, J. Zhou, and D. J. Sellmyer, J. Appl. Phys. **97**, 10J320 (2005).
[46] B. Rellinghaus, S. Stappert, M. Acet, and E. F. Wassermann, J. Magn. Magn. Mater. **266**, 142 (2003).

[47] S. Stappert, B. Rellinghaus, M. Acet, and E. F. Wassermann, J. Cryst. Grow. **252**, 440 (2003).

[48] Y. Xu, M. L. Yan, and D. J. Sellmyer, IEEE Trans. Magn. **40**, 2525 (2004).

[49] D. J. Sellmyer, C. P. Luo, M. L. Yan, and Y. Liu, IEEE Trans. Magn. **37**, 1286 (2001).

[50] M. P. Sharrock, J. Appl. Phys. **76**, 6413 (1994).

[51] Y. Xu, Z. S. Shan, J. P. Wang, and C. T. Chong, J. Magn. Magn. Mater. **232**, 103 (2001).

[52] M. L. Yan, N. Powers, and D. J. Sellmyer, J. Appl. Phys. **93**, 8292 (2003).

[53] H. Zeng, M. L. Yan, N. Powers, and D. J. Sellmyer, Appl. Phys. Lett. **80**, 2350 (2002).

[54] M. L. Yan, H. Zeng, N. Powers, and D. J. Sellmyer, J. Appl. Phys. **91** 8471 (2002).

[55] Y. Shao, M. L. Yan, and D. J. Sellmyer, J. Appl. Phys. **93**, 8152 (2003).

[56] M. L. Yan, X. Z. Li, L. Gao, S. H. Liu, D. J. Sellmyer, R. J. M. van de Veerdonk, and K. W. Wierman, Appl. Phys. Lett. **83**, 3332 (2003).

Chapter 9

SELF-ASSEMBLED NANOMAGNETS

Shouheng Sun

Department of Chemistry,
Brown University,
Providence, RI 02912, USA

Abstract This chapter reviews recent progress in the field of self-assembled nanomagnets. Emphasis is on the fabrication of self-assembled nanomagnets by solution-phase chemical approaches. After outlining the general concepts of self-assembly; we describe several common chemical procedures leading to monodisperse magnetic nanoparticles. Considered examples are Co, iron oxide and FePt nanoparticles. Our review also includes including the preparation of the assemblies, the particle-shape effect on self-assembled structures, interparticle spacing control in self-assembled nanoparticle superlattices, and the magnetic properties of these arrays. The last part of the chapter is devoted to potential applications of the magnetic nanoparticle arrays in information storage, tunneling devices, and permanent magnets.

1. INTRODUCTION

Magnetic nanoparticles have sizes of 20 nm or below in all dimensions. They are especially important for studying nanomagnetism and for technological applications. At this size range, a particle is usually in a single-domain state, with all spins align in one direction [1-3]. As outlined in Chapters 3 and 4, its magnetization reversal occurs through the rotation of its magnetic easy axis. An energy barrier, ΔE, between two orientations of the magnetization determines the relaxation in each orientation, and is proportional to the particle volume, V, and the particle's anisotropy constant, K. As the particle size decreases to a level for which ΔE becomes comparable to thermal energy (k_BT), the particle becomes superparamagnetic and, in the absence of an external field, its overall magnetic moment is randomized to zero due to thermal fluctuations. Such size- and temperature-dependent nanomagnetism has stimulated tremendous interest recently with the hope to approach the scaling limits of magnetic storage technology and understand spin-dependent transport phenomena in nanoscale devices.

To better understand the magnetism of these single-domain magnetic nanoparticles and apply such understanding to technological applications, it is essential to fabricate either 2D or 3D magnetic nanoparticle arrays within which magnetic properties of each particle and interparticle spacing are well-controlled. The popular fabrication methods, such as vacuum deposition, sputtering, melt-spinning, mechanical milling and lithography, are very effective in making thin magnetic granular films, bulk composite materials, or well-ordered magnetic dot arrays. But they have very limited success in producing magnetic dots at sizes of 20 nm or smaller and fabricating the dot array with inter-dot spacing within 2 nm. In contrast to these physical methods, self-assembly of magnetic nanoparticles is, in principle, a versatile process for fabrication of uniform magnetic nanostructures. This approach uses pre-formed monodisperse magnetic nanoparticles as building blocks. By controlling deposition conditions, the nanoparticles can self-organize on a solid substrate into 2D or 3D superlattice arrays. Because nanoparticles can be synthesized with controlled core dimension and surface coating, their self-assembly can yield superlattice structures with tunable interparticle distance, facilitating nanomagnetic studies and nanofabrication for advanced magnetic applications.

This chapter illustrates how self-assembled nanomagnets can be fabricated using solution phase chemical approaches. Section 2 outlines the general concepts of self-assembly; Section 3 describes several common chemical procedures leading to monodisperse magnetic nanoparticles; Section 4 gives a few self-assembly examples of magnetic nanoparticles, and Section 5 discusses potential applications.

2. SELF-ASSEMBLY IN GENERAL

Self-assembly is a naturally occurring process. It generally refers to building blocks forming a structure spontaneously without an additional source of energy [4-8]. The structure in either 2D or 3D is usually determined by Van der Waals, hydrogen bonding, and electric/magnetic dipolar interactions. Self-assembly occurs when the building blocks interact with one another through a balance of attractive and repulsive interactions. The strength of the interactions between the components, therefore, must be comparable to the forces tending to disrupt them. For molecules or small particles, the disrupting forces are generated by thermal motion. Self-assembly is normally carried out in solution or at an interface to allow the required motion of the components. Therefore, the interaction of the components with their environment can strongly influence the course of the process. An evident example of self-assembly is the formation of cell membrane of bipolar lipid molecules in aqueous solvent, as illustrated in Fig. 1 [9]. The typical phospholipid molecule illustrated in Fig. 1(A) has a polar head group and two hydrophobic hydrocarbon tails. When these amphiphatic molecules, building blocks, are surrounded on all sides by an aqueous environment, they tend to bury their hydrophobic tails and leave their hydrophilic heads exposed to water, a process known as self-assembly. They can form spherical micelles, with the tails inward, or bimolecular sheets (bilayers) with the hydrophobic tails sandwiched between the hydrophilic head groups. Such self-assembled structures form essential cell membranes to support important biological functions.

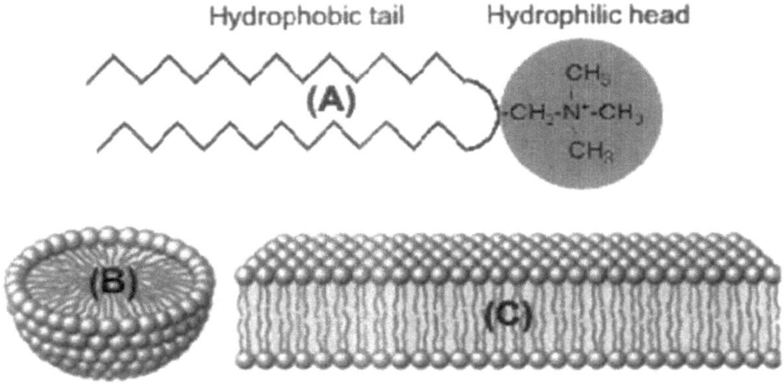

Figure 1. (A) Schematic illustration of a phospholipid molecule with hydrophilic head and hydrophobic tail. Cross-sectional view of (B) the micelle structure and (C) the bilayer sheet structure built through the self-assembly of the lipid molecules (A).

3. MAGNETIC NANOPARTICLES AS BUILDING BLOCKS

3.1. Magnetic Nanoparticle Stabilization

Magnetic nanoparticles contain hundreds to thousands of single atoms/ions and have extremely large surface energy and magnetic interactions. As a result, they tend to undergo uncontrolled aggregation under common preparation conditions. To stabilize these magnetic nanoparticles, repulsive force must be present to counteract the magnetic and surface-related attractions [10, 11]. This stabilization can be achieved via electrostatic and steric repulsion, as indicated in Fig. 2. Coating the particles with ionic compounds will increase electrostatic repulsion when two particles are approaching in close range, Fig. 2(a). But coating the particles with large molecules, such as polymers or surfactants containing long chain hydrocarbon, offers more efficient stabilization, Fig. 2(b). The presence of a hydrocarbon coating layer will greatly enhance the stability of the particles due to the increased steric interactions when two particles come close. The hydrocarbon coating further facilitates the formation of magnetic nanoparticle dispersions in various hydrocarbon solvents, leading to stable magnetic fluids, known as ferrofluids.

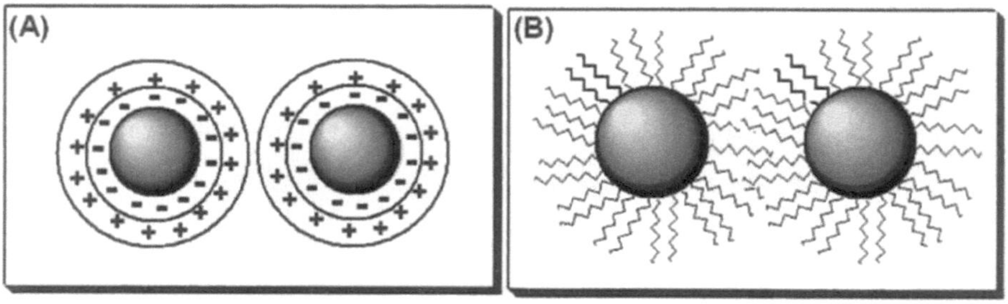

Figure 2. Schematic illustration of particle stabilization *via* (a) electric double layer and (b) steric effect.

3.2. Magnetic Nanoparticle Synthesis

Solution phase chemical synthesis is a convenient way for making surfactant coated magnetic nanoparticles, as described in various reviews [12-18]. Monodisperse Co nanoparticles with standard deviation less than 10% are synthesized by decomposition of $Co_2(CO)_8$ [19-22] or $Co(\eta^3\text{-}C_8H_{13})(\eta^4\text{-}C_8H_{12})$ [23] and reduction of cobalt salt [24, 25] in the presence of oleic acid and trialkylphosphine, or trialkylphosphine oxide, or alkylamine. Monodisperse iron nanoparticles are normally prepared from decomposition of $Fe(CO)_5$ [26-28]. However, metallic iron-based particles from this decomposition procedure are difficult to characterize due to the chemical instability. A recent synthesis using decomposition of $Fe[NSiMe_3)_2]_2$ offers a promising approach to monodisperse Fe nanocrystals [29].

Different from the metallic nanoparticles, iron-oxide nanoparticles are chemically much more stable and readily synthesized. They can be prepared via high-temperature organic-phase decomposition of $FeCup_3$, where Cup represents N-nitrosophenylhydroxylamine, $C_6H_5N(NO)O^-$ [30], decomposition of $Fe(CO)_5$ followed by oxidation [31-33], and decomposition of iron carboxylate [34, 35]. They can also be synthesized from partial reducetion/decomposition of $Fe(acac)_3$ or $Fe(acac)_3$ and $M(acac)_2$ [36-38].

Using similar decomposition/reduction chemistry, multi-component magnetic nanoparticles have been prepared. Specifically, monodisperse FePt nanoparticles are made by decomposition of $Fe(CO)_5$ and reduction of $Pt(acac)_2$ [39-41], or co-reduction of Fe-salt and Pt-salt in the presence of oleic acid and oleylamine [42, 43].

4. SELF-ASSEMBLY OF MAGNETIC NANOPARTICLES

4.1. General Remarks

Monodisperse nanoparticles can form close-packed arrays on a variety of substrates as the solvent from the particle dispersion is allowed to evaporate, as illustrated in Fig. 3. Preparing particles with narrow size distributions is critical to achieving long range order in the assemblies: for size distributions $\sigma \sim 7\%$ short-to-medium-range order is usually observed, reducing σ to ~ 5% enables the formation of assemblies with long-range order known as colloidal crystals or superlattices [44]. However even a sample with $\sigma \sim 5\%$ can yield glassy particle films if the deposition conditions are not appropriately tailored and the solvent evaporation rate is too high.

Figure 3. Schematic illustration of nanoparticle self-assembly *via* solvent evaporation.

To form well-ordered nanostructures using the self-assembly method, three factors have to be considered: (1) the interactions between the solvent and the substrate; (2) the interactions between the particles and the substrate; (3) the interactions between two particles. A nanoparticle surrounded by hydrocarbon molecules is hydrophobic and can be dispersed only in non-polar solvent. Such dispersion can spread well over a hydrophobic surface, such as carbon, or H-Si. By controlling the concentration of the particle dispersion and solvent evaporation rate, large-area 2D or 3D nanoparticle superlattices can be readily formed. If the substrate (such as glass or silicon oxide) is hydrophilic, then the de-wetting nature of the hydrocarbon dispersion of the particles will result in the formation of percolating domains - isolated islands of particles with each island containing either 2D or 3D assemblies. The interparticle interactions can also affect the particle assembly. During the liquid-drying process, attractive molecular interactions between two particles dominate Brownian motion and gravitational precipitation of the particles, causing particles to form percolating domains. This could be a key reason why local disorder and voids are normally present in a compacted monolayer assembly. To maintain a wetting layer on the surface so that nanoparticles can self-assemble into large periodical structures, the evaporation rate of the solvent must be slowed down, as in the case of a crystallization process.

4.2. Self-Assembly Examples

Stabilized with oleic acid, the monodisperse Co particles prepared in solution-phase chemistry are excellent building blocks for constructing 2D and 3D nanoparticle superlattices. Depositing drops of the alkane dispersion of Co particles on a flat carbon surface followed by controlled evaporation of the alkane solvent allows particle self-assemble into superlattice nano-structures. In this case, the evaporation rate is important in controlling the order of the self-assembled structures. Rapid drying (for example <10 seconds) frustrates the formation of long-range order, leaving a disordered structure. Slow evaporation (~ 5 minutes or longer) allows the particles to find energetically favorable lattice sites in a growing assembly, forming a long-range-ordered superlattice. Controlled solvent evaporation can be performed as follows: to cover the substrate with a small container, or to increase the portion of the high-boiling-point solvent in the dispersion. If the evaporation is performed under a covered environment, the evaporation of hexane under the ambient conditions is slowed down (over 5-10 min). As a result, a well-ordered Co nanoparticle superlattice structure is obtained.

Figure 4(A) shows an ensemble of 2D 11 nm Co nanoparticles that has been deposited onto an amorphous carbon surface from hexane under a covered environment [45]. Depositing the same Co nanoparticles from 80% hexane and 20% octane under normal ambient conditions can yield similar nanoparticle superlattices. Here the higher boiling point octane (b.p.126 °C) slows the evaporation rate of the dispersion and contributes to the formation of the superlattice. The use of concentrated dispersion of Co nanoparticles in higher boiling solvents like dodecane (b.p 216 °C) allows slower evaporation of the solvent at higher temperatures, facilitating the formation of a long-range-ordered 3D superlattice [12, 24]. Figure 4(B) shows an 8 nm Co nanoparticle superlattice assembly prepared from the dodecane dispersion at 60°C [12]. The added thermal energy permits the particles to diffuse into their lowest energy superlattice sites during the solvent evaporation, producing a well-defined 3D hcp superlattice structure. Similar to cobalt nanoparticles, both MFe_2O_4 and FePt based nanoparticles stabilized by oleic acid and oleylamine can also self-assemble into iron oxide or FePt nanoparticle superlattices.

Figure 4. TEM images of (A) a 2D assembly of the 11 nm Co nanoparticles, and (B) a 3D assembly of the 8 nm cobalt nanoparticles on an amorphous carbon surface.

4.3. Nanoparticle Shape Effect on Self-Assembled Nanostructures

As the structure of a nanoparticle assembly depends mostly on interparticle interactions, the shape of a particle will alter such interactions and affect the position of the nanoparticles in a superlattice structure. HRTEM analysis on a 11 nm Co nanoparticle assembly shows that the particles in the assembly have three different types of shapes: square-like, hexagon-like and square [45]. Evaporation of the hexane solvent from the dispersion results in an assembly that is different from the spherical ones. These different shapes break the normal six-fold symmetry in a typical hcp superlattice assembly, resulting in two popular defects, twins and stacking faults.

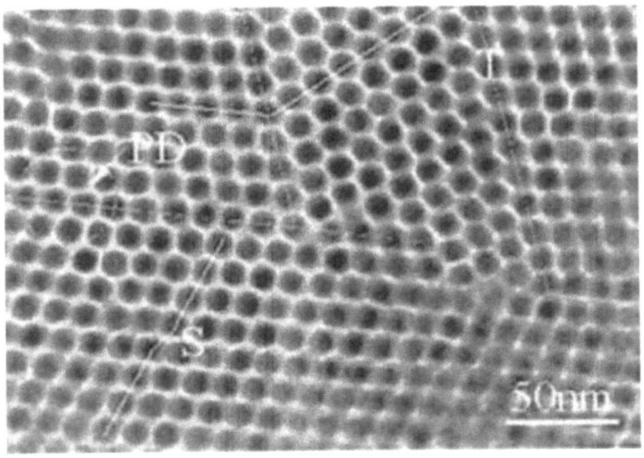

Figure 5. TEM image of 11 nm polyhedral shaped cobalt nanoparticle assembly with the co-existence of stacking faults (S), twins (T) and particle dislocations (PD) [45].

Figure 5 shows the co-existence of twins, stacking faults as well as a partial dislocation in the 11 nm Co nanoparticle assembly. The "strain" induced by the dislocations and defects result in relative misalignments among the particles. These stacking faults and twins totally disturb the local ordering of the self-assembly.

On the other hand, if the shape of the particles can be controlled, self-assembly of these shaped particles can lead to crystal orientation of each particle in a self-assembled superlattice. For example, $MnFe_2O_4$ nanoparticles have been made in cube-like and polyhedron shapes, as shown in Fig. 6 [46]. Controlled evaporation of the carrier solvent from the hexane dispersion (about 2 mg/mL) of the particles led to $MnFe_2O_4$ nanoparticle superlattices.

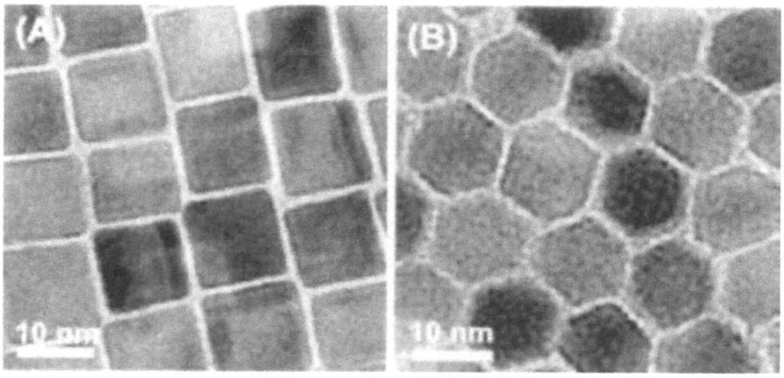

Figure 6. (A) TEM images of the as synthesized (A) 12 nm cube-like and (B) 12 nm polyhedron-shaped $MnFe_2O_4$ nanoparticles [46].

Figure 7(A) shows the superlattice assembly from the cube-like particles, while Fig. 7(B) is the assembly from the polyhedron-shaped particles. The fast Fourier transformation (FFT) of these two images reveals that both assemblies have a cubic packing. But the different shapes possessed by each group of particles affect the crystal orientation of individual particles within the superlattices. XRD of the self-assembled cube-like particles on Si (100) substrate shows the intensified (400) peak, Fig. 7(C), and that of polyhedron-shaped particles reveals the strong reflections of (220), Fig. 7(D). These are markedly different from that of a 3D randomly oriented spinel structured $MnFe_2O_4$ nanoparticle assembly, which shows a strong (311) peak. These indicate that each of the cube-like particles in the cubic assembly has preferred crystal orientation with {100} planes parallel to the Si substrate while for the polyhedron-shaped particle assembly, the {110} planes are parallel to the substrate.

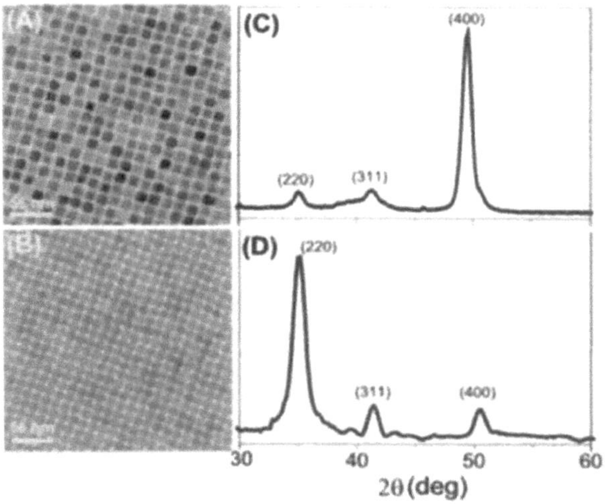

Figure 7. TEM images of 12 nm $MnFe_2O_4$ nanoparticle superlattices of (A) cube-like and (B) polyhedron-shaped nanoparticles. XRD (Co Kα λ = 1.788965 Å) of (C) cube-like and (D) polyhedron-shaped nanoparticle superlattice on Si(100) substrates [46].

4.4. Interparticle Spacing Control in a Self-Assembled Structure

Since the nanoparticles are coated with a layer of surfactants, the separation of any two particles in a self-assembled superlattice is dependent on the thickness of the coating layer. However, the interparticle distance is not a simple addition of two thicknesses as certain degree of intercalation between the particles is present. Nevertheless, by controlling the length of the surfactant, interparticle spacing can be adjusted. Chemically, the surfactant length can be varied by attaching different surfactant molecules to the particle surface during the synthesis, or by replacing the original surfactant molecules with the new ones via surfactant exchange reaction.

One example of interparticle spacing control via surfactant exchange is illustrated in Fig. 8. Here the surface of the Co nanoparticles was modified with Pd to facilitate the replacement of the original RCOO- layer with the new NC-R layer, Fig. 8(A). The p-xylene solution of dicyanobenzene and hexane dispersion of the Pd-modified Co nanoparticles were mixed at room temperature. After an hour, the solvent was allowed to slowly evaporate, giving a superlattice that was different from the one from the original nanoparticles. As shown in Fig. 8(B) and (C), the interparticle spacing is changed from about 4 nm, Fig. 8(B), to about 1 nm, Fig. 8(C). This 1 nm spacing is consistent with the presence of dicyanobenzene between two particles, indicating that, in solution, dicyanobenzene has replaced the oleate molecules, leading to particle assembly with high packing density.

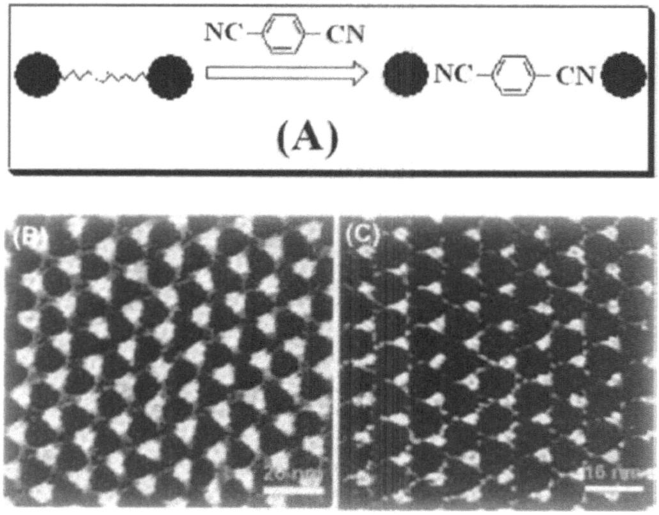

Figure 8. (A) Schematic illustration of surfactant exchange reaction; (B) TEM image of 8 nm nanoparticle superlattice with each particle being surrounded by oleate molecules and interparticle spacing at ~ 4 nm; (C) TEM images of 8 nm nanoparticle superlattice after surfactant exchange with dicyanobenzene. The spacing between two particles is at 1nm.

An alternative approach to the control of interparticle spacing is via thermal annealing. Organic molecules around each nanoparticle are thermally unstable, subject to decomposition/evaporation under high temperature annealing conditions. By controlling the decomposition and evaporation, interparticle spacing can be tuned. Figure 9 shows the interparticle spacing change for an 8 nm Co nanoparticle superlattice before and after thermal annealing [47]. The interparticle spacing in the superlattice of the as-synthesized Co nanoparticles is about 4 nm due to the presence of oleate layer (~ 2.5 nm), Fig. 9(A). After thermal annealing at 350 °C under vacuum for 3 h, interparticle spacing is reduced from about 4 to about 2 nm, Fig. 9(B). Clearly, desorption of the oleate molecules occurs during the annealing process, leading to a more closely packed array. Some coherent strain is observed in the array due to this shrinkage. It is worth to noting that excessive heating can remove all organic coating, making particles sinter into a larger aggregate. Figure 9(C) shows a TEM image of the 6 nm FePt nanoparticle assembly annealed at 600 °C under vacuum for one hour [48]. Detailed HRTEM studies (the inset of Fig. 9C) shows that the coalesced particles do not form single crystals; instead, two originally separated particles form a twin structure with the grain boundaries being easily visible.

This control is the key to form exchange-coupled nanocomposites (see Section 5.3).

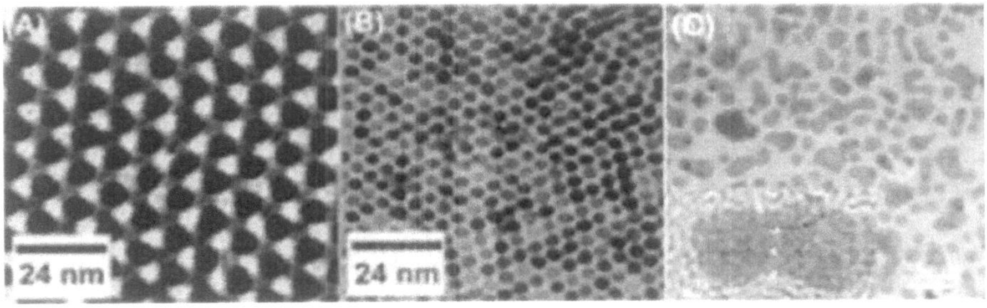

Figure 9. TEM images of 8 nm cobalt nanoparticle assemblies of (A) before and (B) after thermal annealing at 350 °C for 3h. (C) TEM image of 6 nm FePt nanoparticle assemblies annealed 600 °C under vacuum for one hour. Inset: HRTEM of two aggregated FePt particles.

4.5. Polymer-Mediated Self-Assembly

A substrate functionalized with proper molecules can be used to anchor particles on its surface via surface exchange reaction, leading to controlled assembly of the particles. This self-assembly technique is known as molecule-mediated self-assembly and is commonly used for constructing various composite nanostructures [49-55]. Due to their excellent adhesion capability to various substrates, multifunctional polymers are routinely applied as templates to mediate the assembly of the particles. The assembly is carried out as follows: a substrate is immersed into a polymer solution, and then rinsed, leading to a functionalized substrate. Subsequently, this substrate is dipped into the nanoparticle dispersion and then rinsed, leaving one layer of nanoparticles on the substrate surface. By repeating this simple two-step process in a cyclic fashion, a layer-by-layer assembled polymer/nanoparticle multilayer can be obtained.

One assembly example is polyethylenamine (PEI)-mediated self-assembly of FePt nanoparticles [56]. PEI is an all -NH-based polymer that can replace oleate/oleylamine molecules around FePt nanoparticles and attach to hydrophilic glass or silicon oxide surface through ionic interactions [52]. A PEI/FePt assembly is readily fabricated by dipping the substrate alternately into PEI solution and FePt nanoparticle dispersion. Figure 10 shows the assembly process and TEM images of the 4 nm $Fe_{58}Pt_{42}$ nanoparticle self-assemblies on silicon oxide surfaces. Characterizations of the layered structures with X-ray reflectivity and AFM indicate that PEI-mediated FePt assemblies have controlled thickness and the surfaces of the assemblies are smooth with root mean square roughness less than 2 nm.

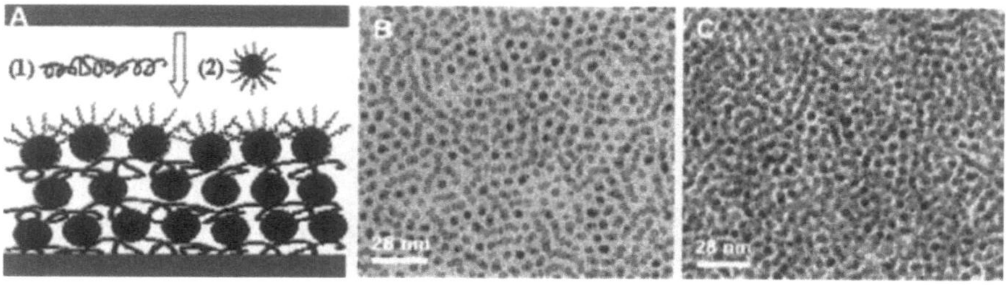

Figure 10. (A) Schematic illustration of PEI-mediated self-assembly of FePt nanoparticles by alternately adsorbing a layer of PEI and a layer of nanoparticles on a solid surface; and TEM images of PEI-mediated assembly of 4 nm $Fe_{58}Pt_{42}$ nanoparticles on silicon oxide surface: (B) one layer of assembly and (C) three layers of assembly [56].

4.6. Magnetic Properties of Self-Assembled Magnetic Nanoparticle Arrays

Due to the KV and k_BT relationship in single-domain particles, magnetic properties of a self-assembled magnetic nanoparticle array are usually size and temperature dependent. A typical zero-field-cooled (ZFC) and field-cooled (FC) scans of the 11 nm Co nanoparticle assembly is shown in Fig. 11(A) [12], in which magnetization is measured as a function of temperature.

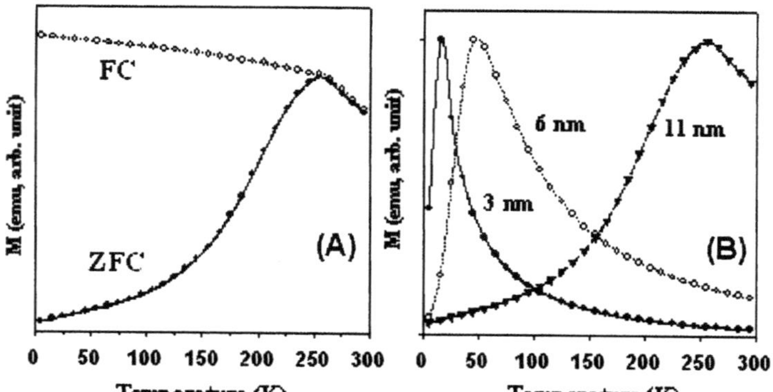

Figure 11. (A) ZFC-FC scans of magnetization over temperature for 11 nm hcp Co nanoparticle arrays, and (B) ZFC scans of the magnetization over temperature for 3 nm, 6 nm and 11 nm Co nanoparticles.

In ZFC scans, the sample is cooled in a zero applied field, and the magnetization is recorded as the temperature is increased in the presence of a small field (10 Oe). The rise of temperature unpins the frozen 3D random magnetization and allows partial magnetic alignment along the applied field, leading to the increase in sample's net moment. At a critical temperature, the thermal energy becomes comparable to the magnetic anisotropy energy of the particles. As a result, thermal fluctuation causes magnetic randomization and reduction of the net magnetic moment. The temperature related to this transition is called superparamagnetic blocking temperature, T_B.

Above T_B, the particles are superparamagnetic and below it, the superparamagnetism is blocked and the particles are ferromagnetic. In FC scans, the particles are cooled in a small field (10 Oe) and the net moment generated under this small field is frozen at low temperature, giving a higher moment, as indicated in Fig. 11(A). When the particles are warmed up, thermal fluctuation of the particles leads to slight moment reduction. Above T_B, the thermal energy overcomes the magnetic anisotropy energy, leading to a sharp drop in moment. The nanoparticle size effect on the T_B of the Co nanoparticles can be readily predicted from KV/k_BT relation and are clearly indicated in the ZFC curves of the particles, as shown in Fig. 11(B).

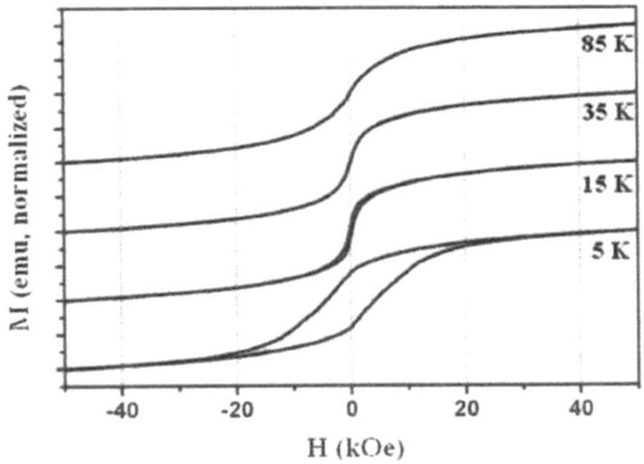

Figure 12. Low temperature hysteresis loops of 4 nm $Fe_{56}Pt_{44}$ nanoparticle assemblies.

Magnetic properties of a self-assembled magnetic nanoparticle array are also dependent on particle structure. The as-synthesized FePt nanoparticles have the chemically disordered fcc structure. Magnetic measurements of the 4 nm $Fe_{56}Pt_{44}$ particles show that they are ferromagnetic only at very low temperature. The coercivity drops sharply as temperature is raised from 5 K to 35 K, above which the particles become superparamagnetic (see Fig. 12). This is consistent with the low magnetocrystalline anisotropy of the fcc structured particles. Thermal annealing converts the internal particle structure from fcc structure to the chemically ordered fct structure and transforms the superparamagnetic FePt nanoparticles into ferromagnetic ones. Studies on these ferromagnetic nanoparticle assemblies reveal that the coercivities of the assemblies are sensitive to the annealing conditions as the fct ordering of the particles is dependent on the annealing history [39, 42, 57].

Interparticle interactions can also influence the magnetic properties of the particles in a self-assembled array. In such an array, each particle is capped with a surfactant layer, and interparticle spacing separated by the surfactant layer essentially eliminates the exchange interactions, leaving the magnetic dipole interactions a dominant factor. As the magnetic dipole interaction energy is proportional to m^2 and $1/d^3$, in which m is magnetic moment of the particle and d is the interparticle spacing, magnetic dipole interaction energy in a close-packed magnetic nanoparticle array with small d can have dramatic effect on the magnetic properties of the array [58-63].

The FePt nanoparticle assembly offers a different system to demonstrate interaction effects on magnetization process of the particles in the assembly [57]. Annealed at 550 °C, the 4 nm $Fe_{58}Pt_{42}$ nanoparticle assembly consists of well-isolated particles. Its hysteresis loop has a remanence ratio of 0.55, Fig. 13(A), that is close to the value of 0.5 predicted for randomly oriented, non-interacting Stoner-Wohlfarth particles. The hysteresis loop of the 600 °C annealed assembly shows a much higher remanence ratio and steeper slope near coercivity, Fig. 13(B). This is typically attributed to the co-operative switching behavior, indicating that the dominant interparticle interaction is exchange coupling. For the 800 °C annealed assembly, however, the initial curve rises steeply at small fields, tends to saturate at lower fields, Fig. 13(C), indicating that the magnetization reversal is controlled mainly by domain nucleation processes. This further reveals that aggregated particles shown in Fig. 9(C) now grow into a large single crystal showing a multi-domain behavior.

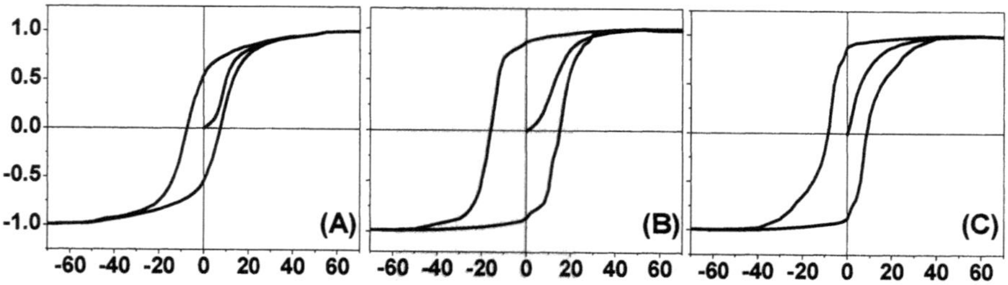

Figure 13. Initial magnetization curves and hysteresis loops for self-assembled $Fe_{58}Pt_{42}$ assemblies annealed at (A) 550°C, (B) 600°C, and (C) 800°C, respectively [57].

5. APPLICATION POTENTIAL OF SELF-ASSEMBLED NANOMAGNETS

5.1. Information Data Storage

Present magnetic disk drives are based on longitudinal recording systems where the magnetization of the recorded bit lies in the plane of the disk. These systems contain recording media that traditionally have a single magnetic storage layer and consist of weakly coupled magnetic grains of CoPtCrX alloy (X = B, Ta) [64, 65]. The fine microstructure of the grains allows writing and storing magnetic transitions at high linear densities with the ultimate density limited by the grain size. The impediment to achieving higher areal densities in magnetic recording is media-dominated system noise that arises from disordered assembly, polydispersed size, and irregular shape of the magnetic grains. Trimming the grain-size distribution and simultaneously reducing the average grain size will allow the maintenance of a larger number of grains per bit and as a result, the increase in recording density.

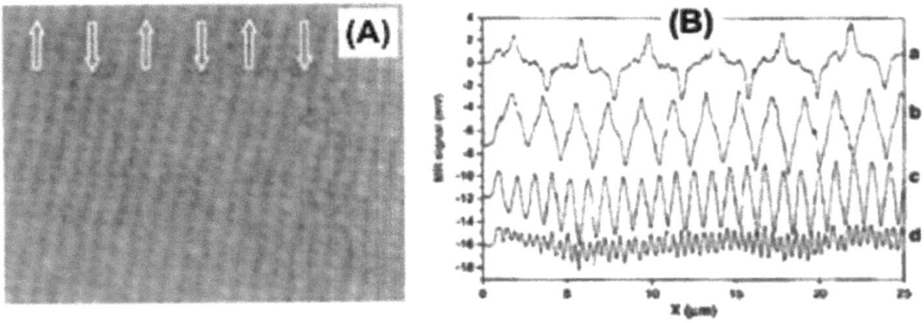

Figure 14. (A) HRSEM image of a ~ 120 nm-thick, ferromagnetic 4-nm $Fe_{48}Pt_{52}$ nanoparticle assembly used for writing experiment. The arrows illustrate the magnetization reversals supported by the assembly. (B) Magneto-resistive (MR) read-back signals from written bit transitions in the assembly of (A). The individual line scans reveal magnetization reversal transitions at linear densities of (a) 500, (b) 1040, (c) 2140, and (d) 5000 fc/mm [39].

Self-assembled FePt nanoparticle arrays are prospective magnetic media candidates [39]. Chemically ordered fct FePt is magnetically hard with anisotropy constant (*K*) reaching ~ 10^8 erg/cc (*K* for current Co-based media is at ~ 10^6 erg/cc). Such large *K* will allow smaller (down to only 3 nm), yet still stable magnetic grains with coercivity high enough to support magnetization transitions at ambient conditions [66]. Preliminary recording experiments on an annealed 120 nm thick 4 nm $Fe_{48}Pt_{52}$ nanoparticle array have been performed at low to medium density [39]. Figure 14(A) is the

SEM image of the 4 nm particle assembly used for the recording demonstration. Figure 14(B) shows read-back signals recorded at linear densities of 500, 1000, 2000 and 5000 fc/mm. Considering the fact that the nanoparticles used for the recording demonstration have only partially ordered fct structure and their magnetic easy axis is 3D randomly oriented, one may expect higher recording densities based on a self-assembled FePt nanoparticle array if each FePt particle can be made ferromagnetic and the magnetic easy axes are aligned.

5.2. Tunneling Devices

In an array containing conducting nanoparticles, electrons can tunnel from one particle to another if two particles are in close proximity [67-70]. The tunneling current is dependent on size and separation of the particles. If the conducting particles are magnetic, the tunneling current can also be controlled by magnetization direction of the particles. Self-assembly is an attractive technique for fabrication of tunneling devices as it has the capability of engineering the position of the nanoparticles. For example, 10 nm Co nanoparticles coated with oleic acid have been self-assembled on a tunnel-junction array of about 100 nm by 100 nm with electrodes having a width of ~ 100-nm, as shown in Fig. 15 [71]. The tunnel junction is made from e-beam lithography and the Co particles are deposited on the device from their octane dispersion. Controlled evaporation of the solvent induces the self-organization of the nanoparticles into an ordered array with interparticle spacing at ~ 4 nm provided by the insulating organic capping group.

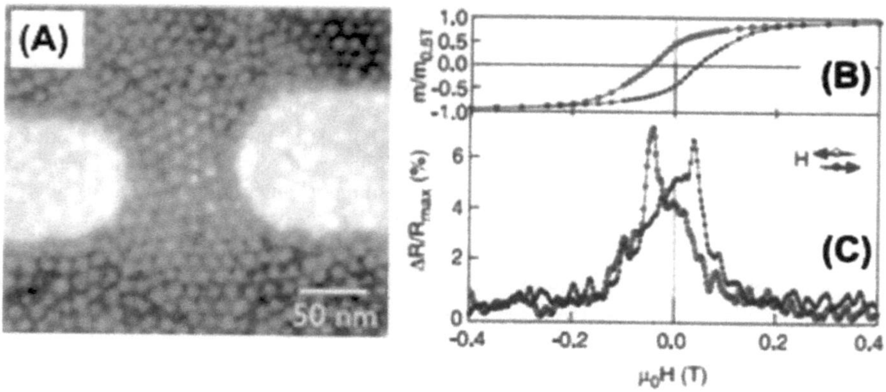

Figure 15. (A) SEM image of self-assembled 10 nm Co nanoparticle array around a tunnel junction; (B) Hysteresis loop of the 10 nm Co nanoparticle assembly at 5 K; (C) Magnetoresistance of the device in (A) at 2 K [71].

Figure 15(A) shows that the shortest current-carrying path in this device contains only about seven nanoparticles in series. The particles form a hexagonal close-packed array over the device dimension of order 100 nm. As deposited, the assembly is electrically insulating ($R > 1$ tera-ohm) due to large interparticle spacing of about 4 nm. Annealing the assembly for 1 hour under reducing conditions partially desorbs the organic coating, brings neighboring particles closer together to ~ 2 nm and decreases the resistivity to 100 kilo-ohm. At temperature lower than 20 K, the *I-V* curve shows Coulomb blockade of current ($R > 1$ tetra-ohm). *M-H* measurements of the particle assembly show a magnetic hysteresis at 5 K with H_c at 0.05 T, Fig. 15(B). When a dc voltage (0.4 V) is applied to overcome the Coulomb blockade of current under an external in-plane field *H*, the device resistance increases as *H* is lowered from saturation to 0.1 T and particle magnetic moments begin to randomize. After *H* passes through zero, the resistance reaches a maximum near H_c, when particle moments are maximally anti-aligned. For fields beyond H_c, the resistance again decreases to the initial value. The device MR ratios are on the order of 10%, approaching the maximum predicted values for ensembles of cobalt islands with randomly oriented magnetic axes.

5.3. Exchange-Spring Magnetic Nanocomposites

An exchange-spring composite contains two modulated phases that are in intimate contact with one being magnetically hard and another magnetically soft. Such a composite system is expected to have higher energy product than the traditional single-phase materials because it has both large coercivity and high magnetic moment [72-74]. To reach the optimum exchange, the size of the soft phase should be around or within 10 nm. Self-assembly is a reliable route to exchange-coupled nanocomposites with controlled nanoscale dimension. In a recent demonstration, FePt and Fe_3O_4 particles are chosen as nanoscale building blocks to form 3D binary assemblies [75].

Figure 16(A) shows the TEM images of the binary composite containing 4 nm Fe_3O_4 nanoparticles and 4 nm $Fe_{58}Pt_{42}$ nanoparticles. It can be seen that both $Fe_{58}Pt_{42}$ and Fe_3O_4 randomly occupy the sites in a hexagonal lattice and form a modulated particle array. Thermal annealing at 650°C under a reducing atmosphere of Ar + 5% H_2 removes the organic coating around each particle in the composite assembly and transforms the composite of FePt-Fe_3O_4 into FePt-Fe_3Pt. Figure 16(B) is a typical high-resolution TEM image of a sintered sample obtained from the 4 nm : 4 nm assembly shown in Fig. 16(A). The coalesced particle is divided into hard magnetic fct FePt phase and soft magnetic fcc Fe_3Pt phase, forming a modulated hard-soft nanocomposites with each phase dimensions on the order of 5 nm.

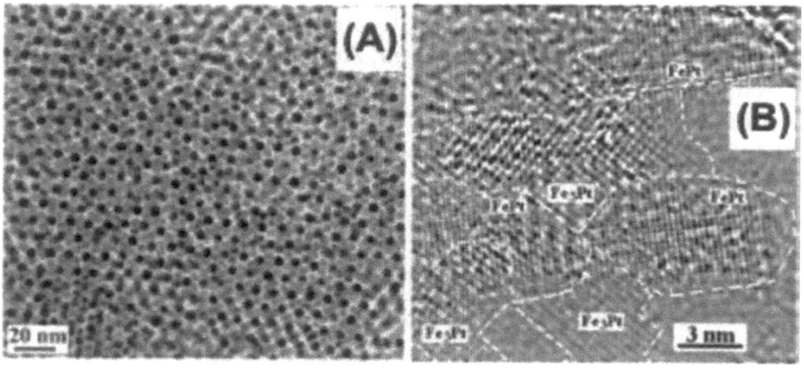

Figure 16. TEM images of binary nanoparticle assemblies of (A) Fe_3O_4(4 nm):$Fe_{58}Pt_{42}$(4 nm) nanoparticles and (B) one sintered FePt-Fe_3Pt composite particle [75].

Magnetic properties of the FePt-Fe_3Pt nanocomposites vary with different initial mass ratios of Fe_3O_4 and $Fe_{58}Pt_{42}$ and particle sizes. Figure 17(A) is a representative hysteresis loop for the composite from the annealed assembly shown in Fig. 16(B). Although the sample consists of both magnetically hard and soft phases, the loop shows that the magnetization changes smoothly with field, similar to the hysteresis behavior of a single-phase material, indicating that effective hard-soft exchange coupling is realized.

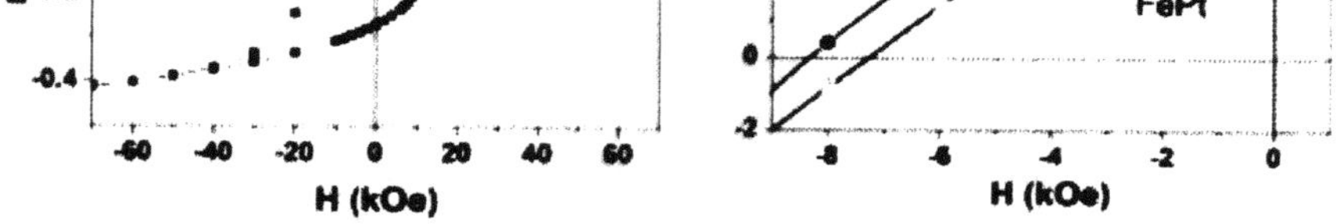

Figure 17. (A) Typical hysteresis loop of a FePt-Fe_3Pt nanocomposites and (B) Second-quadrant B–H curves for the annealed 4 nm $Fe_{58}Pt_{42}$ nanoparticle assembly and hard-soft exchange-coupled FePt–Fe_3Pt nanocomposite [75].

The optimized nanostructures for exchange coupling yield both high remanent magnetization and coercivity, resulting in an enhanced energy product $(BH)_{max}$. Figure 17(B) illustrates two B-H curves: one is from the annealed 4 nm $Fe_{58}Pt_{42}$ nanoparticle assembly and another from the nanocomposite shown in Fig. 16(B). The measured $(BH)_{max}$ for the single-phase $Fe_{58}Pt_{42}$ is 14.7 MGOe while for the nanocomposite, 20.1 MGOe, exceeding the value for single phase $Fe_{58}Pt_{42}$ assemblies by 37%, indicating that exchange-spring system does indeed offer enhanced magnetic properties.

6. CONCLUDING REMARKS

As one of the well-known nanofabrication processes, self-assembly has shown great promise in building high quality magnetic nanostructures. It is a versatile process and can be used to fabricate uniform assembly with controlled structure and interparticle spacing. It is also capable of positioning the nanoparticle building blocks onto a designated area, facilitating the fabrication of single particle devices. Such engineering control on magnetic nanoparticles further allows rational design of new magnetic nanostructures for important nanomagnetic applications. In particular, self-assembled magnetic nanoparticle and composite arrays, as described in this chapter, may evolve into prospective alternative to future ultrahigh density information storage media, highly sensitive magnetic sensors and high-performance permanent magnets.

References

[1] A. H. Morrish, "Physical Principles of Magnetism", Wiley, New York 1965.

[2] B. D. Cullity, "Introduction to Magnetic Materials", Addison-Wesley, Reading, Massachusetts 1972.

[3] C. L. Chien, Mat. Res. Soc. Symp. Proc. **195**, 411 (1990).

[4] A. Ullman, Adv. Mater. **2**, 573 (1990).

[5] G. M. Whitesides, J. P. Mathias, and C. T. Seto, Science **254**, 1312 (1991).

[6] S. I. Stupp, V. LeBonheur, K. Walker, L. S. Li, K. E. Huggins, M. Keser, and A. Amstuz, Science **276**, 384 (1997).

[7] N. C. Seeman, Acc. Chem. Res. **30**, 357 (1997).

[8] G. M. Whitesides and M. Boncheva, PNAS **99**, 4769 (2002).

[9] H. Lodish, A. Berk, S. L. Zipursky, P. Matsudaira, D. Baltimore, and J. Darnell, "Molecular Cell Biology", W. H. Freeman, New York 2000.

[10] J. Shimoiizaka and K. Nakatsuka, "Recent Magnetics for Electronics", **21**, 241 (1985/1986).

[11] D. H. Everett, "Basic Principles of Colloid Science", Royal Society of Chemistry, 1988.
[12] C. B. Murray, S. Sun, H. Doyle, and T. Betley, MRS Bulletin **26**, 985 (2001).
[13] T. Hyeon, Chem. Comm. 927 (2003); M. A. Willard, L. K. Kurihara, E. E. Carpenter, S. Calvin, and V. G. Harris, Int. Mater. Rev. **49**, 125 (2004).
[14] B. L. Cushing, V. L. Kolesnichenko, and C. J. O'Connor, Chem. Rev. **104**, 3893 (2004).
[15] J. Park, E. Lee, N.-M. Hwang, M. Kang, S. C. Kim, Y. Hwang, J.-G. Park, H.-J. Noh, J.-Y. Kim, J.-H. Park, and T. Hyeon, Angew. Chem. Int. Ed. **44**, 2872 (2005).
[16] M. Green, Chem. Commun. 3002 (2005).
[17] D. L. Huber, Small **1**, 482 (2005).
[18] I. Lisiecki, J. Phys. Chem. B **109**, 12231(2005).
[19] S. I. Woods, J. R. Kirtley, S. Sun, and R. H. Koch, Phys. Rev. Lett. **87**, 137205 (2001).
[20] V. F. Puntes, K. M. Krishnan, and A. P. Alivisatos, Science **291**, 2115 (2001).
[21] Y. Gao, Y. Bao, M. Beerman, A. Yasuhara, D. Shindo, and K. M. Krishnan, Appl. Phys. Lett. **84**, 3361 (2004).
[22] Y. Bao, M. Beerman, A. B. Pakhomov, and K. M. Krishnan, J. Phys. Chem. B 109, 7220 (2005).
[23] F. Dumestre, B. Chaudret, C. Amiens, M. Respaud, P. Fejes, P. Renaud, and P. Zurcher, Angew. Chem. Int. Ed. **42**, 5213 (2003).
[24] S. Sun and C. B. Murray, J. App. Phys. **85**, 4325(1999).
[25] C. Petit, A. Taleb, and M. P. Pileni, J. Phys. Chem. B **103**, 1805 (1999).
[26] M. D. Bemtzon, J. van Wonterghem, S. Morup, A. Tholen, and C. J. W. Koch, Phil. Mag. B **60**, 169 (1989).
[27] K. S. Suslick, M. Fang, and T. Hyeon, J. Am. Chem. Soc. **118**, 11960 (1996).
[28] S.-J. Park, S. Kim, S. Lee, Z. G. Khim, K. Char, and T. Hyeon, J. Am. Chem. Soc. **122**, 8581 (2005).
[29] F. Dumestre, B. Chaudret, C. Amiens, P. Renaud, and P. Fejes, Science **303**, 821 (2004).
[30] J. Rockenberger, E. C. Scher, and A. P. Alivisatos, J. Am. Chem. Soc. **121**, 11595 (1999).
[31] T. Hyeon, S. S. Lee, J. Park, Y. Chung, H. B. Na, J. Am. Chem. Soc. **123**, 12798 (2001).
[32] F. X. Redl, K.-S. Cho, C. B. Murray, and S. O'Brien, Nature **423**, 968 (2003).
[33] J. Cheon, N.-J. Kang, S.-M. Lee, J.-H. Lee, J.-H. Yoon, and S. J. Oh, J. Am. Chem. Soc. **126**, 1950 (2004).
[34] J. Park, K. An, Y. Hwang, J.-G. Park, H.-J. Noh, J.-Y. Kim, J.-H. Park, N.-M. Hwang, T. Hyeon, Nature Mater. **3**, 891 (2004).
[35] W. W. Yu, J. C. Falkner, C. T. Yavuz, V. L. Colvin, Chem. Comm., 2306 (2004).
[36] S. Sun and H. Zeng, J. Am. Chem. Soc. **124**, 8204 (2002).
[37] S. Sun, H. Zeng, D. B. Robinson, S. Raoux, P. M. Rice, S. X. Wang, and G. Li, J. Am. Chem. Soc. **126**, 273 (2004).
[38] Q. Song and Z. J. Zhang, J. Am. Chem. Soc. **126**, 6164 (2004).
[39] S. Sun, C. B. Murray, D. Weller, L. Folks, and A. Moser, Science **287**, 1989 (2000).

[40] S. Sun, E. E. Fullerton, D. Weller, and C. B. Murray, IEEE Trans Magn. **37**, 1239 (2005).

[41] M. Chen, J. P. Liu, and S. Sun, J. Am. Chem. Soc. **126**, 8494 (2005).

[42] S. Sun, S. Anders, T. Thomson, J. E. E. Baglin, M. F. Toney, H. F. Hamann, C. B. Murray, and B. D. Terris, J. Phys. Chem. B. **107**, 5419 (2003).

[43] C. Liu, X. Wu, T. Klemmer, N. Shukla, X. Yang, D. Weller, A. G. Roy, M. Tanase, and D. Laughlin, J. Phys. Chem. B. **108**, 6121 (2005).

[44] C. B. Murray, S. Sun, W. Gaschler, H. Doyle, T. Betley, and C. R. Kagan, IBM J. Res. & Dev. **45**, 47 (2001).

[45] Z. L. Wang, D. R. Dai, and S. Sun, Adv. Mater. **12**, 1944 (2000).

[46] H. Zeng, P. M. Rice, S. X. Wang, and S. Sun, J. Am. Chem. Soc. **126**, 11458 (2005).

[47] S. Sun, C. B. Murray, and H. Doyle, Mat. Res. Soc. Symp. Proc. **577**, 385 (1999).

[48] Z. R. Dai, S. Sun, and Z. L. Wang, Nano Lett. **1**, 443 (2001).

[49] G. Decher, Science **277**, 1232 (1997).

[50] Y. Liu, A. Wang, and R. Claus, J. Phys. Chem. B **101**, 1385 (1997).

[51] T. Cassagneau, T. E. Mallouk, and J. H. Fendler, J. Am. Chem. Soc. **120**, 7848 (1998).

[52] J. Schmitt, P. Mächtle, D. Eck, H. Möhwald, and C. A. Helm, Langmuir **15**, 3256 (1999).

[53] N. A. Kotov, MRS Bulletin, 992 (2001).

[54] J. F. Hicks, Y. Seok-Shon, and R. W. Murray, Langmuir **18**, 2288 (2002).

[55] F. Hua, T. Cui, and Y. Lvov, Langmuir **18**, 6712 (2002).

[56] S. Sun, S. Anders, H. Hamann, J.-U. Thiele, J. E. E. Baglin, T. Thomson, E. E. Fullerton, C. B. Murray, and B. D. Terris, J. Am. Chem. Soc. **124**, 2884 (2002).

[57] H. Zeng, S. Sun, T. S. Vedantam, J. P. Liu, Z. R. Dai, and Z. L. Wang, Appl. Phys. Lett. **80**, 2583 (2002).

[58] M. J. Vos, R. L. Brott. J.-G. Zhu, and L. W. Carlson, IEEE Trans. Magn. Magn. **29**, 3652 (1993).

[59] L. M. Malkinski, J.-Q. Wang, J. Dai, J. Tang, and C. J. O'Connor, Appl. Phy. Lett. **75**, 844 (1999).

[60] P. Allia, M. Coisson, M. Knobel, P. Tiberto, and F. Vinai, Phys. Rev. B **60**, 12207 (1999).

[61] H. Rubio and S. Suarez, J. Appl. Phys. **87**, 7415 (2000).

[62] C. R. Pike, A. P. Roberts, and K. L. Verosub, J. Appl. Phys. **88**, 967 (2000).

[63] Y. Sun, M. B. Salamon, K. Garnier, and R. S. Averback, Phys. Rev. Lett. **91**, 167206 (2003).

[64] D. Weller and M. E. Doerner, Annu. Rev. Mater. Sci. **30**, 611 (2000).

[65] A. Moser, K. Takano, D. T. Margulies, M. Albrecht, Y. Sonobe, Y. Ikeda, S. Sun, and E. E. Fullerton, J. Phys. D: Appl. Phys. **35**, R157 (2002).

[66] D. Weller, A. Moser, L. Folks, M. E. Best, W. Lee, M. F. Toney, M. Schwickert, J.-U. Thiele, and M. F. Doerner, IEEE Trans. Magn. Magn. **36**, 10 (2000).

[67] R. P. Andres, J. D. Bielefeld, J. I. Henderson, D. B. Janes, V. R. Kolagunta, C. P. Kubiak, W. J. Mahoney, and R. G. Osifchin, Science **273**, 1690 (1996).

[68] R. P. Andres, T. Bein, M. Dorogi, S. Feng, J. I. Herderson, C. P. Kubiak, W. Mahoney, R. G. Osifchin, and R. Reifenberger, Science **272**, 1323 (1996).

[69] D. L. Klein, R. Roth, A. K. L. Lim, A. P. Alivisatos, and P. L. McEuen, Nature **389**, 699 (1997).
[70] D. J. Schiffrin, MRS Bulletin, 1015 (2001).
[71] C. T. Black, C. B. Murray, R. L. Sandstrom, and S. Sun, Science **290**, 1131 (2000).
[72] E. F. Kneller and R. Hawig, IEEE. Trans. Magn. **27**, 3588 (1991).
[73] R. Skomski and J. M. D. Coey, Phys. Rev. B **48** 15812 (1993).
[74] T. Schrefl, H. Kronmüller, and J. Fidler, J. Magn. Magn. Mater. **127**, L273 (1993).
[75] H. Zeng, J. Li, J. P. Liu, Z. L. Wang, and S. Sun, Nature **420**, 395 (2002).

Chapter 10

PATTERNED NANOMAGNETIC FILMS

J. Cock Lodder
Systems & Materials for Information Storage,
MESA and Institute for Nanotechnology, University of Twente, PO Box 217,
7500AE Enschede, The Netherlands.

Abstract Nanofabrication technologies for patterned structures made from thin films are reviewed. A classification is made to divide the patterning technologies in two groups namely with and without the using a mask. The more traditional methods as well as a few new methods are discussed in relation with the application. As mask-less methods we discussed direct patterning with ions including FIB, nanopatterning with electron beams, interferometric laser annealing and ion beam induced chemical vapor deposition. The methods using masks are ion irradiation and projection, interference lithography, the use of pre-etched substrates and templates from diblock copolymers and imprint technologies. Also a few remarks are given concerning the magnetic properties of patterned films. Nanometer scale magnetic entities (nanoelements, nanodots, nanomagnets) form a fast growing new area of solid-state physics including the new fields of applications. Two of them are summarized namely media for ultra high-density recording and magnetic logic devices.

1. INTRODUCTION

Patterned magnetic nanostructures are researched and developed for fundamental studies and applications in areas such as nanomagnets, ultra high-density storage, spinelectronic devices, memory cells (MRAM) magnetic logic devices. Recently a few overview papers about nanostructuring and related properties have been published [1-3]. Many methods are used to make small magnetic structures from sub micron scale down to nanometer dimensions. They have been made either in single dots or periodic structures such as arrays in rectangular or hexagonal shape depending of the type of study and application. In the literature the fabrication of magnetic nanostructures has been reported either by using the top-down as well as the bottom–up strategy. In this chapter we will focus on the methods following the top down approach. This type of patterned structures can be made by one process step technique (deposition through a shadow mask) or via a multi-process step such as using photoresist, etching, lift-off techniques etc. Patterns with nano-dimensions cannot be prepared by standard-optical lithography and therefore other methods are used such as: electron-beam lithography, X-ray lithography, nanoimprint lithography, and interference lithography. But also focused ion beam techniques, nano-imprint and diblock copolymer template technologies are use for realizing nanostructures in magnetic thin films.

It is well known that the magnetic properties from a ferromagnetic bulk material differ with the same material in thin film form. In addition a comparison between patterned structures will show different magnetic properties from their parent thin film material. In general a magnetic material can change its magnetization either by domain wall motion or by one of the rotation mechanisms. Figure 1 illustrates how the domain structure and in connection the reversal mechanism changes as function of the size. In this case the continuous Co/Pt multilayer film, with a perpendicular anisotropy, shows a typical stripe domain structure. The patterned structures are made by interference lithography (IL) and dry etching. After patterning the dots of 180 nm are still multi-domain (MD) but if they have a size of 70 nm only the single domain (SD) state with was detected in the MFM observations.

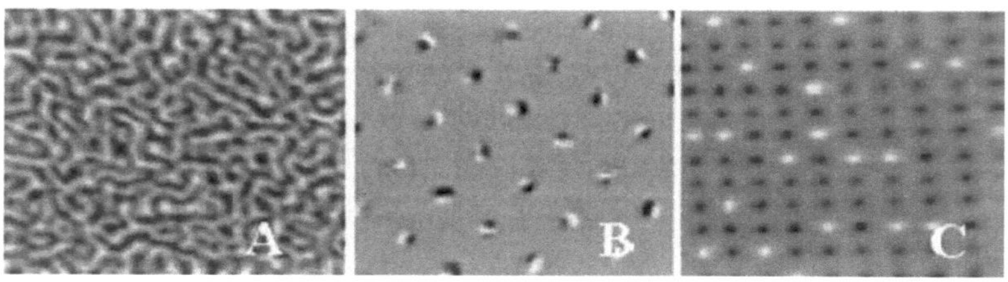

Figure 1. MFM image (4mm x 4mm) from a Co/Pt multilayer continuous film (A), arrays of MD dots of 180 nm (B) and SD dots of 70 nm (C).

Another example showing the influence of the shape on the magnetic behavior of a nanostructure is discussed in [4]. Here experiments and theory have been carried out on various shaped dots of Supermalloy (Ni-Fe-Mo alloy) made by e-beam lithography. They show, as prepared, an in-plane isotropic anisotropy. The various shapes of the magnetic dots (squares, triangle or pentagon) have a great impact on the magnetic anisotropy and related properties. This type of films shows in addition the phenomenon of configurational anisotropy. Configurational anisotropy is related to the role played by small deviations from uniformity in the magnetization field within the nanostructures, which allow unexpected higher-order anisotropy terms to appear. But also the competition that exists between exchange energy and magnetostatic energy plays an important role. This competition determines whether the nanostructures exhibit single domain or incoherent magnetization and also controls the non-uniformities in magnetization, which lead to configurational anisotropy [4].

Hysteresis loops measured form a continuous film and various shapes of nanostructures from Supermalloy are given in Fig. 2. The thickness of the various films is 6nm (a), 5nm (b), 5nm (c) and 3 nm (d). The applied field is assumed to point up the page. Understanding the influence of shape opens the way to designing new nanostructured magnetic materials where the magnetic properties can be tailored to a particular application with a very high degree of precision. It has been determined by that the triangle shape exhibit a 6-fold, square shape a 4-fold and the pentagonal structures a 10-fold anisotropy [4, 5].

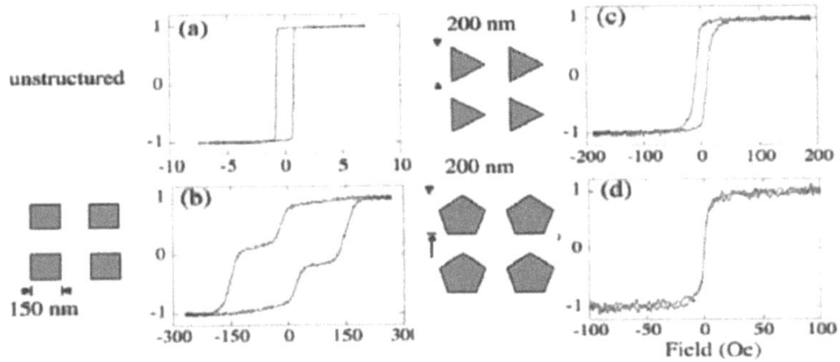

Figure 2. Hysteresis loops as a function of the various shapes squares (b), triangles (c) and pentagons (d) of patterned structures of Supermalloy and the continuous film (a).

2. PATTERNING TECHNOLOGIES FOR MAGNETIC THIN FILMS

It is generally realized that for sub-100 nm structures a new generation of lithography techniques needs to be developed. For the preparation of limited numbers of nanosize structures, electron beam lithography (EBL) is more or less the standard patterning technology. But direct e-beam writing is a very slow process in the case large areas should be covered.

Using EBL or ion beam lithography (IBL) a variation of shapes and dimensions can bee realized even down to 10 nm in size. With EBL large areas cannot be made accurately due to the problems with the long-range coherence of the e-beam writer and also the throughput is very low. But it is a very flexible method for the fabrication of nanostructures of arbitrary shapes for research [4]. Recently it has been shown that EB and IB projection methods can solve the problems with the throughput and mis-coherence

[6]. Other lithography methods such as X-ray lithography and nanoimprint lithography (contact lithography) and the mask-less interference lithography are also used for making large areas of nanostructures. Soft X-rays ($\lambda = 0.5 - 4.0$ nm) lithography is using a thin (about 2 micron) mask just above the sample with the resist layer. The mask is mostly prepared form silicon carbide covered by a metallic pattern (with high absorbing materials such as gold, tantalum or tungsten) with the designed geometry. The advantage is that the mask can be used repeatedly.

Nanoimprint lithography is a fast growing technology and is principally different from the conventional lithographical methods. It has two steps namely imprint and pattern transfer and was proposed in 1995 [8]. The mold with the nanostructures at the surface is pressed into a resist layer on the substrate. The imprint can be carried out by a temperature of about 120 °C at a pressure of 30- 50 bars. The parameters are of course depending on the type of resist used. Sub-10 nm features and 40 nm pitch on Si as well as on Au substrates have been demonstrated [8]. The mask and the master mold used by the two lithographic techniques mentioned above have to be manufactured by using other lithography methods such as e-beam, nanoimprint or interference lithography (IL). In the case of IL a resist layer is exposed by an interference pattern generated by two obliquely incident beams (laser or X-rays). With one exposure resist lines are produced having a period of half the wavelength used. After rotating the sample over 90 degrees a second exposure can be given and after that this illumination dot arrays can be produced.

It is arbitrary to make a classification about the available technologies for making nanostructured magnetic films. In this review we follow two routes namely direct writing and writing by the use of a mask. Direct writing can be carried out by using ion beam, e-beam and laser interference annealing. If a mask is used, methods as ion irradiation, interference lithography, imprint technology and block copolymer templates can be mentioned.

2.1. Mask-Less Thin Film Patterning

Instead of using resist layers for pattern definition, the desired pattern may also be directly written in the magnetic layer or even be directly deposited. Mask less nanopatterning includes ion and electron beam direct irradiation of the film and also a method called laser interference annealing. Irradiation and annealing the film can influence anisotropy or magnetization by changing the phase and or chemical composition of the film locally or over a large area. Also other methods are discussed such as, laser-assisted direct imprint, ion-beam-induced chemical vapor deposition (IBICVD), and ion projection patterning.

2.1.1. Direct patterning with ions

The use of energetic ion irradiation to alter the properties of magnetic films and multilayers has attracted increasing interest in recent years. When ions interact with the surface, many effects occur, such as radiation damage, elastic reflected ions, implantation and ion etching (milling). The magnetic properties such as coercive field [9-11] anisotropy field [12], and exchange biasing [13] can be locally and controllably changed to engineer the behavior of small magnetic devices. Especially multilayer structures are very sensitive for ion irradiation due to the sensitive area of the interface between the ferromagnet and metal layers. A good example is the research carried out on Co/Pt films with perpendicular anisotropy. The multilayers are studied before and after ion irradiation and a model about the mechanism of the destroying interface is given in [12]. The Co atoms moving in the ion direction travel more than one interatomic distance and become an isolated Co-Pt alloy, whereas Co atom moving in opposite direction travel typically only one interatomic distance and contribute to the roughness (local thickness fluctuations). The 'rough' interface will change the magnetic properties a lot. With this technology one can locally change the magnetic state of a sample. Irradiation studies on magnetic multilayers are first published by Traverse *et al.* [14] studying, e.g. Ni/Au multilayers using fluencies between 2×10^{14} and 2×10^{16} ion/cm^2 of 30 keV He^+ ions.

Patterned media (e.g. section 4.1) can be fabricated from magnetic multilayers where ion irradiation is used to differentiate nanoscale magnetic islands in a magnetic matrix, without the removal of material from the multilayers [15, 16]. Arrays of Pt/Co/Pt dots with a dot separation of only 20 nm have been shown by FIB irradiation of 28keV Ga+ ions [17]. The magnetic effects can be influenced by the kinds of ions (He+, Ar+, Ga+, N+), their energies as well as their exposure doses (fluency).

Milling with focused ion beam (FIB)

Patterned structures from films can also be made by removing material by ion-milling/etching. Mostly a Focused Ion Beam (FIB) is used as instrument. Besides, Co/Pt multilayers, perpendicular granular CoCrPt films (20 nm thick) have been prepared using FIB [18]. The patterns are directly made in the film by using 30 keV Ga+ at a beam current of 1 pA. MFM and AFM images of the directly patterned film are given in Fig. 3.

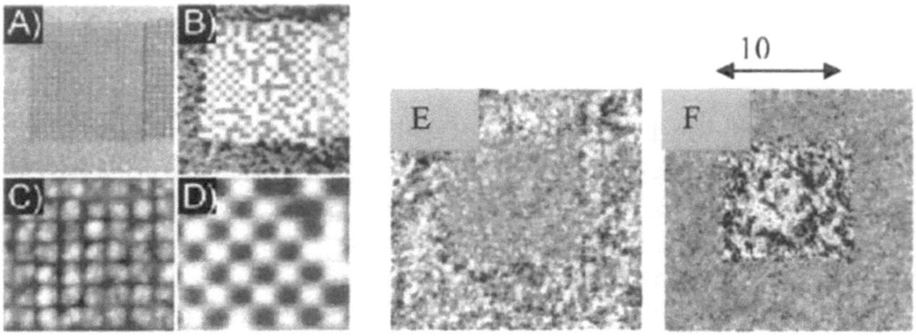

Figure 3. AFM (A and C) and MFM images (B and D) of a CoCrPt thin film medium directly patterned by FIB. Magnetic-domain images taken by spin-SEM after electron bombardment of a centered square for 1200 s; (E) in-plane, and (F) out-of-plane magnetization component.

The pattern trenches are 6 nm deep and square arrays of magnetic isolated islands have been prepared in the range 65-500 nm. Figure 3(A) shows an array of 2x2 mm^2 patterned into a period of 20 x 20 islands with100 nm period and 70 nm side-length. The MFM images (B&D) are made after AC demagnetization procedure of the sample using a perpendicular field from 20 to 100 Oe. With a quasi-static write/read tester, recording experiments have been performed (see section 4.1).

2.1.2. Direct nanopatterning with e-beam

Intense electron beams with keV energy can modify magnetic anisotropies locally. In [19] it has been shown that in very thin epitaxial Co films the easy axis switch from parallel to perpendicular to the surface while the domain size can be enlarged by one order of magnitude. In this paper it has been shown that local magnetic modifications can bee 'written' without adding or removing atoms or molecules. The e-beam of a SEM was used to scan the surface with an operating voltage of 10 keV. A maximum beam current at the sample position of 10 nA was achieved by removing the objective aperture of the microscope. Total exposure times were varied from 10 to 1200 s. The principle of local magnetic modification is illustrated in Fig. 3(E&F). The electron beam scanned an area from 18 × 18 mm^2. The irradiation in the square corresponds to a dose of 2.4 × 10^{10} electrons/mm^2. The spin-SEM image is taken from both the electron-beam-treated area and its surroundings. Within the square, the magnetization direction has switched completely from in-plane to out-of-plane as shown by the uniform gray level in Fig. 3(E) and the black/white contrast in Fig. 3(F).

Another application of direct e-beam irradiation is given in [20]. In this paper, non-ferromagnetic Co-C thin films were magnetically patterned using a focused electron beam with a probe diameter of less than 0.1 μm. The deposited film shows a superparamagnetic behavior but after an anneal process (400-500 °C) the film becomes ferromagnetic. The smallest magnetic dot diameter is about 270 nm. By the use of AFM and MFM characterization the physical dot size and the magnetic dots size can be determined.

In Fig. 4(A) it can be seen that the AFM dot size is almost independent on the dwell time (e-beam irradiation time) while the magnetic dot diameter (MFM) increases almost linearly with the square root of the dwell time.

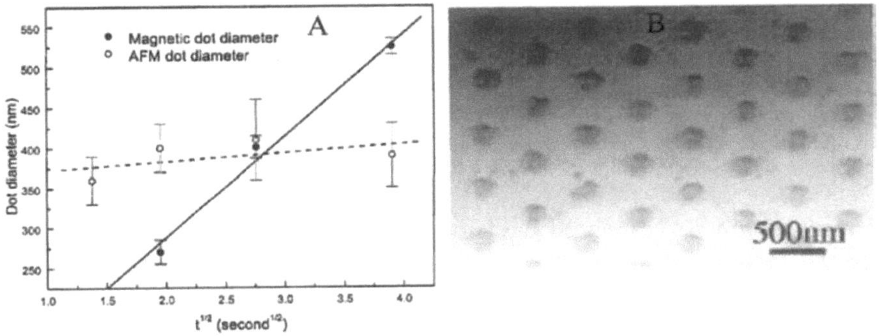

Figure 4. (A) Time dependence of AFM (open circle) and magnetic (solid circle) dot size and (B) TEM micrograph of a patterned Co-C film with direct laser lithography.

The dependence of 'square root of dwell time per dot', means that the magnetic dots are produced by heat-conduction- induced phase change in the film.

2.1.3. Interferometric laser annealing

Laser interference can be used as a lithography method and is particularly suitable for realizing periodic structures in resists. The resist is exposed to a standing wave pattern generated by the interference of two laser beams (see also section 2.2.3.) This interference will give a line pattern with a periodicity Λ given by $\lambda/2\sin\theta$, where λ is the wavelength and θ the angle between the incidence light and the sample normal. By changing θ the periodicity can easily be adjusted from 700 nm down to 100 nm. Hexagonal or cubic patterns are made by double exposure with the sample rotated 60 or 90 degrees between the exposures. Direct laser interference annealing is applied directly on a magnetic thin film. It produces a 2-dimensional hexagonal dot array. This method has been explained in [21] and applied to sputtered non-magnetic Co-C thin films of 40 nm thick. An excimer laser having a wavelength of 308 nm was used and expose a 10-ns-pulse and a flux of 0.17 J/cm^2. The laser beam was split into two beams generating an interference pattern and a periodic modulation of the light intensity. The size and shape of the regions depend on the incident angles of the beam, on the pulse duration, and on the flux. The as-deposited Co-C films are more or less amorphous and in this state the films are not ferromagnetic at RT. After annealing (350-550 °C) the magnetic Co nanoparticles are formed with diameters from 1 to 25 nm. The use of Co-C films was chosen because they form a simple system containing Co carbides which decompose easily into immiscible Co and C upon annealing [22]. Figure 4(B) shows a TEM image of a Co-C film after laser annealing. The dots are 250 nm in diameter and the dot spacing is 650 nm. Dots as well as the matrix are showing a granular structure with 1-2 nm grains with evidence for slightly more order in the dot area [21].

2.1.4. Ion Beam Induced Chemical Vapor Deposition

This method can be compared with Ion Beam Assisted Deposition (IBAD) where the arriving atoms on the substrate receive an additional energy for the growth by ions. In the case of ion beam induced CVD there is a momentum and energy exchange between incident ion beam (Ga^+) and the molecules of a precursor. Besides the 'milling' capacities of the FIB this machine can be used also for deposition of materials in almost arbitrary shape. The principle of FIB deposition is the localized chemical vapor deposition (CVD) by a direct writing technique. The reactions occurring during deposition of metals and insulators are well comparable to, for example, laser-induced CVD [28]. Patterns of deposited areas are fabricated through vector scan rastering. For our application the precursor is suitable for creating a ferromagnetic metal. In the case of Co deposition di-cobalt octacarbonyl has been used. The method is up to now only described for FeCo, CoPt etc materials which can be eventually applied as a magnetic dot array medium for future data storage applications.

The deposition process is illustrated in the left part of Fig. 5. The precursor gases are sprayed on the surface by the 'nozzle', where they are adsorbed. In a second step, the incoming ion beam decomposes the adsorbed precursor gases. Then the volatile reaction products desorb from the surface and are removed through the vacuum system, while the desired reaction products remain fixed on the surface as a thin film. The deposited material is not fully pure however, because organic contaminants as well as Ga ions (from the ion beam) are inevitably included in the deposited film [23].

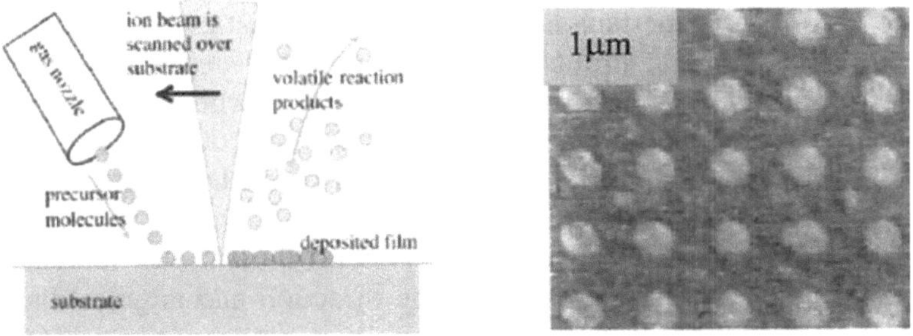

Figure 5. Co dots (800 nm diameter) fabricated by IBICVD technology: the principle of ion–induced deposition (left) and FIB in-situ imaging (right).

The right part of Fig. 5 shows an array of 800 nm Co dots, spaced from each other by 800 nm. Also smaller dot size (150 nm) has been prepared. The improvement is achieved mostly due the selection of a proper substrate that minimizes surface diffusion and mixing within the surface region. Further optimizations may lead to dot sizes with a factor 2-5 smaller [24].

2.2 Patterning with Masks

The most principle idea about a patterning process for a thin film is that the film on the substrate is covered with a spun layer of resist and the pattern is exposed in the resist through a mask. Finally the resist layer or other layers can be used as 'masks' during the following process-step. From the IC-technology it is known that many 'mask' are necessary to final the device. The 2D pattern is transformed into a 3D resist structure and further after processing in the thin film. Developing the resist various profiles are made depending on the developer, the type of resist, the developing time and temperature. Patterning transfer is then realized by etching the thin film. This can be done by wet-chemical as well as by dry-etching methods. Finally a solvent or a dry etching method will strip the remaining resist. An alternative to realize structures is using the lift-off technique. In this case the resist structure is formed on the substrate and the film is deposited through the openings/holes and on the resist surface. While the resist is stripped the material deposit on the resist is lifted off while the film on the substrate remains. Frequently one can find electrodeposition in combination with post-lithographical processes. In this case the substrate should be conductive or should first be coated with a conductive layer.

2.2.1. Ion irradiation through a mask

Here the magnetic properties are modified by irradiation (see also Section 2.1.1.) of the material through a lithographically defined mask. Good examples of the effect of irradiation is given by [25] where ion irradiation is used to alter the direction and angular dispersion of the anisotropy in sputtered soft magnetic NiFe films of 5 nm thickness. During the irradiation in some cases an applied field of 400–700 Oe in a specified direction at the samples was present. The fluency was between 10^{13} and 10^{16} ion/cm^2 while 200 keV Ar^+ ions where used. Anisotropy direction and angular dispersion were determined by angle-dependent remanence measurements [26]. From this the conclusion can be taken that ion irradiation can be used to either increase or decrease dispersion in this film.

The left part of Fig. 6 shows the effect of ion irradiation through a mask for a permalloy thin film. The energy deposited by the ions allows for the movement of the sample atoms, most likely through a series of direct collisions. Much as an applied field during film growth or thermal annealing can be used to affect the bulk magnetic anisotropy of a sample, an applied field during irradiation encourages magnetic ordering along the field direction where ions traverse the film. Thus a film can be anisotropy patterned by irradiation through a mask with the annealing field *set al.*ong any direction. The minimum feature size attainable by this technique will most probably be set by the feature size of the mask (or extent of a focused ion beam), and therefore should be on the order of 10 nm [25].

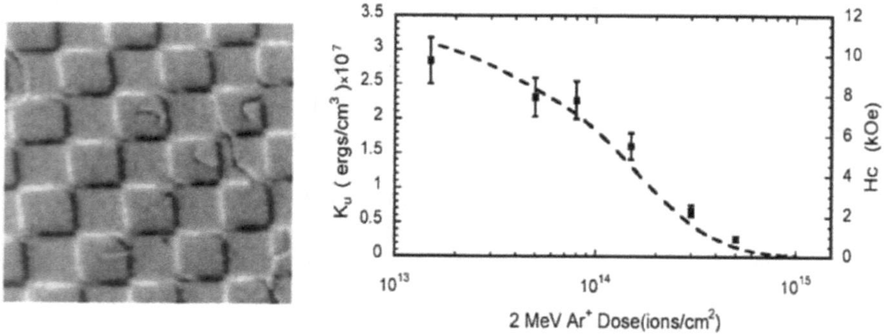

Figure 6. NiFe film with 10 mm square magnetic domains: MFM micrograph (left) and perpendicular anisotropy (K_u) and H_c on exposure to 2 MeV Ar^+ for Co/Pt samples (right).

Many papers have been published about studying this irradiation mechanism for modifying Co/Pt like multilayers with a perpendicular anisotropy (see also Section 2.1.1). In [27], fluencies from 2×10^{14} to 2×10^{16} ions/cm^2 have been used at 30 keV He^+ ions. The He^+ ions do not etch the sample and initial roughness is preserved, but the result is a decrease of magnetic anisotropy, coercivity and Curie temperature. Consequently, in Co/Pt multilayers the anisotropy may change from the perpendicular to the in-plane direction, or the unprotected areas may even change into the paramagnetic state. The mechanism behind this is still in discussion, and in [9, 12] local atomic relaxation (see also Section 2.1.1.) and alloy formation are given. Other authors used 700 keV N^+ ions and found direct evidence for local atomic displacements at the Co/Pt interface, which destroys the anisotropy [28]. The dose required initiating spin reorientation scales with the interface anisotropy K_s of the films used in [28]. For a film with high $K_s \sim 85$ erg/cm^2 they need 1.5 x 10^{15} N^+/cm^2 while for a film $K_s \sim 4$ erg/cm^2, 6 x 10^{14} N^+/cm^2 is necessary. It is roughly equal to the numbers of Co interface atoms per unit interface area contributing to K_s [28]. Irradiation studies of Co/Pt with different energies and ion species (20 keV He^+, 2 MeV H^+, 20 keV Ar^+, 2 MeV Ar^+, 30 keV Ga^+) using doses (10^{11}-10^{17} ions/cm^2) have been carried out [29]. For the same amount of intermixing the dose of He^+ can be replaced by heavier ions such as Ar^+ (100 fold) of the same energy and 400-fold for Xe^+.

In Fig. 6(right) the perpendicular anisotropy K_u (= $K_v + 2\ K_s/t$) are given as a function of the exposure of 2 MeV Ar^+ for a Co/Pt multilayer. In this case the K_u is determined by a contribution from the volume anisotropy (K_v) and the interface anisotropy (K_s) and the thickness of the ferromagnetic layer (t). The conclusion is that the fall of the coercivity with dose reflects a reduction in the perpendicular anisotropy associated with intermixing of the Co/Pt interfaces [29]. In this paper it is concluded that the multilayers are very sensitive to irradiation than expected on the basis of nearest-neighbor coupling and simple ion-beam mixing. It is also stated that the ion irradiation effect strongly depends on the type of samples one made. Moreover, the mask lifetime is a concern, as well as the surface roughness of the irradiated areas. This means that fluency and energy as well as the use of seed layers and substrate must be optimized. The right part of Fig. 6 compares the change of perpendicular anisotropy K_u and coercivity H_c on exposure to 2 MeV Ar^+ for Co/Pt multilayer samples. The data points show the anisotropies as determined with VSM (left scale). The M_s value used to calculate the $2\mu_0 M_s^2$ term was 450 emu/cm^3, except for the point at 5×10^{14} ions/cm^2, where it had fallen to about 250 emu/cm^3. The dashed line follows coercivities obtained for Kerr rotations for the same samples (see right scale in the right part of Fig. 6). Anisotropy also can be engineered to be out-of-plane and in-plane alternatively by using selective epitaxial growth. In [30] an approach to anisotropic patterning for Ni films is reported by using selective epitaxial growth. In this method, a modulated single/polycrystalline substrate surface was used to modify locally the magnetic anisotropy in subsequently deposited magnetic films, which induces the desired artificial magnetic structure. Consequently the Ni film has a spatially varying anisotropy via selective epitaxial growth.

2.2.2. *Ion projection direct structuring (IPDS)*

In the previous section we have seen that irradiation with ions can (locally) modify the magnetic properties of the materials such as Co/Pt multilayers. Direct writing with ion, like e-beam writing, is a slow process. IPDS [6] facilitates a large-area patterning process. It is a non-contact process, which preserve the as-grown surface structures. The mask (mostly SiO_2) consists of all structures necessary for hard disk drives such as bit cells in a circle arrangement and servo (head positioning) structures. IPDS transfer the mask structure with a demagnification factor 4 into the magnetic medium. The exposure was carried out with He+ ions having a dose of 2×10^{15} He^+ /cm^2 at 45 keV in Co/Pt multilayer. The ion energy at the mask is lower than the ion energy at the Co/Pt layer, and moreover the medium is at a significant distance from the masks. This method is proposed for commercialization to fabricate the future hard disks media.

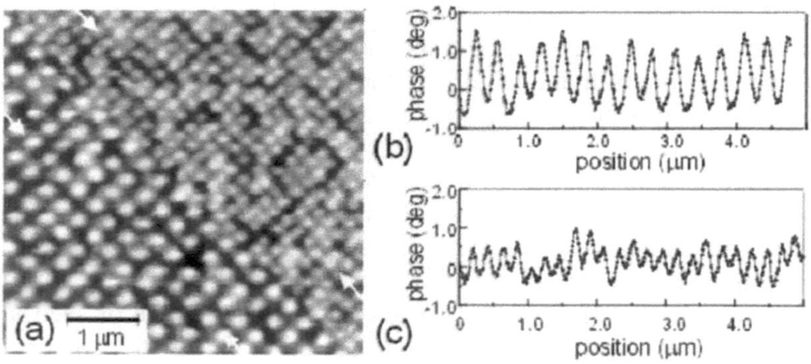

Figure 7. Co/Pt multilayer irradiated through a stencil mask: (a) MFM image, and (b) and (c) line scans from lower left and upper right parts, respectively, along rows of dots as indicated by the white arrows.

Circular tracks of 17 mm in diameter containing data and servo structures were structured [6]. The proposed IPDS technology shows a direct structuring process without any contact with the surface [31]. In Fig. 7 the MFM analysis is given for a Co/Pt multilayer irradiated through a stencil mask with Cartesian dot configuration showing 86 nm dots at a 344 nm pitch in the lower left part and 57 nm dots at 230 nm pitch in the upper right part. The lines scanning from the lower left and the upper right along rows of dots as indicated by white arrows are given in (b) and (c) respectively (see also

[6]). The authors are planning to improve their machine and stated that a storage capacity of 1 Tb/in^2 could realistically be obtained.

2.2.3. *Interference lithography*

With interference lithography (IL) a resist layer is exposed by an interference pattern generated by two obliquely incident laser beams, which is used to expose a photoresist layer without the use of a mask (see section 2.1.3.). Dots can simply be fabricated by a second exposure after rotating the substrate over 90°. The patterned area is determined by the diameter of the two laser beams. For processes with no need for alignment, IL is relatively simple and cheap. The method is already used for more than 20 years for the preparation of submicron gratings for application in integrated optics [32]. In recent years, the possibility of patterning large areas of dots with IL has been recognized. Now, IL is attracting great attention for a wide variety of applications, amongst others: field emission displays, antireflection-structures, photonic crystals, micro filtration and mask making for nanoimprint [41] and X-ray lithography [35] and patterned magnetic media [35-37]. Its capability for patterning large areas is clearly demonstrated by the fabrication of 50 x 50 cm^2 areas of submicron dots [38].

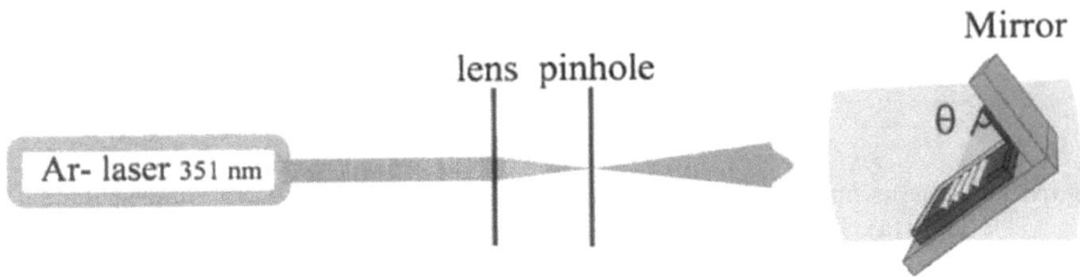

Figure 8. LIL set up for making 2D patterns in a photoresist on top of a thin magnetic film.

Two IL methods have been used for preparing magnetic dots, namely the simple Lloyd's mirror (Fig. 8) consisting of a mirror mounted perpendicular to the substrate [37] and achromatic interference lithography (AIL) [39]. For the preparation of isolated magnetic dots by IL, several processing schemes have been developed and are based on the deposition in resist holes by electroplating [36] or evaporation-and-lift-off [35], or on the selective removal of sputter-deposited material by etching [37]. Figure 9(a) shows IL resist pattern on top of the Co/Pt multilayer (30 nm thick) using a 351-nm laser. A TEM cross-section is shown in Fig. 9(b), after etching the resist and the Co/Pt. The etching has been stopped in the SiO_2/Si substrate. The obtained dots looks like SD with a diameter of 70 nm dots and a period of 200 nm (see Fig. 1(C)). IL was also shown with a laser wavelength of 257 nm and application of deep UV-resists with high contrast [40]. Lasers with even smaller wavelength have poor temporal and spatial coherence, and are therefore not suitable for laser IL. However, achromatic interference lithography with ArF excimer lasers at 193 nm has the capability of patterning resist structures with periods of 100 nm [41]. The principle of the technique is the use of an exposure set-up with vanishing small difference in the length of the two light paths. It also has been shown that the proposed technique could be expanded for the patterning of structures with 50-nm period [41]. In order to produce the smallest possible interference patterns, there should be a compromise between wavelength and stability of the laser source. Pulsed lasers have a short wavelength, but suffer from a poor coherency length. For our LIL setup we chose a highly stable continuous wave (CW) YAG laser with a basic wavelength of 1064 nm.

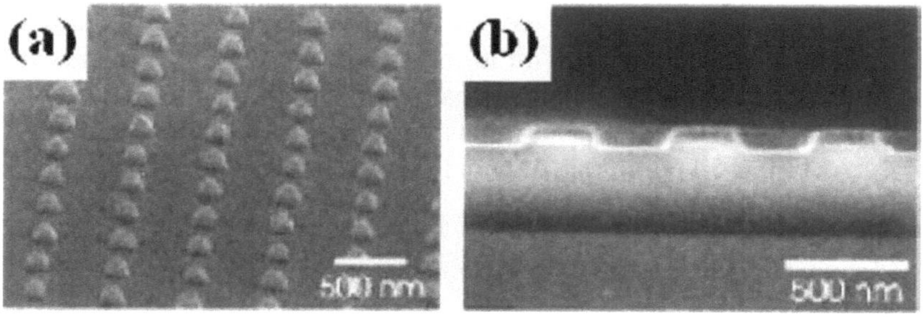

Figure 9. Patterns made by IL using Co/Pt multilayers cover with resist top of the Co/Pt layer (a), TEM cross section of the ion etched structure through the Co/Pt in the Si substrate (b).

This wavelength is frequency-doubled to 532 nm in an intra-cavity. By using an external frequency doubler this wavelength is again halved to λ = 266nm with a small bandwidth (1 MHz). This UV laser beam is optically filtered with an 11 mm lens and a 5 mm pinhole to achieve a clean near-Gaussian beam. This diverging beam travels 2.25 m over an optical table to a Lloyd's Mirror Interferometer mounted on a rotation table. The UV beam illuminates both the mirror and the sample. Part of the light reflects on the mirror and interferes with the portion of the beam that is directly illuminating the sample. To prevent vibrations, which could disturb the interference pattern, the whole setup is built on an actively damped optical table of 3 x 3.6 m. The Lloyd's Mirror Interferometer is placed in a closed cabinet to avoid air movements, which could affect the stability of the interference pattern. With this setup highly regular patterns can be produced over areas of about 2 x 2 cm^2 [42].

2.2.4. *Using pre-etched substrates*

Depositing the magnetic material on a patterned Si wafer can also make patterned media. Patterning can be carried out by e-beam lithography followed by RIE and by nano-imprint technologies (see Section 2.2.5). In Fig. 10 the principle of the deposition of a pre-etched Si wafer is given. The TEM cross section shows the covering with the deposited multilayer of Co/Pt. In this case the Co is deposited normal to the surface and the Pt at oblique incidence from the right. Consequently, the sidewall B consists of Pt grains with a thickness close to the nominal. The sidewall A is covered with a very thin layer of Co-oxide [43]. In the groove opposite to the Pt-source the Si can see a discontinuity on place-C due to the shadowing of Pt. It can be concluded that both sidewalls are non-magnetic, consequently no direct exchange between the top and the bottom of the grooves [43]. Areas of dots with pitch of 60 nm were prepared by using nanoimprint technology.

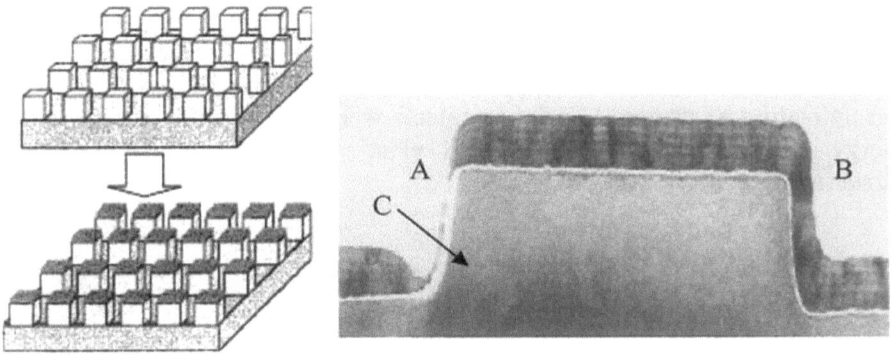

Figure 10. The principle of using pre-etched Si substrates (left top) deposited with a multilayer (bottom left). The TEM cross-section image (right) shows how the multilayer covers the Si structure.

Co/Pt dots do have very high coercivity, which is much larger than the maximum magnetostatic field; consequently the dots are very stable. Write and readback are investigated and it has been demonstrated that it is possible to address individual dots in these arrays, both in the down-track and cross-track directions.

2.2.5. *Imprint technology*

Contact lithography can offer a high resolution and is in principle less expensive than the optical lithographic methods. It uses a method similar as in optical disk fabrication. It has been demonstrated that imprint technology using compression molding of thermoplastic polymers can produce vias and trenches with 25-nm minimum feature size and 100-nm depth in thin polymer [7]. Nanoimprint is using a mold (master) for deforming the resist. This master can be made either by Interference lithography, E-beam lithography or X-ray lithography. Before starting the imprint a polymer is spin coated on the substrate (covered or not with a thin film). This monolayer is about 30 nm thick.. After heating, by hot emboss technology or UV-irradiation, one type of the polymer becomes hard and the other soft for reactive ion etching. It is also possible to use the technology without heating but than by using a very high pressure at room temperature [44].

An important aspect in this technology is of course the reliability and accuracy of the imprint stamp (master). A uniform imprint of 100 nm wide lines has been realized on a 6-in Si wafer [45]. Another technology of mold lithography using a photo-curing polymer shows that sub-100 nm structures can be realized [46]; an automated tool for step and flash imprint lithography has been constructed for using 200 mm wafers [47]. Sub-10 nm structures have been shown from metal dots with a period of 40 nm and a diameter of 10 nm. This tool allows an operator to run automated imprinting experiments without human intervention, except for installation of templates, and loading and unloading of wafers. Imprint templates were treated with a low surface energy self-assembled monolayer to aid selective release at the template-etch barrier interface [47].

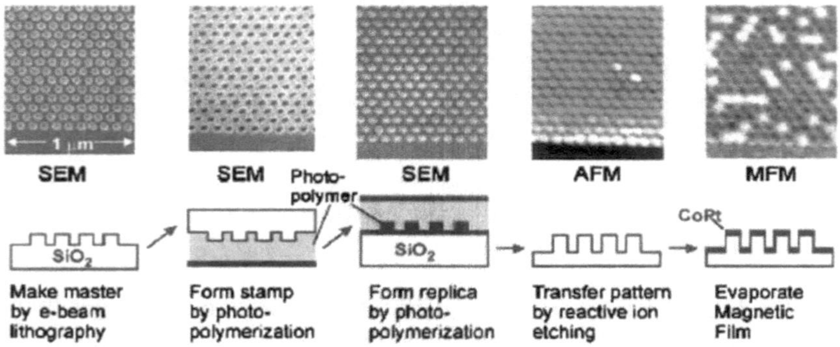

Figure 11. Total process van imprint technology started with mastering, first stamp, replacing the stamp into a flexible stamp, transferring the pattern and finally depositing the Co/Pt multilayer.

These imprint technologies can be combined with various techniques such as electroplating into holes as demonstrated in [33], showing the quantized magnetic disk having perpendicular anisotropy by Ni pillars (diameter = 110 nm) in a non-magnetic matrix. Large area magnetic patterning is also described. In [48] a nanomolding process for producing 55-nm-diameter magnetic islands over 3-cm-wide areas is described. The fabrication steps are given in Fig. 12 consisting of mastering by EBL, stamping, replication, patterned transferring and deposition of the magnetic layer. Finally, an 11-nm-CoPt multilayer is deposited onto a substrate consisting of a pattern of pillars 28 nm high. Magnetic force microscopy reveals that the film on top of each pillar is a magnetically isolated single domain that switches independently (see Fig. 11). With this process [48], a nearly unlimited number of stamps can be made from a single master, because no hard surface comes in contact with the master during stamp formation, and any residue can be removed nondestructively. In addition to accommodating substrate curvature and roughness, the use of a flexible stamp gives a process less sensitive to localized contamination. If this process can be optimized to form a large number of replicas from each stamp, it may be suitable for large-scale manufacturing.

The combination EBL and imprint technology has been optimized by using PMMA (Poly Methyl MethAcrylate) resist, lift-off technique and reactive ion etching (REI) [49]. The authors were able to fabricate dot arrays pitches and periods down to 30 and 40 nm, respectively. Their SiO_2 pillars can be used to fabricate robust molds for imprint lithography up to densities of 700 Gbpsi. These molds have been tested for imprinting experiments, and good replication was achieved for arrays down to a pitch of 50 nm.

Imprint lithography also have several limitations for many applications because it can in principle only be used for small periodic structures. A general fabrication technology for all kinds of applications should be able to realize various combinations and distributions of large as well as small features [50]. Beside the imprint technology it is also possible to use the mold as an 'ink' stamp (Nano Imprint Lithography (NIL). The ink is a chemical solution with certain properties and is applied to the substrate. Consequently it left the chemical on the contact area and after that it can be used as a mask.

Laser Assisted Direct Imprint (LADI). - This is an imprint technology that directly prints the mold in Si with the help of a laser beam. The mask is pressed into the molten layer. When the mask is removed, an accurate imprint of the component design remains in the silicon, with detail rendered at the 10-nm level. This technique proposed by [50] is called 'Laser-assisted direct imprint' (LADI). The process is given in Fig. 12. In this figure laser light passes through a quartz mask above the silicon surface (A and B). A single excimer laser pulse melts a thin surface layer of silicon, and a mould is embossed into the resulting liquid layer. The mask is pressed into the molten layer, Fig. 12(B). When the mask is removed, an accurate imprint of the component design remains in the silicon, with detail rendered at the 10-nm. The embossing time is less than 250 ns. In Fig. 12(D) and (E) a SEM image is given from a cross-section of the quartz mold (Fig. 12D) and the Si imprint (Fig. 12 E). A uniform 300 nm period silicon grating is patterned and has a 140 nm line width and is 110 nm deep.

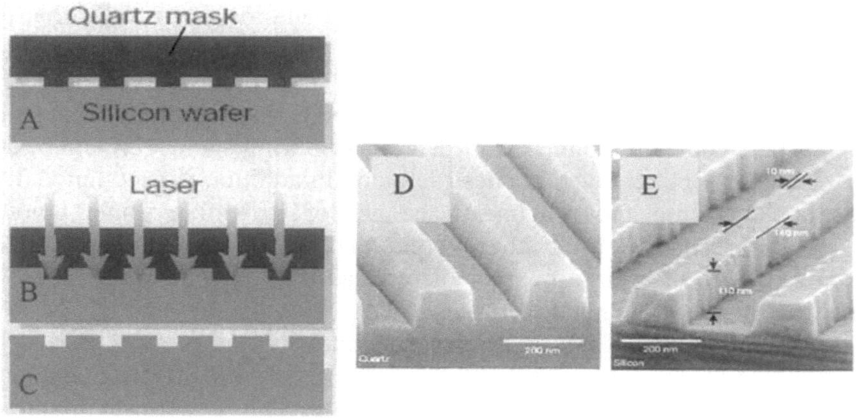

Figure 12. Laser-assisted direct imprint quartz mask is pressed into the molten Si layer heated by the laser light (A,B & C). SEM images of the quartz mold (D) and the imprinted patterns in silicon (E).

The mould after the two LADI processes showing no visible damage. The LADI area could be as large as a whole wafer (4 inch or 8 inch diameter). This method eliminates the costs because no expensive focusing optics and resist is needed.

2.2.6. Block copolymer template technology

Block copolymers consist of polymer chains made from two chemically distinct polymeric materials. Phase separation of the blocks leads to the formation of periodic nanoscale areas with a very uniform distribution of size and shape. Diblock copolymers spontaneously self organized themselves into micro domains. Such micro domain morphology can be used as a template for the formation of large area arrays of p articles finally used as a template for nanostructures. Special devices such as patterned media, waveguides require not only a long range ordering but also accurate pattern registration on a substrate without defects in their order. Various papers have been published to overcome the disorder such as directional solidification, adding electric fields and topographic substrates creating by etching or chemical modification of the surface.

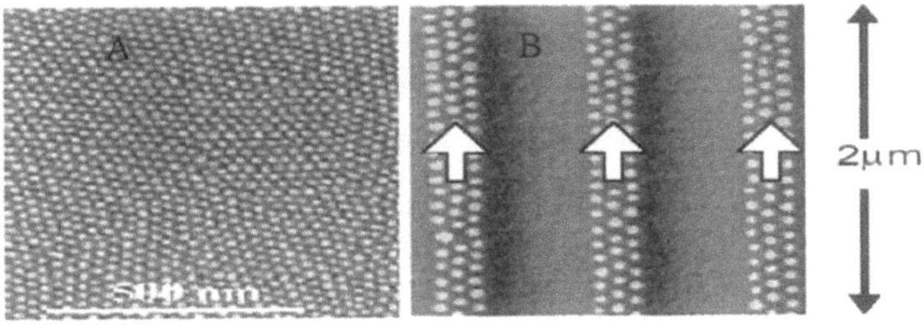

Figure 13. Self-assembled PS-PMMA island with 'domain boundaries' spinned on a smooth substrate (A) and in a grooved substrate the dots are aligned.

In Fig. 13(A) a self-organized block copolymer is spinned on a smooth substrate and defects (domain boundaries) are visible over short distances [51] while the block polymer is spinned over a grooved substrate, Fig. 13(B), the PMMA dots are now aligned along the groove direction [51-53]. Another question is what the switching behavior is of large arrays of particles made by this method. Co particles have been made with 50-nm period arrays with 35 nm widths and 5-20 nm height [54]. The particles have in-plane anisotropy due to their shape.

In [55] a large-area fabrication of hexagonally ordered metal dot arrays with an area density of $\sim 10^{11}/cm^2$ was demonstrated. The metal dots were produced by an electron beam evaporation followed by a lift-off process. The dots size was 20 nm dots with a 40 nm period by combining block copolymer nanolithography and a trilayer resist technique. A self-assembled spherical-phase block copolymer top layer spontaneously generated the pattern, acting as a template. The pattern was first transferred to a silicon nitride middle layer by reactive ion etch, producing holes. The nitride layer was then used as a mask to further etch into a polyamide bottom layer.

2.2.7. Chemical ordering with ion irradiation

It is well known that chemical composition, crystal structure, grainsizes and their distribution are determining the magnetic behavior of thin films and dots. In thin films chemical ordered ferromagnetic phases such as FePt, FePd, and CoPt are very important for applications. This type of materials show the largest magnetic anisotropy in comparison with other ferromagnetic films and that is due to the special chemical ordering in the $L1_0$ phase. To obtain chemical ordering high substrate temperatures are necessary (e.g. FePt 700K) [56]. High substrate temperatures are not suitable for production and consequently alternatives are generated such as the addition of third element, layer-by-layer growth, and also light ion irradiation. For instance He ions (20-150keV) can enhance the chemical ordering. Small energy transfers minimize recoil displacements to 1-2 interatomic distances [57]. It has been observed [58] in co-sputtered FePt films the long-range ordering parameter S (completely order ($S = 1$) totally disorder material ($S = 0$)) increases with the fluency (left part of Fig. 14).

In contrary ion irradiation can also be used to destroy the chemical ordering in alloys. A nice example is given in $CrPt_3$ thin films [59]. In the chemically ordered $CrPt_3$ with $L1_2$ phase (Cu_3Au) are the Cr atoms site at the corners and Pt atoms at the faces of a unit cell. This $CrPt_3$ phase is ferromagnetic, with the localized moments for Cr = 2.33 μ_B and Pt = -0.27 μ_B. The disordered phase is non magnetic and has a FCC structure and the Cr and Pt sites are randomly substituted. Consequently ion irradiation can locally transform chemically ordered $CrPt_3$ into the chemically disordered FCC phase, which means that the ferromagnetic behavior is suppressed.

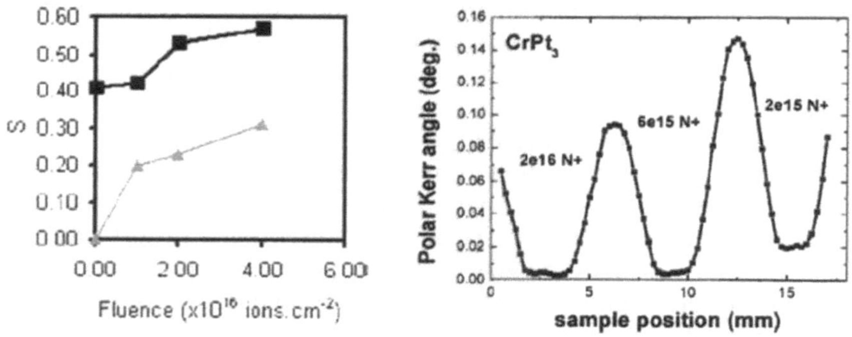

Figure 14. Irradiation and disorder: (left) long-range order parameter S as a function of the fluency for a disorder FePt (50 nm)-Pt (001) thin film (starting S=0) and particularly ordered one (starting S=0.4) and (right) Polar Kerr measurements of three irradiated $CrPt_3$ areas with N^+ ions with a fluency in the range $2x10^{15}$ - $2x10^{16}$ ions/cm^2.

Figure 14(right) shows the polar Kerr measurements of three irradiated $CrPt_3$ areas with N+ ions with a fluency in the range $2 \times 10^{15} - 2 \times 10^{16}$ ions/cm^2. The irradiated arias are 4 x 4 mm^2. The lowest dose is not sufficient to completely suppress the magnetic moment observed by the Kerr rotation while above 6x 10^{15} ions/cm^2 lead to a non-magnetic behavior. This opens the possibility for patterning magnetic and non-magnetic areas in a magnetic film.

3. MAGNETIC PROPERTIES OF PATTERNED THIN FILMS

It has been shown that size, shape, anisotropy and other properties determine strongly the magnetic behavior of a patterned thin film. Magnetization reversal processes of nanosize structures have been studied extensively from both of technical and scientific point of view. MOKE and MFM can observe reversal behavior of a collection of magnetic dots. A special high field MFM operating in vacuum and RT was used to analyze dots having perpendicular anisotropy. It is mounted in between the pole shoes of conventional electro-magnet with a maximum field of 1353 kA/m (1.7T). The magnetic signal is detected at constant height, by means of phase measurement using a lock-in amplifier. A series of measurement was made with fields ranging from -1353 kA/m to 192 kA/m. We used here patterned CoNi/Pt multilayer films having a perpendicular anisotropy [37, 42].

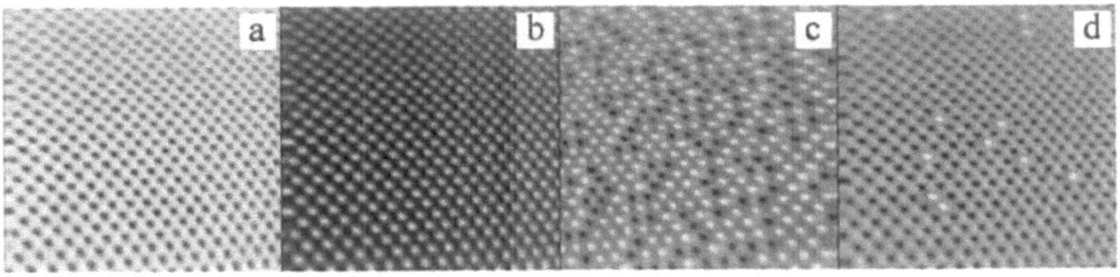

Figure 15. MFM images of a patterned CoNi/Pt multilayer after application of –1350, +4, +127 and +183 kA/m.

An example is shown in Fig. 15. In this case for each image the field is increased to a certain value and taken back to zero, in order to minimize the influence of the tip on the sample. The first image of Fig. 15(a) shows the magnetization after applying -1353 kA/m perpendicular to the sample. The tip and medium are saturated in the same direction so only attractive forces occur (black dots). When field is increased to +4 kA/m, the tip reverses magnetization and all forces become repulsive, so the contrast reverses (white dots). At 127 kA/m almost 50% of the dots is reversed (Fig. 15(c)) and at 183 kA/m almost all the dots are reversed (Fig. 15(d)). The first dots switch already at 80 kA/m, but the last at 192 kA/m.

Studying the reversal mechanism of a single nano-dot is very difficult because the sensitivity of commercial magnetic analysis equipment is not large enough. A few techniques aiming quantitative analysis need a complicated fabrication process and/or their working temperature is limited at low temperature. Analyses with Anomalous Hall Effect (AHE) for films with a perpendicular anisotropy have been proposed already for a long time [60, 61] and it has been shown that this method belongs to the most sensitive measuring methods. The AHE gives information about the magnetic state of a ferromagnetic material, as it is proportional to the perpendicular component of the magnetization in the film or particle. Therefore data of relevant parameters such as Hc, Mr, S, Hk etc can be obtained by recording the hystheresis curves by plotting the AHE-voltage vs. the applied field. The sensitivity of the AHE measurements is very high in relation with SQUID and VSM –magnetometory (~ 10^{-15} emu).

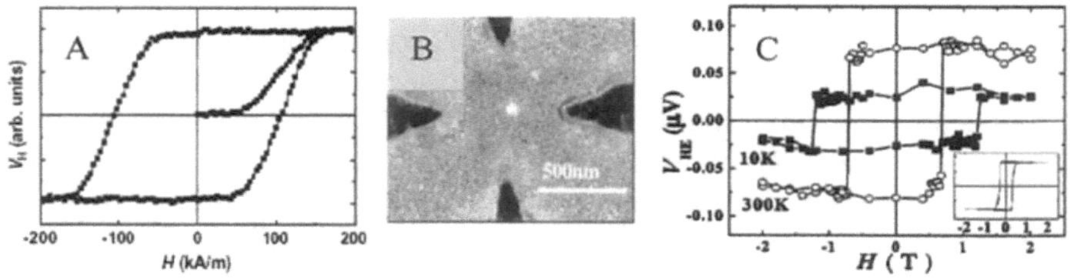

Figure 16. (A). The virgin curve and full AHE curve of a patterned Co/Pt multilayer with a dot size is 120 nm. The loop is measured form about 200 dots in the area of the Hall cross, (B) SEM image of a 60 nm FePt dot (bright spot) covered with a Pt cross-shaped electrode, (C) AHE loops for 10 and 300 K with an inset of the unpatterned FePt film at 300 K.

In a recent paper, the magnetization of an isolated dot prepared by e-beam lithography, from an $L1_0$-FePt (001) film with a diameter of 60 nm has been measured by AHE, and moreover the analysis were carried out as function of temperature [62]. AHE measurement can be adapted to wide range of samples, from a single dot to an array of a few cm^2 and as well as for very thin continuous films. Dot arrays made of polycrystalline Co/Pt multilayer with average grain size of 20 nm are fabricated [63]. Dot patterns of 200 nm thick positive resist for etching mask were exposed on the films by using Interference Lithography (IL) [37, 42, 63, 64] and the patterns were transferred into the Co/Pt multilayers by using Ar+ ion beam etching (IBE). Arrays of circular dots of 120 nm and 200 nm in diameter were fabricated. In Fig.16(A) the virgin curve and full AHE hysteresis curve of a IL patterned Co/Pt multilayer [Pt (2 nm)/Co (0.5 nm)]5 / Pt (4 nm) consisting of 200 dots with a diameter of 120nm is presented. The energy barrier height of the dots has been estimated as 4.0×10^{-12} erg, which corresponding to anisotropy energy of a column of 18 nm in diameter. The activation volume is comparable with the grain size measured with AFM. Recently [80] dot arrays of sub-micron diameter are fabricated from epitaxial grown $Co_{80}Pt_{20}$ film, which has large perpendicular anisotropy. Sweep rate dependent measurements has been carried out as well as angular dependent measurements.

The energy barrier height of reversal is estimated as 1.0×10^{-10} erg. The energy corresponds to the switching volume comparable with the volume of a physical grain in the dots. For application high anisotropy dots are very important. A promising material is the special $L1_0$ structure of FePt. In Fig. 16(C) the result of AHE measurements on a single dot (60 nm in diameter) is given (see Fig. 16(B)). Over the whole temperature range, the dot exhibit perfect rectangular magnetization loops with coercivity almost constant (Fig. 16(C)). The activation energy has been evaluated to be about 4×10^{-19} J, equivalent to the domain-wall energy times the square of the domain-wall thickness, suggesting that the magnetization reversals are initiated by nucleation of reversed nucleus with the dimension of the exchange length. Magnetic imaging and micromagnetic simulations have shown that non-uniform magnetization configurations in patterned structures can considerably affect the magnetization reversal [65].

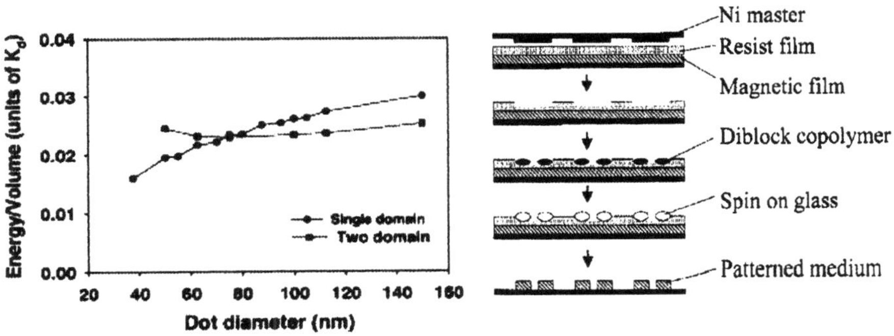

Figure 17. Micromagnetic simulations for a SD and MD state of CoNi/Pt dots with 30 nm thickness (left) and (right) the AASA process for a 2.5 inch disk.

In [66] results are presented from an investigation of the domain structure and magnetization reversal of small circular elements results by use of an existing micromagnetic simulation package. Such a study can contribute to understand the behavior of patterned media for high-density recording and also, for non-volatile magnetic memories [67]. Later we use the micromagnetic simulation to study the critical size at which the energy of single domain state (SD) equals that of the two-domain state (TD) or multi-domain (MD) state [68]. We also started to simulate the switching field distribution of such SD dots. By investigating stable configurations from different initial configurations and comparing the energies as a function of cell size, a set of equilibrium states are found. Using this procedure, the energy of dots of height 30 nm in SD and TD states are calculated as a function of dot diameter. Figure 17(left) shows the calculated total minimum energy density as a function of dot diameter for the SD and MD states. Experimental results give more or less the same value for the critical size [37]. During the fabrication of nanostructures by using dry etching methods a very critical point is the damage that the ions made into the structures. Beside etch damage also imperfect layer growth might influence the switching field. Systematic influences of this kind of factors are discussed in [68]. An example is the slanting of the sidewalls of the dots, resulting from the imperfect resist pattern shape. Magnetization reversals of CoNi/Pt multilayer dots with (a) cylindrical, (b) pointed cone, and (c) truncated cone shape have been simulated and show all different hysteresis properties.

4. APPLICATIONS OF PATTERNED MAGNETIC FILMS

Patterned magnetic thin films are not only interesting from the physics point of view but have also many different applications in the field of magnetism. Data storage, magnetic field sensing, logical elements and in bio-detection one can find different alternatives for applications. In this section we briefly discussed only the application of magnetic logic and high density magnetic recording.

4.1. Media for Ultrahigh-Density Magnetic Storage

The stability of written data becomes more and more a major concern in hard-disk design. The areal bit density of magnetic disk recording has increased enormously over the last decades. Extrapolation leads to bit densities not likely to be achieved without changing the way that the data is stored in the present media. Recently, laboratory demos have shown more than 160 Gbpsi. So far, scaling the critical physical dimensions of the hard disk was sufficient to double the density on an annual basis. Experiment and theory have shown that there is a limit to this scaling. Many alternative technologies have been proposed (see e.g. Ch. 11) to overcome the problems of continuous media. Patterned media are suggested as one of the potential ways to further increase the areal recording density.

It becomes clear if one compares bits in a continuous polycrystalline film (500-1000 crystals per bit) with bits in a patterned medium (1bit = 1 dot) that the highest density is present in the patterned medium. Moreover, the super-paramagnetic bit density limit of this type of medium is much higher than that of present thin film media. Also, the constraints on the writing and reading process itself are strongly reduced. Due to the single-domain nature of every dot, the writing of such a bit is an all or-nothing event and the head does not have to be positioned exactly above the bit. In contrast to continuous medium, however, a write synchronization scheme has to be implemented. For the design of a prototype patterned medium for recording several requirements are given in e.g. [3].

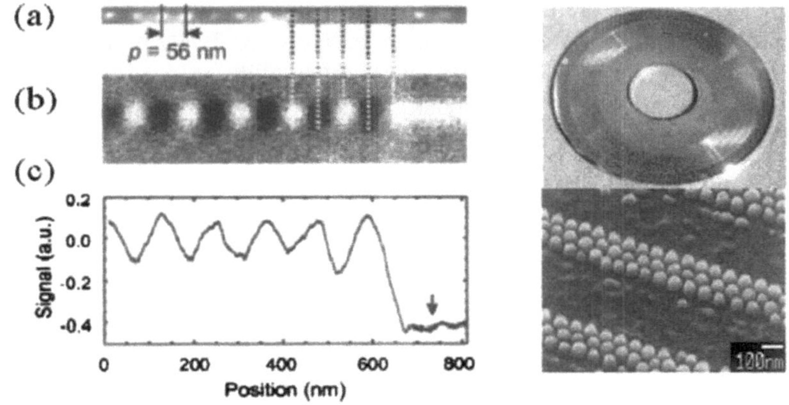

Figure 18. (left) GMR readback signals (c) from a single array of CoPtCr dots made by FIB. The AFM (a) image shows dots (period: 56nm and size:26nm) and MFM images (b) gives the magnetic bit image and (right) resist imprint on a disk with the aligned magnetic CoCrPt particles (at the bottom).

Besides Co/Pt multilayers, perpendicular granular CoCrPt thin films (20 nm thick), suitable for a recording medium, has also been prepared by using FIB [95]. The patterns are directly made in the film by using 30 keV Ga+ at a beam current of 1pA.The pattern trenches are 6 nm deep and square arrays of magnetic isolated islands have been prepared in the range 65–500 nm. Based on MFM measurements it was concluded [69] that below 130nm diameter only single domain behavior was obtained.

With a quasi-static write/read tester, recording experiments have been performed [70, 71]. In Fig. 18(a, b) AFM and MFM data are given from a recorded single row of square dots. The GMR readback signal is given in Fig. 18(c). The obtained density was larger than 200 Gbpsi. In continuous thin film media bits can be placed at any area of the disk. In the case of patterned media the write pulse must be synchronized to the patterning [72]. For the right synchronization requirements, the switching field of distribution of the media and the head field gradients are important. Perpendicular single pole heads in combination with a soft underlayer are also suggested [73].

To increase the density of a rotated disk in [52], artificially assisted self-assembling method (AASA) is introduced. This method is a combination of nanoimprint technology and the use of a diblock copolymer. The result is shown in Fig. 18(right). Here it can be clearly seen that an aligned array of small CoCrPt particles (40nm in diameter) are aligned in the groove of a hard disk (2.5 inch). Figure 17(right) shows a schematic illustration of the preparation method. Ni master possessing spiral patterns with 60-250 nm width lands and 400 nm width grooves was pressed into a resist film on a magnetic layer. Then the diblock copolymer (PMMA-PS) was cast into the grooves and annealed. After removing selectively the PMMA by an oxygen plasma treatment the resulting holes were filled with spin on glass (SOG). The SOG dots are used as a mask for dry etching the magnetic film (CoCrPt).

4.2. Nanoscale Magnetic Elements and Logic Devices

In [74, 75] micron sized NiFe ring-shaped sensors have been fabricated by standard photolithography and proposed as a bio-detection device. Such sensor can detect the radial component of the dipolar-fringing field from a single, partially magnetized micron sized NiFe sphere mounted on an AFM cantilever. When the sphere is above the center of the ring the AMR response to this fringing field is strongly peaked and is minimized when the sphere is moved beside the ring.

Ferromagnetic rings are also fabricated by using electron-beam lithography with a lift-off process for pattern transfer [76]. These nanoscale ferromagnetic rings have a smallest outer diameter of 90 nm, inner diameter of 30 nm and thickness of 10 nm.

Generally in computing data is stored in one place and processing in another. Various groups are proposing magnetic logic elements in which data can be stored and data can be processed [77]. Their proposal is based on MRAM elements that has two independent input lines and it could, in principle operate as any one of four different logic operations - AND, OR, NAND or NOR gates - and lead to increased computational efficiency.

Such type of device has many advantages over transistor-based logic and that it could lead to programmable magneto-logic circuits in the near future. Such system of interacting magnetic nano-cells can be produced today in the order of 100nm and as expected in the future in much smaller size (at least a factor 10 smaller) for the individual building blocks. Moreover such a system has a high speed (switching in magnetic films can be in the GHz) and non-volatility and can operate at room temperature.

A so-called quantum-dot cellular automata (QCA) was proposed by [78], which employs arrays of coupled quantum dots to implement Boolean logic functions [79]. The advantage of QCA lies in the extremely high packing densities possible due to the small size of the dots, the simplified interconnection, and the extremely low power-delay product. Using QCA cells with dots of 20 nm, an entire full adder can be placed within 1 mm^2 [80]. A processing method based on magnetism is also given in [81]. Here networks of sub-micrometer magnetic dots are used to perform logic operations and propagate information. The logic states are signaled by the magnetization direction of the single-domain dots and the dots are coupled to their nearest neighbors through magnetostatic interaction. In [81] QCA-network is shown and consist of single elongated input dot followed by a chain of 69 circular dots (110nm in diameter) [81]. The dots are 10 nm thick and made from Supermalloy and fabricated with e-beam lithography.

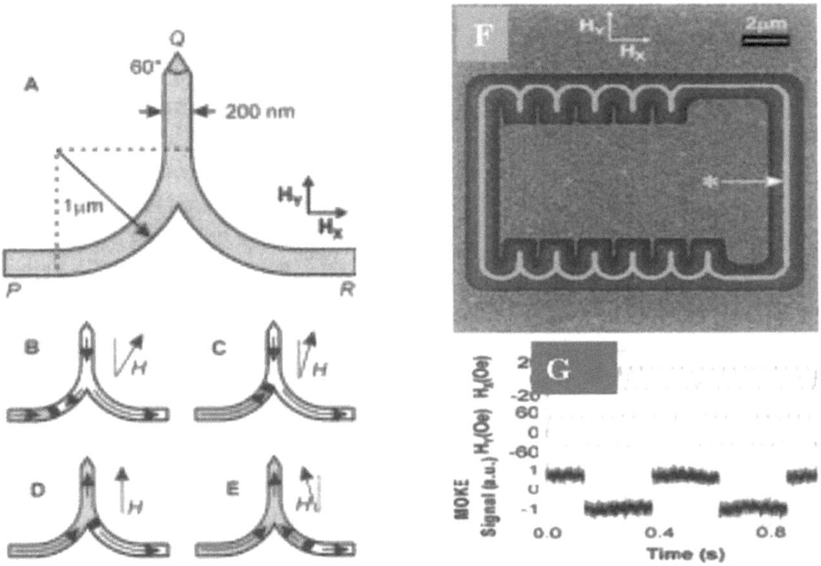

Figure 19. Schematics of the hairpin NOT gate (A) and the mechanism of the domain wall injection (B to E). In (F) the 11 NOT junctions are shown and in G the comparison of the MOKE analysis and the applied field components Hx and Hy.

Magnetic solitons carry information through the networks, and an applied oscillating magnetic field feeds energy into the system and serves as a clock. These networks offer a several thousand-fold increase in integration density and a hundredfold reduction in power dissipation over current micro-electronic technology. The fact that the entire network can be fabricated on a single plane means that many planes could in principle be stacked on top of each other, thus giving a way to realize 3D hardware. In addition also the building blocks for a logic systems such as a NOT gate are studied [82, 83]. An all-metallic sub micrometer device is demonstrated experimentally at room temperature that performs logical NOT operations on magnetic logic signals. When this two-terminal ferromagnetic structure is incorporated into a magnetic feedback loop, the junction performs a frequency division operation on an applied oscillating magnetic field. Up to 11 of these junctions are then directly linked together to create a magnetic shift register. The device is fabricated with a FIB from 5 nm thick Permalloy and a full description of the device is given in [82].

In Fig. 19(A) first of all the metallic ferromagnetic NOT gate is shown with line width of 200nm and a hairpin shape. The operation concept of the NOT gate magnetization reversal is shown in Fig. B to E with the magnetization directions (arrows) in the various parts of the hairpin and the domain wall (thick line) motion as a consequence of rotating applied magnetic field by components H_x and H_y. In Fig. 19(F) the FIB image of a magnetic ring including 11 NOT junctions, is given with the asterisk indicating the position of subsequent MOKE analysis which is given in (G) The MOKE analysis gives an identical structure with the clockwise-rotating magnetic field components.

Domain wall control is an intensive subject of study because it can lead to new logic and memory devices. Domain walls can be injected into planar magnetic nanowires either from magnetically soft large-area "pads" or by using the local magnetic field from an overlying current-carrying wire. The domain wall resistance, magneto resistance, domain wall displacement are measured. A complete nano-magnetic logic system can only work if it is interfaced with conventional electronics. Data can be written in magnetite structures by a current-carrying strip line and data can be read out by incurporating spin tunnel junctions or spin valves into magnetic tracks or alternatively by measuring domain wall resistance.

For non-volatile memory devices it would be a great advantage if a magnetic bit could be switches by the current passing through it, rather than by some external magnetic field. If so the design of high-density non-volatile memory would be greatly simplified. In [84] spin-polarized current switching of a Co thin film nanomagnets (60 nm x 130 nm) is discussed.

5. CONCLUSION

Patterned magnetic nanostructures are researched and developed for fundamental studies and applications. It is clear that there is a growing interest in making well-defined structures of magnetic thin films with nanodimensions. Especially the fields of magnetic recording, sensors, logic systems, MRAM and spin-electronics have shown great interest in nanosize structures. There is a tradeoff from the area of patterning and the accuracy, the size of the individual dots or wires and the price of the equipment and the related technologies. With some of the techniques it is possible to realize structures smaller than 10 nm over a relatively large area. For future applications the price will be a dominant factor.

The properties are always influenced by the structural properties of the continuous film and moreover also the fabrication technologies can have a large influence on the nanosize structure. Important hereby are the applications of the dry etching (implantation, roughening of the sidewalls etc.) and redeposition at the side walls due to the etching.

It has been shown that the size and shape of nanostructures plays an essential role in determining the magnetic properties. One critical size of a structure is the SD or MD state. In certain materials the magnetic anisotropy can be influenced by size, shape and configurational anisotropy and in others the crystalline perfection determines the anisotropy.

The analysis of magnetic nanoelement is not simple due to their small volume. We need sensitive methods. MFM, MOKE and SQUID measurements can give us help in understanding the magnetic behavior of such a small element. Micromagnetic simulations are of great help to understand the behavior and to help with designing the structure. For understanding the individual particle we need to develop new analysis method. In our application particles with perpendicular anisotropy can be analyzed by the AHE and it has been shown that this method is very sensitive because the hystheresis loop of a single FePt dot of 60 nm has been measured. For application it is also important that we use array of particles. In that case we have to deal with the interaction between the dots due to the dipolar fields. This is very important for application because these effects will influence the Hc and switching field distribution of the array. It has been shown that small patterns can principally be made in circular patterns, which is interesting for hard disk application. Alternatively by using the AASA-method have shown the circumferential magnetic patterned hard disk media with 40 nm diameter on a 2,5 inch disk. This method uses the nanoimprint technique of making land-groove spirals using self-assembling diblockpolymer aligned in the groove. The most fascinating area of study and application can be found in the area where different disciplines are meeting each other such as magnetism with, biology, superconductivity, molecules, semiconductors etc.

From the technology point of we are able to realize a variety of structure and dimensions using different kind of materials. But the technology effects in relation with scaling down the dimensions means that we are far of the complete understanding of the properties of the nanosize structures. One of the most important tasks is to understand the magnetization reversal of nanosize dots and the interaction between dots. Dynamic measurements of the magnetic properties are extremely important for application. A good example is here the application in the field of high data rates in relation with logic and computing. The requirement of patterning large areas of regularly spaced uniform dots at nano-dimensions puts strong demands on the patterning technology. Even in the R&D laboratories such technologies are generally not readily available, not to speak of technology for the manufacturing of a patterned hard disk, which is one of the possible application.

Acknowledgment

The author thanks the SMI-group members for their contributions but like to mention specially the cooperation in this field with Leon Abelmann, N. Kikuchi, Nguyen Dao, Martin Siekman and Rogelio Murillo. The Dutch Technology Foundation (STW) supports this work financially.

References

[1] J. P. Wang and T. J. Zhou, in "Encyclopedia of Nanoscience and Nanotechnolgy", Vol.8, Ed. H. S. Nalwa, American Scientific Publishers 2004, p. 415.

[2] J. I. Martin, J. Nogues, Kai Liu, J. L. Vicent, and Ivan K. Schuller J. Magn. Magn. Mater. **256**, 449 (2003).

[3] J. C. Lodder, J. Magn. Magn. Mater. **272-276**, 1692 (2004).

[4] R. P. Cowburn, J. Phys D: Appl. Phys. **33**, R1 (2000).

[5] R. P. Cowburn, J. Magn. Magn. Mater. **242-245**, 505 (2002).

[6] A. Dietzel, R. Berger, H. Grimm, W. H. Bruenger, C. Dzionk, F. Letzkus, R. Springer, H. Loeschner, E. Platzgummer, G. Stengl, Z. Z. Bandic, and B. D. Terris, IEEE Trans. Magn. **38**, 1952 (2002).

[7] S. Y. Chou, P. R. Krauss, and P. J. Renstrom, Science **272**, 85-7 (1996).

[8] S. Y. Chou, P. R. Krauss, W. Zhang, L. Guo, and L. Zuang, J. Vac. Sci. Technol. B **15** (6), 2897 (1997).

[9] C. Chappert, H. Bernas, J. Ferre, V. Kottler, J.-P. Jamet, Y. Chen, E. Cambril, T. Devolder, F. Rousseaux, V. Mathet, and H. Launois, Science **280**, 1919 (1998).

[10] T. Devolder, J. Ferre, C. Chappert, H. Bernas, J.-P. Jamet, and V. Mathet, Phys. Rev. B **64**, 064415 (2001).

[11] E. E. Baglin, M. H. Tabacniks, R. Fontana, A. J. Kellock, and T. T. Bardin, Mater. Sci. Forum **248-249**, 87 (1997).

[12] T. Devolder, Phys. Rev. B **62**, 5794 (2000).

[13] A. Mougin, T. Mewes, M. Jung, D. Engel, A. Ehresmann, H. Schmoranzer, J. Fassbender, and B. Hillebrands, Phys. Rev. B **63**, 060409 (2001).

[14] A. Traverse, M. G. Boite, and G. Martin, Europhys. Lett. **8** (7), 633 (1989).

[15] T. Devolder, C. Chappert, Y. Chen, E. Cambril, H. Bernas, J. P. Jamet, and J. Ferre, Appl. Phys. Lett. **74**, 3383 (1999).

[16] B. D. Terris, D. Weller, L. Folks, J. E. E. Baglin, and A. J. Kellock, J. Appl. Phys. **87**, 7004 (2000).

[17] T. Aign, P. Meyer, S. Lemerle, J. P. Jamet, J. Ferré, V. Mathet, C. Chappert, J. Gierak, and C. Vieu, Phys. Rev. Lett. **81** (25), 5656 (1998).

[18] C. T. Rettner, M. A. Best, and B. D. Terris, IEEE Trans. Magn. **37** (4), 1649 (2001).

[19] R. Allenspach, A. Bischof, U. Durig, and P Grutter,Appl. Phys. Lett. **73** (24), 3598 (1998).

[20] T. J. Zhou, Y. Zhao, J. P. Wang, T. C. Chong, and J. Y. L.Thong, IEEE. Magn. **38** (5), 1970 (2002).

[21] M. Zheng, M. Yu, Y. Liu, R. Skomski, S. H. Liou, D. J. Sellmyer, V. N. Petryakov, Yu. K. Verevkin, N. I. Polushkin, and N. N. Salashchenko, Appl. Phys. Lett. **79**, 2606 (2001).

[22] M. Yu, Y. Liu, and D. J. Sellmyer, J. Appl. Phys. **85**, 4319 (1999).

[23] G. Thornell and S. Johansson, J. Micromech. Microeng. **8**, 251 (1998).

[24] A. Lipicki, K. Kang, and T. Suzuki, IEEE Trans. Magn. **38** (5), 32589 (2002).

[25] S. I. Woods, S. Ingvarsson, J. R. Kirtley, H. F. Hamann, and R. H. Koch, Appl. Phys. Lett. **81** (7), 1267 (2002).

[26] E. O. Samwel, D. B. Bijl, J. C. Lodder, and Th. J. A. Popma, J. Magn. Magn. Mater. **155**, 292 (1996).

[27] J. Ferre, C. Chappert, H. Bernas, J.-P. Jamet, P. Meyer, O. Kaitasov, S. Lemerle, V. Mathet, F. Rousseaux, and H. Launois, J. Magn. Magn. Mater. **198-199**, 191 (1999).

[28] D. Weller, J. E. E. Baglin, A. J. Kellock, K. A. Hannibal, M. F. Toney, G. Kusinski, S. Lang, L. Folks, M. E. Best, and B. D. Terris, J. Appl. Phys. **87** (9), 5768 (2000).

[29] C. T. Rettner, S. Anders, J. E. E. Baglin, T. Thomson, and B. D. Terris, Appl. Phys. Lett. **80** (2), 279 (2002).

[30] S. P. Li, W. S. Lew, J. A. C. Bland, L. Lopez-Diaz, M. Natali, C. A. F. Vaz , and Y. Chen, Nature **415**, 600 (2002).

[31] H. Loescher, E. J. Fantner, R. Kornter, E. Platzgummer, G. Stengl, M. Zeininger, J. E. E. Baglin, R. Berger, W. H. Bruenger, A Dietzel, M. I. Baraton, and L. Merhari, Proc. MRS **739**, 3 (2003).

[32] L. F. Johnson, G. W. Kammlott, and K. A. Ingersoll, Appl. Opt. **17**, 1165 (1978).

[33] Wei Wu, Bo Cui, Xiao-yun Sun, Wei Zhang, Lei Zhuang, Linshu Kong, Stephen Y. Chou, J. Vac. Sci. Technol. B **16**, 3825 (1998).

[34] M. L. Schattenburg, C. R. Canizares, D. Dewey, K. A. Flanagan, A. Hammett, A. M. Levine, K. S. K. Lam, R. Manikkalingam, T. Markert, and H. I. Smith, Opt. Eng. **30**, 1590 (1991).

[35] E. F. Wassermann, M. Thielen, S. Kirsch, A. Pollmann, H. Weinforth, and A. Carl, J. Appl. Phys. **83**, 1753 (1998).

[36] M. Farhoud, M. Hwang, H. I. Smith, M. L. Schattenburg, J. M. Bae, K. Youcef-Toumi, and C. A. Ross, IEEE Trans. Magn. **34**, 1087 (1998).

[37] M. A. M. Haast, J. R. Schuurhuis, L. Abelmann, J. C. Lodder, Th.J., IEEE Trans. Magn. **34**, 1006 (1998).

[38] J. P. Spallas, R. D. Boyd, J. A. Britten, A. Fernandez, A. M. Hawryluk, M. D. Perry, and D. R. Kania, J. Vac. Sci. Technol. B **14**, 2005 (1996).

[39] T. A. Savas, M. L. Schattenburg, J. M. Carter, and H. I. Smith, J. Vac. Technol. B **14**, 4167 (1996).

[40] W. Hinsberg, F. A. Houle, J. Hoffnagle, M. Sanchez, G. Wallraff, M. Morrison, S. Frank, J. Vac. Sci. Technol. B **16**, 3689 (1998).

[41] T. A. Savas, S. N. Shah, M. L. Schattenburg, J. M. Carter, H. I. Smith, J. Vac. Sci. Technol. B **13**, 2732 (1995).

[42] R. Murillo, H. A. van Wolferen, L. Abelmann, and J. C. Lodder, Micr. Eng. in press (2005).

[43] J. Moritz, L. Buda, B. Dieny, J. P. Nozie`res, R. J. M. van de Veerdonk, T. M. Crawford, and D. Weller, App. Phys. Lett. **84** (1), 1519 (2004).

[44] S. Matsui, Y. Igaku, H. Ishigak, J. Fujita, M. Ishida, Y. Ochiai, M. Komuro, and H. Hiroshima, J. Vac. Sci. Techn. B **12**, 2801 (2001).
[45] Babak Heidari, Ivan Maximov, and Lars Montelius, J. Vac. Sci. Technol. B **18** (6), 3557 (2000).
[46] Jan Haisma, Martin Verheijen, Kees van den Heuvel, Jan van den Berg, J. Vac. Sci. Technol. B **14** (6), 4124 (1996).
[47] T. Bailey, B. J. Choi, M. Colburn, M. Meissl, S. Shaya, J. G. Ekerdt, S. V. Sreenivasan, and C. G. Willson, J. Vac. Sci. Technol. B **18**, 63572 (1996).
[48] Gary M. McClelland, Mark W. Hart, Charles T. Rettner, Margaret E. Best, Kenneth R. Carter, and Bruce D. Terris, Appl. Phys. Lett. **81** (8), 1483 (2002).
[49] F. Carcenac, C. Vieu, A. Lebib, Y. Chen, L. Manin-Ferlazzo, and H. Launois, Microelct. Eng. **53**, 163 (2000).
[50] L. Jay Guo, J. Phys. D: Appl. Phys. **37**, R132 (2004).
[51] J. Y. Cheng, C. A. Ross, E. L. Thomas, H. I. Smith, and G. J. Vancso, Adv. Mater. **15**, 1599 (2003).
[52] K. Naito, H. Hieda, M. Sakurai, Y. Kamata, and K. Asakawa, IEEE. Trans. Magn. **38** (5), 1949 (2002).
[53] A. Kikutsi, Y. Kamata, H. Hieda, M. Sakurai, K. Asakawa, and K. Naito, Trans. Magn. Soc. Japan **4**, 1 (2004).
[54] J. Y. Cheng, C. A. Ross, E. L. Thomas, H. I. Smith, R. G. H. Lammertink, G. J. Vancso IEEE Trans. Magn. **38** (5), 2541 (2002).
[55] M. Park, P. M. Chain, R. A. Register, D. H. Adamson, Appl. Phys. Lett. **79** (2), 257 (2001).
[56] R. F. C. Farrow, D. Weller, R. F. Marks, M. F. Toney. S. Hom, G. R. Harp, and A. Cebollada, Appl. Phys. Lett. **69**, 1166 (1996).
[57] J. Fassbender, D. Ravelosona, and Y. Samson, J. Phys. D: Appl. Phys. **37**, R179 (2004).
[58] D. Ravelosona, C Chapper, V. Mathet, and H. Bernas, Appl. Phys. Lett. **76**, 236 (2002).
[59] Olav Hellwig, D. Weller, A. J. Kellock, J. E. E. Baglin, Eric E. Fullerton, Appl. Phys. Lett **79**(8), 1151 (2001).
[60] S. de Haan and J. C. Lodder , J. Magn. Magn. Mater. **155** (1-3), 193 (1996).
[61] B. C. Webb, S. Schulz, and S. B. Oseroff, J. Appl. Phys. **63** (8), 2923 (1988).
[62] N. Kikuchi, S. Okamoto, O. Kitakami, Y. Shimada, and K. Fuakmichi, Appl. Phys. Lett. **82**, 4313 (2003).
[63] N. Kikuchi, R. Murillo, and J. C. Lodder, accepted for publication JMMM 2005.]
[64] N. Kikuchi, R. Murillo, J. C. Lodder, K. Mitsuzuka, and T. Shimatsu, Digest Intermag, Nagoya, March, 2005.
[65] J. Shi, T. Zhu, M. Durlam, E. Chen, S. Tehrani, Y. F. Zheng, and J.-G. Zhu, IEEE Trans. Magn. **34**, 997 (1998).
[66] K. Ramstock, J. J. M. Ruigrok, and J. C. Lodder, Sensors & Actuators A **81** (1–3), 359 (2000).
[67] G. A. Prinz, Science **282**, 1660 (1998).
[68] S. Sindhu, M. A. M. Haast, K. Ramstock, L. Abelmann, and J. C. Lodder, J. Magn. Magn. Mater. **238**, 246 (2002).

[69] C.T. Rettner, M.E. Best, B.D. Terris, IEEE Trans. Magn. **37** (4) 1649, (2001).

[70] J. Lohau, A. Moser, C. T. Rettner, M. E. Best, and B. D. Terris, Appl. Phys. Lett. **78** (7), 990 (2001).

[71] M. Albrecht, C. T. Rettner, A. Moser, M. E. Best, and B. D. Terris, Appl. Phys. Lett. **80** (18), 3409 (2002).

[72] M. Albrecht, C. T. Rettner, A. Moser, M. E. Best, and B. D. Terris, Appl. Phys. Lett. **81** (15), 2875 (2002).

[73] G. F. Hughes, IEEE Trans. Magn. **36** (2), 521 (2000).

[74] G. A. Prinz, J. Magn. Magn. Mater. **200**, 57 (1999).

[75] M. M. Miller, G. A. Prinz, S. -F. Cheng, and S. Bounnak, Appl. Phys. Lett. **81**, 2211 (2002).

[76] L. J. Heyderman, C. David, M. Klauui, C. A. F. Vaz, and J. A. C. Bland, J. Appl. Phys. **93** (12), 10011 (2003).

[77] A. Ney, C. Pampuch, R. Koch, and K. H. Ploog, Nature **425**,·485 (2003).

[78] C. S. Lent, P. D. Tougaw, W. Porod, and G. H. Bernstein, Nanotechnology **4**, 49 (1993).

[79] P. D. Tougaw and C. S. Lent, J. Appl. Phys. **75**, 1818 (1994).

[80] G. L. Snider, A. O.Orlov, I. Amelani, X. Zuo, G. H. Bernstein, C. S. Lent, J. L. Mertz, and W. Porod, J. Appl. Phys. **85** (8), 4283 (1999).

[81] R. P. Cowburn and M. E. Welland, Science **287**, 1466 (2000).

[82] D. A. Allwood, G. Xiong, M. D. Cooke, C. C. Faulkner, D. Atkinson, N. Vernier, and R. P. Cowburn, Science **296**, 2003 (2002).

[83] D. A. Allwood, Gang Xiong, and R. P. Cowburn, Appl. Phys. Lett. **85**, 2848 (2004).

[84] F. J. Albert, J. A. Katine, R. A. Buhrman, and D. C. Ralph, Appl. Phys. Lett. **77**, 3809 (2000).

Chapter 11

MEDIA FOR EXTREMELY HIGH DENSITY RECORDING

D. Weller
Seagate Recording Media Operations
47010 Kato Road
Fremont, CA 94538, USA

T. McDaniel
Seagate Research
1251 Waterfront Place
Pittsburgh, PA 15222, USA

Abstract The development and engineering of the magnetic material that stores information largely determines the progress of magnetic storage technology, as epitomized by magnetic hard disks. In this chapter, we consider the ultimate capability for the storage of digital information. Noise performance and spatial resolution are key parameters in recording media and an ongoing challenge in advancing the areal density. The dominant media noise source today is transition jitter noise, or the uncertainty in positioning neighboring bit transitions. In sputtered media, it reflects the finite size, random positioning and dispersions in size, orientation and magnetic properties of the fine grains that comprise the media. Highly anisotropic materials, combined with heat-assisted magnetic recording (HAMR), promise significant reductions in the average, thermally stable grain size from currently about 7-9 nm in Co-alloys to about 2-3 nm in FePt-based media. In addition, self-organized magnetic arrays (SOMA) promise a significant reduction in jitter noise, because they yield nearly monodisperse magnetic nanoparticles such as FePt and Co. SOMA media may serve not only as conventional media with reduced dispersions and media with bit-transitions defined by rows of particles, but may also be scaled to the ultimate goal of single-particle-per-bit recording. In this last scenario, the eventual areal density is governed by the minimal thermally stable size and by the center-to-center spacing of neighboring particles. Ultimately, subject to the mastering of all writability, signal retrieval, bit-addressability and spacing issues, areal densities of the order of 40-50 terabit per square inch may be reached.

1. INTRODUCTION

Magnetic recording is a mature technology, which is over one hundred years old. Performance in commercial tape and hard disk drives (HDDs) has been improving exponentially for over fifty years. A "Moore's law"-type growth [1] of the areal density (AD), the leading performance metric in magnetic recording, has been tracked for nearly a half-century now, a longer period than that for the namesake law. The average compound average growth rate (CAGR) for HDDs over this period is about 45% (comparable to that for semi-conductors), with recent bursts persisting over many years reaching the 60-120% range. As the technology now reaches the upper portion of the classic "*S*-curve" [2], characteristic of the growth of all technologies, the CAGR has begun to diminish. This suggests that technological infusions are needed to allow the historical exponential growth to continue toward hard physical limits.

This chapter summarizes this situation from the vantage point of storage medium materials, citing the current situation and the advances that can be foreseen. We will discuss emerging technologies that show promise for providing a boost to the growth rate AD. At present, the HDD industry is experiencing the last stages of the extension of *longitudinal recording*, in which recorded magnetization lies in the plane of a layer of memory material, be it a spun coating or sputter-deposited thin film (see Fig. 1). This classic orientation of the recorded magnetic information has persisted from the earliest days of the technology until now. Longitudinal recording is approaching limits on the further reduction in size of the recorded "bit cell", due mainly to the phenomenon of thermal instability of the grains of magnetic material comprising the imprinted permanent magnet [3, 4]. Further increase in the material magnetic anisotropy energy density in order to retain the recorded magnetization in the face of thermal agitation (the historical strategy for coping with magnetic particle volume reduction) results in a commensurate increase of the necessary switching field to record the information. The dimensional scaling arguments necessitating this particle volume reduction are discussed in a subsequent section. This strategy has nearly outrun the available fringing magnetic field strength below the gap of the inductive writing element (dictated by available magnetic head materials) shown in Fig. 1.

The currently emerging technology shift in magnetic media is the reorientation of the stored magnetization to the direction perpendicular to the plane of the medium, and consequently it is called *perpendicular recording* (see Fig. 2). As explained in the figure caption, this is a strategy for increasing the usable magnetic field of the writing head. In addition, since the imprinted permanent magnet is "stood on its end", a larger effective grain volume can be achieved relative to its surface area, and this is a favorable metric for increasing AD. As we show below, achievable AD is roughly proportional to the medium anisotropy energy density, as is the media switching field. Consequently, if the effective recording field could be increased by roughly 50-75 percent in the perpendicular configuration, then we might realize a similar AD gain with this change. Thus, the leverage for AD gain with the transition from longitudinal to perpendicular recording is worthwhile, but it is not huge.

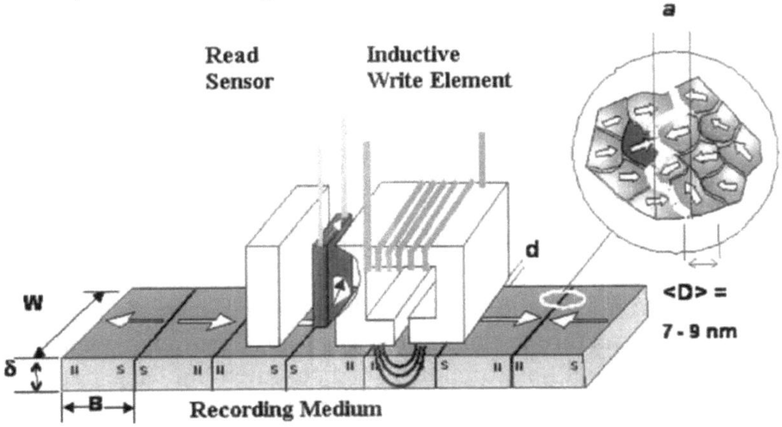

Figure 1. Schematic illustration of the longitudinal magnetic recording configuration (from Ref. [4]). W=recorded track width (track pitch); d=Head-to-Media separation (HMS); δ=medium thickness; B=Bit-length; <D>: average magnetic grain size (7-9 nm today); *a*=transition width parameter. The recording medium is comprised of 80-100 magnetically isolated grains per bit cell at today's commercial areal densities of 100 Gbit/in^2. The track width *W* is determined by the magnetic width of the write element (write field contour). The track in the recording medium is imprinted in a continuous layer of a granular magnetic material, so it is not a finite width structure as the drawing may suggest.

Beyond perpendicular recording, what technological shifts involving recording media are possible that may yield more leverage than the transition to perpendicular? The two most frequently discussed possibilities at present are *heat-assisted magnetic recording* (HAMR) and *bit-patterned media* (BPM), with the latter eventually incorporating the *self-organized magnetic array* (SOMA) technology. We will argue that HAMR alone has the potential to buy a tenfold increase of areal density, simply on the basis of utilization of much higher anisotropy media that is known to exist. HAMR is a technology that directly addresses the writability limitations of inductive heads by using local media heating during the recording process to temporarily and drastically lower the media switching field requirements. This can be done with media of the highest available anisotropy, thus taking care of the thermal stability issues at much reduced grain volumes. [4-8]

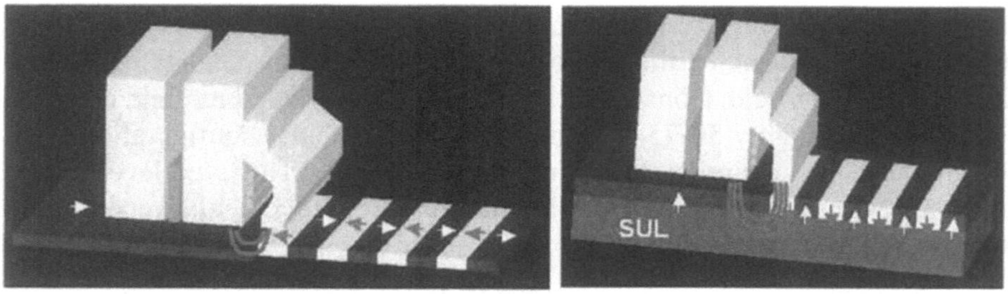

Figure 2. Schematic illustration of the geometrical transformation from the longitudinal to perpendicular magnetic recording configuration. The perpendicular arrangement allows realization of significantly higher applied fields from the head, as it places the recording medium directly in the "gap" of a head whose mirror image is created in the soft magnetic underlayer material (SUL).

BPM with SOMA takes the important step of moving away from reliance on averaging over random medium structures for achievement of an adequate signal-to-noise ratio (SNR) in the information readout process. As seen in the grain structure in the inset in Fig. 1, the particles comprising the storage medium layer are randomly distributed in size and shape, contributing in part to statistical distributions of magnetic properties. All commercial magnetic media constructions to the present have relied on a recorded magnet (or digital bit) being composed of N random particles, with N being a number "large enough" to hold property variance below some threshold through averaging. In such a situation, a simple analysis shows that a linear measure of *power* SNR is essentially N [9]. It is this paradigm that drove recording engineers to the brink of the "superparamagnetic limit" in the past decade. As AD increased, $V_{bit} = N \cdot V_{grain}$ was driven downward, but the need to maintain high N for SNR forced $V_{grain} = V_{bit}/N$ toward thermal instability as the ratio $K \cdot V_{grain}/k_B T$ became too small for the smallest grains in a random distribution to resist spontaneous reversal due to thermal perturbation.

We will see that patterning bits and diminishing randomness from the arrangement of the constituent magnetic particles moves one out of the requirements of large N and statistical averaging. In this situation, variance in the medium configuration arises from different considerations, such as the precision and accuracy of bit patterning or particle self organization. Simple geometrical analysis will show us that a relation $AD \propto 1/N$ based on thermal stability can, in well engineered BPM in which SNR is decoupled from N, be used down to the minimum limit of N, namely $N = 1$. This implies that the extreme limit of BPM has the potential of improving AD by a factor $N_{granular}/N_{BPM}$, which can be of the order of 10^2 ($N \sim$ 80-100 in today's state-of-the-art longitudinal or perpendicular media). Consequently, we will argue that, at least on the basis of known limits of physical laws, ultimately combining HAMR with BPM and SOMA could provide an AD increase of the order of $10^1 \cdot 10^2 = 10^3$. This claim ignores other possible engineering limitations that may arise.

2. FUTURE ADVANCES IN AREAL DENSITY

2.1. Getting to 1 Terabit per Square Inch – The Near Term

Improving media noise performance requires continued reduction in grain-size, grain size distributions and magnetic dispersions, and control of magnetic grain isolation. Continued grain-size reduction, however, will eventually cause thermal destabilization of the media grains. This effect fundamentally limits the attainable areal density. To postpone the onset of thermal destabilization and to push out the so-called *superparamagnetic limit*, it is generally necessary to enhance the media magnetic anisotropy and along with it the write coercivity, which enables smaller, thermally stable grains [3, 4, 10]. Recording on such smaller grains will therefore require enhanced write fields. While more than ten-fold higher anisotropy media materials are known, the saturation moment density in today's write pole materials, which determines the maximum possible write field, is already near the known FeCo-alloy limit of about 2.5 Tesla. The thermal stability problem in magnetic recording is hence a write-field problem and advancements are largely linked to improvements in write-head and field-delivery technology.

Near term advancements are realized in perpendicular recording (PMR), where a single-pole write head is combined with a double layered, hard/soft magnetic medium. The presence of the soft magnetic underlayer (SUL) permits flux closure through the recording layer into the SUL, enabling larger fields than are possible in longitudinal recording (LMR). Respective heads and media have been developed and areal densities up to about 230 Gbit/in^2 have been demonstrated. The media are comprised of well oriented nanocrystalline structures with exchange isolated grains of about 7 nm in diameter, large perpendicular magnetic anisotropy, full remanence squareness and negative onset of magnetization reversal at large fields. Both perpendicular CoPtCr-based alloys and CoX/Pd(Pt)-based multilayers are being considered. SUL approaches include Fe or Co based nanocrystalline materials, laminated exchange biased structures, and films with strong radial anisotropy. In the latter cases, the SUL is kept in a single domain state during read-back, thereby efficiently eliminating SUL related read-back noise.

To reach ultimate densities of 500-1000 Gbit/in^2 in PMR the average grain diameter needs to be reduced to about 5 nm and dispersions need to be narrowed significantly [11, 12]. Other enhancements include optimizing the write field angle with respect to the media anisotropy direction via new head architectures and/or tilting the media easy axis direction, which should further ease the write coercivity [13-16]. Side shielded heads are being pursued to concentrate the write field in the cross track direction thereby limiting the effect of side-track erasure. Details about various such designs and current issues and integration challenges with PMR can be found in the recent literature [17-19].

2.1.1 Moving beyond longitudinal recording

The recent history of CAGR in AD is summarized in Fig. 3 – see the figure caption for more information. Although more or less standardized recording demonstrations ("demos") have been practiced since the early 1990's, experience has shown that they can serve as precursors for the AD evolution of the products that follow in time. The high CAGR for the early perpendicular demonstrations is actually an artifact of the introduction of a new technology, and is more reflective of advances associated with the early part of a learning curve ("catching up with longitudinal") rather than a measure of faster intrinsic advance.

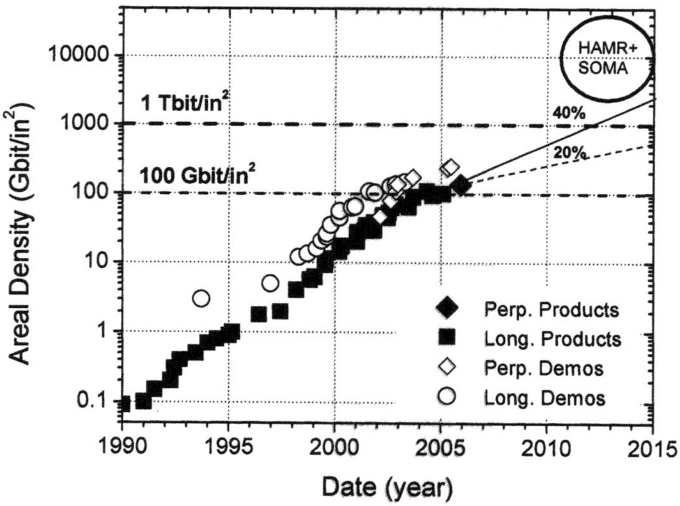

Figure 3. Areal density progression data for the HDD industry since 1990. Two future CAGR scenarios are projected from the last available datum. Notice that the slope of the trends in historical data varies for different periods reflecting variations in CAGR. The period 1998-2002 was notable for the highest values of CAGR ever seen, up to 120%. Clearly CAGR has decreased markedly since 2002-2003 for both products and demos.

We are just now (mid-2005) experiencing the first announcements and introductions of perpendicular recording products at about 130 Gbit/in^2 areal density. Again, we note that this transition to perpendicular recording is a response to the marked slowdown in progress for the very mature longitudinal recording technology, denoted by the recent arrival at the upper portion of an "S-curve" (the 'Long. Products' points in Fig. 3). The specific cause of this slowdown in longitudinal progress is fundamentally due to the collision of application of a remedy for increasing thermal decay of small grain media (namely, higher anisotropy media material), and the onset of writability difficulties due to the unavailability of head materials to provide the required increased switching fields for higher anisotropy media.

As also noted in the introduction, perpendicular recording has become the consensus candidate technology to address this issue in the short term. By making the seemingly straightforward transition of reorienting the medium magnetization from in-plane, down-track to perpendicular to the plane of the memory layer, one has the potential of recovering roughly 50-75% higher recording field from the same head materials *if* the medium is redesigned to include a relatively thick soft magnetic underlayer below the vertically oriented memory layer (Fig. 2). The perpendicular orientation of the storage bits may offer other secondary advantages that could contribute to media SNR gains, thus possibly pushing the AD gain potential of perpendicular relative to longitudinal recording into the 2- to 4-times range. Principal among these are (1) the superior "orientation ratio" (degree of anisotropy axis alignment), (2) the potential for relatively larger grain volumes (and hence greater thermal stability and perhaps readout signal) for perpendicular media, and (3) superior head field gradients (in addition to higher fields) from the perpendicular head configuration. The grain volume effect is due to "standing bits on end" (see again Fig. 2) and then thickening the media layer, a configuration that is more favorable at high linear density from the magnetic energy (demagnetization fields) standpoint than shortening and fattening bits (grains) in a longitudinal orientation.

Although this projected AD gain is not an enormous leverage for technology extendibility on the face of it, at least in historical terms for HDDs, recording technologists believe that the perpendicular orientation of magnetic media is appropriate as staging for follow-on technologies such as HAMR and BPM. The origin of this belief appears to lie in the facts that (a) the highest anisotropy candidate materials for media *can be* vertically oriented, and (b) the composite of the higher head fields along with the secondary advantages (1)-(3) enumerated above. Viewed in this light, the significant investment of the HDD industry in the transition to perpendicular technology is easier to understand from an economic viewpoint. In short, there is much larger future AD growth potential beyond the modest improvement of "conventional" perpendicular recording, particularly if that intermediate step helps prepare the way for the higher payoff technologies of HAMR and BPM.

2.1.2. Noise and dimensional scaling in granular media

In general, to advance AD (areal density) the overall system SNR in the disc drive needs to be maintained or improved at ever-smaller bit size, i.e. higher AD. The overall system SNR requirement is dictated by the allowable bit-error rate or sector failure rate and depends on the channel electronics among other factors [20]. A typical requirement, for advanced channels to properly operate, is $SNR \sim 20$ dB, assuming jitter noise domination and a relationship $SNR \sim 20 \times \log(B/\sigma_j)$ *(B = minimum bit length)*. The jitter, σ_j needs to be of the order of 10% of B or better to meet this requirement. At 100 Gbit/in^2 (today's products), we have $\sigma_j \leq 2.5\text{-}3$ *nm* and at 1 Tbit/in^2 we will need $\sigma_j \leq$ *1 nm,* assuming dimensional scaling (fixed aspect ratio BAR = W:B ~ 6-8 of the bit cell). The main strategy to get there, from the media perspective, is to reduce the magnetic grain size $<D>$. In the above scenario, a reduction of σ_j by 3 × would require a commensurate reduction of $<D>$ by 3 times, based on an averaging relationship $\sigma_j \propto \sqrt{<D>^3 / W_{read}}$, which assumes that the transition width or *a*-parameter, as e.g. indicated in Fig. 1, is proportional to <D> [21]. Such a reduction, from 7-9 nm today to 2.3 -3 nm in <D> poses an immense fabrication and microstructural control challenge and is not likely to occur in the timeframe indicated in Fig. 3 (1 Tbit/in^2 products in 2012-2013 at 40% CAGR). It is generally agreed upon today that other strategies are needed to advance the AD.

Figure 4 shows the development of the average physical grain size and grain size distribution in sputtered longitudinal media over a period of about 6-7 years. The mean grain size has been reduced from 15 to about 7-9 nm, while the distributions have come down from 33 to about 20%. Fig. 4(b) shows these data as a function of the areal density. Also included in Fig. 4(b) are extrapolated requirements for 1 Tbit/in^2 perpendicular recording ($<D> \sim$ 5 nm) and 10 Tbit/in^2 HAMR recording (<D> ~ 2-3 nm). Table 1 summarizes respective numerical values including demo dates.

As the figure shows, the physical media properties have been improved by less than a factor of two over the indicated period of 6-7 years, while AD has grown more than 15-fold. Current perpendicular recording models suggest that 1 Tbit/in^2 is only possible if the grain-count-per-bit is drastically reduced, from currently about 80 (100 Gbit/in^2: 35 nm × 180 nm bit cell; 750 kfci × 150 ktpi; 10-15% channel overhead) to as low as 5-6 (500 Gbit/in^2: 11 nm × 47 nm bit cell; 1850 kfci × 540 ktpi; >35% channel overhead) [11-13]. The suggested approach assumes higher relative media noise and the availability of significantly improved channel detectors, with associated larger error-correction overhead. The media grain size still needs to be reduced to <D> ~ 5 nm (center-to-center spacing) and the grain size distribution needs to be trimmed significantly below the currently achieved value of about 20%. Even under such somewhat relaxed requirements, media scaling remains a major challenge, given the historical trends.

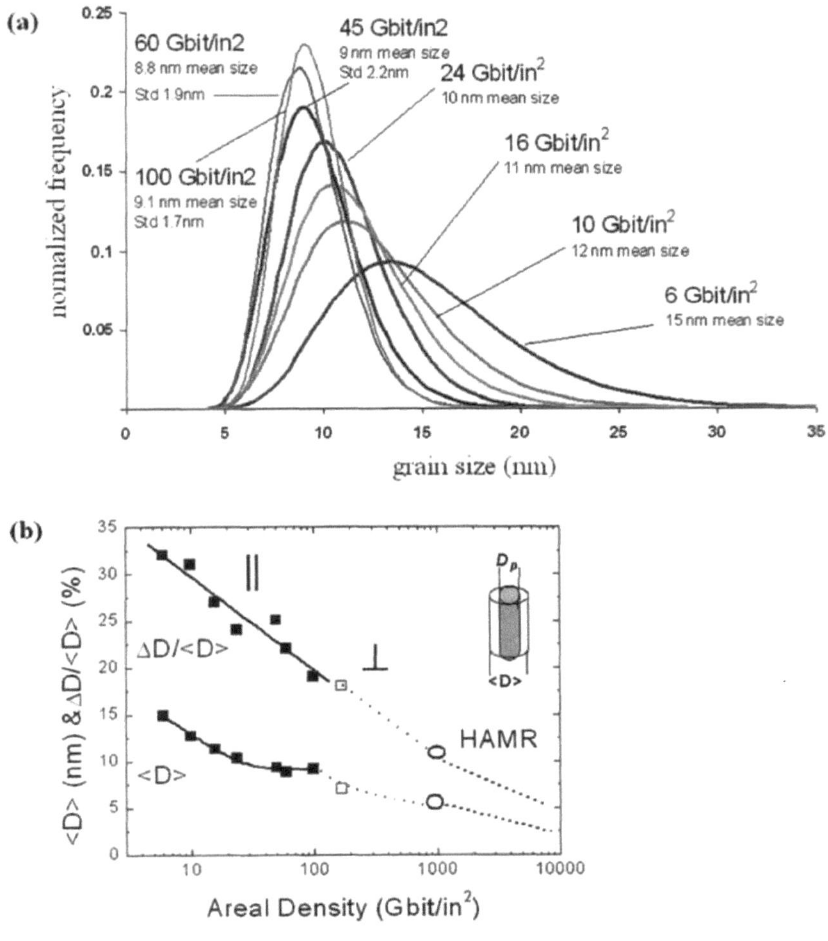

Figure 4. Seagate longitudinal media of 6 Gbit/in^2 and beyond (courtesy of Tom Nolan, Rajiv Ranjan, Gary Rauch): (a) development of mean grain size and grain size distribution in. The curves are fits to the data using a lognormal size distribution as is common for sputtered granular media (see Eq. (1) below (b) average grain size <*D*> and grain size dispersion *ΔD*/<*D*> as a function of areal density including trends and targets for 1 Tbit/in^2 (perpendicular) and beyond (HAMR, SOMA). Note that the average grain size included the grain boundary (see inset).

Table 1: Evolution of mean grain size <D> and grain size distribution ΔD/<D> for Seagate laboratory demonstration media of 6 Gbit/in^2 and above. The data point at 170 Gbit/in^2 is for perpendicular media. Trends at higher densities are estimated.

density (Gbit/in^2)	10,000	1,000	170	100	60	45	24	16	10	6
demo dates	?	~2010	2003	2001	2000	2000	1999	1999	1998	1997
<D> (nm)	~2	~4-5	~7-8	9.1	8.8	9.3	10.3	11.3	12.7	14.9
std. dev./mean	~0.05	~0.10	~0.18	0.19	0.22	0.25	0.24	0.27	0.31	0.32

Note, that *<D>* is a magnetic grain or cluster size, which in the limit of negligible interactions equals the physical grain size (center-to-center spacing). High-resolution magnetic force microscopy studies [22] and soft x-ray characterization [24] of state-of-the-art perpendicular granular oxide media indicate a significant disparity of 2-3 × between both quantities, implying that considerable improvements of media are possible, if magnetic (dipolar and exchange) interactions between adjacent grains can be reduced further. However, any such improvements will have to preserve thermal stability, which will limit the ability to reduce the physical grain size.

Regardless, however, of whether intergranular interactions are lowered or the physical grain size is reduced, magnetically harder media materials and larger write fields will be required in the future, in order to meet thermal stability requirements [3, 4, 7, 10].

Thermal stability is a function of the grain size and grain size distribution [24]. To estimate the minimal thermally stable grain core size $D_{p,}$ one needs to make assumptions about the magnetization reversal and thermal decay mechanisms. Thermal decay is described by an exponential Arrhénius law, which relates the time constant τ for storage to the ratio of a reversal energy barrier E_B and thermal energy k_BT (T = absolute temperature in Kelvin) (Néel-Brown model [25-27]):

$$\tau = f_0^{-1} e^{E_B / k_B T} \tag{1}$$

To maintain thermal stability, hence a condition $E_B/k_BT = ln\ (f_0\tau)$ needs to be fulfilled. For τ = 10 years storage, $f_0 \sim 10^9$-10^{11} Hz [28] and ignoring dispersions, i.e. assuming monodisperse particles, this becomes $E_B/k_BT = 40$-45. Reversal for isolated, well-decoupled grains to first order can be described by coherent rotation over E_B. This simple model, as first discussed by Stoner and Wohlfarth in 1948 [29], considers only intrinsic anisotropy and external field (Zeeman) energy terms. For perpendicular geometry one obtains the following expression:

$$E_B = K_u V \left(1 - \frac{4\pi N_d M_S}{H_K}\right)^2 . \tag{2}$$

The pre-factor in Eq.(2) is the intrinsic energy barrier given by the anisotropy energy density K_u (materials constant) multiplied by the grain core volume V. The extrinsic, external field term in Eq.(2) is given by the demagnetization field $4\pi N_d M_S$ normalized to the anisotropy field H_K. Here, for generality we use a demagnetization factor of N_d, which can vary from 0 to 1, but which will tend toward 1 for long bits in perpendicular recording [30], M_S is the saturation moment density and K_u can be expressed as $K_u = H_K M_S/2$.

Eq. (2), together with assumptions about the grain shape (typically cylinders with constant height δ), the demagnetization field and the actual E_B/k_BT requirement, can be used to estimate the minimal thermally stable grain size D_p. Before calculating D_p for a variety of media materials, we first consider the effect of physical grain dispersions on the E_B/k_BT requirement.

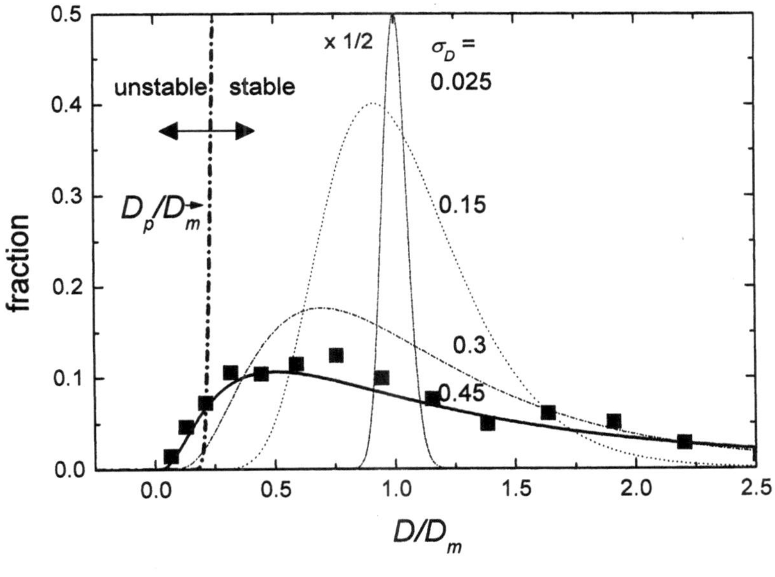

Figure 5. Illustration of superparamagnetic cut-off as function of grain core size for three different widths σ_D of log-normal grain size distributions. Small grains cause thermal decay and large grains compromise the media noise performance. Reducing σ_D avoids thermal decay and improves SNR. The horizontal axis is normalized to the median grain size and the vertical axis shows the fraction of grains with a given diameter. The data points for the wide distribution are taken from earlier longitudinal recording media (similar to Ref. [24] but converted to grain size D instead of grain area A).

An illustration of this effect on thermal stability is given in Fig. 5. The figure shows calculated lognormal grain size distributions *G(D)* for different dispersion widths σ_D ranging from 0.45 (earlier longitudinal media) to 0.025 (hypothetical, nearly ideal monodisperse media). *G(D)* and σ_D are defined in Eq.(3). Grains, which fall below the stability boundary D_p/D_m are thermally unstable. Grains larger than D_p/D_m are thermally stable. D_p denotes the minimal stable magnetic grain core diameter (excluding boundary) and D_m denotes the median of the distribution (for a more detailed discussion see [4, 31]).

$$G(D) = \frac{1}{\sqrt{2\pi}\sigma_D D} \exp\left\{ -\frac{1}{2} \frac{\left[\ln(D/D_C)\right]^2}{\sigma_D{}^2} \right\} \tag{3}$$

Reducing σ_D from 0.45 to 0.3 in this example removes all small particles from the thermally unstable regime. The benefit of reducing σ_D is two-fold: (1) direct SNR gains due to reduced dispersions (a gain of order of +2.5 dB may be expected by reducing σ_d from 0.45 to 0.025 [32]) and (2) ability to lower the mean grain size without compromising thermal stability (results in additional small SNR gains).

Next, we estimate potential AD gains, based on the above discussion and assuming isolated grains with constant height δ. Based on purely geometrical arguments, the maximum AD should scale as

$$AD_{\max} = \frac{1}{A_{bit}} = \frac{1}{N \cdot \langle A_g \rangle} \propto \frac{p}{N \cdot \langle D_p \rangle^2}, \tag{4}$$

where A_{bit} is the minimum bit area, $\langle A_g \rangle$ is the average grain area (at the medium top surface and including the grain boundary), N is the average number of grains per minimum sized bit and $\langle D_p \rangle$ is the average stable grain core size as defined before. The packing fraction p is defined as the ratio of the average grain core area (excluding grain boundary width w) to the total grain area (including boundary w):

$$p = \frac{\langle D_p \rangle^2}{(\langle D_p \rangle + w)^2} \tag{5}$$

Here $\langle D \rangle = \langle D_p + w \rangle = \sqrt{\langle A_g \rangle}$ for grains with arbitrary shape. Figure 6 shows the impact of reducing $\langle D_p \rangle$ for a range non-magnetic grain boundary widths w. Advanced perpendicular media today, have $w \approx 1$ nm for $\langle D \rangle = 7$ nm, corresponding to a packing fraction of about 75%. It is seen that as one scales the average grain size down to smaller sizes the packing fraction reduces. Assuming $w \approx 1$ nm as a reasonable grain boundary width, one ends up with a packing fraction of order of 60-70% for the currently targeted Tbit/in^2 perpendicular design. The packing fraction limits the maximum read-back signal. It has an impact as well on dc-noise and it is crucial to reduce the grain boundary width as the grain diameter is reduced [33]. The goal is to keep p as high as possible, which will be a significant fabrication and processing challenge for future recording media approaches using HAMR and SOMA, where the grain size target is 2-3 nm.

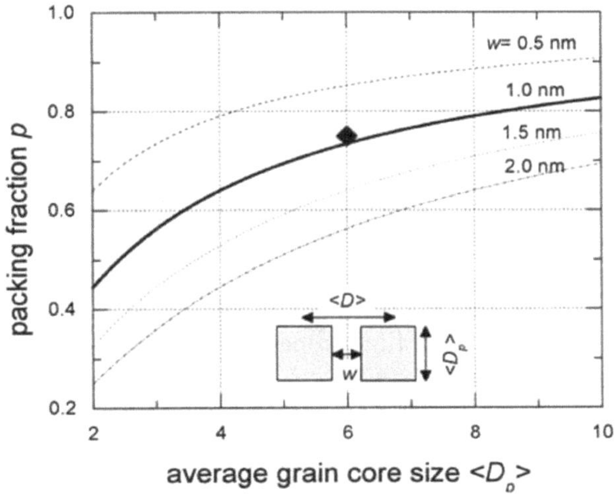

Figure 6. Packing fraction p according to Eq. (5) as a function of magnetic core size $<D_p> = <D> - w$, with $<D>$ the average diameter or width of the grain's magnetic core plus non-magnetic shell and w the non-magnetic separation between grains (grain boundary width). The figure inset defines the simple model of grain geometry employed here. Today's state-of-the-art perpendicular media have $<D> \approx$ 7nm and $w \approx 1$ nm, hence $p \approx 0.75$, as indicated y the data point.

Using the thermal stability criterion $E_B/ k_BT = r_k$, Eqs. (2) to (4) we can be re-written to yield the following relation for AD_{max}:

$$AD_{\max} \approx \frac{p}{N}\left[\left(\delta M_S H_K \big/ 2r_k k_B T\right)\left(1 - 4\pi N_d M_S \big/ H_K\right)^2\right] \tag{6}$$

Eq. (6) provides general design guidelines for increasing areal density. The three main options are to

(i) maintain a large packing fraction p, implying scaling down $<D>$ as well as w.

(ii) use higher H_K media materials, which requires increasing the write field capability.

(iii) reduce the grain count N, which requires operation at lower SNR and/or lithographic patterning of bits ($N = 1$).

It is generally advantageous to optimize the magnetization thickness product $M_S \cdot \delta$ and to keep $4\pi M_S/H_K$ small. $M_S \cdot \delta$ needs to be compatible with the read head capability and should generally be high, but at the same time the ratio of M_S and H_K needs to be relatively low (about 1/3 [34]) to avoid destabilization due to demagnetization fields. The choice of the film thickness δ needs to be optimized to achieve maximum write field gradient (small δ) and high thermal stability (large δ). The grain aspect ratio δ/D_p cannot exceed values of about 2.5-4, above which the magnetization tends to switch incoherently [35]. The stability ratio r_K in Eq.(6) cannot be smaller than about 40-45 and should be of the order of $r_K \approx 50$ to allow for small dispersions $\Delta E_B/E_B \approx 10\%$.

We now turn back to the question of what the minimal stable grain size for various perpendicular media materials candidates is. Table 2 summarizes a variety of intrinsic and derived properties for such candidate materials. The data are based on a similar table for longitudinal media [4, 10, 36]. They represent intrinsic bulk and thin film properties. The last three columns show the minimal thermally stable grain size dimensions D_p calculated based on the stability criterion $r_K = 50$: (a) diameter for cylindrical grains of constant height $\delta = 10$ nm, (b) size for cubic grains with $D_p = \delta$, (c) diameter for spherical grains and (d) size for elongated cubes with constant grain aspect ratio $\delta/D_p = 4$. The dimension D_p is calculated at storage temperature $T_S = 350$ K ($k_B T_S = 5 \times 10^{-14}\, erg$) as follows:

$$D_p = \left(k \frac{2 \cdot r_K \cdot k_B T_S}{H_K M_S (1 - \frac{4\pi M_S}{H_K})^2} \right)^n \tag{7}$$

where $n = 1/2$ for cylinders (case a) and $n = 1/3$ for cubes and spheres (cases b-d). The geometric prefactor k is (a) $k = 4/\pi\delta$ ($\delta = 10$ nm) for cylinders, (b) $k = 1$ for cubes, (c) $k = 6/\pi$ for spheres and (d) $k = D_p/\delta = 1/4$ for elongated cubes.

Grain sizes of the order of $D_p = 2$-3 nm become possible for the highest anisotropy materials such as CoPt or FePt-$L1_0$ phases. The last row also shows a value for the high anisotropy Co_5Sm phase, indicating thermal stability down to $D_p = 1.3$ nm (10 nm tall needles), 2.4 nm (cubes), 2.9 nm (spheres) or 1.5 nm (elongated cubes). Those represent the most aggressive values; however, chemical stability may be a major issue for rare-earth containing materials. Also, at such small dimensions, the ratio of surface to volume of the particles increases significantly (~ 25 % for 3 nm diameter spheres) and it is to be expected that magnetic properties will be strongly affected by surface effects and in particular the chemical environment of the grains (grain boundary material). From this perspective, chemically ordered CoPt or FePt $L1_0$ compounds, with close to equiatomic composition are more likely candidates for ultimate areal densities in magnetic recording than, e.g. Co_5Sm systems [37-39].

Table 2: Magnetic properties and theoretical minimal grain diameters of various perpendicular media candidates including the most advanced chemically ordered Co_3Pt [36] and $(CoCr)_3Pt$-alloys. The minimal physical grain diameter is estimated using maximum demagnetization $4\pi M_S$ ($N_d = 1$) and the four different grain geometries (a-d) explained below the table. Definitions: $H_K = 2K_1/M_S$ anisotropy field; M_s and K_1 are measured in emu/cm^3 and 10^7 erg/cm^3 respectively.

Alloy System	Material	K_1	M_S	H_K (kOe)	T_C (K)	$D_p^{(a)}$ (nm)	$D_p^{(b)}$ (nm)	$D_p^{(c)}$ (nm)	$D_p^{(d)}$ (nm)
Co-alloys	$CoCr_{20}Pt_{15}$	0.25	330	15.2		15.5	12.4	15.3	7.8
	Co_3Pt	2	1100	36.4	1200	6.4	6.9	8.5	4.3
	$(CoCr)_3Pt$	0.39	410	19		12.4	10.6	13.2	6.7
	$CoPt_3$	0.5	300	33.3	600	9.0	8.6	10.7	5.4
CoX/Pt(Pd) multilayers	Co2/Pt9	1	360	55.6	500	6.1	6.7	8.3	4.2
	Co2/Pd9	0.6	360	33.3	500	8.4	8.2	10.2	5.2
$L1_0$ phases	FePd	1.8	1100	32.7	760	7.3	7.5	9.3	4.7
	FePt	7	1140	122.8	750	2.4	3.6	4.4	2.3
	CoPt	4.9	800	122.5	840	2.8	3.9	4.9	2.5
	MnAl	1.7	560	60.7	650	4.9	5.7	7.1	3.6
rare-earth alloys	$Fe_{14}Nd_2B$	4.6	1270	72.4	585	3.4	4.5	5.5	2.8
	$SmCo_5$	20	910	439.6	1000	1.3	2.4	2.9	1.5

(a) cylindrical grains of constant height $\delta = 10$ nm
(b) cubic grains with $\delta = D_p$
(c) spherical grains
(d) elongated cubes with constant grain aspect ratio $\delta/D_p = 4$.

Note again, that Table 2 contains intrinsic materials properties, representative of homogeneous bulk or thin film materials. Addition of third or fourth elements for grain decoupling or columnar growth will likely lower anisotropy and magnetization values and hence compromise the ultimately achievable areal density.

2.2. Finding Another 100 to 1000 Times Increase in Density Growth – New Technologies

Densities beyond 0.5-1 Tbit/in^2 will likely require more drastic departures from conventional recording. In *heat-assisted magnetic recording* (HAMR), the write coercivity is temporally lowered via rapid local heating. Recording occurs during subsequent cooling in the presence of the head field [5, 7, 41-43]. If heat delivery systems capable of reaching the Curie temperature of candidate HAMR media materials such as $Fe_{50}Pt_{50}$-based $L1_0$ structures (T_C = 480 °C, Table 2) can be built, then stable grains as small as 2.4 nm in diameter and 10 nm tall should be possible. HAMR technology, however, faces many challenges, not the least of which is fabricating and assembling such small elongated grains with nearly *monodisperse* properties. Many efforts are ongoing to realize such materials using advanced sputtering combined with *in-situ* or *ex-situ* heating and significant progress has been made to fabricate actual perpendicular granular FePt based media [44, 45]. Other efforts are concentrating on lowering the write temperature requirement by doping e.g. Ni into FePt [46] or by composite media approaches. Thiele *et al.* [47, 48] for example showed that a dual layer structure of antiferromagnetic (AFM) FeRh and ferromagnetic (FM) FePt can significantly lower the write coercivity at temperatures as low as 400-500 K, corresponding to the AFM-FM phase transition in FeRh, which can be induced by laser heating on time scales as short as a few picoseconds [49]. Such efforts are crucial to reduce head-disc interface heating, which could be a major problem.

Regardless of the material combination, departures from conventional sputter deposition, such as cluster beam methods with particle size selection, shaping and heating capabilities independent of the substrate, may be necessary to achieve the desired granular microstructure [50-53].

2.2.1. *The potential of heat-assisted magnetic recording (*HAMR*)*

The current-write anisotropy-field cut-off is believed to be about $H_{K,max}$ = 20-30 kOe for perpendicular recording, which requires most advanced writer head technology and includes enhancements such as tilted [15, 16] and/or recently proposed exchange-coupled composite media [54, 55]. Hence, most of the materials listed in Table 2 are excluded from perpendicular media applications. Co-rich Co-Pt based alloys and Co/Pd type multilayers look promising from the perspective of near term extensions to higher write coercivities. Future HAMR media candidates are mainly Co/Pt based multilayers (D_p ~ 6.1-8.3 nm) (entry level) and ultimately FePt-$L1_0$-compounds (D_p ~ 2.4-4.4 nm).

A graphical depiction of the data in Table 2 is shown in Fig. 7. Plotted are the anisotropy fields H_K versus minimum stable cylindrical grain diameter D_p for fixed media thickness δ = 10 nm. The dashed line indicates the trend toward higher H_K materials with D_p dropping. Write fields H_0 larger 100 kOe are required to write the smallest grain diameters of $D_p \cong$ 2.4

nm for chemically ordered FePt- $L1_0$-compounds. $H_0 > 200$ kOe would be required for Co_5Sm.

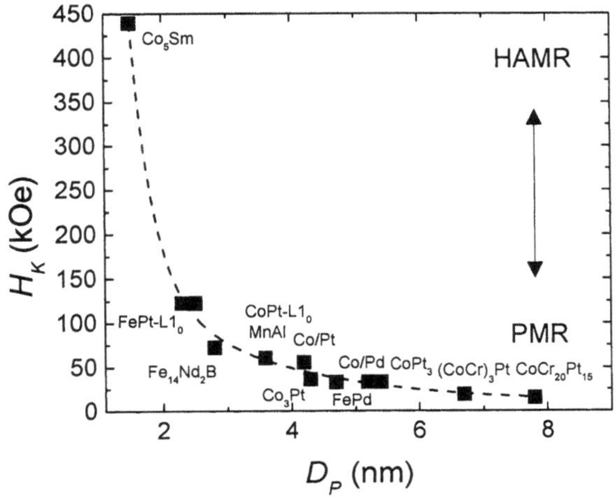

Figure 7. Anisotropy field H_K versus minimal stable grain diameter D_p for cubic grains ($\delta = D_p$) (last column in Table 1). The dotted line represents today's write field limit, without thermal assist (the dashed line is a guide to the eye).

To depict the areal density capability of HAMR we can re-write Eq. (6) as follows

$$AD_{\max} \approx K_u(T_S)\frac{p}{N}\left[\left(\delta \big/ r_k k_B T_S\right)\left(1 - 4\pi N_d M_S \big/ H_K\right)^2\right] \quad (8)$$

According to Eq. (8), AD scales largely linearly with the anisotropy energy density K_u at storage temperature T_S. It can be concluded, that on a relative scale, i.e. assuming purely geometrical scaling, FePt should provide an areal density advantage of roughly a factor of ten. For Co_5Sm the gain factor would even be higher (> 20×). The data of Table 2 are re-plotted in Fig. 8 to express that areal density gain potential and approximately linear relationship of AD_{max} with $K_u(T_S)$ expected from the leading term in Eq.(8). The vertical scale is in relative units with a base of 250. It can be readily converted into an absolute scale of Gbit/in^2 if one identifies the base with the highest AD demonstrated in perpendicular recording to date. Accordingly, densities of about 2 Tbit/in^2 and 5 Tbit/in^2 become possible for FePt-$L1_0$ and Co_5Sm, respectively, if p and N are kept constant. Reductions in N (number of grains per bit), e.g. enabled by channel gains, will enhance the AD potential of HAMR further.

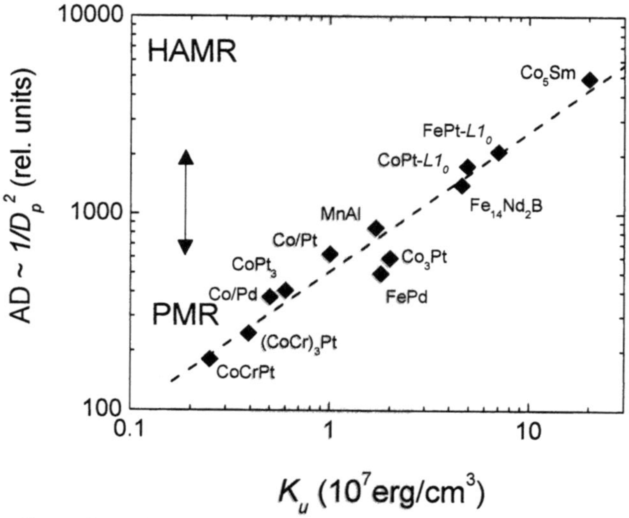

Figure 8. Scaling of areal density (AD) with anisotropy energy density K_u using room temperature bulk data from Table 2 and elongated cubes with constant grain aspect ratio $\delta/D_p = 4$ (case d) (the dashed line is a linear fit of the data to the power one).

2.2.2. The benefit of patterning bits in the recording medium

Even though HAMR technology is appealing from the perspective of continuing to scale media to smaller grain sizes without compromising thermal stability, there is significant risk and uncertainty associated with such a new technology. Inability to control the microstructure at such small length scales, inability to generate a reliable head-disk interface at ever smaller head-media spacing with the added heating requirement, inability to heat and cool the media on rapid time scales of sub-nanosecond are just some examples of such risks. More fundamentally, it is unknown, whether bulk media properties, as assumed in Fig. 8 and Table 2, can be retained for the tiny grains, that would be required. For example, the ratio of surface-to-volume atoms increases to about 40% for FePt and 60% for Co_5Sm for the most aggressive scenarios (case (d) in Table 2).

As an alternative to HAMR, forms of patterned media are being considered as a means to lower media noise by lithographic definition of tracks or the recording transitions. The basic concept is to operate with fewer grains per bit and keep the grains larger, which avoids problems discussed above. Likewise, patterned media will face many fundamental and technological challenges, not the least of which are pattern replication with the required precision at densities larger than about 500 Gbit/in^2 and manufacturability at low cost.

Ultimate recording densities of 40-50 Tbit/in^2 are conceivable in *self-organized magnetic array* media (SOMA), in which guided self-assembly processes are exploited to go beyond the limitations of lithographic patterning [56, 57]. Terris and Thomson have recently reviewed the field of patterned media, including bit-patterned media (BPM) and discrete-track media (*DTM*) based on nano-imprint as well as guided self-assembly techniques [58].

In the present chapter in the following we give a brief review of one form of SOMA, which could overcome processing limitations in conventional sputtering and serve as HAMR, BPM and ultimately both.

Self-ordered magnetic array (SOMA) media. - Distribution sigmas below 0.15 (in linear size dimension) to our knowledge have not been achieved in conventional thin film deposition processes. Chemical synthesis routes on the other hand can generate nearly monodisperse ($\sigma_D < 0.05$) particle arrays [56, 57]. Two examples of such tight distributions are shown in Fig. 9, which displays plan view 100 nm ×100 nm TEM images of 6.4 nm and 8.2 nm-FePt nanoparticle arrays [56]. Such FePt nanoparticle arrays with mean diameters in the 3-10 nm range and with room-temperature coercivities larger than 1 Tesla were first successfully reported by Shouheng Sun *et al.* in 2000 and are now routinely fabricated in many industrial and university laboratories [56, 59].

To use such SOMA media for either conventional (longitudinal or perpendicular) or BPM, a number of obstacles have to be overcome.

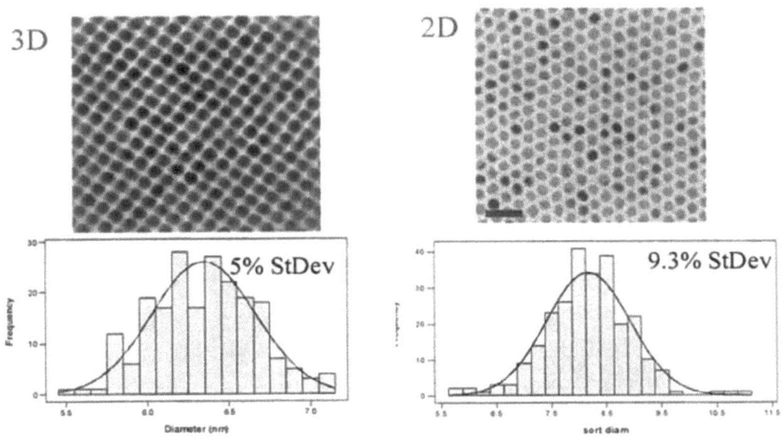

Mean: 6.4 nm StDev: 0.32

Mean: 8.2 nm StDev: 0.76

Figure 9. Size distribution analysis of plan view 100 nm × 100 nm TEM images of 3D and 2D FePt SOMA structures (see also Ch. 9).

First, while self-organization and size distributions of better than 5% have been found in as-deposited FePt SOMA structures, annealing to temperatures of the order of 600-700 °C, as required to fully convert the FePt material into the desired magnetically hard face-centered tetragonal (*fct*) or $L1_0$ phase, has been found to destroy the array order and size distribution via metallurgical sintering processes [60, 61]. This effect is illustrated in Fig. 10. Proposals to avoid sintering include synthesis of core-shell structures with harder shells (diffusion barriers) and enhanced adhesion of the particles to the substrate surface via chemical modification of substrates as well as particle ligand shells. For example, it has been shown that sandwiching FePt nanoparticle layers between carbon films may avoid sintering after heating to 700-800 °C. Kodama *et al.* [62] from Fujitsu demonstrated recording on such media, which comprised a 2.5 inch disk substrate coated with SiO-Si / 3 layers of [$Fe_{53}Pt_{47}$/ carbon] and a 5 nm lubricant layer. The 3-4 nm FePt nanoparticles were deposited by spin-coating. Other efforts are concentrating on lowering the required annealing temperature for the $L1_0$ phase transformation, e.g. by alloying, without at the same time inducing grain growth [61, 63-65].

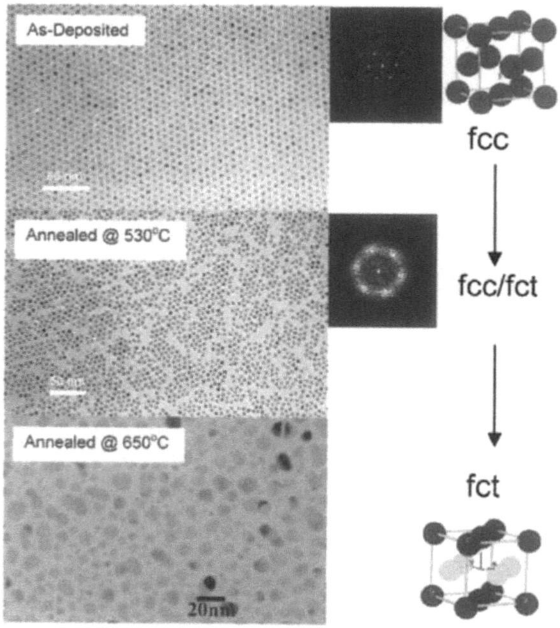

Figure 10. Effect of annealing on FePt SOMA structures. TEM images indicate that sintering of small FePt nanoparticles takes place at temperatures above about 530-650°C. Thermogravimetric analysis (TGA) indicates breakdown of the surfactant shell at temperatures above 350 °C (see Ref. [61]).

Second, all FePt SOMA structures reported to-date have more or less random easy-axis orientation, which is not acceptable for future high density media as it will result in extra media noise and reduced the signal strength. This appears to be one of the most difficult unsolved problems today. Attempts to anneal SOMA structures in large fields or deposit in the presence of a magnetic field so far, to our knowledge, have failed or produced only marginally oriented magnetic particles [66, 67]. The most promising approach to date appear to be to pursue shaped FePt particles that can be crystallographically and subsequently magnetically oriented, e.g. by deposition or annealing in an external field [68-70].

Third, FePt SOMA nanoparticles today are approximately spherical (truncated octahedral before anneal). The surfactant shell coating is between 1-2 nm thick (oleic acid, oleylamine). Scaling, under these conditions is much more unfavorable as e.g. for cylindrical grains as discussed in conjunction with PMR or HAMR media. Figure 11 shows the effect of packing fraction as function of molecule chain length, $l = w/2$, for 2D arrays of 3 nm FePt particles in spherical, cylindrical or cubic shapes [71]. For a hexagonal close packed two-dimensional array of spheres and a chain length of $l = 1$ nm the packing density is only 23%. Even if l could be reduced to 0.5 nm (optimistic case; $2l = w = 1$ nm) p would only be about 33%, which is far less than what is needed to avoid *dc*-noise dominance in read-back [33].

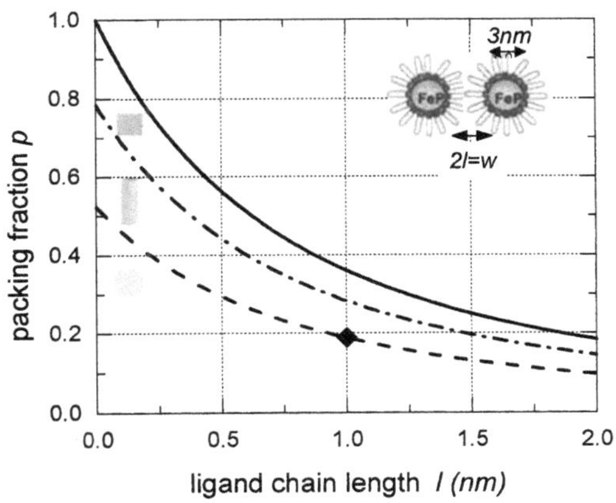

Figure 11. Effect of packing density on ligand chain length of surfactant molecules for 3 nm diameter FePt spheres, cylinders and cubes. Current state-of-the-art spherical 3nm diameter particles use $l = 1$nm oleic acid surfactant molecules and have a packing fraction of only $p = 19\%$. Cubic or cylindrical structures with shorter chains are needed to increase p to acceptable levels.

The question is whether there are surfactant or other boundary materials approaches that enable higher packing densities than those typically reported in the literature so far ($p \sim 20$-30%).

Finally, we point to some recent discussion in the literature pertaining to a size dependence of the thermodynamic $L1_0$ critical temperature [72, 73], which will impede the chemical ordering process for the smallest particle sizes.

Using lithography and self-organization (guided self-assembly). - Finally, self-organization is limited to dimensions of order one to ten micrometer. To overcome these limitations two-step processes involving lithographic pre-patterning of substrates and subsequent nanoparticle assembly on the smaller scale of 100 nm – 10 μm have been proposed [74-76]. This concept, which could be used to obtain extend the self-assembly order over the full size of a recording disk, is summarized in Fig. 12. In rotating storage one would align the lithographically pre-patterned regions into a circumferential structure. For alternative probe storage devices one could choose an x-y pattern. Both topographic as well as chemical pre-patterning schemes have been proposed.

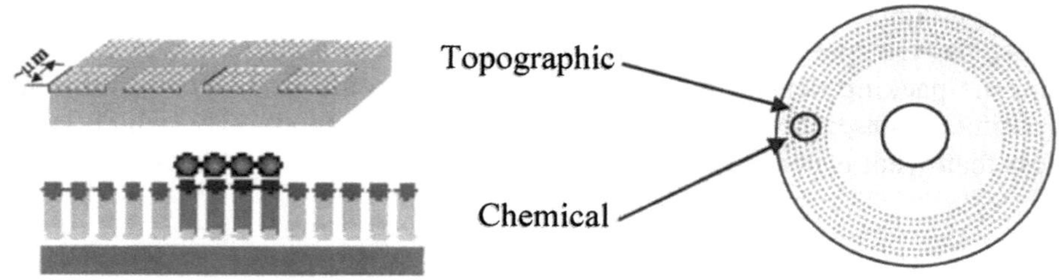

Figure 12. Dual-Patterning or guided self-assembly approaches to SOMA disks. The array ordering length scale of 1-10 μm is replicated around the circumference of a disk by topographic or chemical guide patterns.

SOMA ultimately represents an approach toward single-grain-per-bit bit-patterned magnetic recording. Figure 13 is used to summarize the potential of this approach.

Considering thermal stability alone we argue that 40-50 Tbit/in^2 AD might become possible in single particle per bit SOMA patterned media recording using FePt. Assuming cylindrical grains of diameter D_p = 4.0 nm and a surfactant chain length of l = 1 nm gives about 18 Tbit/in^2. For D_p = 2.4 nm, as quoted in *Table 2* for FePt and with l = 1 nm chain length one obtains 33 Tbit/in^2. Even higher hypothetical particle densities can be estimated on purely geometrical grounds, e.g. if one succeed with shaping particles and reducing the chain length. It is clear, however, that all those estimates ignore overhead e.g. due to dual patterning or to side fringing of the write field or read-back signal.

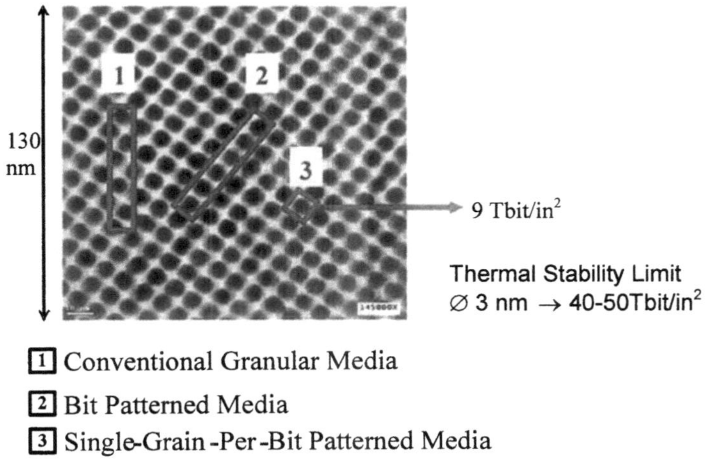

Figure 13. FePt SOMA structure with 6.4 nm particles (from Fig. 9) and different media application scenarios. Case 1: conventional media: the particles are randomly placed in relation to the bit transition; case 2: bit-patterned media with several particles per bit allowing for bit-aspect ratio > 1: the bit transition is pre-located into the transition line between rows of particles; case 3: single-particle-per-bit patterned media recording: as case 2 but with bit-aspect ratio = 1. For 6.4 nm particle size a maximum density of 9 Tbit/in^2 is estimated. If the particle size is reduced to 3 nm, 40-50 Tbit/in^2 could be possible.

2.2.3. Combining technologies

To reach the thermal stability limit of extremely high anisotropy SOMA media in the AD range of ~ 40 Tbit/in^2 implies that HAMR is in use as well. Clearly, the highest H_k media in Table 2 require switching fields far above the presumed limit of inductive head materials, which is around 25 kOe, and so thermally-assisted writing will be necessary. We believe that, as indicated in the upper right of Fig. 3, when AD approaches and moves beyond ~ 1-5 Tbit/in^2 that HAMR will be in use, whether BPM is or not. At some point, HAMR and SOMA, as a more mature form of BPM, will combine to carry magnetic recording toward ultimate physical limits.

Our earlier discussion surrounding Fig. 3 indicates that a transition from longitudinal to perpendicular recording is occurring somewhere above 120 Gbit/in^2. Perpendicular recording, without additional heat assist should enable densities in the 500 Gbit/in^2 – 1 Tbit/in^2 range. HAMR and BPM with SOMA are technology options beyond 1 Tbit/in^2.

2.3. Approaching the Ultimate Limits of Magnetic Storage

2.3.1. Physical limits of magnetic media

We previously developed one of the main results of physical limit analysis for this chapter, which is embodied in Eqs. (4) or (6). This is an expression for AD based on simple geometry combined with the basic criterion of small magnetic particle thermal stability (or avoidance of superparamagnetism). Whenever information is stored magnetically in a single domain particle comprised of atoms, these equations will be valid. (These equations would not pertain to some type of storage of magnetic information at the atomic level.) The underlying physical picture is that a particle of condensed atomic matter has sufficiently high uniaxial magnetocrystalline anisotropy (although in principle the particle could be amorphous rather than crystalline), that its anisotropy energy K_uV is sufficiently large relative to the environmental thermal energy k_BT, and that the probability of the composite magnetization orientation of the ferromagnetically ordered spins coherently reversing its component along the anisotropy axis is "sufficiently small" as to render the magnetization "stable". This means that in an ensemble of tens or hundreds of magnetically oriented such particles, the probability of more than x percent of them (typically $x \leq 10$) reversing in 10 years at storage temperature is effectively zero.

Possible physical limits associated with media may be inherent in the HAMR process as discussed by Lyberatos *et al.* [77]. The issue here is that in heating media to effect HAMR, one is purposely inducing extreme thermal instability, even to the point of initial erasure prior to reforming the magnetization and stabilizing it by rapid cooling. The detailed physics of how this occurs, particularly when the peak medium temperature exceeds the Curie temperature, is not yet well developed. It may be there will exist conditions of diminishing returns when trying to confine media heating, magnetization collapse, and magnetization re-formation. Lyberatos *et al.* [78] have analyzed this problem within the context of more classical theories of thermally-induced magnetic stability of uniaxial, single domain particles treated *below* the Curie temperature. In [78], the issue of adjacent track erasure in HAMR is considered, and here it seems clear that AD will be directly limited by the degree of thermal localization that is possible in HAMR.

2.3.2. Other system limits in physics and engineering

There are other physical limits surrounding magnetic storage technology, and many of them are linked to values of fundamental physical constants of known materials. For example, the output magnetic fields on inductive or

permanent magnet recording heads cannot significantly exceed the saturation induction B_s of the head material. For known soft magnetic materials, this value is 2.43 T for $Fe_{65}Co_{35}$, in agreement with the Slater-Pauling curve.
In present day magnetic recording systems, readout of stored magnetic information is done by detecting weak fringing fields near the medium surface with a magnetoresistor scanned over that surface. The useful signal from magnetoresistance (*MR*) is effectively an electrical potential drop due to current passing through the field-dependent resistor, $V = IR(H)$. Since SNR must be sufficient, there is ongoing pressure to increase the signal V relative to the shrinking volume of the bits. The only levers to do so are to push the current (density) upward and/or to increase the inherent sensitivity of magnetoresistors *R(H)*. Much progress has been made on the latter tack, but there probably exists an upper limit on improvement of *R(H)* imposed by available materials. Meanwhile, the Joule heating in the resistor $I^2R(H)$ produces useless thermal energy (and Johnson noise) in the ever-decreasing volume of the magnetoresistor. Consequently, there will be an ultimate limit on magnetoresistor SNR when waste thermal energy overcomes the *MR* signal.

These examples of physical limits do not mean that magnetic recording technology has hit a brick wall when they are encountered. They do mean that "workarounds" or workable alternative approaches must be identified. "Alternative approaches" cannot mean overcoming the physical limit directly, since this is impossible by definition. Workarounds amount to engineering innovation, which identify a different way of achieving necessary ends.

There may be situations in which a physical limit combines with engineering difficulties to block progress. One example is the head-medium interface in recording systems. The magnetic spacing between the head and medium is well-known to be a critical parameter in the writing and reading processes. This spacing parameter has decreased steadily toward zero (a physical limit!) over the fifty-year history of HDD*s*. But can the rate of decrease in logarithmic terms continue indefinitely, or is there a practical minimum value of this spacing set by surface roughness, lubricant and overcoat effective thickness, and the like? If there is a hard lower bound set by practical or physical considerations, then downward physical scaling will cease, for a fundamental condition of scaling is that *all* relevant dimensions and parameters must continue to scale in concert. If one parameter ceases to change, this limits the process and scaling cannot be recovered. Examples of the difficulty of maintaining rigorous scaling are already observed in present day HDD technology. Finding ways to overcome these limitations can bring engineers and scientists face to face with either physical or engineering limits, or both.

3. CONCLUDING REMARKS

Magnetic recording has in the past fifty years shown remarkable technological advance. Areal density has increased by a factor of nearly 10^8, from the initial HDD product introduced in 1956 with an AD of 2 kbit/in^2 up to present day products that are rapidly advancing toward 200 Gbit/in^2. This march is in every way as impressive as the better-known Moore's law advance that governs the evolution of semiconductor technology. In this chapter, we have discussed issues associated with the components of the HDD system, particularly the storage media, which are central to the support of this progress.

We have identified the new technologies on the horizon that have the potential to carry magnetic recording forward toward fundamental physical limits. These include perpendicular recording (PMR), heat assisted magnetic recording (HAMR), and bit patterned media (BPM) (including guided assembly of self-organized magnetic arrays (SOMA)). Simple physical analyses suggest that another three orders of magnitude of AD growth of the one hundred year old magnetic recording concept is foreseeable before these limits are reached. There are also practical engineering issues that could intervene to block the progress toward physical limits, but these cannot be anticipated. The history of technology advancement strongly suggests that engineering limits tend to be pushed back by unforeseen innovation, and this process distinguishes these kinds of barriers from those imposed by physics.

In summary, the future of magnetic recording is still bright, in spite of brushes with maturing technological S-curves that have already been experienced. If true, this situation should buy enough time (a few decades?) for truly revolutionary future forms of information storage to be conceived and developed into viable technologies.

Acknowlegdement

The authors acknowledge contributions by and discussions with E. Gage, G. Ju, T. Klemmer, M. Kryder B. Lu, T. Rausch, S. Sun, and E. Svedberg.

References

[1] R.R. Schaller, IEEE Spectrum 34(6), **52** (1997).

[2] M. Nieto, F. Lopez, and F. Cruz, Technovation **18**(6-7), 439 (1998); C.M. Christensen, Production and Operations Management **1**(4), 334 (1992); "Part II: Architectural Technologies", op.cit. **358** (1992).

[3] S. H. Charap, Pu Ling Lu, and Yanjun He, IEEE Trans. Magn. **33**, 978 (1997).

[4] D. Weller and A. Moser, IEEE Trans. Magn. **35**, 4423 (1999).

[5] M. Alex, A. Tselikov, T.McDaniel, N. Deeman, T. Valet, and D. Chen, IEEE Trans. Magn. **37**, 1244 (2001).

[6] H. Sukeda, H. Saga, H. Nemoto, Y. Itou, C. Haginoya, and F. Matsumoto, IEEE Trans. Magn. **37**, 1234 (2001).

[7] J. J. M. Ruigrok, R. Coehoorn, S. R. Cumpson, and H. W. Kesteren, J. Appl. Phys. **87**, 5398 (2000).

[8] H. J. Richter, "Dynamic Effects in High-Density Recording Media", in: The Physics of Ultra-High-Density Magnetic Recording, Eds. Plumer, van Ek, Weller, Springer 2001, chapter 6.

[9] JC. Mallinson, "The Foundations of Magnetic Recording", 2nd edition, Academic Press, 1993.

[10] D. Weller, A. Moser, L. Folks, M. E. Best, W. Lee, M. F. Toney, M. Schwickert, J.-U. Thiele, and M. F. Doerner, IEEE Trans. Magn. **36**, (2000).

[11] R. Wood, IEEE Trans. Magn. **36** (1), 36 (2000).

[12] R. H. Victora, K. Senanan, and J. Xue, IEEE Trans. Magn. **38** (5), 1886 (2002).

[13] M. Mallary, A. Torabi, and M. Benakli, IEEE Trans. Magn. **38** (4), 1719 (2002).

[14] H. J. Richter and A. Yu. Dobin, J. Magn. Magn. Mater. **287**, 41 (2005).

[15] K. Gao and H. N. Bertram, IEEE Trans. Magn. **38** (6), 3675 (2002).

[16] Y. Y. Zou, J. P. Wang, C. H. Hee, and T. C. Chong, Appl. Phys. Lett. **82**, 2473 (2003).

[17] M. H. Kryder and R. W. Gustafson, J. Magn. Magn. Mater. **287**, 449 (2005).

[18] D. Gurarisco, Y. Wu, P. Luo, B.E. Higgins, and K. Saito, J. Magn. Magn. Mater. **287**, 459 (2005).

[19] Y. Tanaka, J. Magn. Magn. Mater. **287**, 468 (2005).

[20] Zheng Zhang, Tolga M. Duman and Erozan M. Kurtas, "Information Theory of Magnetic Recording Channels," Coding and Signal Processing for Magnetic Recording Systems, Eds. Bane Vasic and Erozan M. Kurtas, CRC Press, 2004, pp. 12.1-12.20.

[21] In a micro-track model, $\sigma_j \sim a\ (<D>/W_{read})^{1/2}$, where a is the transition-width parameter. In the limit of negligible exchange $a \approx <D>$, hence $\sigma_j \sim a\ (<D>^3/W_{read})^{1/2}$; see e.g. H. N. Bertram, H. Zhou, and R. Gustafson, IEEE Trans. Magn. **34**, 1845 (1998).

[22] J. Ahner (unpublished data, 2004); J. Ahner *et al.* (to be published).

[23] E. E. Fullerton, O. Hellwig, Y. Ikeda, B. Lengsfield, K. Takano, and J. B. Kortright, IEEE Trans. Magn. **38**, 1693 (2002).

[24] D. Weller and M. F. Doerner, Ann Rev Mat Sci **30**, 611 (2000).

[25] L. Néel, Ann. Geophys. 5, **99** (1949).

[26] W. T. Coffey, D. S. F. Crothers, J. L. Dorman, L. J. Geoghegan, and E. C. Kennedy, Phys. Rev. B **58**, 3249 (1998).

[27] W.F. Brown, Jr. Phys. Rev. **130**, 1677 (1963).

[28] P. Gaunt, J. Appl. Phys. **78**, 3470 (1977).

[29] E. C. Stoner and E. P. Wohlfarth, Trans. Roy. Soc. **A240**, 599 (1948).

[30] Hong Zhou (private communication): Demag = $8\ \arctan[kBW\delta^{-1}((Bk)^2+W^2+\delta^2)^{-1/2}] \approx 4\pi$ for realistic conditions ($k \geq 8$:max number of bits before head flux reversal, B=bit-

length ≅ 40 nm @ 100 Gbit/in^2, *W*=bit-width ≅140 nm @ 100 Gbit/in^2, δ=medium thickness ≅10 nm.

[31] A. Moser and D. Weller, "Thermal Effects in High Density Recording Media", in The Physics of Ultra-High-Density Magnetic Recording, Eds. Plumer, van Ek, Weller, Springer 2001, chapter 5.

[32] H. Zhou, H. N. Bertram, M. F. Doerner, and M. Mirzamaani, IEEE Trans. Magn. **35**, 1239 (1999).

[33] H. N. Bertram, M. Marrow, J. Ohno, and J. K. Wolf, IEEE Trans. Magn. **40**, 2311 (2004).

[34] H. J. Richter, "Dynamic Effects in High Density Recording Media", "Dynamic Effects in High-Density Recording Media", in The Physics of Ultra-High-Density Magnetic Recording, Eds. Plumer, van Ek, Weller, Springer 2001, chapter 6, page 201, eq. (6.24).

[35] X.-W. Wu *et al.* J. Magn. Magn. Mater., to be published (2005)

[36] T. Klemmer, D. Hoydick, H. Okumura, B. Zhang, and W. A. Soffa, Scripta Met. Mater. **33**, 1793 (1995).

[37] D. J. Sellmyer, J. Zhou, Y. Liu, and R. Skomski, "Magnetism of Sputtered Sm-Co-Based Thin Films", in Rare Earth Magnets and Their Applications, Eds. G.C. Hadjipanayis and M.J. Bonder, Rinton Press 2002, p. 712.

[38] K. J. Strnat, in "Ferromagnetic Materials", Eds. E. P. Wohlfarth, K. H. J. Buschow, vol. 4, North-Holland, Amsterdam 1998, pp. 131-209 and references cited therein.

[39] J. Sayama, K. Mizutani, T. Asahi, J. Ariake, K. Ouchi, S. Matsunuma, and T. Osaka, J. Magn. Magn. Mater. **287**, 239 (2005).

[40] Y. Yamada, T. Suzuki, H. Kanazawa, and J.C. Österman, J. Appl. Phys. 85, 5094 (1999) and refs. therein.

[41] H. Saga, H. Nemoto, H. Sukeda, and M. Takahashi, Jpn. J. Appl. Phys. **38**, 1839 (1999).

[42] H. Takano, Y. Nishida, A. Kuroda, H. Sawaguchi, T. Kawabe, A. Ishikawa, H. Aoi, H. Muraoka, Y. Nakamura, and K. Ouchi, "A practical approach for realizing high-recording density hard disk drives", paper CA-01, MMM/Intermag 2001, San Antonio, TX, January 7-11.

[43] T. W. McDaniel, W. A. Challener, and K. Sendur, IEEE Trans. Magn., **39**, 1972 (2003).

[44] Z. Zhang, A. K. Singh, J. Yin, A. Perumal, and T. Suzuki, J. Magn. Magn. Mater. **287**, 224 (2005).

[45] M. L. Yan, X. Z. Li, L. Gao, S. H. Liou, and D. J. Sellmyer, Appl. Phys. Lett. **83**, 3332 (2003).

[46] J.-U. Thiele, K. R. Coffey, M. F. Toney, J. A. Hedstrom, and A. J. Kellock, J.Appl. Phys. **91**, 6595 (2002).

[47] J.-U. Thiele, S. Maat, and E. E. Fullerton, Appl. Phys. Lett. **82**, 2859 (2003).

[48] J.-U. Thiele, S. Maat, J. Lee Robertson, and E. E. Fullerton, IEEE Trans. Magn. **40**, 2537 (2004).

[49] G. Ju, J. Hohlfeld, B Bergman, R. J. M. van de Veerdonk, O. N. Mryasov, J.-Y. Kim, X. Wu, D. Weller, and B. Koopmans, Phys. Rev. Lett. **9**, 197403 (2004).

[50] Y. Qiang, R. F. Sabirianov, S. S. Jaswal, H. Haberland, and D. J. Sellmyer, Phys. Rev. B **66**, 064404 (2002).
[51] Y. Zhang, J. Wan, M. J. Bonder, G.C. Hadjipanayis, and D. Weller, J. Appl. Phys. **93**, 7175 (2003).
[52] J. P. Wang and T. J. Zhou, Patterned Magnetic Nanostructures (invited chapter) in Encyclopedia of Nanoscience and Nanotechnology, American Scientific Publishers, in press.
[53] J.-M. Qiu, J. H. Judy, D. Weller, and J.-P. Wang, J. Appl. Phys. 97, 10J319 (2005).
[54] R. H. Victora and Xiao Shen, "Exchange Coupled Composite media for Perpendicular Recording", paper BB01, Intermag 2005, Nagoya, Japan (IEEE Trans. Magn. (2005) to be published).
[55] D. Suess, T. Schrefl, M. Kirschner, G. Hrkac, and J. Fidler, "Optimization of Exchange Spring Perpendicular Recording Media", paper GB07, Intermag 2005, Nagoya, Japan (IEEE Trans. Magn. (2005) to be published).
[56] S. Sun, C. B. Murray, D. Weller, A. Moser, and L. Folks, Science **287**, 1989 (2000).
[57] S. Sun, D. Weller, C. Murray, in "The Physics of Ultra-High-Density Magnetic Recording", Eds. Plumer, van Ek, Weller, Springer 2001, Ch. 9.
[58] B. D. Terris and T. Thomson, J. Phys. D: Appl. Phys. **38**, R199 (2005).
[59] C. Liu, X. W. Wu, T. Klemmer, N. Shukla, X. M. Yang, D. Weller, A. Roy, M. Tanase, and D. E. Laughlin, J. Phys. Chem. B **108**, 6121 (2004).
[60] R. Chantrell, D. Weller, T. Klemmer, E. Fullerton, and S. Sun, MMM Seattle, November 2001, J. Appl. Phys. (to be published).
[61] T. J. Klemmer, C. Liu, N. Shukla, X. W. Wu, D. Weller, M. Tanase, D. E. Laughlin, and W. A. Soffa, J. Magn. Magn. Mater. **266**, 79 (2003).
[62] H. Kodama, S. Momose, T. Sugimoto, T. Uzaumaki, and A. Tanaka, IEEE Trans. Magn. **41**, 665 (2005).
[63] S. Kang, Z. Jia, D. Nikles, and J. W. Harrell, IEEE Trans. Magn. **39**, 2753 (2003).
[64] S. Kang, Z. Jia, D. Nikles, and J. W. Harrell, J. Appl. Phys. **93**, 7178 (2003).
[65] C. L. Platt, K. W. Wierman, E. B. Svedberg, R. van de Veerdonk, J. K. Howard, A. G. Roy, and D. E. Laughlin, J. Appl. Phys. **92**, 6104 (2002).
[66] S. Kang, Z. Jia, S. Shi, D.E. Nikles, J.W. Harrell, Appl. Phys. Lett. **86**, 062503 (2005).
[67] S. Kang, Z. Jia, D. E. Nikles, J.W. Harrell, J. Appl. Phys. **97**, 10J318(2005).
[68] Z. L. Wang, J. Phys. Chem. B **104**, 1153 (2000).
[69] H. Zeng, P. M. Rice, S. X. Wang, and S. Sun, J. Am. Chem. Soc. **126**, 11458 (2004).
[70] M. Tanase *et al.* (to be published)
[71] $p = p_{max}\,[1/(1+2l/D)]^2$, with the maximum packing ($l = 0$ nm) $p_{max} = 1$ for cubes, $p_{max} = \pi/4 = 0.785$ for cylinders and $p_{max} = \pi/6 = 0.524$ for spheres. D = particle core diameter, l = surfactant molecule chain length with $2l = w$.
[72] Y. K. Takahashi, T. Ohkubo, M. Ohnuma, and K. Hono, J. Appl. Phys. **93**, 7166 (2003).
[73] B. Yang, M. Asta, O. N. Mryasov, T. Klemmer, and R. W. Chantrell, Acta Metallurgica, (accepted for publication, 2005).
[74] K. Naito, H. Hieda, M. Sakurai, Y. Kamata, and K. Asakawa, IEEE Trans. Magn. **38**, 1949 (2002).

[75] J. Y. Cheng, C. A. Ross, E. L. Thomas, H. I. Smith, R. G. H. Lammertink, and G. J. Vancso, IEEE Trans. Magn. **38**, 2541 (2002).

[76] X. M. Yang, C. Liu, J. Ahner, J. Yu, T. Klemmer, E. C. Johns, and D. Weller, J. Vac. Sci and Tech. B **22**, 31 (2004).

[77] A. Lyberatos and K. Yu. Guslienko, J. Appl. Phys. **94**, 1119 (2003).

[78] A. Lyberatos and J. Hohlfeld, J. Appl. Phys. **95**, 1949 (2004).

Chapter 12

HARD-MAGNETIC NANOSTRUCTURES

S. Rivoirard
CRETA (CNRS)
25 avenue des Martyrs
BP 166, 38042 Grenoble Cedex 9, France

D. Givord
Laboratoire Louis Néel (CNRS)
25 avenue des Martyrs
BP 166, 38042 Grenoble Cedex 9, France

Abstract The hard- magnetic nanostructures discussed in this chapter consists of assembled nanosize particles, having diameters of typical less than 100 nm. They are based on the same hard phases which are involved in bulk Rare-Earth Transition-Metal hard magnetic materials or on the hard $L1_0$ FePt or CoPt phases. The main processes used for the preparation of these materials are described. The original magnetic properties observed are consequences of the crystallite size reduction, when it approaches the domain wall thickness, δ. In exchange-decoupled particle systems, reversal by coherent rotation tends to be favored with respect to nucleation/pinning-depinning. Remanence enhancement characterizes magnetic nanostructures made of exchange-coupled hard nanoparticles. Exchange-spring mechanism characterizes nanostructures which incorporates hard and soft nanoparticles.

1. INTRODUCTION

Our knowledge of the magnetism of matter indicates that the hard magnetic properties of the Rare-Earth Transition Metal (RE-TM) compounds, such as $Nd_2Fe_{14}B$ and $Sm(Co,Cu,Fe,Zr)_{7\text{-}8}$, are close to the optimum properties which can be shown by high anisotropy compounds. Thus, it appears unlikely that new materials with significantly improved properties will emerge in the future from the conventional research routes of preparing new alloys and compounds. Nanostructuring of magnetic materials opens an alternative and vast field of research of new materials. Specific properties may exist in these systems because their dimensions are of the order or below the domain wall width, dimension defining the characteristic length-scale of magnetization processes.

Hard magnetic nanostructures, discussed in the present chapter, assemble nanosize particles, below typically 100 nm diameter. They are based on the same hard phases which are involved in bulk hard magnetic systems ($Nd_2Fe_{14}B$, $SmCo_5$ and hard hexaferrites [1] or on the hard $L1_0$ FePt or CoPt phases [2]). Two main categories of hard nanostructures may be distinguished, depending whether the constituting nanocrystallites are exchange-coupled or not. Single-phase and nanocomposites constitute further divisions of exchange-coupled systems.

The study of these systems have become possible thanks to the development of various preparation routes, from sophisticated routes for the preparation of model materials with controlled nanostructures to industrial routes for the production of large quantity of materials. It has benefited as well from the development of new experimental techniques, allowing the properties of matter to be quantitatively examined at the nanometre scale. These include Hall micro-probe [3] or micro-SQUID magnetometry [4], XMCD at synchrotron radiation facilities [5] and scanning probe microscopes [6]. This is not the topic of this chapter to describe in detail these various techniques. They are only quoted in the following sections. The reader may find in the associated references the detailed technical descriptions that he may need.

In the section 2 of this chapter, an introduction is given to the main concepts required for the understanding of coercivity in hard magnetic materials. The various types of hard nanostructures and the procedures used for their preparation are described in section 3. The specific properties of hard nanostructure magnetism are described in section 4. In section 5, the properties of industrial materials are presented and some applications are briefly summarised.

2. COERCIVITY IN HARD MAGNETIC MATERIALS

2.1. Coercivity

2.1.1. *Definition of coercivity*

Let consider a ferromagnetic or ferrimagnetic material with spontaneous magnetization, M_s, and uniaxial magnetocrystalline anisotropy; the easy magnetization direction is z and K is the second order anisotropy constant. The magnetization, **M**, is assumed to be saturated along z, in one of the two favoured orientations. Under an applied magnetic field, $\mathbf{H}_{app}$, antiparallel to **M**, the magnetization should reverse in order to align back along the field and minimize the Zeeman energy. However, magnetization reversal does not occur as the applied field is reversed but it requires that the field reaches a certain critical value. This resistance to magnetization reversal defines coercivity. Magnetic materials are characterized by their hysteresis cycle. The value of the applied field at which M reverses is the coercive field, H_c and the value of the magnetization in zero applied field is the remanent magnetization, M_r. In modern hard magnets, $\mu_0 M_r$ is of the order of 0.5 - 1 T and $\mu_0 H_c$ is of the order of 1 T or above.

2.1.2. Stoner-Wohlfarth Model

The resistance to magnetization reversal indicates that an energy barrier separates the initial and the final magnetic states. This energy barrier is a consequence of magnetic anisotropy. This can be illustrated within the so-called Stoner-Wohlfarth (SW) model, in which reversal is assumed to occur by in-phase rotation of all moments (coherent rotation) [7]. For $\mathbf{H}_{app}$, antiparallel to **M**, the energy may be expressed as:

$$E = K \sin^2\Phi + M H_{app} \cos\Phi \qquad (1)$$

where Φ is the angle between **M** and z ($\Phi = 0$ in the initial state) (see Fig. 1, inset). In Eq. (1), the first term represents the anisotropy energy and the second term represents the Zeeman energy. From energy minimization, it is found that the barrier against reversal vanishes when H_{app} reaches the value of the so-called anisotropy field $H_A = 2\ K/M_s$ (by definition, H_A is the field required to align the moment along a field applied perpendicular to z).

Assuming that the field is applied at an angle θ from z ($\theta = 0°$ when H_{app} is exactly antiparallel to M), the coercive field becomes [7]:

$$H_c(\theta) = H_{SW}(\theta) = \frac{H_A}{(\sin^{2/3}\theta + \cos^{2/3}\theta)^{3/2}} \qquad (2)$$

where H_{sw} represents the coercive field in the SW approach; H_{sw} decreases from $H_{SW}(0°) = H_A$ to a minimum value $H_{SW}(45°) = H_A/2$, it increases again at larger θ values up to $H_{SW}(90°) = H_A$ (Fig. 1).

2.2. Coercivity in Real Systems

2.2.1. Other processes than coherent rotation

Other reversal processes than coherent rotation may be envisaged and, thus, the validity of the above approach be questioned. However, in the case where the coercivity of a small particle originates from anisotropy of magnetocrystalline origin, Brown [8] has shown that the energy barrier associated to coherent rotation is of minimum height and thus reversal should occur necessarily by coherent rotation. This may be used to approximate modern hard magnetic materials, which are based on Fe or Co compounds ($SmCo_5$, $Nd_2Fe_{14}B$, FePt).

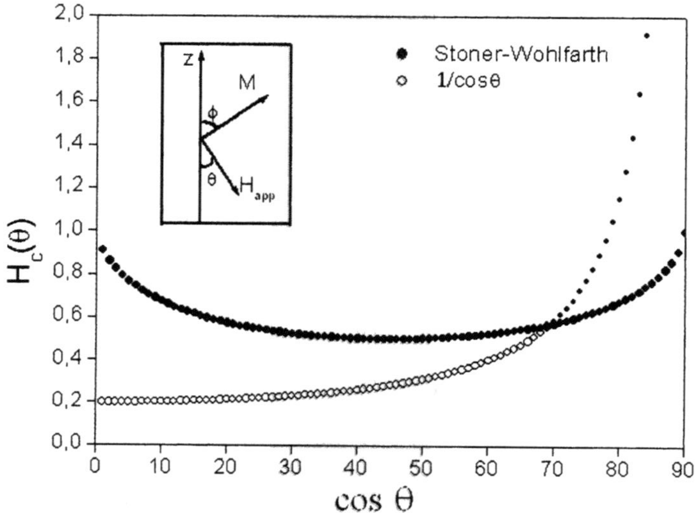

Figure 1. Angular variation of the normalized coercive field $H_c(\theta)/H_A$: (•) Stoner-Wohlfarth model and (○) $1/\cos\theta$ dependence, approximately observed in usual hard magnets. The value of $H_c(\theta)$ is arbitrary, $H_c(\theta)/H_A = 0.2$ has been assumed. At large θ values, when $H_c(\theta) > H_{SW}(\theta)$, coherent rotation is favoured again. Inset: definitions of the various angles involved in Eqs. 1-3.

Yet, in all these materials, coherent rotation is not observed and the coercive field is much weaker than the anisotropy field [9]. Actually, reversal is nucleated at local defects, a process which we call true nucleation. A nucleus is formed in which the magnetization is not fully aligned along z. The moment configuration within the nucleus has necessarily similarities with a magnetic wall since a wall constitutes the non-uniform moment configuration allowing magnetization reversal at minimum energy cost. As the nucleus grows and enters into the main phase, the wall energy, γ, (passage mechanism) and its surface area (expansion mechanism) vary. Experimentally, it is almost impossible to distinguish between true nucleation, passage and expansion. These three mechanisms may be gathered under the generic term of nucleation; they are distinct from another possible mechanism, domain wall pinning (or depinning), which may occur when the magnetic properties vary locally within the main phase and thus the wall energy may vary from place to place. The effective coercive field is defined by the process to which is associated the highest energy barrier. It depends critically on the material microstructure, i.e. coercivity is an extrinsic physical properties. Typically, H_c is one order of magnitude weaker than H_A.

Additionally to the reduction of the coercive field with respect to the anisotropy field, the occurrence of nucleation or pinning-depinning may be characterized by the fact that the angular dependence of the coercive field is very different from the one corresponding to coherent rotation. Rather it may be expressed as [[1]0]:

$$H_c(\theta) \approx \frac{H_c(0)}{\cos\theta} \tag{3}$$

where $H_c(0)$ is the coercive field obtained for $\theta = 0$ (see Fig. 1). Equation (3) expresses the fact that only the projection of the applied field along the moment direction is active in the reversal process.

2.2.2. Activation volume and thermal activation

Whatever the reversal mechanism, the critical phenomenon which determines reversal occurs within a local critical volume under an effective critical field. Together with the experimental coercive field, the critical volume, associated with the determinant magnetization reversal process, is another important parameter characterizing coercivity [9, 11, 12]. The size of this critical volume may be derived from the analysis of thermal activation effects and we are thus led to discuss these effects briefly.

To first approximation, thermal activation may be described in terms of the Arrhenius law :

$$\tau = \tau_o \exp(\Delta E/kT) \tag{4}$$

where τ is the characteristic time to overcome the energy barrier ΔE at temperature T and τ_0 is the associated intrinsic time, typically of the order of 10^{-9} s. In the common case where a large distribution of energy barriers exists, the resulting magnetization variation is proportional to the logarithm of time; one defines the magnetic viscosity $S = \partial M / \partial Ln(t)$ and the magnetic viscosity parameter $S_v = S/\chi_{irr}$, where χ_{irr} is the irreversible susceptibility. Due to the extremely fast variation of τ with ΔE, the time dependence of the magnetization at time t is dominated by the particles characterized by $\tau \approx t$. It results that activation effects may be described in terms of an equivalent magnetic field, the fluctuation field, $H_f = S_v \ln(t/\tau_0)$ which is active over the activation volume, equal to $v_{act} = k_B T/\mu_0 S_v M_s$. In practice, the time t is defined by the measuring technique used. For $t \approx 1$s, $Ln(t/\tau_0) \approx 25$.

In usual hard magnetic materials, thermal activation effects are the source of a coercive field decrease by typically 10% at room temperature. Importantly, they provide an experimental access to the critical volume for reversal, which may be identified with the activation volume, v_{act}.

2.2.3. *Experimental analysis of $H_c(T)$ – demagnetizing field contribution*

The temperature dependence of the coercive field, $H_c(T)$, is often used as a tool to relate the experimentally observed coercive field to intrinsic physical parameters characteristic of the hard magnetic phase, i.e. the anisotropy field, H_A [13, 14] or $\gamma/v_{act}^{1/3}$ [15], a parameter related to the domain wall energy. This type of analysis is based on the simplified assumption that H_A or γ in the activation volume may be expressed as a function of the same parameters in the main hard phase. Under the hypotheses that the anisotropy in the activation volume is dominated by second order anisotropy, the coercive field may be related to a parameter in which the main phase anisotropy is not involved, $A/v_{act}^{2/3}$, where A is the exchange constant in the activation volume, assumed to be equal to the main phase exchange constant [16]. Additionally, the analysis of $H_c(T)$ reveals that a demagnetizing field contribution, represented by the demagnetizing field, H_D, systematically leads to a further reduction in the experimentally observed coercive field [9, 17]. H_D may be relatively large, of the order of M_s or above.

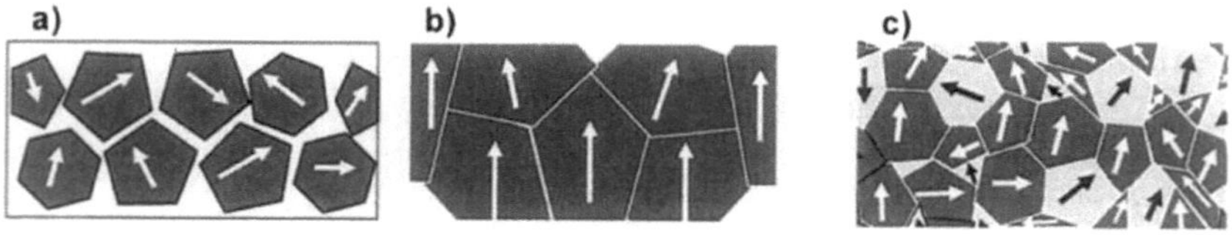

Figure 2. Schematic microstructures of the three main types of nanocrystalline magnets a) exchange-decoupled nanostructures (the hard magnetic grains are surrounded by a non magnetic phase), b) single-phase exchange coupled nanostructures, and c) nanocomposite (exchange coupling between the hard and the soft magnetic nanograins).

3. PREPARATION OF HARD MAGNETIC NANOSTRUCTURES

The preparation methods and the essential properties of the three main types of hard nanostructures are presented in this section: (i) in exchange-decoupled nanostructures, high coercivities are achieved, (ii) single-phase exchange-coupled nanostructures show enhanced remanence, (iii) and hard nanocomposites associate soft and hard nanograins in such a way that the soft grains develop a certain coercivity. The specific properties of each of these systems are determined by their microstructure, schematically shown in Fig. 2.

3.1. Exchange-Decoupled Hard Nanostructures

As described in Section 2, the coercivity of hard magnetic materials is generally determined by undesired structural defects. As the size of hard particles is progressively reduced, the coercive field tends to increase. This is explained by considering that the impact of a given defect extends to the volume of the concerned particle. It is reduced as the particle size is reduced. However, below a certain critical size, a reduction in coercivity is observed. The process used to reduce particle size inevitably introduces additional defects, which become dominant.

This argument shows that the processes used for the production of very small particles must be carefully controlled in order that a minimum number of additional defects is introduced. Various sophisticated processes have been developed which allow nanoparticles to be prepared, showing more less controlled nanostructures. These processes may involve size reduction of larger particles (top-down approach) or direct nanoparticle growth (bottom-up approach). The most important of these processes are described in the following sections.

3.1.1. HDDR powders

Hydrogen decomposition desorption recombination (HDDR) process is the only top-down industrial process used for the preparation of coercive nano-particles. This process applied to rare-earth transition-metal (RE-TM) alloys consists in heating the concerned alloy under hydrogen until it decomposes into a fine mixture of RE-hydride and TM. The hard magnetic phase is recombined with a much finer microstructure. This process was first developed to convert 100 microns sized cast $Nd_2Fe_{14}B$ grains into 200-300 nm crystallites [18, 19]. Later, it has been applied to other RE-TM alloys [20, 21]. Recently, a new variation of this process has been developed towards developing texture in the final materials [22]. It is briefly described below.

Anisotropic $Nd_2Fe_{14}B$ HDDR powder. - In the so called dynamic HDDR (d-HDDR) process, the control of the reaction rate between NdFeB and hydrogen is much tighter than in the conventional one [23]. It is believed that the texture is induced by an orientation relationship between the newly formed $Nd_2Fe_{14}B$ grains and the original parent grains. Gutfleisch *et al.*

suggest that the iron-boride phase acts as the anisotropy-mediating phase throughout the different stages of the process [24]. The nucleation and growth of this phase is controlled by the hydrogen pressure value during disproportionation.

HDDR processed Sm_2Fe_{17} compounds. - The HDDR process has also been applied to $Sm_2Fe_{17}N_x$ [20]. The HDDR $Sm_2Fe_{17}N_3$ crystallites (500 nm) exhibit $\mu_0 H_c = 1.5$ T after nitrogenation. Milling the $SmFe_{17}$ powder prior to the HDDR treatment allows the coercivity to be increased from 1.5 to 3.5 T [25]. No texture could be developed either by conventional or modified HDDR treatments. This has been ascribed to the lack of intermediate phase in this system.

3.1.2. Melt spun materials

Melt spinning is by excellence the bottom-up approach to prepare nanoparticles with typical grain size around 50 nm.

$Nd_2Fe_{14}B$-type magnets. - Rapidly solidified melt-spun Nd-Fe-B alloys form the basis for the entire bonded magnet industry. Depending on the processing parameters, the ribbons obtained can be amorphous or directly nanocrystallised. In the former case, final annealing is required to obtain nanoparticles. When the alloys are overstoichiometric in Nd, a thin Nd-rich intergranular layer ensures exchange decoupling between the grains with typical grain size around 30-60 nm.

Hot deformation of $Nd_2Fe_{14}B$-based melt-spun ribbons: Die-upsetting. - Hot deformation, such as die-upsetting is used to produce bulk nanocrystalline NdFeB and PrFeB magnets from melt spun ribbons, with excellent magnetic properties [26]. Solution precipitation creep is thought to be the basis for this deformation process [27]. A material flow occurs perpendicular to the applied stress. Thanks to the anisotropy in shape and elastic behaviour of the $Nd_2Fe_{14}B$ phase, a texture is produced with the magnetically easy direction (c-axis) parallel to the deformation direction. The microstructure of deformed magnets consists of $Nd_2Fe_{14}B$ platelet-shape grains of about 50 nm thickness and 100 to 400 nm^2 surface area. The platelets are continuously coated by the Nd-rich amorphous paramagnetic phase of approximately 1 nm thickness.

3.1.3. Deposition techniques

Nanoparticle systems can be prepared by various deposition techniques (sputtering, cluster deposition) and in particular, serve as model systems to study magnetisation processes at the nanoscale. In addition, the magnetic recording industry requires the use of materials in thin film forms, which could constitute new recording media or elements of MEMS devices. The FePt and CoPt hard materials are directly competing with the RE-TM alloys. The high cost of the raw materials is compensated by their extremely good resistance to corrosion.

FePt (CoPt). - The Fe-Pt phase diagram (or Co-Pt) reveals the presence of various phases: Fe_3Pt, FePt and $FePt_3$, each of them existing in an ordered and a disordered state. The most important of these is the ordered $L1_0$ FePt phase, which is made of an alternate stacking of Fe and Pt planes. The spontaneous magnetisation, $\mu_0 M_s$ reaches 1.43 T at 300 K, the magnetocrystalline anisotropy is high (7 MJ/m^3) and $T_c = 750$ K.

In view of obtaining decoupled FePt (or CoPt) nanoparticles, films can be prepared by magnetron sputtering from two targets (tandem method): a FePt target and a non magnetic one. A non magnetic layer is deposited between the FePt layers which crystallises in the fcc (face-center-cubic) disordered phase. Upon annealing the fcc phase transforms into the fct (face-center-tetragonal) hard phase and intermixing between the two layers leads to the formation of hard FePt grains embedded in a non magnetic matrix. Elements such as Pt itself, Ag [28], C [29], BN [30], SiO_2 [31] or B_2O_3 [32] may be used to isolate the FePt grains.

For most of the samples studied, the size of the particles (around 10 nm) is well below the single domain particle size (340 nm for the $Fe_{50}Pt_{50}$ phase [33]). The coercivity is found to depend critically on the layer thickness, on the annealing temperature and on the additives-content, parameters which determine the particle size and the interparticles separation.
The major drawback of room temperature sputtering is the need for high temperature post-annealing of the films. The annealing process modifies the initial morphology of the sample, and the resulting particle growth and matrix structural changes are essentially uncontrolled. Much smaller particle size are obtained by direct sputtering from FePt and C targets on heated substrates; C impedes FePt grain growth, $\mu_0 H_c = 3.4$ T was obtained for particles size around 15 nm [34].

Sputtering [35] or pulsed laser deposition [36] on heated Si or MgO substrates leads to the growth of highly textured films with island morphology and the c-axes of the fct FePt phase perpendicular to the film plane. Each island is a single crystalline FePt grain. Extremely high coercivities are obtained ($\mu_0 H_c > 4$ T).

Alternative procedures to obtain FePt nanoparticles use direct cluster deposition with a particle gun [37]. Under normal conditions, the FePt clusters have the fcc structure. In this original development, the particles passing through halogen lamps are annealed before deposition. The nanoparticles are embedded in a carbon matrix with typical size of 6 nm. $\mu_0 H_c =$ 0.3 T is obtained.

RE-TM-based films. - The deposition techniques (sputtering or PLD – pulsed laser deposition) have been used successfully to prepare SmCo5 and Nd2Fe14B nanoparticles [38, 39]. After room temperature deposition, the nanoparticles are amorphous. The desired crystal structure and magnetic properties are obtained through application of a post-deposition annealing treatment. Due to the high reactivity of the materials, this constitutes a non trivial task.

In the case of Nd-Fe-B films, a common approach is to deposit a Nd-overstoichiometric film at room temperature. The hard $Nd_2Fe_{14}B$ grains which crystallise during post-deposition annealing become directly isolated by the non-magnetic matrix. Coercivity develops at this stage, but the films are isotropic. The same type of nanostructure may be obtained by direct sputtering on heated substrate with the benefit that the films are highly textured [40].

Recently, 300 nm thick $Nd_2Fe_{14}B$ grains were epitaxially grown by PLD on a Ta (110) buffer itself grown at the surface of single crystalline (0001) Al_2O_3 heated substrates. The high value of the coercive field reached (2T) was attributed to the high crystallinity of the isolated $Nd_2Fe_{14}B$ grains, which constituted the film. The tetragonal c-axis was almost perfectly aligned perpendicular to the film plane. The resulting remanence to saturation ratio reached 0.97 [41].

In view of further isolating the $Nd_2Fe_{14}B$ grains, Panagiotopoulos *et al.* [42] prepared Nd-Fe-B/W multilayers with typically 28 nm thicknesses. 10 nm $Nd_2Fe_{14}B$ nanoparticles embedded in a W matrix were obtained through annealing at 700 °C thus leading to $\mu_0 H_c \approx 2$ T.

In the case of the Sm-Co system, exchange decoupled grains can not be obtained by sputtering from a Sm-rich target, since the intergranular phase is still magnetic. Stoichiometric $SmCo_5$ grains have been pulsed laser deposited together with a non magnetic matrix (Al), which was sputtered simultaneously. $\mu_0 H_c = 1.5$ T was obtained. These films show very high corrosion resistance [43].Very recently, $SmCo_5$ 8 nm particles embedded in a Cu-rich matrix were prepared by a multilayer (Sm-Co/Ti-Cu) sputtering technique and showed coercivity close to 5 T after annealing [44].

Stoyanov *et al.* have demonstrated the feasibility to grow Sm-Co nanoparticles without the need for post-deposition annealing, using the cluster gun described above [45]. However, the 7 nm nanoparticles embedded in a carbon matrix shown a low coercivity ($\mu_0 H_c = 0.07$ T).

3.1.4. *Chemical synthesis*

Recently, chemicals routes were used for the preparation of high anisotropy FePt and CoPt nanoparticles with excellent control of the particle size and morphology [46] and the first preparation of rare-earth transition-metal hard magnetic materials by such route was announced [47].

The preparation of 2 nm diameter FePt or CoPt nanoparticles by Sun *et al.* constituted a spectacular achievement which was obtained through high temperature (297 °C) decomposition of $Fe(CO)_5$ and $Pt(acac)_2$ phase reduction [46]. Quasi-monodispersed nanoparticles assemblies offer the unique possibility to self-organise on a substrate. This effect is influenced by the nature of the interactions between the particles and the substrate and among the particles themselves. Final thermal annealing was required to transform the as-produced fcc particles into the ordered fct $L1_0$ structure. However, particles aggregation and particle growth occurred at the expense of coercivity and self-organisation. Recently, a rapid thermal annealing treatment was shown to limit particle growth thus leading to coercive fields of 1.8 T [48].

The self assemblies are difficult to obtain and, up to now, the patterns extend to only a few hundred of nanometers in diameter. Therefore, the use of structured substrate has been proposed as a way to obtain long distances particle ordering. As an example, lithography patterning [49, 50] allowed to obtain FePt nanoparticles arrays on Si wafer, with $\mu_0 H_c$ = 1.2 T after annealing.

3.2. Single-phase exchange-coupled nanostructures

Two main techniques are used for the preparation of RE-TM exchange-coupled hard nanostructures: melt spinning and mechanical alloying. Alloys close to the stoichiometric composition of the hard magnetic phase are used. Exchange between nanoparticles results from the absence of secondary non-magnetic phases. Fairly uniform microstructures are generally obtained, consisting of randomly oriented grains of the hard magnetic phase. High coercivities can easily be obtained for grain sizes below 80 nm. At finer grain sizes (< 40 nm), a tendency to coercivity decrease is found and remanence enhancement occurs.

3.2.1. Exchange coupled Nd-Fe-B materials

As first observed by McCallum *et al.* [51], melt spun alloys of composition close to the $Nd_2Fe_{14}B$ stoichiometry and grain size below typically 20 nm exhibit a remanence to magnetisation ratio, *S*, higher than 0.5, the maximum value expected for uniaxial decoupled grains. This phenomenon known as remanence 0enhancement was confirmed by many other studies on both melt spun and mechanically alloyed samples [52-54]. Remanence enhancement constitutes a signature for the presence of significant exchange between the grains. It may be attributed to the fact that in alloys close to the $Nd_2Fe_{14}B$

stoichiometric composition, many grains are in direct contact. Enhanced remanence increases as the grain size decreases [55]. Due to the short range character of exchange coupling, the effect is significant over thicknesses of the order of the domain wall width (see section 4.2.). Concomitantly, the coercivity decreases with the grain size, since the tendency to common alignment of the grain magnetisations leads to averaging out the anisotropy in the individual grains. The preservation of coercive field around 1 T together with $S \approx 0.55$ - 0.6 requires very well controlled metallurgical processing. Direct crystallisation during melt spinning allows more uniform grain size to be obtained than crystallisation by annealing of overquenched amorphous ribbons.

In the Nd-Fe-B system, remanence enhancement may also be obtained by mechanical alloying [54]. The as-milled samples are amorphous. Crystallization of the $Nd_2Fe_{14}B$ phase is obtained by subsequent annealing. As for melt spun ribbons, an optimisation of the metallurgical process is required. At low annealing temperatures, the $Nd_2Fe_{14}B$ phase formation is incomplete, some coarse free iron grains remain which are detrimental to both coercivity and remanence enhancement. At high annealing temperatures, grain growth occurs which again leads to reduction in both H_c and S values. The optimum properties approach those obtained with ribbons ($\mu_0 H_c \approx 1.2$ T and $S \approx 0.60$) [54].

3.2.2. *Exchange-coupled SmCo-based materials*

Single phase materials based on the Sm-Co system have essentially been prepared by mechanical alloying. The as-milled powder is usually amorphous. The microstructure and coercivity are optimized during a subsequent annealing. Good properties were obtained with materials based on both the $SmCo_5$ and the Sm_2Co_{17} phases [56-58]. In both cases, the remanence to saturation ratio obtained is close to 0.7 with coercive fields higher than 5 T for the $SmCo_5$-type materials.

Rhen *et al.* [59] reported the successful stabilisation of the SmCoZrTi 1:7 phase at high temperature (up to 400 °C) with a tiny amount of ZrB_2 addition. This additive, which decomposes during milling, prevents excessive grain growth during the annealing step. Coercivity of 0.5 T at 400°C and a remanence ratio of about 0.65 are obtained. These compounds could constitute a good alternative for high temperature applications.

Exchange-coupled behaviour was also found in Sm_2Co_{17} alloys processed by reactive milling under hydrogen [60]. This process leads to disproportionation of the very stable 2:17 compound into an intimate mixture of Sm-hydride and Co grains with grain size significantly smaller (< 10 nm) than those of conventionally HDDR materials. During recombinetion, the grain size remains below 30 nm. $S = 0.93$ and $\mu_0 H_c = 0.56$ T were obtained.

Sm-Co films have been obtained by various deposition techniques [61, 62]. $SmCo_5$ or $SmCo_7$ nanograins with characteristic single phase behaviour are grown on heated substrates, with typical grain size of 20 nm. The c-easy –axis of magnetization lies in the film plane for all compositions. In general, pulsed laser deposition leads to slightly higher coercive fields ($\mu_0 H_c \approx 2$ T) [63] than magnetron sputtering ($\mu_0 H_c \approx 1.5$ T) [64].

3.2.3. Exchange coupled $Sm_2Fe_{17}N_x$ ($Sm_2Fe_{17}C_x$) and 1:12 compounds

Mechanical alloying was the first successful method for making highly coercive Sm-Fe-N magnets [65]. Powders of Sm and Fe are cold worked in a ball mill to induce an amorphous Sm-Fe alloy. The Sm_2Fe_{17} phase crystallises during subsequent annealing. The powder is nitrogenated (or carbonated) by a solid-gas reaction under N_2 or C_2H_2 atmospheres. The grain size of the hard phase is around 30 nm and $\mu_0 H_c \approx 3$ T [65]. The nanograins form micron size coarse agglomerates from which the magnets are made. Due to the isotropic distribution of the nanograins within the powder particles, the poor texture and the presence of additional non magnetic phases (particularly Zn) no definite evidence for remanence enhancement can be obtained [25, 66].

The same type of phenomena is found in nitrogenated melt spun Sm_2Fe_{17} ribbons. However, the nitrogenation time required is longer than in the case of mechanically alloyed materials. Partial decomposition of the nitride can occur which leads to the formation of a non negligible amount of free α-Fe. The coercive field ($\mu_0 H_c \approx 2$ T) is lower than in mechanically alloyed materials [67].

Another category of hard nanostructures was obtained by nitrogenation of the $RFe_{12-X}M_X$ compounds (R = Nd, Pr; M = V, Ti, Nb, Ta, Mo, Cr, Mn, W). Before nitrogenation, the easy magnetisation direction of the compounds of interest lies in the basal plane of the tetragonal structure. Upon nitrogenation, the easy direction becomes the c-axis, with an anisotropy field reaching 12 T in $NdFe_{12-X}V_XN$ [68]. This large anisotropy combines with the high spontaneous magnetisation to provide potentially excellent hard magnetic properties. However, mastering the metallurgical processes remains a challenge. Higher coercivity develops in the $NdFe_{12-X}V_X$ compounds obtained by melt spinning ($\mu_0 H_c \approx 1$ T) [69] than in the same compounds prepared by mechanical alloying ($\mu_0 H_c \approx 0.6$ T) [70].

3.2.4. Exchange coupled FePt compounds

FePt–based hard nanostructures have been obtained by film deposition techniques and severe cold deformation.

FePt films were sputtered onto single crystalline MgO (001) heated substrates [35]. With increasing the film thickness, the size of the FePt particle increases up to typically 100 nm and particle coalescence is observed, finally leading to percolation. Most generally, the c-axis of the $L1_0$ phase is oriented perpendicular to the film plane. The coercivity is large ($\mu_0 H_c \approx 2.5$ T) for film thickness up to typically 50 nm but it decreases beyond this value [35].

FePt based magnets have also been obtained by annealing nanostructured Fe/Pt bulk multilayers produced by cyclic mechanical deformation [71]. The full process typically consists of four stages. At each stage, Fe and Pt foils are stacked periodically, inserted into a stainless steel tube and sealed. A number of rolling passes between 50 and 100 is applied and the thickness is gradually reduced by a cumulated factor of 10. After extraction from the stainless steel sheath, the sample is cut into new platelets ready for a new stacking/rolling cycle. At the end of the process, the Fe/Pt multilayer is typically 100 microns thick, with individual layers thicknesses of about 20 to 50 nm. Upon annealing, the formation of $L1_0$ FePt is diffusion limited, owing to the multi-layer of the starting material. The coercive field obtained approaches 0.75 T. No texture has been developed so far with this technique.

3.3. Nanocomposites

Since the successful preparations of $Nd_2Fe_{14}B/Fe_3B$ and $Nd_2Fe_{14}B/Fe$ nanocomposites magnets were first reported by the Philips group in the late 1980s [72], the RE-TM nanocomposites have been extensively studied [73-75]. Until recently, these materials were prepared under the form of ribbons or powders only. They were isotropic. Recent research focused on the preparation of textured materials [76] and of other potential materials, in particular based on the FePt system [77].

In nanocomposites, exchange coupling occurs between hard and soft magnetic nanograins. The hard phase provides coercivity and the soft phase contributes to magnetisation enhancement (see section 4.3). The grain size of both phases must be of the order of the hard phase domain wall thickness. The crystallisation behaviour plays an important role in the microstructural control of the magnetic properties.

3.3.1. Rare-earth iron boron nanocomposite magnets

Nd-Fe-B and Pr-Fe-B nanocomposites are rare-earth deficient with respect to the $R_2Fe_{14}B$ stoichiometry. The soft phase which is formed is Fe_3B or α-Fe. The control of the soft grain size is crucial to preserve a significant coercivity. This can be obtained by crystallisation of amorphous ribbons. The annealing temperature must be high enough to allow crystallization of the desired phases and the annealing time must be short enough to avoid grain growth. However, coarse grains of the soft phase are usually observed. Additional substitutions of minor elements are often used to hinder grain growth [78]. Nanostructured materials may also be obtained directly from the melt using lower wheel speed. The nanoparticle size is larger than in overquenched ribbons but the particle size distribution is more uniform. The properties obtained with the $Nd_2Fe_{14}B/\alpha\text{-}Fe$ type nanocomposites ($\mu_0 M_r$ = 0.98 T, $\mu_0 H_c$ = 0.7 T) are more suitable for bonded magnets applications than those obtained with the $Nd_2Fe_{14}B/Fe_3B$ nanocomposites ($\mu_0 M_r$ = 0.7 T, $\mu_0 H_c$ = 0.9 T) [79]. The finer and more uniform $Nd_2Fe_{14}B/\alpha\text{-}Fe$ nanostructure allows a better squareness of the demagnetization curve to be obtained.

Skomski and Coey [80] have suggested that a multilayer arrangement of alternating hard and soft magnetic layers, could provide a means to better control the nanostructure. Additionally, it could allow crystallographic alignment of the hard phase, thus higher remanence values. This has stimulated the preparation of model materials based on $Nd_2Fe_{14}B/Fe$ multilayers by various groups [81-83]. Yet, no texture has been obtained, and the coercivity could never be increased above 0.8 T.

In terms of applications, most significant success towards texturing was achieved by hot deformation of NdFeB melt spun ribbons [84]. A very rapid hot compaction process (2 minutes total duration) was used, thus avoiding excessive grain growth. Hot deformed nanocomposite magnets contain elongated nanograins of which the c-axis is aligned along the deformation direction. However, the volume fraction of the soft phase is 4% only. The best properties achieved are $\mu_0 H_c$ = 1.026 T and B_r = 1.48 T. Increasing the amount of the soft phase results in a sharp drop in coercivity and the remanence is also reduced due to a decrease in the crystallographic texture.

3.3.2. Sm-Co and Sm-Fe-N-based nanocomposites

Magnetic hardened Sm_2Co_{17}-type of magnets. - As pointed out by Hadjypanayis [85], the microstructure of precipitation-hardened 2:17 magnets show similarities with that of textured hard-soft nanocomposites. It consists of a 2:17 matrix with a cell size of 40 - 100 nm surrounded by the Cu-rich 1:5 cell boundary phase with a thickness of about 10 nm and a Zr-rich lamellar phase. Magnets can be obtained either by sintering of powdered cast ingots or of melt-spun materials. Details of the metallurgical processes have been described in several papers [86-88]. The best properties are obtained in sintered magnets with a remanence of the order of 1.2 T and coercivity in excess of 3.5 T. By contrast, magnets made from melt spun materials are isotropic in general. Appropriate element addition or control of the cooling rate during melt spinning may allow the c-axis of the crystallites to be aligned within the ribbon planes [89]. The remanence ratio is strongly improved reaching a value of 0.9.

It is thought that the coercivity development is due to the diffusion of copper from the 2:17 phase into the 1:5 cell boundary phase. The resulting modification of the magnetic properties of the 1:5 phase leads to a significant domain wall energy difference between the 1:5 and 2:17 phases. Recently, it has been suggested that the coercivity is controlled in fine by the Cu-distribution inside the cell boundary phase [90]. An original behaviour found in these materials is the possible increase of coercivity with temperature in a certain temperature range. This abnormal temperature dependence critically depends on the thermal treatment applied [88, 91].

Other SmCo-based nanocomposites systems. - Chen *et al.* showed that partial substitution of Cu for Co in the stoichiometric Sm_2Co_{17} compounds leads to the joint crystallization of 1:5 and 2:17 nanograins, without formation of the cellular nanostructure characteristic of the Sm_2Co_{17} hardened magnets [92]. Similar results were reproduced by Yan *et al.* by mechanical alloying of elemental $SmCo_5$ and Sm_2Co_{17} powders [93]. As the composition is varied from 1:5 to 2:17, the coercivity decreases from 3.6 T in the 1:5 single phase materials to 0.72 T in materials which contain 80% of the Sm_2Co_{17} phase. This effect is related to the large difference in the anisotropy field of both phases. The optimum properties are obtained in alloys made of 40% $SmCo_5$ and 60% Sm_2Co_{17}.

Nanocomposites have also been obtained based on Sm_2Co_{17}/Co or FeCo. High coercivity values were obtained in Sm_2Co_{17}/Co by mechanical alloying [94].Very recently, high remanence $SmCo_7$/Co nanostructures have also been produced by melt spinning the appropriate Sm-Co alloy composition with carefully selected additives [84, 95]. An isotropic grain structure consisting of 10-15 nm fcc Co grains embedded into 100-200 nm $SmCo_7$ grains has been obtained with the Co phase amounting to 2 -4%vol only [96]. A remanence ratio of 0.7 was reached which is remarkable for an isotropic material, with a coercivity approaching 1 T. Lee *et al.* prepared bulk isotropic $(Sm,Gd)_2(Co,Fe)_{17}$/Co,Fe nanocomposites magnets by hot compaction with $H_c = 1.04$ T, $B_r = 0.76$ T [84].

Exchange coupled behaviour was found in bilayer films of SmCo/CoFe, nanocomposites [97, 98]. Liu *et al.* [99] reported high magnetic properties in Sm-Co/Co and Pr-Co/Co sputtered multilayers ($\mu_0 H_c = 4.5$ T). Adjusting the fraction of the soft phase was realised by varying the Co-layer thickness.

Finally, remanence enhancement has been achieved by producing a coherent nanocrystalline structure in mechanically alloyed isotropic Sm_7Fe_{93}-nitride powders consisting of α-Fe and $Sm_2Fe_{17}N_3$ [100].

3.3.3. FePt-based nanocomposites

Several studies focused recently on the preparation of FePt-based nanocomposites [101]. Rapid annealing of Fe/Pt multilayers leads to the formation of highly textured FePt/α-Fe nanocomposites with magnetization perpendicular to the film plane. Despite the short annealing time, FePt crystallizes in the $L1_0$ phase structure, the resulting high anisotropy allowed a coercivity of 1.3 T to be obtained.

FePt/Fe-B nanocomposite materials have been prepared by Chang *et al.*, using ribbons as precursors [77]. After annealing the as quenched ribbons, the hard magnetic fct phase formed together with the Fe_3B or Fe_2B soft magnetic phase. The addition of a sufficient boron content into the alloy was needed to promote the formation of the fct phase and to prevent grain growth during annealing. Both the remanence and the coercivity values compares well with those obtained in Nd-Fe-B nanocomposites (see section 5).

FePt prepared by mechanical alloying. - FePt based nanocomposites have also been prepared by mechanical alloying. The milling must be performed at liquid nitrogen temperature in order to allow the mechanically soft Fe and Pt to become brittle. A lamellar structure is obtained with Fe and Pt having thicknesses of about 20-250 nm. A proprotion of the disordered fcc FePt is also present in the alloys, with grain diameter around 5 nm. Various phases of the Fe-Pt phase diagram form during subsequent annealing. Optimum properties are found in an alloy annealed 48 hours at 450°C, which contains 85% of the $Fe_{50}Pt_{50}$ phase. Various combinations of phase concentrations in the alloy lead to various magnetic properties. In the $Fe_{40}Pt_{60}$ alloys, $\mu_0 H_c$ = 0.95 T, $\mu_0 M_r$ = 0.4 T. The large coercivity obtained is attributed to the decoupling of $L1_0$ lamellae by the paramagnetic $L1_2$ phase. In $Fe_{60.5}Pt_{39.5}$ alloys, the two phases are $L1_0$ and the magnetically soft Fe_3Pt phase. A high remanence ($\mu_0 M_r$ = 0.8 T) is obtained together with $\mu_0 H_c$ = 0.33 T [102].

3.4. Hexaferrites

The Ba and Sr hexaferrites constitute the only family of hard magnetic materials which is not based on metallic systems. Most generally, the preparation involves annealing of the material at temperatures around 700°C. This leads to grain growth and the final materials are not nanocrystallised. Only a limited number of processes allow nanostructures materials to be prepared. Yu *et al.* prepared $BaFe_{12}O_{19}$ nanoparticles using a process which combined the citrate precursor method and spray technique [103]. Another process involves wet chemical synthesis through gel-to-crystallite (G–C) conversion method [104]. In these processes, the materials obtained are exchange decoupled. The properties approach at best those found in larger particle size systems.

4. HARD-NANOSTRUCTURE MAGNETISM

The original magnetic properties of hard nanostructures, described in the above sections, are direct consequences of the crystallite particle size reduction, when this size approaches the domain wall thickness, δ. In this section, we apply and extend the findings of Chapters 3 and 4 to hard-magnetic nanostructures. We discuss (i) systems of exchange decoupled particles, and (ii) systems of exchange-coupled particles, themselves divided into single-phase systems and nanocomposites.

4.1. Exchange-Decoupled Nanoparticles

In this section, we concentrate on the fundamental impact of particle size reduction on magnetization processes in individual particles. Although not directly related to coercivity, the classical effect of single domain particle formation is described. At small particle size, reversal by coherent rotation tends to be favoured with respect to nucleation/pinning-depinning; finally thermal activation effects and macroscopic quantum tunnelling are discussed.

4.1.1. Single-domain particles

To describe single-domain particles, it must be considered that the formation of magnetic domains is determined by the competition between domain wall energy lost and demagnetizing field energy gained. The walls are quasi 2D objects and the domain wall energy is thus a surface energy term. As the volume of the particle decreases, the strength of the wall energy is relatively reduced with respect to that of the demagnetizing field energy term, a volume term. At very small sizes, the state of minimum energy is the single domain state. A schematic description of this behaviour was given by Kittel [105] who considered a spherical particle of radius R, characterized by an anisotropy constant K. The energy E_1 in the single domain state is:

$$E_1 = \frac{1}{3} M \frac{4}{3} \pi R^3 \tag{5}$$

When the particle is divided into two domains by a single wall, the value of the demagnetizing field energy is approximately divided by two. The corresponding energy E_2 is :

$$E_2 = 4\sqrt{AK}\pi R^2 + \frac{1}{6}M\frac{4}{3}\pi R^3 \tag{6}$$

where the term in R^2 represents the wall energy [106]. Below a certain critical radius R_c, E_2 becomes less than E_1 and the single domain state is favoured ; R_c is given by:

$$R_c = \frac{36\sqrt{AK}}{\mu_0 M_s^2} \tag{7}$$

For Co particles, $R_c \approx 9$ nm and for $Nd_2Fe_{14}B$ particles, $R_c \approx 15$ nm. Single-domain particle formation implies that the observation of coercivity does not require prior magnetic saturation under applied field, unlike the case of multidomain particles. In itself, it does not give any indication on the magnetization reversal process involved under H_{app}.

4.1.2. Coherent rotation vs. nucleation and propagation

Reversal by nucleation-propagation involves the formation of a non-uniform magnetic state which is similar to a magnetic wall (see Section 2.2). The formation of such a wall is not involved in the coherent rotation process. In the same way that the single-domain state is favored at small particle size, it can be anticipated that coherent rotation is favored as well. To evaluate the critical radius below which coherent rotation will happen necessarily, one may compare the two magnetic configurations of maximum energy which are involved during reversal in a spherical shape sample. For coherent rotation, this corresponds to the magnetic moment being perpendicular to the easy axis, z. For nucleation-propagation, this corresponds to an equatorial wall which defines two equal domains of antiparallel magnetization. The critical radius, $R_c^{'}$, below which coherent rotation dominates is derived to be:

$$R_c^{'} \approx \delta \tag{8}$$

$R_c^{'}$ is of the order of 30 nm for Co metal, but of 5 nm only for a high anisotropy system such as $Nd_2Fe_{14}B$. Whilst the SW model of coherent rotation has been used for many years as a model paradigm to illustrate the phenomenon of coercivity, the first unambiguous experimental evidence for the occurrence of coherent rotation is recent. Magnetization reversal in individual 4 nm diameter Co nanoparticles has been measured thanks to ultra-sensitive μ-SQUID magnetometry. The angular variation of the coercive field found in these nanoparticles corresponds exactly to the prediction of the SW model [107].

Due to the high reactivity of the rare-earth metals, the controlled preparation of rare-earth based hard nanoparticles is difficult. $Nd_2Fe_{14}B$ nanoparticles embedded in a Nd-rich non-magnetic matrix have been prepared by fast annealing of ribbons, with minimum particle volume of the order 100 x 40 x 25 nm^3 [108]. The coercive field, $\mu_0 H_c \approx 2.7$ T at 300 K, reaches a value which is close to the SW calculated value when account is taken of shape anisotropy. This suggests that coherent rotation occurs whilst the particle radius is still an order of magnitude larger than $R_c^{'}$. Due to the random orientation of the nanoparticles within the matrix, the angular dependence of the coercive field has not been measured. FePt isolated nanoparticles may be grown on an insulated MgO substrate, of the order of 20 x 20 nm^2 in-plane particle size and 3 nm thickness [109]. Coercive field values in excess of 6 T were obtained at 300 K, which approach the value of the anisotropy field. This shows that magnetization reversal should resemble coherent rotation in these systems as well.

4.1.3. *Thermal activation and superparamagnetism*

Thermal activation effects become more effective at very small particle size. In particular, they lead to the appearance of superparamagnetism. Consider a single domain uniaxial particle which may reverse its magnetization by coherent rotation. In zero applied field, the barrier against reversal amounts to $\Delta E = KV$, where V is the particle volume. The particle moment may fluctuate above the energy barrier at a frequency equal to $1/\tau$, where τ is given by Eq. 4. In systems made of micron size particles, τ amounts to values longer than any physical time. Thermal activation effects are insignificant. However, when the particle volume is reduced in the nanosize range, the time τ may be reduced to values of the order of the characteristic time of the measurement technique used. For instance, the characteristic time of magnetization measurements is 1s typically; it is reached at the so-called blocking temperature T_B such that $\Delta E/kT_B \approx 25$ (see Section 2.2.2). Above T_B, the magnetization fluctuates above the energy barrier, it behaves similarly as if it was paramagnetic (note that T_B depends on the measurement technique). The phenomenon of superparamagnetism sets a limit to the exploitation of nanoparticle hard magnetic properties. In 4 nm diameter Co metal nanoparticles, $T_B \approx 50$ K; in $Nd_2Fe_{14}B$ particles of the same diameter, it is of the order of 450K.

Even below T_B, the magnetization of nanoparticles may be strongly affected by thermal activation. Very small Co particles showing coherent rotation allowed thermal activation effects to be analyzed quantitatively [110]. From 40 mK up to 12 K, the coercive field was found to be a function of the expression theoretically expected for thermal activation, $Tln(\tau/\tau_0]^{2/3}$.

The volume of the smallest hard $Nd_2Fe_{14}B$ nanoparticles obtained until today is of approximately 10^5 nm^3 (see Section 4.1.2.). This is 100 times more than the size of the activation volume usually measured in NdFeB magnets [12]. Assuming that, in nanoparticles, reversal is governed by the same mechanism as in magnets, no thermal activation effect may be expected. Alternatively, assuming that reversal occurs by coherent rotation, the coercive energy barrier is 0.5 10^{-15} J, 500 times more than $25\ k_BT$, the energy provided by thermal activation. Again, no thermal activation may occur. Thermal activation effects are expected to become substantially enhanced with respect to the effects observed in micron-size particles at particle size below 10 δ^3 ($\approx v_{act}$) only. In rare-earth based and FePt hard nanostructures this corresponds to particle volume below 10^3 nm^3.

4.1.4. Macroscopic quantum tunneling

Under the conditions that the magnetic states on both sides of the energy barrier are not strictly orthogonal, magnetization reversal may occur by quantum tunneling, without overcoming the barrier which characterizes coercivity. The first experimental evidence for such effect was obtained in molecular clusters, containing ten atoms typically [111, 112]. In magnetic nanoparticles, the first attempts to reveal the influence of quantum effects were based on the analysis of the temperature dependence of the coercive field, searching for low temperature deviation from the relation $H_c \propto Tln(\tau/\tau_0]^{2/3}$ (see Section 4.1.4). This type of analysis led to ambiguous results due to the unavoidable presence of a distribution of particle sizes. To lift this ambiguity, Wernsdorfer *et al.* measured the time dependence of the magnetization in a unique 10 nm hard hexaferrite particle [113]. In the study, the field was applied at various angles away from the easy magnetization direction, which constitutes a stringent test for the possible occurrence of quantum effects. The cross-over temperature, at which thermal activation effects equal quantum effects, was found to be in excellent agreement with theory. This is strong indication for the occurrence of reversal by quantum tunneling.

In 3d metallic nanoparticles, quantum effects have not been seen yet. Most often the particles are embedded in a matrix to avoid oxidation. Due to the itinerant character of the magnetic d electrons, it may be suggested that the actual reference volume encompasses the particle volume.

4.2. Exchange-Coupled Single-Phase Materials

Remanence enhancement and exchange-spring mechanism are the two characteristic properties of magnetic nanostructures which are made of exchange-coupled nanoparticles. These properties are described in the present section. About 10 years ago, when the exchanges-spring phenomenon was discovered, it was hoped that it could offer a novel route towards the preparation of new families of hard magnetic materials, with properties superior to those of already known materials. This expectation has not been realized yet. The prospect for obtaining high-performance hard nanocomposites is critically examined at the end of this section.

4.2.1. Coercivity in exchange-coupled nanosystems

As shown by Herzer [114], exchange interactions between nanoparticles are expected to be the source of a reduction in the coercive field strength. Let characterize the interparticle interactions by the exchange constant A. The interparticle exchange interactions, which favour a common parallel alignment of the moments, compete with the anisotropy which favours their individual alignment along the easy magnetization direction of each crystallite. The crystallites are exchange coupled through their surface, i.e. exchange interactions constitute a surface term whereas the anisotropy is a volume term as usual. Over a volume V which contains N crystallite and in which the moments are assumed to be aligned, the mean anisotropy may be expressed as :

$$\overline{K} = \frac{K}{\sqrt{N}} = \frac{K\,v^{1/2}}{V^{1/2}} \tag{9}$$

where v is the individual crystallite volume. The correlation length, δ_{corr}, may be defined as the associated domain wall width:

$$\delta_{corr} = \pi\sqrt{A/\overline{K}} \tag{10}$$

$\overline{K}$ depends on V through N, whereas V depends on $\overline{K}$ through δ_{corr}. The correlation volume ($V_{corr} = \delta^3_{corr}$) is derived self-consistently. One obtains

$$V_{corr} = \frac{A^6}{K^6\,v^3} \tag{11}$$

$\overline{K}$, evaluated over V_{corr}, is a function of the ratio between the crystallite diameter, *D*, and the wall width, δ:

$$\overline{K} = K\frac{D^6}{\delta^6} \tag{12}$$

In soft magnetic materials, the correlation volume deduced is of macrocopic size and the anisotropy is vanishingly small ($\overline{K} \approx 10^2$ J/cm^3). In high anisotropy uniaxial compounds ($Nd_2Fe_{14}B$, $SmCo_5$, $Sm_2Fe_{17}N$ or FePt) the magnetocrystalline anisotropy constant reaches values $K \approx 10^7$J/m^3. The deduced correlation volume V_{corr} ($\approx 10^3$ nm^3) is of the order of the crystallite size. Thus, the present model fails. To first order, the crystallite magnetic moments remain aligned along the local easy direction of each crystallite. This explains why large coercivity may be preserved in assemblies of exchange-coupled nanoparticles, as observed experimentally. However, a certain deviation from the local easy magnetization direction occurs close to the grain boundary, when two grains are interacting. This effect leads to remanence-enhancement as discussed in the next section. It becomes significant at particle size below typically 20 nm. Simultaneously, it is the source of a reduction in the magnetic anisotropy and thus the coercive field value, of the same origin but much reduced strength than in soft materials. These effects have been extensively confirmed by numerical modeling [115, 116].

At larger particle size, several experimental results suggest the existence of a moderate coercivity increase as the particle size is decreased [54]. This unexpected behavior was examined by numerical modeling. When more than two particles are involved in reversal [116], a competition may exist between the various local simultaneous processes, thus accounting for the observed effect.

4.2.2. Remanence enhancement

The most striking property observed in exchange-coupled nanoparticles is remanence-enhancement. In an assembly of randomly oriented non-interacting uniaxial particles which have been initially saturated along H_{app}, the expected value of the remanent magnetization, M_r, is 0.5 M_s. Remanence enhancement corresponds to M_r values higher than 0.5 M_s. Qualitatively, this behavior indicates directly that the particles are coupled through the interface moments. As a result of the favored common alignment, the interface moments rotate out of their local easy direction. They orient along a direction *z'*, which is intermediate between the easy directions of the interacting particles. The *z'* directions are distributed within a solid angle which is more restricted than the z directions and this gives direct account for the observed effect.

To provide a physical picture for this phenomenon, the nanoparticles may be arbitrarily separated into a core, in which the magnetization is along the grain easy magnetization direction and a shell, of thickness δ, in which the moments may deviate from the local easy axis. Thus, the remanent magnetization may be expressed as:

$$M_r = 0.5 M_s(1 - f(r)) + \beta f(r) \tag{13}$$

In this equation, $r = R/\delta$ where R is the nanoparticle radius and $f(r) = V_{shell}/V_{particle} = 3/r - 3/r^2 + 1/r^3$. The curve shown in Figure 3 was obtained for $\beta = 0.85$.

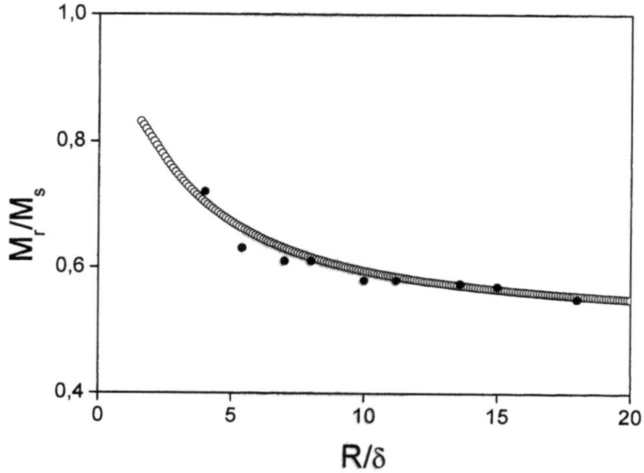

Figure 3. Remanence enhancement in exchange-coupled single –phase nanostructures – (•) Numerical modeling [117, 118] and (o) calculation according to Eq. 13.

More quantitative values are obtained by numerical modeling [117, 118]. The agreement with the simple Eq. 13 is good. The calculated remanences tends to be higher than the experimental ones. This suggests that a certain exchange reduction exists at the interface.

4.3. Soft-Hard Magnetic Nanocomposites

4.3.1. Introduction to hard-soft hardening

The discovery by Coehoorn *et al.* [119, 120] of a significant coercivity in nanocomposite magnetic materials which associate exchange coupled soft and hard magnetic nanograins, constitutes a spectacular properties of magnetic nanomaterials. This so-called spring-magnet behaviour was first described in its one-dimensional form by Kneller and Hawig [121], although there are earlier precursors (Ch. 3).. The essential features can be reproduced within the simple picture of a soft layer sandwiched between two hard layers of infinite anisotropy, in which the easy axes z are supposed to be parallel. Let M_{soft}, A_{soft} and d_{soft} be the soft layer magnetization, exchange constant and thickness respectively. Under field, the moment configuration within the soft layer is determined by the balance between exchange energy and Zeeman energy. The moments rotate progressively, as in a domain wall, from $\theta = 0$ at the interface to $\theta = \theta_{max}$ at the centre of the soft layer (see Fig. 4).

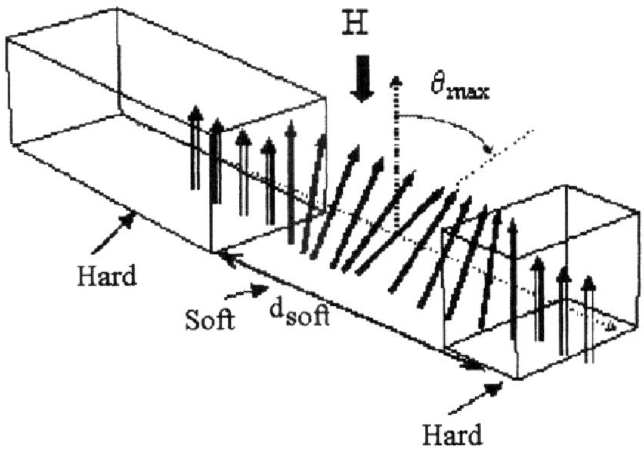

Figure 4. Schematic representation of hard-soft nanocomposites.

In a small applied magnetic field, the soft layer remains fully aligned along the uniaxial hard magnetization direction as a result of exchange coupling through the interfaces. The nucleation field, H_n, at which first departure from saturation takes place, amounts to [75]:

$$H_n = \frac{12\, A_{soft}}{M_{soft}\, d_{soft}^2} \tag{14}$$

When the anisotropy energy within the hard layer cannot be considered as infinite as compared to the Zeeman energy, the nucleation field depends on the hard layer magnetic properties [122]. However, as long as $d_{hard} > \approx 3\delta_{hard}$ (d_{hard} and δ_{hard} are the hard layer thickness and domain wall width respectively), H_n does not depend much on d_{hard}. For $d_{soft} \approx 10$ nm, the room temperature nucleation field $\mu_0 H_n$ is typically 1 T.

As the applied field is increased further, the domain wall formed in the soft layer becomes progressively compressed at the soft/hard interface, until it penetrates into the hard layer and full reversal occurs. For sharp interface and assuming that the magnetisation and exchange constant are the same in both types of layers, the propagation field, H_p, has been calculated by Aharoni [123]. At small d_{hard}, H_p is very small, it increases to a maximum value $H_p \approx 0.5\ H_A$ (H_A is the hard phase anisotropy field) at $d_{hard} \approx \delta_{hard}$ and decreases further to 0.25 H_A as $d_{hard} \rightarrow \infty$. When account is taken of the different magnetisation and exchange constant values in the soft and hard layers, the reduced propagation field $h_p = H_p/H_A$ becomes at large d_{hard} [124]:

$$h_p = \lambda/\left(1+\sqrt{\lambda}\right)^2 \quad (15)$$

with $\lambda = M_hA_h/M_sA_s$. A typical experimental value is $h_p \approx 0.15$ (see §3), i.e. the propagation field is significantly smaller than calculated. Real interfaces are not ideally sharp; the propagation field is directly proportional to the derivative of the domain wall energy across the wall, thus qualitatively accounting for the observed effect.

4.3.2. *Analyses of nanocomposite magnetic properties*

Several semi-quantitative experimental studies of exchange–spring behaviour were performed on R-Fe-B nanocomposites [120]. At 300 K, magnetization reversal appears to take place in one step, with $\mu_0H_p \approx \mu_0H_n \approx$ 0.4 - 0.5 T. However, using the element sensitivity of XMCD measurements, soft and hard phase magnetisation reversal could be separated [125]. It was found that (reversible) soft phase reversal occurs at μ_0H_{app} values which are approximately 0.1 T less than those at which (irreversible) hard phase reversal takes place. Assuming grain sizes around 20 nm and $A_s \approx 1.5 \times 10^{-11}$ J/m (mean value between Fe_3B (1.3×10^{-11} J/m) and α-Fe (1.8×10^{-11} J/m), the values of H_n were of the order of magnitude of calculated values. Typical values for the propagation field (in Pr-Fe-B, $\mu_0H_p = 0.5$ T at 300 K and 2.5 T at 20 K) are strongly reduced with respect to the values of H_A. This is in agreement with the discussion presented in Section 4.3.1.

The individual measurements of the hard grain hysteresis cycles by XMCD have revealed that the hard phase magnetisation show a significant remanence enhancement ($M_r/M_s \approx 0.6$) [123]. This observation implies that the soft phase magnetisation shows remanence enhancement as well. This is in agreement with experimental data on other nanocomposite systems.

SmCo/Fe multilayers, epitaxially grown on sapphire, constitute model systems on which quantitative analyses of reversal were performed [126]. The first reversible step in the hysteresis cycle may be ascribed to nucleation of reversal in the soft Fe layer. The second step is irreversible and corresponds to Sm-Co reversal. $\mu_0 H_n = 0.22$ T (0.09 T) and $\mu_0 H_p = 0.6$ T (0.7 T) are obtained in the 10 nm (20 nm) Fe film. Reversible magnetisation reversal was quantitatively fitted to a simple numerical model, assuming that the magnetic properties of the Sm-Co films are identical to those of individual films and those of the Fe films correspond to bulk Fe. By contrast, hard layer reversal occurs at field values which are smaller than calculated values. It is a common fact that the experimental values of the coercive field are much lower than calculated values. Structural features, not considered in the calculation, are the source of strong coercive field reduction.

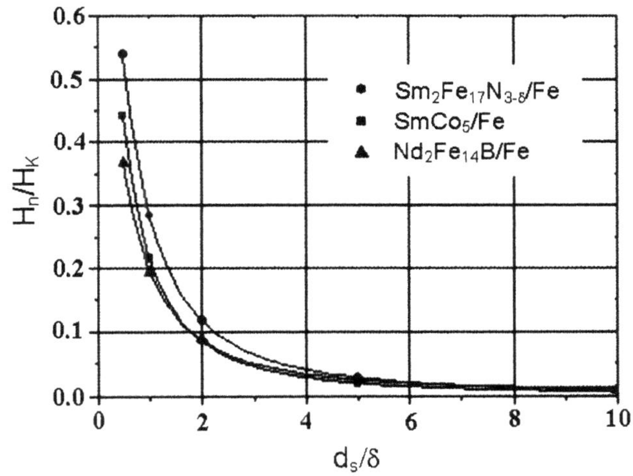

Figure 5. Nucleation field calculated for a series of model multilayered systems.

The properties of nanocomposite systems, whose microstructures aim at reproducing real systems, have been examined in various numerical modelling studies [127, 128]. In general, the essential features of the hysteresis cycles may be satisfactory reproduced. In particular, soft layer reversal is quantitatively accounted for, which is expected for reversible phenomena. By contrast, the calculated high-field irreversible reversal of the hard phase magnetization is not reproduced in general. Such discrepancy illustrates the already mentioned difficulty to describe irreversible processes.

4.3.3. Prospect for improving nanocomposite magnetic properties

We discuss in this section whether routes may be envisaged towards nanocomposite materials with magnetic properties significantly superior to present properties. The target remanent magnetisation value is estimated to be $\mu_0 M_r = 1.2$ T in isotropic systems and $\mu_0 M_r = 1.5$ T in anisotropic systems. Considering that $\mu_0 H_c$ should be larger than $\mu_0 M_r/2$ to benefit from remanence, $\mu_0 H_n \approx \mu_0 H_p \approx 0.8 - 1$ T is deduced.

To reach the remanent magnetization values quoted here, the soft layer spontaneous magnetization must be higher than typically 1.6 T. Such values imply that the soft layer be a Fe based alloy. The important adjustable parameter to be considered is the layer thickness, d_{soft}. Based on the relation established by Skomski and Coey [80], the d_{soft} dependence of $h_n = H_n/H_K$ is plotted in Fig. 5 for $Sm_2Fe_{17}N_{3-\delta}$ /Fe, $SmCo_5$/Fe and $Nd_2Fe_{14}B$/Fe (d_{hard} was assumed to be large and under these conditions, it does not enter into the calculation). Other relevant parameters are listed in Tab. I. The d_{soft} values at which $\mu_0 H_n = 1$ T are found to be 10 nm in $Sm_2Fe_{17}N_{3-\delta}$ /Fe and $SmCo_5$/Fe and 7.5 nm in $Nd_2Fe_{14}B$/Fe. A real nanocomposite material will contain 50 % of both the hard and the soft phase, and thus $d_{hard} \approx d_{soft.}$ The reduced nucleation fields, h_n calculated for such hard layer thickness do not differ by more than a few percent from h_n calculated for $d_{hard} \rightarrow \infty$.

Soft layer thicknesses of about 10 nm are significantly smaller than those obtained in nanocomposites prepared by usual metallurgical routes, such as ribbon melt-quenching or powder metallurgy. Only thin film preparation procedures have allowed films of such small thicknesses to be prepared. For industrial development alternative routes should be developed. Mechanical deformation routes such as rolling might be a possibility.

Furthermore, as the layer thickness will become smaller, exchange interactions between the particles will favour parallel alignment of the moments throughout the whole material. In a non-textured material, this will occur at the expense of coercivity. It may be expected that large nucleation fields can only be obtained with textured materials. The development of the corresponding preparation procedures constitutes a challenge for material scientists.

Table 1. Physical parameters used in the calculation of H_n (Fig. 5) (see text).

	Fe	$SmCo_5$	$Sm_2Fe_{17}N_{3-\delta}$	$Nd_2Fe_{14}B$
$\mu_0 M_s$ (T)	2.2	1.2	1.55	1.6
A (10^{-11} J/m)	2.00	1.73	1.48	1.13
δ (nm)		3.2	3.4	4.0

Assuming thin layer thicknesses and large H_n, it is still necessary that the hard layer resists reversal. From the discussion of propagation in Section 4.3.1., the question arises whether $\mu_0 H_p \approx 1$ T may be attained. The best properties will require sharp interface. When a close relationship exists between hard phase and soft phase crystal structures, it may be anticipated that the formation of diffuse interfaces will be favoured. This is expected to favour propagation and thus be detrimental to coercivity. In other systems, such as NdFeB/Fe, no simple relationship exists between the crystal structures of the hard and soft phase and TEM images reveal sharp interfaces. However, in this case, due to the poor structural matching between the two phases involved, there is a risk that a significant exchange discontinuity exists at the interfaces.

An alternative approach would be to use pinning-type magnets for the hard layer. Domain walls would be pinned within the hard phase, thus hindering reversal. Potentially, spring magnets based on such materials (e.g. $Sm(FeCoCuZr)_{7-8}$) would be able to maintain large coercivity in soft/hard mixtures. We suggest that they constitute promising materials for spring-magnets applications, provided appropriate metallurgical routes are found.

4.4 Demagnetizing Field in Nanostructured Materials

It is well known that the magnetization processes of magnetic materials do not directly depend on the applied magnetic field H_{app}, but rather on the internal field, H_{int}, existing within matter. H_{int} is the sum of the applied field, H_{app}, and the demagnetizing field, H_{D}, itself created by the magnetization. In the only case where the magnetisation is uniform and the sample under study has the shape of a second order ellipsoid, H_{D} is uniform; the demagnetizing field may be expressed as

$$H_D = -N\overline{M} \tag{16}$$

where N is the demagnetizing field coefficient along the magnetization direction.

In other cases, the demagnetizing field has a local character. In principle, it should be calculated in each point within matter. However, this constitutes a formidable task and a uniform demagnetizing field is still assumed most generally, with the demagnetizing field being expressed as in Eq. 16 and N being evaluated by approximating the sample shape to that of an ellipsoid.

In hard nanostructures, the magnetization is by nature highly heterogeneous and the validity of the usual demagnetizing field corrections may thus be questioned [129]. Let separate the demagnetizing field H_{D} acting on a given nanoparticle, (i), into three contributions:

$$H_D = H_D^{bulk} + H_{cav} + H_D^i \tag{17}$$

where H_D^{bulk} is the bulk demagnetizing field created by the surface magnetic charges, H_{cav} is the cavity field created on a given particles by the surrounding magnetic charges and H_D^i is the self demagnetizing field of particle (i). When the magnetization reverses within particle (i), it flips from saturation along one direction to saturation in the opposite direction. Thus, the field H^i_D does not depend on the magnetization state of the system. It may be subtracted from H_D and considered as a contribution to the coercive field.

The bulk demagnetizing field, H_D^{bulk} and the cavity field, H_{cav}, remains to be evaluated. Let first assume that the material is made of exchange decoupled nanoparticles. Magnetization reversal within individual grains may be assumed to be independent of reversal in the other grains and the magnetization within matter as homogeneous, equal to $\overline{M}$. The bulk demagnetizing field is equal to $-N\overline{M}$ and the cavity field becomes $N_i\overline{M}$, where N_i is the relevant demagnetizing field coefficient of particle (i). The field H_D becomes:

$$H_D = -N\overline{M} - N_i\overline{M} = -(N - N_i)\overline{M} . \tag{18}$$

Of particular interest is the case where $N = 1$ (field applied perpendicular to the plane in a film or a foil) and N_i = 1/3 (spherical particles). The demagnetizing field slope instead of being equal to $1/N \approx 1$ becomes $1/(N-N_i) \approx 1.5$. It differs from the usual demagnetizing field slope.

Alternatively, let us consider an assembly of exchange-coupled particles. Reversal is expected to occur essentially by growth of the reversed regions into the regions in which the magnetization is not reversed yet. For any considered grains, reversal takes place in a region in which the magnetization is still saturated. The cavity field proportional to $N_i M_s$ is constant. Only the bulk demagnetizing field needs to be considered in demagnetizing field corrections. It may be thought that the usual demagnetizing field corrections thus apply. However, starting from the saturated state, a small number of domains are nucleated in general. As a result, significant demagnetizing field energy remains in the magnetic state of zero magnetization. Let expresses the remaining energy ½ $N' M_s^2$, where N' may be considered as a purely phenomenological parameter. As the field is varied and the magnetization progressively grows along the field direction, the demagnetizing field energy increases until it reaches the maximum value ½ $N M_s^2$. From zero magnetization to saturation, the variation in the demagnetizing field energy is ½$(N-N')M_s^2$. To fist approximation, the slope of the magnetization variation is $1/(N-N')$ instead of the usual slope $1/N$.

This simple discussion of the demagnetizing field effects in hard nanostructures show that the usual corrections should not be applied without careful analysis of the system behavior. In particular, it may be suggested that the impossibility to apply the $1/N$ corrections found in several FePt hard magnetic systems [130, 131] is due to the inappropriateness of these corrections.

5. APPLICATIONS

The main application of hard nanostructures is for the preparation of bonded magnets. Although economically marginal today, MEMS constitute a specific domain of applications, since the magnetic properties of nano-structured materials are maintained at small magnet dimensions. Finally it is realised that hard nanostructured materials could find applications as high density recording media.

5.1. Bonded Magnets

One can find high performance bonded magnets in our daily lives in various electronic devices, office automation equipment, automotive components and home appliances such as computer hard disk drives, CDs, DVDs. Isotropic powders are most commonly used for the production of bonded magnets. The bonded magnet market throughout the world was over $1 billion in 2001 [132]. While bonded ferrites have still the biggest market by mass, RE magnets are becoming leader in value. Four RE powder families are commercially used: (i) isotropic and anisotropic nanostructured Nd-Fe-B, ribbons (ii) "lean neo" $Nd_2Fe_{14}B/Fe/Fe_3B$ nanocomposites, (iii) $SmCo_5$ and Sm_2Co_{17} and finally (iv) Sm-Fe-N.

Thermo set as well as thermoplastic and elastomer binders are used in compression, injection and calendaring process respectively. Extrusion moulding uses elastomers as well as thermoplastic materials such as PVC. Compression and injection moulding processes make only rigid magnets, calendaring makes flexible magnets and extrusion can lead to both states.

In most of these magnets, isotropic NdFeB powders are used which are produced by the melt spinning process. The high corrosion resistance of the ribbons together with the flexibility in shape and magnetic properties largely open the field of applications of these materials. Moreover, the melt spinning process is significantly less expensive than powder metallurgy.

The highest remanence available is achieved in $Nd_2Fe_{14}B/\alpha$-Fe nano-composite materials with typical grain size 20 – 30 nm: $\mu_0 M_r = 0.98$ T, $\mu_0 H_c = 1.1$ T, and $BH_{max} = 127$ kJ/m^3 [133]. However, due to the difficulty in material processing, the commercial development of these materials remains very limited. Anisotropic NdFeB powders are produced by the HDDR process. Nanosize grains form within micron sized powder particles and they are therefore protected against corrosion. Thanks to the induced texture obtained, large energy product may be reached (342 kJ/m^3, in the powder) [134].

Sm-Co anisotropic powders have been developed to prepare magnets meeting high operating temperature requirements [135]. These powders are produced by pulverization of bulk ingots which have been submitted to the precipitation hardening heat treatment.

$Sm_2Fe_{17}N_x$ powders are produced commercially since 1999 by the reduction and diffusion process [136]. The currently mass-produced powder has a BH_{max} of 290 kJ/m^3 with μ_0M_r = 1.35 T and μ_0H_c = 1.1 T. Milling and surface treatment have been optimized to improve the squareness of the demagnetization curve and resistance to corrosion [137].

5.2. Recording Media

Magnetic data storage constitutes a large potential market for hard magnetic films (Ch. 11). While the bit density increases by about 60% each year, the need appears for materials with higher coercivities. Compared to others applications, the amount of material required is much less. The media consist of films of typical thickness below 50 nm, themselves made of hard magnetic nanoparticles embedded in a non magnetic matrix. The use of high anisotropy materials allows the signal to noise ratio to be reduced, while preserving thermal stability. FePt and CoPt films are good candidates for this type of applications, since high corrosion resistance is needed additionally. Efforts are directed at optimizing the nanostructure with uniformity in particle size and space arrangement. Self-organisation is expected. A high perpendicular magnetic anisotropy can be obtained in NdFeB thin films by various physical deposition techniques. In particular, epitaxial growth is a promising way to obtain high coercivities in highly textured films [41] (see section 3). In comparison to other deposition techniques, chemical synthesis is inexpensive, easy to use and it may lead to self-organised nanoparticles [46]. However, laboratory production for each synthesis run by the chemical method is still limited to less than 1 gram. The main problem is the scaling up of the process. Many grams of FePt nanopowders with coercivity of 2 T after annealing were already produced in one batch [47].

5.3. MEMS Applications

The application of actuators in micro-electro-mechanical systems (MEMS) requires the miniaturization (100 to 500 microns) of the permanent-magnet components while preserving as much as possible their bulk magnetic properties. In many instances, a severe degradation in coercivity occurs when the magnet size is reduced at such small dimensions. This is particularly true for Nd-Fe-B magnets. The use of nanosize hard magnetic nanostructures allows coercivity to be preserved in micron-sized powders.

Screen-printing, bonding and injection moulding are well adapted to the fabrication of thick films for MEMS. Most often, isotropic ribbons are used, and the dilution of the powders in the non magnetic binder limits further the value of the remanent magnetisation.

Die-upsetting of fully dense hot compacted melt spun NdFeB powders produces axially anisotropic sheet-shaped magnets of about 300 microns thickness. The following properties were obtained $\mu_0 M_r = 1.25$ T, $\mu_0 H_c = 1.33$ T and $(BH)_{max} = 290$ kJ/m^3 together with good mechanical integrity [138].

With the continuous increase in the demand for smaller and smaller systems, film preparation emerges as a very promising route towards MEMS applications [139]. Plasma spraying is a high rate deposition technique (several μ/min) suited to the fabrication of thick films over large surface area [140]. Other possible techniques are sputtering or PLD. When magnetic films are deposited with these techniques, texture control determines not only the remanence achieved in the magnet itself, but also the relative orientation of magnetisation with respect to the film plane and thus the applicability of the film for a given system. Though most of the sputtering deposition rates are still too low for industrial thick film deposition, NdFeB films have been deposited at rates up to 50 μ/h over surfaces of 25 cm^2 [141]. The corresponding magnetic properties of the films range from $BH_{max} = 195$ kJ/m^3 (with $\mu_0 M_r = 1$ T and $\mu_0 H_c = 3.2$ T) to $BH_{max} = 352$ kJ/m^3 (with $\mu_0 M_r = 1.35$ T and $\mu_0 H_c = 0.8$ T). High deposition rates have also been reported for pulsed laser deposition but over small surface areas only (1 cm^2) [142]. Finally electro-deposition is a high rate deposition technique but its applicability is restricted to a limited number of systems. FePt films can be electro-deposited at rates up to a few μ/h but the magnetic properties remain low [143].

6. CONCLUSION

The magnetic properties of hard nanostructures have been reviewed in this chapter. Various processes may be used for the preparation of these systems and specific properties arise which are directly associated to the nanometer scale of the structures. Our analysis suggests that a significantly enhanced performance could be achieved by an improved control of the nanostructure. The impressive development of innovative physical and chemical preparation routes that occurred during the last ten years lead to think that new routes will appear in the coming years allowing even better nanostructure control and thus further progress in the hard magnetic properties of these systems. The goal of obtaining a material with energy product of the order of 0.8 MJ/m^3 does not appear unrealistic. Beside this, a need exists for new materials for magnetic recording and MEMS applications. The specific processes required for the preparation of nanostructured systems can meet nanotechnology requirements. It can be hoped that these highly challenging domains will open to hard magnetic materials.

Acknowledgments

We acknowledge Nora Dempsey for very helpful discussion from the initial stage in the preparation of the manuscript to the realisation of the final version.

References

[1] R. Skomski and J. M. D. Coey, "Permanent Magnetism", I.O.P., Bristol 1999, p. 23.

[2] R. A Ristau, K. Barmak, L. H. Lewis, K. R. Coffey, and J. K. Howard, J. Appl. Phys. **86**, 4527 (1999).

[3] A. D. Kent, S. von Molnar, S. Gider, and D. D. Awschalom, J. Appl. Phys. **76**, 6656 (1994).

[4] W.Wernsdorfer *et al.,*. Phys. Rev. Lett. **77**, 1873 (1996).

[5] "Magnetism and Synchrotron radiation", Ed. E. Beaurepaire, F. Scheurer, G. Krill, and J.P. Kappler, Springer, Berlin 2001.

[6] I. R. Gentle, Materials Forum **22**, 123 (1998).

[7] E. C. Stoner and E. P. Wohlfarth, Philo. Trans. R. Soc. London A **240**, 599 (1948).

[8] W. J. Brown Jr. "Micromagnetics", Interscience Publishers 1963, p. 68.

[9] D. Givord and M. F. Rossignol, in "Rare-Earth Iron Permanent Magnets", Ed. J. M. D. Coey, Clarendon Press, Oxford 1996, p. 218.

[10] D. Elbaz, D. Givord, S. Hirosawa, F. P. Missel, M. F. Rossignol, and V. Villas-Boas, J. Appl. Phys. **69**, 5492 (1991).

[11] E. P. Wohlfarth, J. Phys. F, **14**, L155 (1984).

[12] D. Givord, A. Liénard, P. Tenaud, and T. Viadieu. J. Magn. Magn. Mat. **67**, L 281(1987).

[13] M. Sagawa, S. Hirosawa, H. Yamamoto, S. Fujimura, and Y. Matsuura, Jap. J. of Appl. Phys. **26**, 785 (1987).

[14] H. Kronmüller, K. D. Durst, S. Hock, and G. Martinek, J. Magn. Magn. Mater. **69**, 149 (1987).

[15] D. Givord, P. Tenaud, and T. Viadieu, I.E.E.E. Trans. Mag. **24**, 1921 (1988).

[16] D. Givord, M. F. Rossignol, and V. M. T. S. Barthem, J. Magn. Magn. Mater. **258-259,** 1 (2003).

[17] D. Givord, Q. Lu, F. P. Missell, M. F. Rossignol, D. W. Taylor, and V. Villas-Boas, J. Magn. Magn. Mater. **104-107,** 1129 (1992).

[18] P. J. Mc Guiness, X. J. Zhang, X. J. Yin, and I. R. Harris, J. Less Common Metals **158**, 379 (1990).
[19] T. Takeshita, R. Nakayama, Proceedings of the 10th International Workshop on Rare Earth Magnets and their Applications, Kyoto, 551 (1989).
[20] H. Nakamura, S. Sugimoto, M. Okada, and M. Homma, Mater. Chem. Phys. **32**, 280 (1992).
[21] M. Kubis, O. Gutfleisch, K. H. Müller, I. R. Harris, and L. Schulz, J. Appl. Phys. **83** 6905 (1998).
[22] Y. Honkura, C. Mishima, and H. Mitarai, Tokaihei 10-135019 (japanese patent) (1996).
[23] Y. Honkura, Proceedings of the 17th International Workshop on Rare Earth Magnets and their Applications, Delaware, 52 (2002).
[24] O. Gutfleisch, K. Khlopkov, A. Teresiak, K. H. Müller, G. Drazic, C. Mishima, and Y. Honkura, IEEE Trans. Magn. **39**, 2926 (2003).
[25] K. H. Müller, L. Cao, N. M. Dempsey, and P. A. P. Wendhausen, J. Appl. Phys. **79** (8), 5045 (1996).
[26] A. Kirchner, J. Thomas, O. Gutfleisch, D. Hinz, K. H. Müller, and L. Schulz, J. Alloys and Comp. **365**, 286 (2004).
[27] W. Grünberger, D. Hinz, A. Kirchner, K. H. Müller, and L. Schulz, J. Alloys and Comp. **257**, 293 (1997).
[28] S. C. Chen, P. C. Kuo, A. C. Sun, C. T. Lie, and W. C. Hsu, Mater. Science and Engineering B **88**, 91 (2002).
[29] J. A. Christodoulides, Y. Huang, Y. Zhang, G. C. Hadjypanayis, I. Panagiotopoulos, and D. Niarchos, J. Applied Physics **87** (9), 6938 (2000).
[30] M. Daniil, P. A. Farber, H. Okumura, G. C. Hadjypanayis, and D. Weller, J. Magn. Magn. Mat. **246**, 297 (2002).
[31] C. P. Luo and D. J. Sellmyer, Appl. Phys. Letters **75**, 3162 (1999).
[32] C. M. Kuo and P. C. Kuo, J. Appl. Phys. **87**, 419 (2000).
[33] T. Klemmer, D. Hoydick, H. Okomura, B. Zhang, and W. A. Soffa, Scripta Metal. Mater. **33**, 1793 (1995).
[34] S. C. Chen, P. C. Kuo, S. T. Kuo, A. C. Sun, C. T. Lie, and C. Y. Chou, Mater. Science and Engineering B **98**, 244 (2003).
[35] T. Shima, K. Takanashi, Y. K. Takahashi, and K. Hono, Appl. Phys. Letters **81** (6), 1050 (2002).
[36] M. Weisheit, L. Schulz, and S. Fähler, J. Magn. Magn. Mater., in press (2005).
[37] S. Stoyanov, Y. Huang, Y. Zhang, V. Skumryev, G. C. Hadjypanayis, and D. Weller, Appl. Phys. Letters **93** (10), 7190 (2003).
[38] F. J. Cadieu, T. D. Cheung, and L. Wickramasekara, J. Magn. Magn. Mat. **54**, 535 (1986).
[39] V. Neu, U. Hannemann, S. Fähler, B. Holzapfel, and L. Schulz, J. Appl. Phys. **91** (10), 8180 (2002).
[40] L. Castaldi, M. R. J. Gibbs, and H. A. Davies, J. Appl. Phys. **93** (11), 9165 (2003).
[41] U. Hannemann, S. Fähler, V. Neu, B. Holzapfel, and L. Schulz, Appl. Phys. Letters **82** (21), 3710 (2003).

[42] I. Panagiotopoulos, X. Meng-Burany, and G. C. Hadjypanayis, J. Magn. Magn. Mater. **172**, 225 (1997).
[43] L. Chen, B. Li, C. F. Cadieu, C. Theodoropoulos, and F. J.Cadieu, J. Appl. Phys. **87** (9), 6128 (2000).
[44] J. Zhou, R. Skomski, A. Kashyap, K. D. Sorge, Y. Sui, M. Daniil, L. Gao, M.L. Yan, S. H. Liou, R. D. Kirby, and D. J. Sellmyer, J. Magn. Magn. Mater. **290-291**, 227 (2005).
[45] S. Stoyanov, V. Skumryev, Y. Zhang, Y. Huang, G. C. Hadjypanayis, and J. Nogues, J. Appl. Phys. **93** (10), 7592 (2003).
[46] S. Sun, C. B. Murray, D. Weller, and L. Folks, A. Moser, Science **287**, 1989 (2000).
[47] J. Ping Liu, Proceedings of the 18th Workshop on High Performance Magnets and their Applications, Annecy, France, 679 (2004).
[48] H. Zeng, S. Sun, R.L. Sandstrom, C.B. Murray, J. Magn. Magn. Mater. **266**, 167 (2003).
[49] C.A. Ross, M. Hwang, M. Shima, H. I. Smith, M. Farhoud, T. A. Savas, W. Schwarzacher, J. Parrochon, W. Escoffier, H. Neal Bertram, F. B. Humphrey, and M. Redjdal, J. Magn. Magn. Mater. **249**, 200 (2002).
[50] M. Chen, D. E. Nickles, H. Yin, S, Wang, J. W. Harrel, and S. A. Majetich, J. Magn. Magn. Mater. **266**, 8 (2003).
[51] R. W. McCallum, A. M. Kadin, G. B. Clemente, and J. E. Keen, J. Appl. Phys. **61** (8), 3577 (1987).
[52] G. B. Clemente, J. E. Keen, and J. P. Bradley, J. Appl. Phys. **63**, 3507 (1988).
[53] A. Manaf, R. A. Buckley, H. A. Davies, and M. Leonowicz, J. Magn. Magn. Mater. **101**, 360 (1991).
[54] V. Neu, L. Schulz, and H. D. Bauer, Nanostructured Mat. **12**, 769 (1999).
[55] J. I. Betancourt and H. A. Davies, J. Magn. Magn. Mater. **246**, 6 (2002).
[56] M. Venkatesan, C. Jiang, and J. M. D. Coey, J. Magn. Magn. Mater. **242-245**, 1350 (2002).
[57] J. Ding, P. J. Mc Cormick, and R. Street, J. All. Comp. **191**, 197 (1993).
[58] J. Wecker, M. Katter, and L. Schulz, J. Appl. Phys. **69** (8), 6058 (1991).
[59] F. M. F. Rhen, M. Ventakesan, I. R. Harris, and J. M. D. Coey, J. Appl. Phys. **93** (10), 8683 (2003).
[60] O. Gutfleisch, A. Bollero, D. Eckert, B. Gebel, M. Kubis, K.-H. Müller, Proc. of 16th Int. Workshop on Rare-Earth Magnets and their Applications, ed. H. Kaneko, M. Homma and M. Okada, Sendai, Japan, 883 (2000).
[61] F. J. Cadieu, R. Rani, T. theodoropoulos, and L. Chen, J. Appl. Phys. **85** (8), 5895 (1999).
[62] T. Speliotis and D. Niarchos, J. Magn. Magn. Mater. In press (2005).
[63] A. Singh, V. neu, S. Fähler, L. Schulz, and B. Holzapfel, J. Magn. Magn. Mater. In press (2005).
[64] E. Pina, M. A. Garcia, I. Carabias, F. J. Palomares, F. Cebollada, A. De Hoyos, R. Almazan, M. I. Verdu, M. T. Montojo, G. Vergara, A. Hernando, and J. M. Gonzales, J. Magn. Magn. Mater. **272**, e833 (2004).
[65] K. Schnitzke, L. Schulz, J. Wecker, and M. Katter, Appl. Phys. Letters **57** (26), 2853 (1990).

[66] C. Kuhrt, K. O'Donnell, M. Katter, J. Wecker, K. Schnitzke, and L. Schulz, Appl. Phys. Letters **60** (26), 3316 (1992).
[67] M. Katter, J. Wecker, L. Schulz, J. Appl. Phys. **70** (6), 3190 (1991).
[68] S. L. Tang, C. P. Yang, B. W. Wang, X. M. Jin, S. Y. Zhang, and Y. W. Du, J. Magn. Magn. Mater. **189**, 341 (1998).
[69] J. Yang, Ph. Oleinek, K.H. Müller, J. Appl. Phys. **88** (2), 988 (2000).
[70] J. Yang, W. Mao, B. Cheng, Y. Yang, H. Xu, B. Han, S. Ge, W. Ku, Appl. Phys. Letters, **71** (22), 3290 (1997).
[71] N. H. Hai, N. M. Dempsey, M. Verdier, M. Veron, and D. Givord, J. Magn. Magn. Mater. **257**, L139 (2003).
[72] K. H. J. Buschow, D. B. de Mooij, and R. Coehoorn, J. Less-Common Metals **145**, 601 (1988).
[73] E. F. Kneller and R. Hawig, IEEE Trans. Magn. **27**, 3588 (1991).
[74] A. Manaf, R. A. Buckley, and H. A. Davies, J. Magn. Magn. Mater. **128**, 302 (1993).
[75] S. David and D. Givord, J. All. Compounds **281**, 6 (1998).
[76] R. Skomski, J. Appl. Phys. **76** (9), 7059 (1994).
[77] C. W. Chang, H. W. Chang, C. H. Chiu, W. C. Chang, S. K. Chen, and A. C.Sun, J. Magn. Magn. Mater. In press (2005).
[78] Z. M. Chen, Y. Zhang, Y. Q. Ding, G. C. Hadjypanayis, Q. Chen, B. M. Ma, and G. C. Hadjypanayis, J. Appl. Phys. **85** (10), 5908 (1999).
[79] P. Campbell, D. N. Brown, Z. M. Chen, P. C. Guschl, D. J. Miller, B. M. Ma, Proceedings of the 18th Workshop on High Performance Magnets and their Applications, Annecy, France, 68 (2004).
[80] R. Skomski and J. M. D. Coey, Phys Rev B **48** 812 (1993).
[81] M. Shindo, M. Ishizone, A. Sakuma, H. Kato, and T. Miyasaki, J. Appl. Phys. **81** (8), 4444 (1997).
[82] W. Liu, Z. D. Zhang, J. P. Liu, B. Z. Cui, X. K. Sun, J. Zhou, and D. J. Sellmyer, J. Appl. Phys. **93** (10), 8131 (2003).
[83] S. Parhofer, G. Gieres, J. Wecker, and L. Schulz, J. Magn. Magn. Mater. **163**, 32 (1996).
[84] D. Lee, C. H. Chen, S. Liu, and M. Q. Huang, Proceedings of the 18th Workshop on High Performance Magnets and their Applications, Annecy, France, 667 (2004).
[85] G. C. Hadjypanayis, J. Magn. Magn. Mater. **200**, 373 (1999).
[86] A. Yan, A. Bollero, K. H. Müller, and O. Gutfleisch, Appl. Phys. Letters **80** (7), 1243 (2002).
[87] J. F. Liu, Y. Zhang, D. Dimitrov, and G. C. Hadjypanayis, J. Appl. Phys. **85** (10), 2800 (1999).
[88] D. Goll, H. Kronmüller, and H. H. Stadelmaier, J. Appl. Phys. **96** (11), 6334 (2004).
[89] W. Y. Zhang, X. D. Zhang, Y. C. Yang, and B. G. Shen, J. All. Comp. **353**, 274 (2003).
[90] A. Yan, O. Gutfleisch, T. Gemming , and K. H. Müller, Appl. Phys. Letters, **83** (11), 2208 (2003).
[91] A. Yan, A. Handstein, T. Gemming, K. H. Müller , and O. Gutfleisch, J. Magn. Magn. Mater., in press (2005).

[92] Z. Chen, X. Meng-Burany, H. Okumura, and G. C. Hadjypanayis, J. Appl. Phys. **87** (7), 3409 (2000).
[93] A. Yan, A. Bollero, O. Gutfleisch, and K.H. Müller, J. Appl. Phys. **91** (4), 2192 (2002).
[94] S. K. Chen, J. L. Tsai, and T. S. Chin, J. Appl. Phys. **79** (8), 5964 (1996).
[95] R. Gopalan, D. H. Ping, and K. Hono, J. Magn. Magn. Mater. **284**, 321 (2004).
[96] J. E. Shield, V. K. Ravindran, S. Aich, S. Hsiao, and L. H. Lewis, Scripta Mater. **52**, 75 (2005).
[97] Al-Omari, D.J. Sellmyer, Phys Rev B **52**, 3441 (1995).
[98] E. E. Fullerton, J. S. Jiang, and S. D. Bader, J. Magn. Magn. Mater. **200**, 392 (1999).
[99] J. P. Liu, R. Skomski, Y. Liu, and D. J. Sellmyer, J. Appl. Phys. **87** (9), 6740 (2000).
[100] J. Ding, P. J. Mc Cormick, and R. Street, J. Magn. Magn. Mater. **124**, 1 (1993).
[101] J. P. Liu, C. P. Luo, Y. Liu, and D.J. Sellmyer, Appl. Phys. Letters **72** (4), 483 (1998).
[102] J. Lyubina, O. Gutfleisch, K. H. Müller, and L. Schulz, J. Magn. Magn. Mater. in press (2005).
[103] Hsuan-Fu Yu and Hsin-Yi Lin, J. Magn. Magn. Mater. **283**, 190 (2004).
[104] C. Sudakar, G. N. Subbanna, and T. R. N. Kutty, **263**, 253 (2003).
[105] C. Kittel , Rev. Mod. Phys. **21,** 541 (1949).
[106] A. H. Morrish, "The Physical Principles of Magnetism", Wiley, New York 1965.
[107] E. Bonet, W. Wernsdorfer, B. Barbara, A. Benoit, D. Mailly, and A. Thiaville, Phys. Rev. Lett. **83**, 4188 (1999).
[108] Er. Girt, Kannan M. Krishnan, G. Thomas, Z. Altounian, and M. Dikeakos, J. Appl. Phys. **88**, 5311 (2000).
[109] T. Shima, K. Takanashi, Y. K. Takahashi, and K. Hono, Appl. Phys. Lett. **85**, 2571 (2004)
[110] W. Wernsdorfer, E. Bonet Orozco, K. Hasselbach, A. Benoit, B. Barbara, N. Demoncy, A. Loiseau, H. Pascard, and D. Mailly, Phys. Rev. Lett. **78**, 1791 (1997).
[111] J. R. Friedman, M. P. Sarachik, J. Tejada, and R. Ziolo, Phys. Rev. Lett. **76**, 3830 (1996).
[112] L. Thomas, F. Lionti, R. Ballou, D. Gatteschi, R. Sessoli, and B. Barbara, Nature (London) **383**, 145 (1996).
[113] W. Wernsdorfer, E. Bonet Orozco, K. Hasselbach, A. Benoit, D. Mailly, O. Kubo, H. Nakano, and B. Barbara, Phys. Rev. Lett. **79**, 4014 (1997).
[114] G. Herzer, IEEE Transactions on Magnetics **26**, 1397 (1990).
[115] R. Fischer, T. Schrefl, H. Kronmüller, and J. Fidler, J. Magn. Magn. Mater. **153**, 35 (1996).
[116] W. Rave and K. Ramstock, J. Magn. Magn. Mater. **171,** 69(1997).
[117] M. K. Griffith, J. E. L. Bishop, J. W. Tucker, and H. A. Davies, J. Magn. Magn. Mater. **183,** 49 (1998).
[118] J. C. Toussaint and B. Kevorkian (unpublished).
[119] R. Coehoorn, D. B. de Mooij, J. P. W. B. Duchateau, and K. H. J. Buschow, J. Phys. (Paris) Colloq **49**, C8-669 (1988).
[120] R. Coehoorn, D. B. de Mooij, and D. de Waard, J. Magn. Magn. Mater. **80**, 101 (1989).
[121] E. Kneller and R. Hawig, I.E.E.E. Trans. Mag. **27,** 3588 (1991).

[122] R. Skomski and J.M.D. Coey, "Permanent Magnetism", I.O.P., Bristol 1999, p. 23.
[123] A. Aharoni, Phys. Rev. **119**, 127 (1960).
[124] J. C. Toussaint (private communication).
[125] S. David, K. Mackay, M. Bonfim, S. Pizzini, A. Fontaine, and D. Givord, Mat. Res. Soc. Symp. Proc. **577** (1999).
[126] E. E. Fullerton, J. S. Jiang, M. Grimsditch, C. W. Sowers and S. D. Bader, Phys. Rev. B **58**, 12 193 (1998).
[127] T. Schrefl and J. Fidler, J. Magn. Magn. Mater. **203,** 28 (1999).
[128] M. K. Griffiths, J. E. L. Bishop, J. W. Tucker, and H. A. Davies, J. Magn. Magn. Mater., **234**, 331 (2001).
[129] F. Ingwiller, V. M. T. S. Barthem, N. H. Hai, and D. Givord (unpublished).
[130] N. H. Hai, N. M. Dempsey, and D. Givord, J. Magn. Magn. Mater, **262**, 353 (2003).
[131] R. Skomski, J. P. Liu, and D. J. Sellmyer, J. Appl. Phys. **87**, 6334 (2000).
[132] V. Panchanathan, D. F. Davies, in "Bonded Magnets", NATO Science Serie , Kluwer Academic Publisher, 45 (2003).
[133] P. Campbell, D. N. Brown, Z. M. Chen, P. C. Guschl, D. J. Miller, and B. M. Ma, Proceedings of the 18th Workshop on High Performance Magnets and their Applications, Annecy, France, 68 (2004).
[134] C. Mishima, N. Hamada, H. Mitaria, and Y. Honkura, IEEE Trans. Magn. **37**, 2467 (2001).
[135] J. F. Liu, in "The Global Outlook for Polymer Bonded Magnets", Chicago, Illinois, USA (2002).
[136] S. Yoshizawa, T. Ishikawa, I. Kaneko, S. Hayashi, A. Kawamoto, and H. Ohmori, IEEE Trans. Magn. **35**, 3340 (1999).
[137] T. Ishikawa and K. Ohmori, in "Bonded Magnets", NATO Science Serie , Kluwer Academic Publisher, 73 (2003).
[138] D. Hinz, O. Gutfleisch, and K. H. Müller, Proceedings of the 17th Workshop on High Performance Magnets and their Applications, Annecy, France, 797 (2004).
[139] N. M. Dempsey, N. V. Kornilov, and O. Cugat, Proceedings of the 18th Workshop on High Performance Magnets and their Applications, Annecy, France, 779 (2004).
[140] G. Rieger, J. Appl. Phys. **87** (9), 5329 (2000).
[141] Y. L. Linetsky and N. Kornilov, J. Mat. Eng. Performance **4** (2), 188 (1995).
[142] M. Nakano, Proceedings of the 18th Workshop on Rare-Earth Magnets and their Applications, Delaware, USA, 457 (2002).
[143] G. Zangari and I. Zana, Proceedings of the 17th Workshop on High Performance Magnets and their Applications, Annecy, France, 817 (2004).

Chapter 13

SOFT MAGNETIC NANOSTRUCTURES AND APPLICATIONS

K. Suzuki
School of Physics and Materials Engineering
Monash University
Clayton, VIC 3800, Australia

G. Herzer
Vacuumschmelze GmbH & Co. KG
D-63450 Hanau, Germany

Abstract The focus of this chapter is primarily directed towards nanocrystalline soft magnetic materials prepared by crystallization of amorphous precursors. The key elements involved in the development of this class of materials are three-fold: (i) theoretical models for magnetic softness in nanostructures (ii) nanostructure-property relationships and (iii) nanostructural formation mechanisms. This chapter surveys recent research on these three areas with emphasis placed on the principles underlying alloy design in soft magnetic nanostructures.

1. INTRODUCTION

Reducing the structural correlation length of solids (*e.g.* crystallite or device sizes) on a nanoscale often brings about dramatic changes in their physical properties which are usually unpredictable from the classical point of view. Magnetic softening in Fe-based nanostructures [1-3] is one of the examples of such size effects. The magnetic softening effect in nanostructures has opened the possibility of promising exploration for material developments in soft magnetic materials for practical use.

Two major alloy systems in the family of soft magnetic nanostructures are Fe-Si-B-Nb-Cu [1, 4-6] and Fe-*M*-B-(Cu) (M = Zr, Hf or Nb) [3, 7-9], commercially known as FINEMET (or VITROPERM) and NANOPERM, respectively. They are produced by primary crystallization of amorphous ribbons. Hence, the nanostructural formation upon crystallization as well as the mechanism of magnetic softening has been actively studied for these two alloy systems. These topics have already been addressed in some of the reviews in the literature. Herzer [10] reviewed the magnetic properties of Fe-Si-B-Nb-Cu alloys along with a brief summary of the random anisotropy model [2, 11-13], currently the most commonly referred theoretical framework in the field. Hono *et al.* [14] summarized the nanostructural formation mechanisms in Fe-based systems with emphasis on research performed by atom probe field ion microscopy (APFIM). McHenry *et al.* [15] focused on α'-$Fe_{50}Co_{50}$-based alloys in their review. In the present chapter we must regrettably exclude much interesting fundamental research in magnetism which does not contain information on technical magnetic properties. Useful data obtained from vapour deposited thin films or mechanical alloyed powders are also omitted.

Following this introductory section, we will overview the development of random anisotropy models and discuss the origin of the magnetic softness in nanostructures. The nanostructural formation process and alloy development in the Fe-*M*-B-(Cu) alloys to which less attention has been addressed in the previous reviews, will be another focal point in this chapter.

1.1. Historical Development of Soft Magnetic Nanostructures

Historically, the effect of nanoscale grain sizes on the technical magnetic properties was first studied for Ni-Fe (permalloy) thin films in the 1960s. The magnetic properties of Ni-Fe thin films were studied extensively owing to their potential application to non-volatile magnetic memory devices. The major technological problem with this application was the presence of local magnetic anisotropy dispersions in the Ni-Fe thin films. The magnetization fine structure induced by the local anisotropy dispersions, the so called magnetic ripple, was observed experimentally by means of Lorentz microscopy [16]. In addition, micromagnetic models were developed by Rother [17] and Hoffmann [18, 19] to describe the effect of microstructural properties on the local anisotropy dispersions. Hoffmann [19] predicted in his ripple theory that the soft magnetic properties of polycrystalline Ni-Fe films could be improved through the suppression of the effect of K_1 by grain refinement. However, K_1 of permalloy is by nature small enough to realize good magnetic softness and hence, the earlier ripple theory did not seem to have an impact large enough to stimulate attempts to reduce the large effect of K_1 in Fe-rich alloys by nano-scale grain refinement.

The study of nanostructured thin films was revitalized in early 1980s by the requirement of soft magnetic thin films for the metal-in-gap (MIG) recording heads. The purpose of this MIG configuration was to enhance the saturation magnetic flux density at the gap and hence, high saturation magnetization was the primary requirement for the soft magnetic thin films. Thus, much attention was paid to Fe-rich single- and multi-layer soft magnetic thin films [20, 21]. Although, there was a growing awareness about the significance of grain refinement even in Fe-rich films [22] where the intrinsic K_1 is substantially larger than that of permalloy, the study of soft magnetic nanostructures remained in thin film materials in the early to mid 1980s.

The era of intense research on soft magnetic nanostructures has opened after the development of nanocrystalline $Fe_{74.5}Si_{13.5}B_9Nb_3Cu_1$ by Yoshizawa *et al.* [1]. This new material seems to have originated from research aimed at the improvement of soft magnetic properties in amorphous alloys. Although stress-relief annealing was a standard process in the optimization of soft magnetic properties for amorphous alloys, it was widely believed that the magnetic softness of any amorphous alloy deteriorates by increasing the annealing temperature far beyond the crystallization temperature. Yoshizawa *et al.* demonstrated that this common belief was wrong. They showed that combined addition of Nb and Cu to amorphous Fe-Si-B alloys could alter the microstructure after crystallization into nano-scale and the resultant nano-structure is magnetically soft. This discovery was soon followed by many researchers and the use of amorphous ribbons as precursors to nanostructures has become a standard approach in material development.

The excellent soft-magnetic characteristics of rapidly solidified soft magnetic nanostructures have stimulated fundamental research on the mechanism of the softness in magnetic nanostructures. In 1989, Herzer [2] analysed the scaling behaviours of the magnetocrystalline anisotropy energy density of nanocrystalline materials based on a theoretical approach, the so called random anisotropy model (RAM), which was originally proposed by Alben *et al.* [23] in 1978 for amorphous systems. Herzer's random anisotropy model pointed out that the coercivity in nanocrystalline materials should scale as the 6th power of the grain size (D). This model explained experiments well, and nanoscale grain refinement has become a guiding principle in alloy development for soft magnetic materials.

1.2. Technological Advantage of Nanostructures

Magnetically soft materials are used as magnetic cores for the purpose of magnetic flux multiplication. Major requirements for the materials include high saturation magnetization ($J_s = \mu_0 M_s$), small coercivity (H_c) and high permeability (μ).

The technical magnetic properties such as H_c and μ are primarily a function of the magnetocrystalline anisotropy constant (K_1) of the material. But once the magnetocrystalline anisotropy has been made small, the soft magnetic properties are still limited by magneto-elastic anisotropies due to internal mechanical stress. Hence, materials development has focused on compositions and microstructures where both K_1 and the saturation magnetostriction (λ_s) are reduced towards zero. The principal advantage of soft magnetic nanostructures is that the reduction of the effects of these two major sources of magnetic anisotropy energy can be realized without sacrificing the concentration of ferromagnetic elements in the alloys.

A wide range of soft magnetic materials developed to date may be classified into three groups in terms of the approaches used for reducing the effects of K_1 and λ_s; these are (1) conventional crystalline materials, (2) amorphous materials and (3) nanocrystalline materials. The key approaches employed in these three types of materials are summarized in Table 1. Until relatively recently, most of the development was performed by optimizing the chemical composition and heat treatment of alloys prepared by conventional melting and casting. A considerably large amount of additives is required for realizing small K_1 and λ_s in this conventional approach and the saturation magnetization (J_s) is severely reduced.

Table 1. Approaches used for reducing the effects of magnetocrystalline anisotropy (K_1) and magnetostriction (λ_s) in bulk soft magnetic materials.

Material families	Preparation methods	Approaches $K_1 \to 0$	$\lambda_s \to 0$
Conventional materials	Melting and casting	Composition	Composition
Amorphous alloys	Rapid solidification (amorphization)	Exchange interaction	Composition
Nanocrystalline alloys	Rapid solidification (amorphization) and crystallization	Exchange interaction	Composition and proportion of phases

In the 1970s it was discovered that the magnetocrystalline anisotropy in transition metal based alloys could be dramatically reduced by amorphization [24]. The physics behind is that the random atomic scale anisotropies are averaged out by exchange interaction [23], and this very mechanism is also responsible for the suppression of K_1 in nanocrystalline alloys [2]. This has set the materials development free from the quest for alloy compositions having $K_1 \sim 0$. Still, the said effect of exchange interaction alone is not enough in reducing the saturation magnetostriction λ_s, which still is a matter of composition. However, unlike conventional crystalline materials, the magnetostrictive behaviour in soft magnetic amorphous or nanocrystalline alloys is isotropic and characterized by a single magnetostriction coefficient. This consequence of exchange interaction results in true stress insensitivity of the magnetic properties for compositions with $\lambda_s = 0$. This is hardly the case in conventional crystalline materials where the various magnetostriction coefficients such as λ_{100} and λ_{111} are barely zero at the same time when they combine to a nominally zero magnetostriction constant $\lambda_s = (2\ \lambda_{100}+3\ \lambda_{111})/5$ Still, in nanocrystalline systems exchange interaction helps us to combine the magnetic properties of structural phases which expands the versatility as compared to the more conventional property adjustment by the "atomic scale composition". This has opened the way to new compositions with $\lambda_s = 0$ based on inexpensive raw materials like iron. Hence, small effective values of K_1 and λ_s are realized in Fe-rich nanocrystalline soft magnetic alloys, having led to a new class of soft magnetic materials possessing both high μ and high J_s. In Fig. 1, we show the relationship between the saturation magnetization and the relative permeability at 1 kHz for various soft magnetic materials so far developed [1, 6, 8, 25]. These new alloys exhibit a high saturation magnetization of 1 to 1.7 T as well as superior soft magnetic properties (initial permeability as high as 2 to 15 × 10^4). These alloys are clearly the most advanced family of soft magnetic materials developed to date.

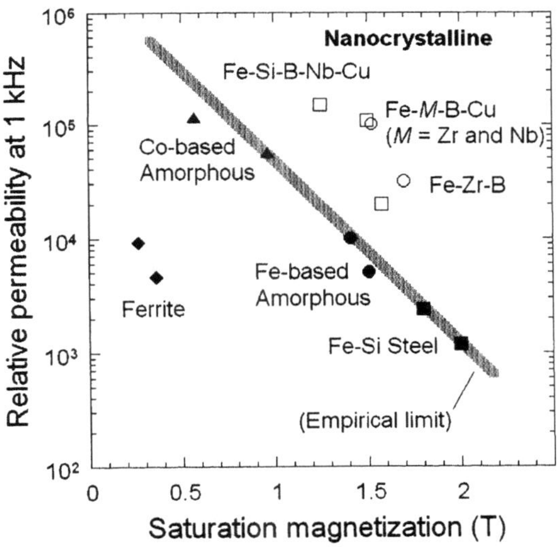

Figure 1. Saturation magnetization and initial permeability of various soft magnetic materials. The open squares and circles denote FINEMET [1, 6] and NANOPERM [8, 25], respectively.

2. MODELS FOR MAGNETIC SOFTNESS IN NANO-STRUCTURES

The primary focus of this section is to summarize the current theoretical understanding of the origin of magnetic softness in nanostructured ferromagnetic materials. The original single-phase random anisotropy model [2, 11, 12] and extended models [13, 26-28] will be overviewed, and the applicability and limits of these models will be discussed. Prior to this discussion, we will briefly go over some basic features of phenomenology in soft-magnetic materials.

2.1. Coercivity and Anisotropy

Two important concepts relevant to the magnetization process of ferromagnetic materials are magnetic domains and domain walls. The change in the magnetic polarization reflects the processes of domain wall motion and domain rotation. Thus, ease of domain wall motion is the key to realizing magnetically soft materials. Consequently, small H_c and high permeability (μ) are obtained for a material with a small potential energy barrier to domain wall movement. The potential energy barrier of the wall motion is defined by the spatial fluctuation amplitude of the anisotropy energy (ΔK) and the coercivity due to the hindrance of the wall motion is approximately given by the well-known pinning expression

$$H_c = \frac{1}{2J_s} \overline{\left| \frac{\partial \gamma}{\partial x} \right|}_{\max} \approx \frac{\Delta K}{J_s} \frac{L_{ex}}{l_K} \tag{1}$$

where γ is the domain wall energy, L_{ex} the exchange correlation length (which basically defines the domain wall width) and l_K the wavelength of the anisotropy fluctuation[11, 28]. The reduction of the fluctuation amplitude ΔK of the anisotropy energy is the key to a magnetically soft material. However, the effectiveness of ΔK as the energy barrier to the wall motion also depends on the wavelength of the anisotropy fluctuation, *i.e.* on the coherence of the magnetic anisotropy energy in the sample. Thus, for $l_k > L_{ex}$, the domain walls move the easier the longer the wavelength of the anisotropy fluctuations. The lower bound for l_k is given by L_{ex} which is the scale below which the magnetic moments are largely aligned parallel due to exchange interaction. In this case Eq. (1) simplifies to $H_c \approx \Delta K/J_s$ which essentially is the theoretical result for nucleating a reverse domain by domain rotation.

The most prominent contribution to the magnetic anisotropy energy is determined by the symmetry axis of the local atomic structure (Ch. 3). The free energy density e_K of this magnetocrystalline anisotropy can be expressed, for example, by

$$e_K(\theta) = K_0 + K_1 \sin^2\theta + K_2 \sin^4 \theta ... \quad (2)$$

for a uniaxial system. Here, θ is the angle between the magnetization direction and the magnetic easy axis of the crystal and K_m (m = 0, 1, 2 ...) are the magnetocrystalline anisotropy constants (Ch. 4). The θ independent part, *i.e.* K_0, is an arbitrary and ultimately irrelevant constant. As for the angular dependence it is often sufficient to consider the first order contribution, *i.e.* K_1, only.

The anisotropy fluctuations (ΔK) discussed above typically result from fluctuating orientations of the magnetic easy axis which varies from one grain to another in conventional polycrystalline microstructures, though Eq. (1) is not limited to this mechanism of anisotropy fluctuations. Hence, for conventional polycrystalline materials where $l_k > L_{ex}$, the parameter ΔK introduced into Eq. (1) can be well approximated by the magnetocrystalline anisotropy constant K_1. However, as we discuss in the subsequent section, the approximation of ΔK by K_1 is no longer applicable for small structural correlation lengths.

2.2. Random Anisotropies in Nanostructured Materials

2.2.1. Basic concepts

The most appreciable difference between classical and nanostructured ferromagnetic systems lies in the effect of the grain size on the effective anisotropy relevant to the magnetization process. In conventional polycrystalline materials, the magnetization will follow the individual easy magnetic directions of the structural units. The magnetization process, thus, is determined by the local magnetocrystalline anisotropy constant K_1 of the grains. For small structural correlation lengths, however, ferromagnetic exchange interaction forces the magnetic moments more and more to align parallel, thus impeding the magnetization to follow the easy axis of each individual structural unit. As a consequence the effective anisotropy for the magnetic behaviour will be an average over several structural units and, thus, be reduced in magnitude. The critical scale below which this averaging mechanism takes place is given by

$$L_0 = \varphi_0 \sqrt{A/|K_1|} \tag{3}$$

where A is the exchange stiffness constant and φ_0 is a dimensionless parameter in the order of one. For Fe-base alloys, this *natural* or *basic* exchange correlation length is typically 30 - 50 nm.

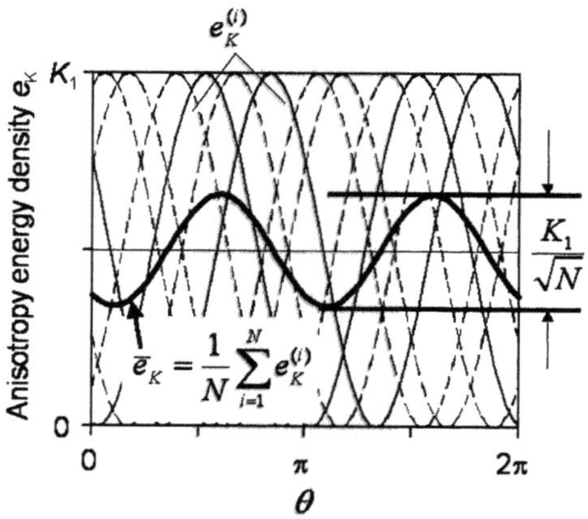

Figure 2. Random average of anisotropies.

Figure 2 schematically illustrates how the local anisotropies are averaged out for $D < L_0$. In this illustration we have modelled the local anisotropy energy density for each individual grain by

$$e_K^{(i)}(\theta) = K_1 \sin^2(\theta - \psi_i) \tag{4}$$

The angle ψ_i denotes the orientation of the local easy axis and is randomly fluctuating from grain to grain. The angle θ is the orientation of the magnetization. It is kept constant by exchange interaction over the N grains within the exchange coupled volume V_{ex}. The average anisotropy energy density over V_{ex} is, accordingly, given by

$$\bar{e}_k(\theta) = K_1 \frac{1}{V_{ex}} \int_{V_{ex}} \sin^2(\theta - \psi_i) dV = K_1 \frac{1}{N} \sum_{i=1}^{N} \sin^2(\theta - \psi_i) \tag{5a}$$

where we have assumed that each grain has the same size and the same local anisotropy constant. The sum over the random anisotropy orientations ψ_i evaluates like in a random walk process, that is, it is proportional to $N^{1/2}$, and we finally arrive at (*cf.* [28])

$$\bar{e}_k(\theta) \approx \frac{K_1}{\sqrt{N}}\sin^2(\theta - \Psi) + \frac{K_1}{2}\left(1 - 1/\sqrt{N}\right) \tag{5b}$$

In this argument one finds that the minimum anisotropy energy (*i.e.* the energy in the easiest direction) for $D < L_0$ is greater than that for $D > L_0$ since the magnetization does not follow the individual easy axes of the exchange-coupled grains. Yet, this mainly affects the "ground state energy" which is independent of the magnetization orientation θ and, hence, irrelevant to the magnetization reversal. The relevant contribution is solely given by the angular variation whose amplitude defines the average magnetocrystalline anisotropy constant, *i.e.*

$$\langle K_1 \rangle = \frac{K_1}{\sqrt{N}} \tag{6}$$

The angle Ψ in Eq. (5b) represents the resulting easiest axis of the N exchange coupled grains. It is still a random variable which fluctuates on the scale of L_{ex} from one set of N coupled grains to the other. Accordingly, the conditions $\Delta K \sim \langle K_1 \rangle$ and $l_K \sim L_{ex}$ are relevant to the relationship in Eq. (1) for $D < L_0$ and hence, the coercivity H_c is expected to be proportional to $\langle K_1 \rangle$.

The number N of exchanged coupled grains now is

$$N = \left(L_{ex}/D\right)^3 \tag{7}$$

where the exchange length L_{ex} is obtained by substituting $\langle K_1 \rangle$ for K_1 in Eq. (3), that is,

$$L_{ex} = \varphi_0 \sqrt{A/\langle K_1 \rangle} \tag{8}$$

This renormalization means that the exchange correlation length expands at the same time as the anisotropy is averaged out. By combining Eqs. (6) to (8) we finally get [2, 11]

$$\langle K_1 \rangle = K_1 \left(D/L_0\right)^6 \tag{9}$$

where L_0 is the basic exchange length of the local anisotropy K_1 as introduced in Eq (3). Accordingly the coercivity is expected to vary proportional to D^6 for $D < L_0$. By definition, the power law of Eq. (9) is limited to large values of L_0 and should not be used to judge other regions.

The prefactor φ_0 introduced in Eqs. (3) and (8) remains the open parameter within the above scaling analysis. A theoretical determination of φ_0 would ultimately require a by far more complex micromagnetic analysis of the problem. It is therefore best to estimate φ_0 from experiment from the upper critical grain size where coercivity no longer follows a D^6 law and tends to reach its maximum (*cf.* Fig. 4). This yields $\varphi_0 \approx 1.5$ for Fe-Si-B-Nb-Cu type alloys and is consistent with rough theoretical estimates [28].

2.2.2. *Extended random-anisotropy model*

The arguments discussed above were based on a single phase system. In real materials, however, we deal with various structural phases. Thus, in typical soft magnetic nanocrystalline materials, randomly oriented crystallites of about 10 nm in size are embedded in an amorphous matrix. The latter is made up again of structural units with magnetic easy axes randomly fluctuating on the much smaller scale of atomic distances. Moreover, real materials reveal additional anisotropies, such as magneto-elastic and field anisotropies (see below), which are uniform on a scale much larger than the exchange length. Such long-range anisotropies ultimately determine the soft magnetic properties of optimized nanocrystalline alloys where the contribution of the random anisotropies tends to become negligible. The original model has been extended correspondingly [13, 26-28]. As a result, the average anisotropy constant $\langle K \rangle$ of a *coupled* 3-D multi-phase system with anisotropies randomly oriented on a scale smaller than a magnetic correlation length L_{ex} can be described by [28]

$$\langle K \rangle = \sqrt{K_u^2 + \sum_\nu x_\nu \beta_\nu^2 K_{1,\nu}^2 \cdot \left(D_\nu / L_{ex}\right)^3} \tag{10}$$

where K_u denotes a uniaxial anisotropy which is *uniform* on a scale much larger than L_{ex}. The random contributions are represented by the local anisotropy constants $K_{1,\nu}$, the grain sizes D_ν and the volume fractions x_ν of the individual structural phases labelled by the index ν. The parameters β_ν mainly involve conventions used for defining the anisotropy constants for different symmetries, but also include some statistical corrections in the order of 10% to 20%. Numerical simulations for single phase systems [28] result in $\beta \approx 1$ for uniaxial and $\beta \approx 0.4$ for cubic symmetry. The rather distinct value of β for the *cubic* case is largely a consequence of common conventions for the anisotropy energy. The latter result in $\Delta e_K = |K_1|/3$ for cubic and $\Delta e_K = |K_1|$ for uniaxial anisotropies where $\Delta e_K = e_K^{max} - e_K^{min}$ is the energy difference between the hardest and easiest axis. The average anisotropy constant <*K*> is defined as the difference between the maximum and minimum of the average anisotropy energy density.

The above result is valid as long as the average number of *coupled* grains $N_\nu = x_\nu (L_{ex}/D_\nu)^3$ is larger than one for each individual phase. For the derivation it is only necessary to assume that the magnetization is parallel within a volume defined by a correlation length L_{ex} without specifying the precise coupling mechanism.

If the coupling mechanism is dominated by *exchange interaction*, the correlation length L_{ex} is self-consistently related to the total average anisotropy constant <*K*> by

$$L_{ex} = \varphi\sqrt{A/\langle K\rangle} \tag{11}$$

where φ is a prefactor in the order of unity.

In the general case, the average anisotropy <*K*> has to be determined from Eq. (10) and Eq. (11) by numerical iteration. Explicit solutions can be obtained in the limiting cases of a vanishing or dominating macroscopic anisotropy K_u. The results are

$$\langle K\rangle = \left(\sum_\nu x_\nu\sqrt{\beta_\nu\left|K_{1,\nu}\right|}\left(D_\nu/L_{0,\nu}\right)^3\right)^2 \tag{12a}$$

for $K_u = 0$, and

$$\langle K\rangle \approx K_u + \tfrac{1}{2}\sum_\nu x_\nu\sqrt{\beta_\nu\left|K_{1,\nu}\right|K_u}\,\left(D_\nu/L_{0,\nu}\right)^3 \tag{12b}$$

if the uniform anisotropy is dominating over the random contributions (*i.e.*, $K_u >> \langle K_1\rangle$). In the above relations

$$L_{0,\nu} := \varphi_{0,\nu}\sqrt{A/\left|K_{1,\nu}\right|} \text{ with } \varphi_{0,\nu} := \varphi/\sqrt{\beta_\nu} \tag{13}$$

define the *basic* exchange lengths related to the *local* anisotropies of the individual structural phases [*cf.* Eq. (3)]. These length scales should not be confused with the *renormalized* exchange length L_{ex} of Eq. (11) which is self-consistently related to the *average* anisotropy <*K*>.

Figure 3 illustrates the expected grain-size dependence of the average random anisotropy for the material parameters of optimized nanocrystalline Fe-Si-B-Nb-Cu alloys [29]. We hereby have included the contribution of the random atomic scale anisotropy of the amorphous matrix as well as the case of a small uniform anisotropy K_u.

In the absence of long-range anisotropies, the average anisotropy <*K*> is scaling with D^6 down to grain sizes of about 5 nm below which the random anisotropy of the residual amorphous phase becomes significant relative to $<K_1>$. Although the atomic scale anisotropy of the amorphous phase in this simulation is almost two orders of magnitude higher than that of the bcc crystallites, its average contribution is virtually negligible in this regime since the structural anisotropies are fluctuating on the much shorter scale of atomic distances ($D_{am.} \approx 0.5$ nm).

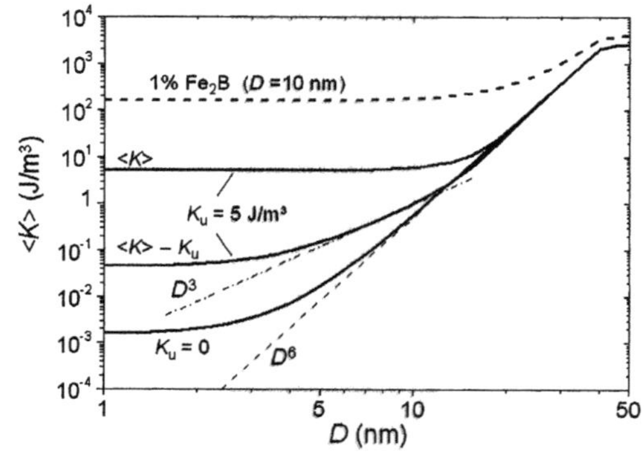

Figure 3. Theoretical estimate of the average anisotropy <*K*> for a system of randomly oriented crystallites of bcc $Fe_{80}Si_{20}$ (K_1 = 8.2 kJ/m³) with grain size *D* and embedded in an amorphous matrix with a volume fraction *x* = 0.75 [29]. The atomic-scale anisotropy constant of the amorphous phase was taken as K_1 = 430 kJ/m³, which is the value for Fe_2B and which can be looked upon as an upper bound. Further materials parameters are found in Ref. [28].

In the regime where the contribution of the bcc crystallites is dominating <*K*>, Eq. (10) simplifies drastically, so that

$$\langle K \rangle := \beta_{cr} \langle K_1 \rangle = \beta_{cr} |K_1| \cdot x_{cr}^2 \left(D/L_0 \right)^6 \tag{14}$$

This relation corresponds to exchange-coupled crystallites diluted in an ideally soft magnetic matrix. The only modification made over the original single phase model is the inclusion of the crystalline volume fraction x_{cr}. It should be noted that <*K*> is scaling with the *volume* fraction in the same way as with the crystalline *volume* D^3.

The random anisotropy of the amorphous matrix becomes only visible for very small grains, resulting in a grain-size independent anisotropy. However, the related coercivity (H_c ~ 0.001 A/m) is so small, that the situation shown for smallest grain sizes in Fig. 3 remains academic.

In real materials, whether amorphous or nanocrystalline, the minimum H_c is ultimately determined by longer-range anisotropies. To illustrate this realistic situation we have assumed the presence of a small uniform or 'coherent' anisotropy K_u. Figure 3 shows that the average anisotropy constant $<K>$ is almost totally determined by K_u for grain sizes below about 10 - 15 nm. The grain size dependence of the random contribution, $\delta K = <K> - K_u$, changes from a D^6-dependence to a D^3-dependence and, finally, becomes grain size independent due to the random anisotropy of the amorphous phase. The latter is larger than that for $K_u = 0$, because the maximum value for the renormalized exchange length is limited by $L_{ex} = \varphi (A/K_u)^{1/2}$. As a consequence the random anisotropies of the amorphous phase are less effectively averaged out.

The discussion so far assumed that the superimposed uniaxial anisotropy is perfectly uniform. This is rather the exception than the rule in reality, mostly due to internal stresses and/or surface defects. The typical fluctuation wavelengths are much larger than L_{ex} and range from a few to about 100 μm. Such K_u fluctuations ultimately provide the limiting factor for the soft magnetic properties in amorphous and optimized crystalline (conventional or nanocrystalline) alloys.

A very small volume fraction of a phase with high anisotropy can finally change the picture totally. This is illustrated in Fig. 3 assuming a 1% fraction of Fe_2B precipitates with 10 nm grain size. The example explains at least qualitatively the finding in experiment that Fe_2B precipitates can significantly degrade the soft-magnetic properties even though the grain size of the bcc crystallites remains unchanged.

2.2.3. Exchange interaction in multiphase systems

Adding up anisotropies is only one aspect of extending the random anisotropy model to multi-phase systems. Another most challenging problem is how to relate the effective exchange interaction between the grains to the local material parameters. The exchange constant A has ultimately to be understood as an effective average value on the scale of the exchange length L_{ex}. However, as clearly demonstrated by experiment, the effective A is not a simple volume average. It is rather determined by the 'weakest link' in the exchange chain which, for example, is the amorphous intergranular phase in typical nanocrystallized materials [2]. Thus, the soft magnetic properties degrade drastically when the measuring temperature exceeds the Curie temperature of the amorphous matrix and the exchange interaction between the nanocrystallites is largely interrupted.

The problem has been tackled by analysing the behaviour of the tilting angle between exchange coupled local magnetizations (or spins) where the local exchange stiffness is changing [26]. It was assumed that the tilting per unit length is inversely proportional to a local exchange length $(A_{loc}/<K>)^{1/2}$, *i.e.* the tilting of the spins becomes more rapid when the local exchange constant A_{loc} is smaller. As a result the "average" exchange stiffness A can be related to the local exchange constants of the amorphous matrix (A_{am}) and of the bcc grains (A_{cr}) as

$$\frac{D+\Lambda}{\sqrt{A}} = \frac{D}{\sqrt{A_{cr}}} + \frac{\Lambda}{\sqrt{A_{am}}} \tag{15}$$

where thickness Λ of the amorphous phase can be finally approximated by $\Lambda \approx D(x_{cr}^{-1/3} - 1)$. This behaviour is qualitatively similar to the exact solution of the micromagnetic problem of two adjacent grains separated by a matrix phase of reduced exchange stiffness (Ch. 3). Hence, the effective exchange stiffness A results from an 'inverse averaging' of the local exchange constants, similar to the way how parallel resistors are adding.

2.2.4. *Thin films*

The discussion so far assumed that the exchange interaction is expanding into three dimensions. However, care must be taken when the smallest sample dimension is comparable to the exchange length. By replacing the material parameters of Fe(Si)-based nanocrystalline soft magnetic alloys; D = 10 nm, K_1 = 10 kJ/m^3 and $A = 10^{-11}$ J/m, into Eq. (8), the L_{ex} in a typical nanocrystalline sample is estimated to be about 1 μm. This estimate is verified by the domain observation of nanocrystalline $Fe_{73.5}Si_{13.5}B_9Nb_3Cu_1$ by Schäfer *et al.* [30]. Hence, the three-dimensional expansion of L_{ex} is well justified for nanocrystalline ribbons where the smallest dimension, *i.e.* the thickness (typically 20 μm) is much larger than L_{ex}. This is no longer the case for thin films with a thickness comparable to or less than L_{ex}. The random anisotropy model can be easily adapted to this situation (*cf.* [12]). For example, Eq (10) simply modifies to

$$\langle K \rangle = \sqrt{K_u^2 + \sum_\nu x_\nu \beta_\nu^2 K_{1,\nu}^2 \cdot \left(D_\nu / L_{ex}\right)^n} \tag{16}$$

when the exchange interaction expands into n dimensions. L_{ex} again is given by Eq. (11). For two dimensions (n = 2), Eqs. (16) and (11) can be explicitly resolved for $\langle K \rangle$ which yields

$$\langle K \rangle = \frac{1}{2}\left(\delta K + \sqrt{\delta K^2 + 4K_u^2}\right) \tag{17a}$$

with

$$\delta K := \sum_{\nu} x_{\nu} \beta_{\nu} K_{1,\nu} \cdot \left(D_{\nu} / L_{0,\nu}\right)^2 \tag{17b}$$

where $L_{0,\nu}$ is given by Eq. (13). We, thus, arrive at a D^2 grain size dependence for δK and, hence, for H_c. Interestingly, the power law is independent of K_u in two-dimensions.

It should be noted that a D^2 grain size dependence for H_c is also predicted by Hoffmann's ripple theory [19], which is a more detailed micromagnetic analysis for a simplified, two dimensional model system.

3. MICROSTRUCTURE-PROPERTY RELATIONSHIPS

The survey in the previous section suggests that the most significant physical parameters in soft magnetic nanostructures would be the mean grain size D, the crystalline fraction x_{cr}, the intrinsic magnetocrystalline anisotropy K_1, the exchange stiffness A as well as more long range uniaxial anisotropies K_u. In this section, the effects of these parameters on the soft magnetic properties of actual magnetic nanostructures are discussed.

3.1. Grain Size

Figures 4(a) and 4(b) show the relationship between the average grain size and the coercivity in various Fe-based nanocrystalline soft magnetic alloys prepared by crystallization of amorphous precursors (For details, see Herzer [13], Yoshizawa [31], Müller and Mattern [32], Fujii *et al.* [33], and Suzuki *et al.* [34, 35]). As shown in Fig. 4(a), the coercivity H_c of the nanocrystalline Fe-Si-B-*M*-Cu (M = IVa to VIa metal) alloys follows the predicted D^6 dependence in a D range below L0 (~ 30 to 40 nm for this alloy system) although the plots deviate from the predicted D^6 law in the range below H_c ~ 1 A/m where the effect of grain refinement on $\langle K \rangle$ is overshadowed by magneto-elastic and annealing induced anisotropies. Hence, the experiments are better described by $H_c \approx [a^2 + (bD^6)^2]^{1/2}$, where $a \approx 1$ A/m is the coercivity originating from the induced anisotropies. The grain size dependence of H_c in Fig. 4(a) suggests that the grain refinement below 10 - 15 nm in this alloy system may not be beneficial to better magnetic softness unless care is taken over the reduction of the effect of the induced anisotropies.

In Fig. 4(b) we show the grain size dependence of H_c for the nanocrystalline Fe-Zr-B-(Cu) and Fe-P-C-Ge-Si-Cu alloys. The coercivity follows a simpler D-power law with an exponent of about 3 and no clear deviation of the plots from the power law is evident, probably due to the limited range of grain sizes. The uninterrupted D^3 dependence indicates that

H_c is governed by the grain size dependent part of $\langle K \rangle$ (*i.e.* $\langle K_1 \rangle$) and the spatial fluctuation amplitude of K_u is insignificant relative to $\langle K_1 \rangle$. The remanence to saturation ratio (J_r/J_s) of the $Fe_{91}Zr_7B_2$ samples shown in Fig. 4(b) is around 0.5 [27], substantially lower than the theoretical value for cubic Stoner-Wohlfarth particles (0.83). Thus, the dominant anisotropy in the $Fe_{91}Zr_7B_2$ samples appears to have a uniaxial symmetry rather than the cubic symmetry of $\langle K_1 \rangle$. Therefore, a large coherent K_u is evident in the $Fe_{91}Zr_7B_2$ samples. As we discussed in the derivation of Eq. (12), the effect of such a large K_u on the exchange correlation length alters the scaling properties of RAM significantly and the observed D^3 dependence can be fully attributed to the presence of the large coherent K_u ($\langle K_1 \rangle \ll K_u$).

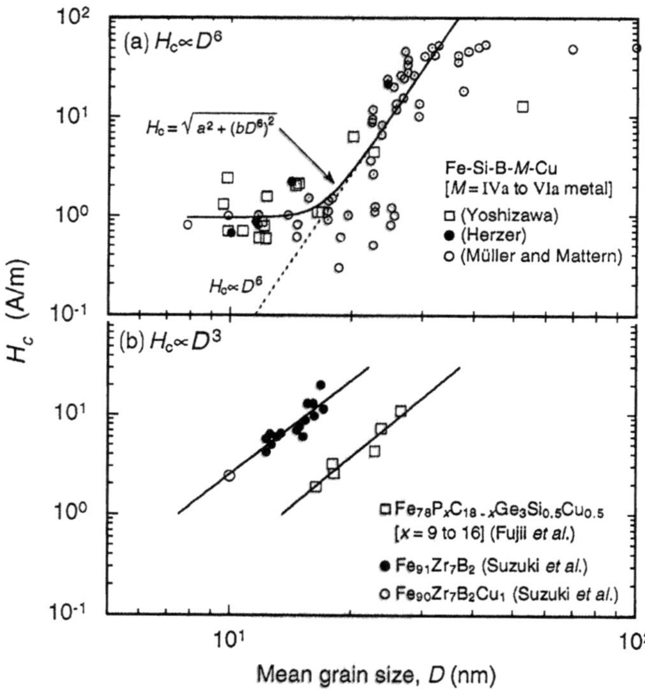

Figure 4. Grain size dependence of coercivity (H_c) for (a) Fe-Si-B-*M*-Cu (*M* = IVa to VIa metal) [13, 31, 32] and (b) Fe-P-C-Ge-Si-Cu [33] and Fe-Zr-B-(Cu) [34, 35] alloys.

The effect of a large coherent K_u on the D scaling property was tested by means of numerical simulations and the change over from D^6 to D^3 has been clearly seen [28].

It is important to note here that the experimental variation of the grain size for the nanocrystallized material is not completely unambiguous. It inevitably requires changes of the alloy composition and/or the annealing conditions which both changes the volume fraction and composition of the precipitated crystallites and the residual matrix. As a consequence the local magnetocrystalline anisotropy constant, K_1, and the exchange interaction, A, between the grains change simultaneously which causes the broad scatter of the data in Fig. 4.

3.2. Local Magneto-Crystalline Anisotropy

It is not straightforward to examine the effect of the local magnetocrystalline anisotropy K_1 on the averaged random magnetocrystalline anisotropy $\langle K_1 \rangle$ in experiment. This is because alteration of K_1 requires a modification of chemical compositions which concurrently alters other influential parameters. Such concurrent alterations overshadow the effect of K_1 on $\langle K_1 \rangle$ in experiment. This has made an explicit discussion on the scaling property of K_1 in soft magnetic nanostructures challenging.

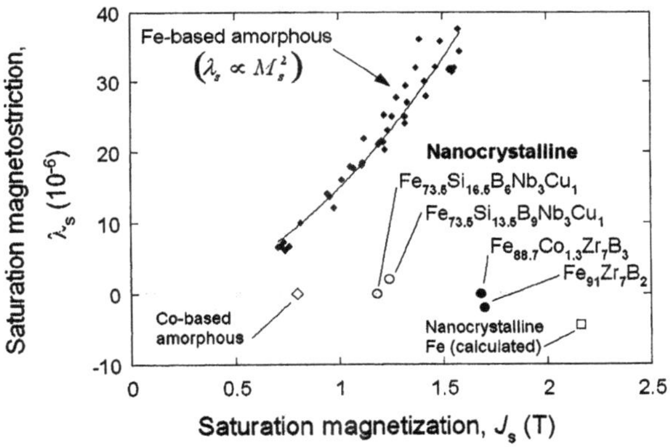

Figure 5. Relation between saturation magnetostriction and saturation magnetization for amorphous [41] and nanocrystalline [1, 39, 40] soft magnetic materials.

However, qualitative discussion is possible. For example, the coercivity at comparable grain size is lower in the high Si-content alloys than in nanocrystalline Fe-Zr-B alloys since Si considerably reduces the magnetocrystalline anisotropy of α-Fe (Fig. 4). Another example of the importance of K_1 was given by Hasegawa and Saito [36], who studied the compositional dependence of the coercivity in nanocrystalline soft-magnetic thin films having the composition (Fe, Co, Ni)$_x$Ta$_y$C$_z$ (x = 75 - 85, y = 6 - 9, z = 10 - 16). The alloys exhibit a α-Fe/TaC two-phase microstructure and have been prepared in a thin-film form only. The authors showed that nanocrystalline Co-base alloys typically reveal a 5 to 10 times higher coercivity than Fe-base alloys owing to the high local K_1 value of Co and, hence, the Co-base alloys are less suited for soft magnetic applications. They also showed that the lowest H_c in the nanocrystalline (Fe, Co, Ni)$_x$Ta$_y$C$_z$ alloys appears to lie along the compositions where $K_1 = 0$ is confirmed in the bulk Fe-Co-Ni ternary system. This clearly demonstrates that to try reducing the intrinsic K_1 of the bcc-Fe nanocrystallites is an effective approach for better soft magnetic properties in magnetic nanostructures when $\langle K_1 \rangle$ remains the dominant magnetic anisotropy.

The effect of alloying elements on K_1 for Fe-based crystalline thin films has been reviewed by Hayashi *et al.* [37]. It was summarized that $K_1 = 0$ is expected in Fe-Ga-Si, Fe-Al-Ge, Fe-Co-Ga-Si and Fe-Co-Al-Ge, along with classical Fe-Ni and Fe-Al-Si. Consequently, the addition of Al, Si, Ga and Ge to Fe-based magnetic nanostructures may result in a further reduction of $\langle K_1 \rangle$. However, a substantial amount of these additives is accompanied by a serious decrease in the saturation magnetization [38], making the Fe-based soft magnetic nanostructures featureless from the viewpoint of technological applications.

3.3. Magneto-Elastic Anisotropy

Magneto-elastic anisotropies are the key to understanding the advantage of the soft magnetic nanostructures over Fe-based amorphous soft magnetic alloys. Figure 5 shows the relationship between the saturation magnetization (J_s) and saturation magnetostriction (λ_s) for various nanocrystalline [1, 39, 40] and amorphous [41] materials. The saturation magnetostriction of Fe-based amorphous alloys varies as the square of the spontaneous magnetization [42] and hence, the combination of small λ_s and high J_s cannot be achieved in Fe-rich amorphous materials. The magneto-elastic K_u is given by $-3/2\lambda_s\sigma$, where σ is the residual stress in the sample (typically ~ MPa) and thus, the Fe-based amorphous alloys with $\lambda_s \sim 10^{-5}$ could have K_u of a few 10 J/m^3. By contrast, zero-magnetostrictive alloys with very high saturation magnetizations of 1 to 1.7 T are obtained in nanocrystalline soft magnetic materials.

The behaviour of λ_s can be understood from the balance of magnetostriction among the structural phases present in the nanocrystalline state, *i.e.* [12]

$$\lambda_s^{cr} = x_{cr}\lambda_s^{cr} + (1 - x_{cr})\lambda_s^{am} \tag{18}$$

where λ_s^{cr} and λ_s^{am} represent the saturation magnetostriction constant of the crystalline and the residual amorphous phases. Thus, near zero magnetostriction in nanocrystalline Fe-base alloys requires a large crystalline volume fraction with negative magnetostriction in order to compensate the high positive value of the amorphous Fe-based matrix. This is achieved either by a high Si-content in the bcc grains ($\lambda_s^{FeSi} \approx -6 \cdot 10^{-6}$ for α-$Fe_{80}Si_{20}$), like in the Fe-Cu-Nb-Si-B system, or if the grains consist of pure α-Fe ($\lambda_s^{Fe} \approx -4 \cdot 10^{-6}$) like in Fe-Zr-B alloys.

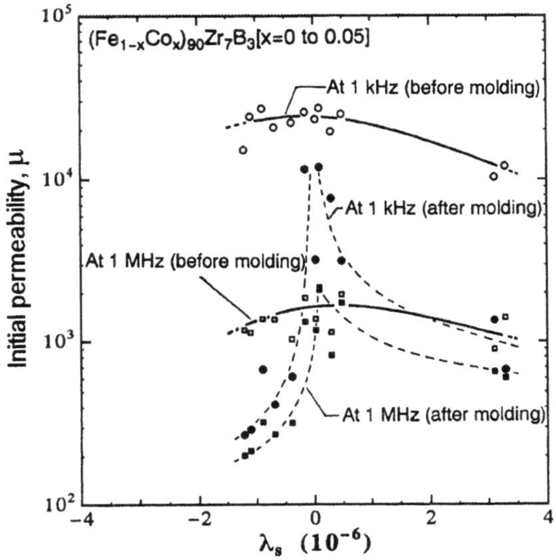

Figure 6. Relation between saturation magnetostriction (λ_s) and μ_e (at 1 kHz and 1 MHz) before and after epoxy resin moulding for nanocrystalline $(Fe_{1-x}Co_x)_{90}Zr_7B_3$ [40].

An important point is that the superposition of the local magnetostriction constants to zero really results in stress-insensitivity of the magnetic properties like in amorphous Co(Fe)-base alloys. This is again a consequence of the smoothing effect of exchange interaction for structural correlation lengths much smaller than the domain wall width. Thus, the nano-scale fluctuations in magneto-elastic anisotropy associated with the locally varying magnetostrictions are randomly averaged out which results in a single isotropic magnetostriction coefficient. The situation contrasts with that for large-grained crystalline systems, where magnetostriction is anisotropic and has to be characterized by at least two magnetostriction coefficients (*e.g.* λ_{100} or λ_{111} in the cubic case). Hence, for conventional systems, an average zero saturation magnetostriction $\lambda_s = (2\ \lambda_{100}+3\ \lambda_{111})/5$ does usually not imply stress-insensitivity of the hysteresis loop.

Figure 6 gives a practical example for the relevance of $\lambda_s \approx 0$ in application. The saturation magnetostriction of the $(Fe_{1-x}Co_x)_{90}Zr_7B_3$ alloy series shown in this Figure has been fine-tuned around $\lambda_s \approx 0$ by adding a small amount of Co. The figure shows the initial permeability μ_e at 1 kHz and 1 MHz for magnetic cores before and after epoxy-resin moulding. The internal mechanical stress caused by the moulding process is about 140 MPa for this example. It may cause large magneto-elastic anisotropy, leading to deterioration of magnetic softness. This effect is minimized for Co contents around $x = 0.01$ to 0.02 [40] where $\lambda_s \approx 0$ and a high μ_e value above 10^4 is obtained even after epoxy resin moulding.

3.4. Annealing-Induced Anisotropies

So far, magnetic anisotropies have been discussed as a rather disturbative factor for soft magnetic properties. However, if properly controlled, they can be a powerful tool in order to tailor the shape of the hysteresis loop according to the demands of application [10].

Of particular importance are "magnetization-induced" uniaxial anisotropies which form along the local direction of the spontaneous magnetization during annealing due to directional atomic ordering. In general, this yields a distribution of uniaxial anisotropies reflecting the local magnetization directions (domain structure) during annealing. Such anisotropy fluctuations are rather unfavourable for the soft magnetic properties. It is therefore advantageous to apply a saturating magnetic field during annealing to induce a homogeneous anisotropy. In fact, the coercivity of field-annealed samples is typically lower than that of those annealed conventionally [10].

Another approach to reducing the anisotropy fluctuations is to minimize the directional atomic ordering during annealing. This can be realized by constantly altering the direction of the applied field during annealing, i.e. rotating field annealing (RFA). In Fig. 7 we show the effect of RFA on the hysteresis loop of nanocrystalline $Fe_{84}Nb_7B_9$ [43]. The squareness of the *B-H* loop is enhanced after RFA, implying a reduction of K_u. In addition, the coercivity of the Fe-Nb-B alloy is reduced by as much as 65% by RFA. Since K_u has a significant effect on the *D*-power scaling exponent, as predicted in RAM, a part of this large reduction in H_c by RFA may well be attributed to the change in the *D*-power scaling exponent induced by the suppression of K_u in the sample.

A systematic study by Herzer [44] on the effect of Si content on the field induced K_u in the nanocrystalline Fe-Si-B-Nb-Cu alloys indicated that K_u is mainly induced in the crystallites. By appropriate choice of alloy composition and annealing conditions, transverse field annealing of nanocrystalline Fe-Si-B-Nb-Cu alloys allows to induce anisotropies in the range of $K_u \approx 5 - 100$ J/m³. This corresponds to initial permeabilities of about $\mu_i \approx 10^4 - 2 \times 10^5$. They perfectly cover the needs of applications like common mode chokes or earth leakage circuit breakers [45] which require high permeabilities.

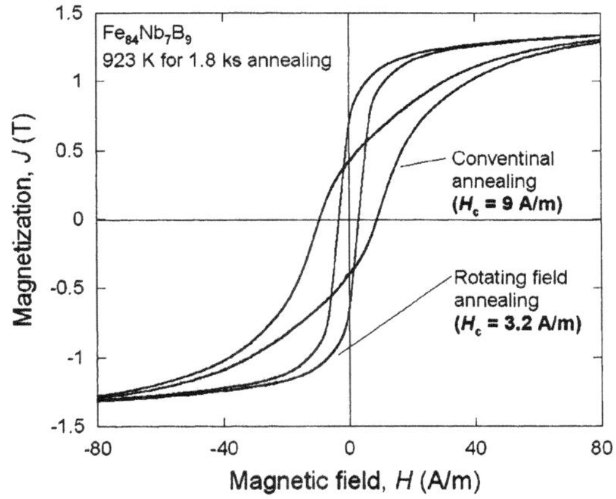

Figure 7. DC hysteresis loops of nanocrystalline $Fe_{84}Nb_7B_9$ annealed under zero field and rotating field (640 kA/m at 1500 rpm) [43].

However, there are also other applications like energy-storage chokes which rather require low permeabilities in the order of several hundreds to several thousands only. This requires relatively high induced anisotropies, which can be achieved by the addition of Co. Yoshizawa *et al.* [46] showed that the annealing-induced uniaxial anisotropy (K_u) in nanocrystalline $Fe_{78.8-x}Co_xSi_9B_9Nb_{2.6}Cu_{0.6}$ ($x = 0$ to 78.8) tends to increase with Co content and that K_u is enhanced from ~ 100 J/m^3 at $x = 0$ to ~ 600 J/m^3 at $x = 50$, reflecting the extent of the pair ordering of Fe and Co atoms. Yet the draw-back is that such large additions of Co increase saturation magnetostriction. Müller *et al.* [47], for example, investigated the magnetic properties of nanocrystalline $Fe_{73.5-x}Co_xSi_{13.5}B_9Nb_3Cu_1$ ($x = 0$ to 73.5) and $Fe_{86-x}Co_xZr_7B_6Cu_1$ ($x = 0$ to 86) and showed that both the magnetostriction and coercivity of these two alloy systems increase more than an order of magnitude when substantial amounts of Co are added. They also showed that the addition of Co to Fe-based nanocrystalline alloys results in severe deterioration of magnetic softness when being annealed without magnetic field. The reason is both the increased magnetostriction as well as the high locally induced anisotropy K_u.

Uniaxial anisotropies can also be induced by applying tensile stresses to ribbon samples during annealing. This creep-induced anisotropy can be as large as a few kJ/m^3 for nanocrystalline $Fe_{73.5}Si_{15.5}B_7Nb_3Cu_1$ [48, 49]. The mechanism of this effect was clarified by means of transmission x-ray diffractometry by Ohnuma *et al.* [50]. They showed that the lattice spacing of Fe(Si) nanocrystallites is elongated along the stress direction after the stress annealing, confirming local magneto-elastic interactions as the direct cause of the large K_u [49]. The advantage of the creep-induced anisotropies is that one can combine low magnetostriction and high anisotropy, as contrasted to the case of Co-additions. However, the practical application of stress annealing is rather limited due to the severe embrittlement after nanocrystallization.

3.5. Residual Amorphous Phase

As we will discuss in this section, the presence of a residual amorphous phase is essential to maintain the metastable phase equilibrium of the nanostructure in both the Fe-Si-B-Nb-Cu and Fe-*M*-B-(Cu) alloys. Hence, there is a layer of the residual amorphous phase in the intergranular region in these alloys. In the temperature range below the Curie temperature of the residual amorphous phase (T_C^{am}), the nanocrystallites are coupled magnetically via the exchange interaction acting over the crystallite-amorphous-crystallite coupling chain. However, this coupling chain is diminished in the temperature range above T_C^{am} and the nanocrystalline alloys behave as an assembly of isolated magnetic particles in which the magnetically hardest domain configuration is expected. Consequently, a dramatic increase in H_c takes place in the temperature range near T_C^{am} [2]. This effect is a possible disadvantage of the nanocrystalline soft magnetic alloys from the application viewpoint.

On the other hand, examining the grain decoupling phenomenon is a sophisticated way of clarifying the relationship between the exchange stiffness of the intergranular phase and the random magnetocrystalline anisotropy. There are at least five different parameters other than the exchange stiffness constant of the intergranular phase that may influence $\langle K \rangle$. However, among these parameters, φ_0, x_{am} and D are all temperature independent. In addition, the temperature dependence of both K_1 and A_{cr} near T_C^{am} is negligibly small if the difference between T_C^{am} and the Curie temperature of the crystalline phase (T_C^{cr}) is large enough. Consequently, the effect of A_{am} can be isolated from other influential parameters when the temperature dependence of H_c near T_C^{am} is examined. Furthermore, a limiting condition $A_{am} \ll A_{cr}$ is valid at $T \sim T_C^{am}$ and the exchange correlation length is governed chiefly by A_{am}, *i.e.* the weakest link in the coupling chain. Hence, $\langle K \rangle$ becomes approximately proportional to A_{am}^{-3} near T_C^{am} when each nanocrystallite is well isolated by the residual amorphous phase. Given the fact that the exchange stiffness constant of ferromagnetic materials varies approximately as the square of the spontaneous magnetization, the temperature dependence of $\langle K \rangle$ in two-phase soft magnetic nanostructures at $T \sim T_C^{am}$ is described by [2]

$$\langle K(T)\rangle \propto A_{\mathrm{am}}(T)^{-3} \propto M_{\mathrm{s}}^{\mathrm{am}}(T)^{-6} \propto \left(1-\frac{T}{T_{\mathrm{C}}^{\mathrm{am}}}\right)^{-6\beta} \tag{19}$$

where β is the critical exponent, which must not be confused with the parameter β introduced previously in Eq. (11). This scaling property of RAM was tested for nanocrystalline $Fe_{91}Zr_7B_2$ [26].

Figure 8(a) shows the temperature dependence of the coercivity for nanocrystalline $Fe_{91}Zr_7B_2$ annealed at 823 K for 60 s ($x_{\mathrm{am}} \approx 0.5$, $D \approx 13$ nm and $J_r/J_s \approx 0.8$ at room temperature). The temperature dependence of the mean ^{57}Fe hyperfine field of the residual amorphous phase ($<B_{\mathrm{hf}}>$) as well as a fit with the Brillouin function for $S = ½$ is also shown in this figure. The coercivity begins to increase drastically at around T_C^{am} due to the grain decoupling. In Fig. 8(b) we show the plots of the coercivity to the power -6β with a typical β value of 0.36. A critical behaviour is evident on the temperature dependence of $H_c^{-6\beta}$ and a critical point of 369(10) K is estimated from the variation of $H_c^{-6\beta}$. This critical point is fully consistent with T_C^{am} in Fig. 8(a), indicating that the effect of the exchange stiffness of the intergranular phase on $<K_1>$ is very well described by Eq. (15).

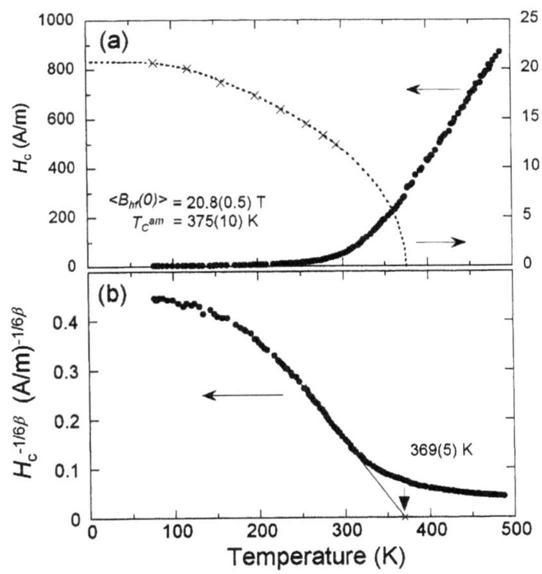

Figure 8. Temperature dependence of (a) coercivity (H_c) and (b) plots of the coercivity to the power -1/6β (β = 0.36) for nanocrystalline $Fe_{91}Zr_7B_2$ annealed at 823 K for 60 s [26].

The critical point in Fig. 8(b) corresponds to the divergence of H_c due to the disappearance of the intergranular exchange coupling at T_C^{am}, however, the experimental H_c in Fig. 8(a) keeps increasing even above this critical temperature. This is a clear indication of additional intergranular magnetic interactions other than the crystallite-amorphous-crystallite exchange coupling. The extent of the extra interactions tends to be greater when the amorphous volume fraction is small. It is still premature at this stage to conclude the origin of the additional interactions, although models based on Fe-exchange field penetration [51], dipolar interaction [13] and direct contact of the grains are considered.

Figure 9 shows some simulations of $\langle K_1 \rangle$, in accordance with Eqs. (14) and (15), for a typical two-phase nanocrystalline system (D = 10 nm, K_1 = 10^4 J/m^3, $A_{cr} = 10^{-11}$ J/m, $\varphi_0 = 1$ and $\beta_{cr} = 1$) for various values of A_{am} and x_{am}. These simulation results indicate that $\langle K_1 \rangle$ decreases with x_{am} when $A_{am} \approx A_{cr}$ whereas it increases with x_{am} when $A_{am} \ll A_{cr}$. Hence, the optimum volume fraction of the residual amorphous phase to realize good magnetic softness should vary depending on alloy systems. Naturally, these simulations are only for the random magnetocrystalline anisotropy ($\langle K_1 \rangle$) while the magnetic softness in practical materials may also be governed by induced anisotropies. An interesting aspect of the simulations in Fig. 9 is that the $\langle K_1 \rangle$ shows a peak when A_{am} is considerably smaller than A_{cr}. Hence, H_c of nanocrystalline alloys with small A_{am} may increase at the initial stage of nanocrystallization. Such a magnetic hardening effect is often seen for Fe-*M*-B-(Cu) alloys (*e.g.* $Fe_{86}Zr_7B_6Cu_1$) [52] as the Curie temperature of the residual amorphous phase in these alloys is close to room temperature.

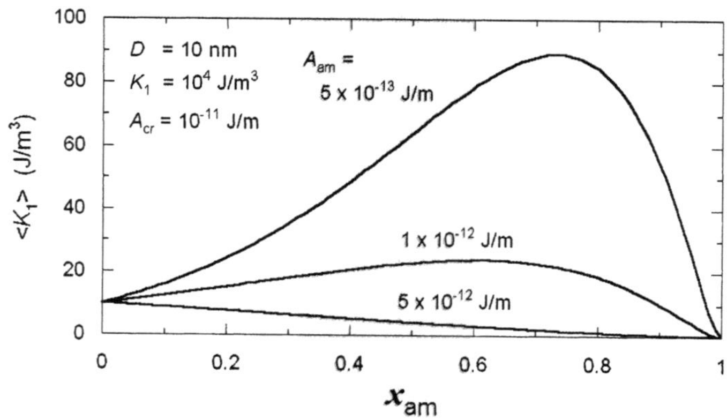

Figure 9. Change in calculated $\langle K_1 \rangle$ at various A_{am} as a function of the volume fraction of the residual amorphous phase (x_{am}).

The soft magnetic properties of nanocrystalline soft-magnetic materials deteriorate drastically above the grain-decoupling temperature and thus, these materials are usually unsuited for operation at elevated temperatures. However, the soft magnetic properties at elevated temperatures can be significantly improved by enhancing the Curie temperature of the residual amorphous phase. A prominent example was given by Willard *et at.* [53] who investigated the magnetic properties of nanocrystalline $Fe_{44}Co_{44}Zr_7B_4Cu_1$ (this α'-FeCo based nanocrystalline alloy is also known to as HITPERM). The Curie temperatures of both the crystalline and the residual amorphous phases in nanocrystalline $Fe_{44}Co_{44}Zr_7B_4Cu_1$ are higher than those of typical Co-free nanocrystalline soft magnetic alloys, making HITPERM a potential candidate of magnetic cores operated at elevated temperatures.

Although the information on the static hysteresis loop or the coercivity of HITPERM at room temperature is limited, Johnson *et al.* [54] showed that the initial permeability (at 1 KHz) of nanocrystalline $Fe_{44}Co_{44}Zr_7B_4Cu_1$ is 650, two orders of magnitude smaller than that (4.8×10^4 at 1 KHz) of nanocrystalline $Fe_{86}Zr_7B_6Cu_1$ (NANOPERM) [52]. Hence, the areas of potential applications for HITPERM are expected to be quite different from those for FINEMET or NANOPERM. McHenry *et al.* [15] suggested the rotor material used in the aircraft power unit as a potential application of HITPERM. Škorvánek *et al.* [55] reported that the coercivity of nanocrystalline $Fe_{44.5}Co_{44.5}Zr_7B_4$ remains below a moderate value of 25 to 35 A/m over a wide temperature range from room temperature to 673 K. Since the magnetization induced K_u is one of the two major causes of the poor soft magnetic properties in the Fe-Co-based nanocrystalline alloys, further improvement of the magnetic softness in the Fe-Co-based nanocrystalline alloys is highly likely when the annealing induced K_u is minimized by rotating field annealing.

4. NANOSTRUCTURAL FORMATION MECHANISMS AND ALLOY DEVELOPMENT

The random anisotropy model predicts that the effective magnetocrystalline anisotropy can be reduced to $K_1(D/L_0)^6$ when the grain size (D) is smaller than the natural exchange length ($L_0 \approx 30$ nm for α-Fe). Hence, the $<K_1>$ value of α-Fe ($K_1 = 47$ kJ/m^3) is expected to be suppressed to about 10 J/m^3 when $D \approx 10$ nm. However, K_1 of Fe-borides or Fe-M (M = early transition metals) intermetallic compounds may remain unchanged even if the grain size is refined to ~ 10 nm because L_0 of these ferromagnetic compounds is usually shorter than that of α-Fe (*e.g.*, L_0 of Fe_2B is 6.5 nm [28]). Thus, to prevent the formation of these ferromagnetic compounds is a crucial condition for realizing magnetically soft nanostructures in Fe-based nanocrystalline soft-magnetic alloys. The focus of this section is, therefore, to discuss how such a compound-free nanostructure has been prepared from amorphous precursors.

4.1. Primary Crystallization

The magnetically soft nanostructure in FINEMET or NANOPERM is a direct result of primary crystallization of amorphous ribbons. Here, we discuss this type of crystallization process with emphasis on the nanostructural formation upon crystallization.

It is instructive to consider the free-energy hierarchy and the metastable phase equilibria when crystallization of an amorphous material is discussed. Köster and Herold [56] discussed these aspects of crystallization and showed that crystallization reactions of amorphous alloys can be classified into the following three types: polymorphic, primary and eutectic crystallization reactions. Among these three types, the slowest crystal growth process is expected for primary crystallization and thus, primary crystallization is ideal for tailoring fine microstructures upon decomposition of amorphous alloys.

Another significant aspect of primary crystallization is that the crystallization products can be free of compounds. In Fig. 10(a) we show a schematic free-energy vs. composition diagram for a hypothetical Fe-based glass forming system. A eutectic alloy system is assumed here, as the formation of an amorphous phase by rapid solidification occurs in the vicinity of a deep eutectic point. The common tangent drawn over the free energy curves of bcc-Fe and amorphous phases represents a metastable equilibrium of these two phases while the tie line between the bcc-Fe and compound curves shows the stable equilibrium. Given these phase equilibria, there are at least two possible crystallization routes in this hypothetical system; one is eutectic crystallization from amorphous to a mixed bcc-Fe and compound phases ($1 \rightarrow 3$) and the other is a two-step reaction ($1 \rightarrow 2$ and then $2 \rightarrow 3$) initiated by primary crystallization of bcc-Fe. The second stage reaction ($2 \rightarrow 3$) is triggered by crystallization of the residual amorphous phase, and if the thermal stability of the residual amorphous phase is sufficiently high, the primary crystallization reaction is fully complete and a bcc-Fe/amorphous two-phase microstructure free of compounds is realized.

The sluggish crystal growth in primary crystallization is explained by taking into account the solute partitioning behaviour. Figure 10(b) shows a schematic view of the compositional profile across a growing bcc-Fe crystallite at an early stage of primary crystallization where the crystallization has little effect on the average solute concentration of the amorphous matrix. When a steady state is established, the partial equilibrium between bcc-Fe (solute content C_{bcc}) and the residual amorphous phase (solute content C_{am}) is maintained at the bcc/amorphous interface. Further growth of the bcc-Fe crystallite requires the diffusion of solute atoms from the interface to the amorphous matrix away from the reaction front. Thus, the growth process is governed by long-range atomic diffusion. The growth kinetics of the nanocrystallites under such a diffusion-limited growth process can be discussed qualitatively although the exact analytical solution for the entire growth kinetics may not be given [57].

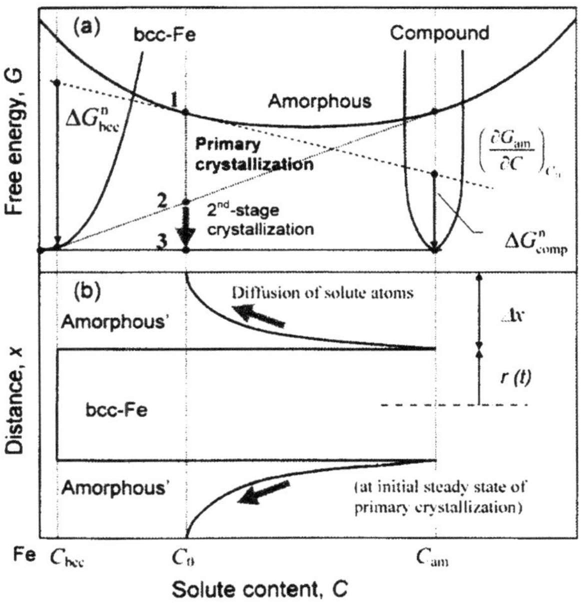

Figure 10. Nanocrystallization: (a) hypothetical free energy composition diagram and (b) composition profile of a primary precipitate for an Fe based glass forming system.

Provided that the diffusion field Δx is approximated by r in Fig. 10(b) and 1-dimensional growth is assumed for simplicity of the argument, the following relationship is expected:

$$r(t) \propto \sqrt{2tD_v\left(\frac{C_{\mathrm{am}} - C_t}{C_{\mathrm{am}} - C_{\mathrm{bcc}}}\right)} \tag{20}$$

Here D_v is the volume diffusion coefficient of the solute atoms and C_t is the solute content of the residual amorphous phase ($C_t \approx C_0$ at an early stage of primary crystallization). Here, dr/dt is proportional to $t^{-1/2}$ and the growth rate decreases progressively with time. An *in-situ* time-resolved x-ray diffractometry study on nanocrystalline $Fe_{73.4}Si_{13.4}B_{9.1}Nb_{3.1}Cu_1$ by Köster *et al.* [58] clearly showed such a parabolic growth of nanocrystallites. Moreover, the crystal growth is severely retarded when the population density of primary crystallites is high because the diffusion fields of neighbouring crystallites tend to overlap each other and C_{am} - C_t decreases drastically. This effect known as soft impingement [59] as well as the parabolic time dependence of the growth rate explains why primary crystallization is preferable for nanostructural formation upon crystallization of amorphous precursors.

4.2. Development of Nanocrystalline Fe-Zr-B-(Cu) Alloys

The earliest investigation on the soft-magnetic properties of nanocrystalline Fe-*M*-B alloys (NANOPERM) is found in a report by Suzuki *et al.* [3]. In Fig. 11(a) we show the compositional dependence of magnetic properties (J_s and μ_e at 1 kHz) for rapidly solidified Fe-Zr-B alloys annealed at 873 K. A high permeability value, above 10^4, is limited to the compositions where the precipitates after primary crystallization are bcc-Fe with a grain size of 10 to 20 nm. This magnetically soft nanostructure is prepared through the first stage (primary) crystallization of the following two-step reaction [60]:

> Amorphous → nanoscale bcc-Fe + residual amorphous → α-Fe + compounds,

where the compounds include Fe_3Zr (or $Fe_{23}Zr_6$), Fe_3B and Fe_2B. The type of compounds forming at the 2nd stage crystallization depends upon the chemical composition. This sequence of phase decomposition is readily understood by considering the free energy hierarchy (1 → 2 → 3) in Fig. 10(a). An atom-probe-field-ion microscopy study by Zhan *et al.* [61] confirmed that the Fe content of the residual amorphous phase in nanocrystalline $Fe_{90}Zr_7B_3$ is about 78 at%, implying that the compound (Fe_3Zr) is formed by polymorphic crystallization of the residual amorphous phase. The high μ_e region in Fig. 11(a) is limited to a narrow Fe-rich composition range as the driving force for the nucleation of the magnetically harder compounds ($\Delta G_{comp}{}^n$) is negatively enhanced in the solute-rich region. On the other hand, the preparation of the amorphous precursor becomes arduous when Fe content exceeds 92 at% and consequently, a high initial permeability above 10^4 in the Fe-Zr-B system is limited in a narrow composition range of 5 to 7 at% Zr and 2 to 6 at% B.

It is worth mentioning here that a compound-free bcc-Fe nanostructure with grain sizes smaller than L_0 can be obtained even in a binary $Fe_{92}Zr_8$ alloy [3, 60] although the initial permeability of nanocrystalline $Fe_{92}Zr_8$ (2×10^3 at 1 kHz) is similar to that of magnetically soft Fe-Si steels. Weak intergranular magnetic coupling was suggested for the primary reason for the poor magnetic softness of nanocrystalline $Fe_{92}Zr_8$ as the maximum Curie temperature in the amorphous Fe-Zr binary alloys is usually limited to below room temperature.

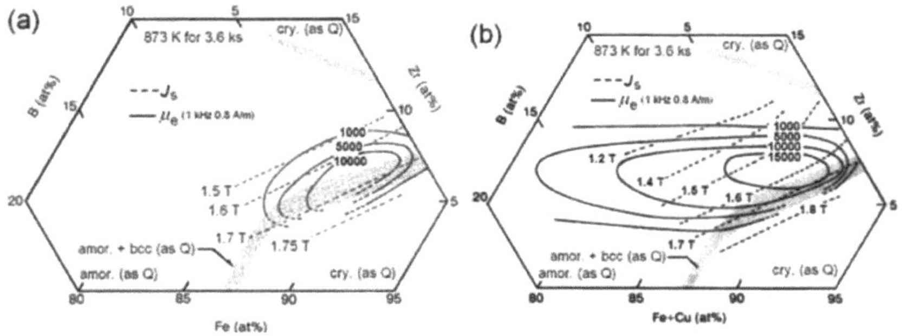

Figure 11. Compositional dependences of initial permeability (μ_e at 1 kHz) and saturation magnetization (J_s) for nanocrystalline (a) Fe-Zr-B [3] and (b) Fe-Zr-B-Cu [34] alloys.

As we discussed in Section 3.5, the intergranular magnetic coupling is often strengthened by choosing higher annealing temperatures where the volume fraction of the amorphous phase is minimized. However, this option is not available in the $Fe_{92}Zr_8$ alloy because the thermal stability of the residual amorphous phase is so low that the Fe_3Zr phase starts to precipitate before the primary crystallization process is complete. This problem is resolved by boron addition. The onset temperature of the second stage crystallization in amorphous $Fe_{91}Zr_7B_2$ is higher by 140 K than that of amorphous $Fe_{92}Zr_8$ and the thermal stability of the residual amorphous phase is enhanced quite efficiently by a small amount of boron. This is again due to the preferential partitioning of solute elements to the residual amorphous phase upon primary crystallization. A later investigation by Mössbauer spectroscopy showed that the Curie temperature of the residual amorphous phase (T_C^{am}) in nanocrystalline $Fe_{92}Zr_8$ is 330(10) K [62], which is possibly enhanced by the effect of neighbouring ferromagnetic nanocrystallites [51] yet still lower than the T_C^{am} value (about 370 K) of nanocrystalline $Fe_{91}Zr_7B_2$ (Fig. 8). The addition of boron is therefore effective in enhancing the exchange stiffness of the residual amorphous phase in the Fe-Zr based nanocrystalline alloys. Hence, boron has become an essential element in NANOPERM.

Since the deterioration of the soft magnetic properties on the solute rich side in Fig. 11(a) is principally due to the formation of magnetically hard compounds, the high μ_e region can be extended toward the solute rich side if the formation of compounds is prevented upon primary crystallization. This was demonstrated by substituting only 1 at% of Fe with Cu [34]. Figure 11(b) shows the compositional dependence of magnetic properties for the Cu-added Fe-Zr-B alloys. The formation rage of the compound-free nanostructure is extended towards the solute rich side and the high μ_e region appears to be considerably wider in comparison with that of the Cu-free Fe-Zr-B alloys. The effect of Cu addition on preventing the formation of compound phases upon crystallization of Fe-based amorphous alloys had already been known prior to the development of NANOPERM. Yoshizawa and Yamauchi [63] showed that the crystallization products of amorphous $Fe_{74.5}Si_{13.5}B_9Nb_3$ (Cu-free FINEMET) changes from iron borides to bcc-Fe(Si) by substituting 1 at% of Fe with Cu. The microalloying of Cu is vital to this original FINEM*Et al.*loy as the formation of the compound-free nanostructure is difficult without the aid in the nucleation of bcc-Fe(Si). Contrarily, the Fe-rich Fe-Zr-B alloys have a natural tendency to form the preferable bcc-Fe phase presumably because of their large driving force for the bcc-Fe nucleation (*i.e.*, $|\Delta G_{bcc}{}^{n}| >> |\Delta G_{comp}{}^{n}|$) due to their low solute contents. Hence, Cu addition in NANOPERM is only needed when the solute content is high and $\Delta G_{comp}{}^{n}$ becomes negatively greater.

Hono *et al.* [64, 65] confirmed that Cu in both amorphous $Fe_{74.5}Si_{13.5}B_9Nb_3Cu_1$ and $Fe_{89}Zr_7B_3Cu_1$ alloys forms clusters prior to primary crystallization and the Cu clusters act as heterogeneous nucleation sites for bcc-Fe(Si) precipitates. They also confirmed orientation relationships between the Cu clusters and bcc-Fe crystals and concluded that the heterogeneous nucleation of the bcc-Fe phase is due to the lower interfacial free energy for nucleation [66].

The role of Cu in FINEMET has been discussed almost solely from the viewpoint of microstructural refinement. However, it is equally important to discuss the effect of Cu in suppressing the formation of magnetically hard compounds upon crystallization of the amorphous precursors. Prevention of the magnetically harder compounds from forming upon crystallization is surely the primary role of Cu in NANOPERM.

Still the addition of Cu becomes ineffectual for alloys containing a substantial amount of Co. For example, Ping *et al.* [67] who investigated the nanostructural formation process of nanocrystalline $Fe_{44}Co_{44}Zr_7B_4Cu_1$ (HITPERM) using APFIM found that the well-known Cu cluster formation does not take place in this alloy. These authors concluded that Cu in HITPERM is ineffective for microstructural refinement. In fact, Škorvánek *et al.* [55] reported that the coercivity of nanocrystalline $Fe_{44.5}Co_{44.5}Zr_7B_4$ is lower than that of nanocrystalline $Fe_{44}Co_{44}Zr_7B_4Cu_1$. Similar effect of Co on the Cu clustering behaviour was also confirmed for FINEMET by Ohnuma *et al.* [68]. It was found that the Cu cluster formation prior to primary crystallization in amorphous $Fe_{78.8-x}Co_xSi_9B_9Nb_{2.6}Cu_{0.6}$ (x = 5 to 60) alloys is spoiled when 20 at% or more of Co is added and the grain size after primary crystallization showed a tendency to increase with increasing Co content. These authors estimated the driving force for the formation of Cu clusters in Fe-Cu and (Fe, Co)-Cu model systems using the CALPHAD method and showed that the driving force could be reduced by Co addition. Although the thermodynamics of actual supercooled melts in the $Fe_{78.8-x}Co_xSi_9B_9Nb_{2.6}Cu_{0.6}$ alloys may be far more complicated, this estimate in the model systems explains the effect of Co on the clustering behaviour of Cu well.

4.3. Substitution of Zr with Other IVa to VIa Metals

A systematic investigation on the decomposition behaviours of amorphous $Fe_{93-x}M_7B_x$ (M = IVa to VIa metal) ternary alloys has been carried out by Suzuki *et al.* [69]. The M content of 7 at% in their study reflects the compositional dependence of soft magnetic properties in the nanocrystalline Fe-Zr-B-(Cu) alloys where the highest permeability is obtained at Zr = 7 at%. Since the formation of compound-free nanostructures is most likely to take place at Fe-rich compositions, the lowest boron content required for the formation of an amorphous phase by rapid solidification was chosen for each M element. The alloys studied were amorphous $Fe_{83}Ti_7B_{10}$, $Fe_{91}Zr_7B_2$, $Fe_{91}Hf_7B_2$, $Fe_{79}V_7B_{14}$, $Fe_{85}Nb_7B_8$, $Fe_{83}Ta_7B_{10}$, $Fe_{79}Cr_7B_{14}$, $Fe_{81}Mo_7B_{12}$ and $Fe_{81}W_7B_{12}$. The amorphous samples after primary crystallization were identified by means of x-ray diffractometry and transmission electron microscopy and the following reactions have been confirmed:

for M = Zr, Hf, V, Nb, Cr and Mo: amorphous → bcc-Fe + residual amorphous;

for M = Ti, Ta and W: amorphous → bcc-Fe, tetragonal-Fe_3B + residual amorphous.

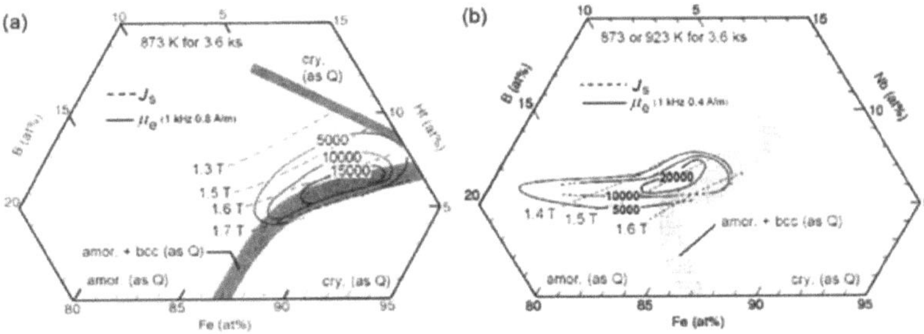

Figure 12. Compositional dependences of initial permeability (μ_e at 1 kHz) and saturation magnetization (J_s) for nanocrystalline (a) Fe-Hf-B and (b) Fe-Nb-B alloys [69].

Hence, the primary crystallization products can be free of compounds for amorphous $Fe_{93-x}M_7B_x$ (M = Zr, Hf, V, Nb and Cr) alloys. Among these alloys, a small grain size of 10 to 15 nm is confirmed for M = Hf and Nb as well as Zr. Consequently, a high initial permeability (μ_e) above 10^4 at 1 kHz is obtained for the nanocrystalline Fe-M-B (M = Zr, Hf and Nb) ternary alloys as shown in Figs. 11(a), 12(a) and 12(b).

The full substitution of Zr by Nb is important since the oxygen reactivity of Nb is lower than that of Zr or Hf , and because the amorphous precursors of some Fe-Nb-B based alloys can be prepared under ambient atmosphere [70], making NANOPERM mass producible.

In spite of the compound-free microstructural evolution upon primary crystallization in amorphous $Fe_{79}V_7B_{14}$ and $Fe_{79}Cr_7B_{14}$, a magnetically soft nanostructure is unrealizable in these alloys as their grain sizes are comparable to the natural exchange length. The coarse microstructure of these two alloys is attributed to the effect of the M element on the grain-growth kinetics.

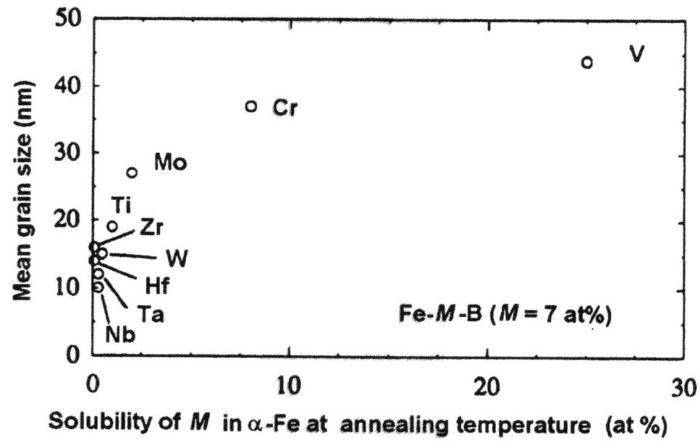

Figure 13. Relation between the mean grain size and the solubility of the M (IVa to VIa metal) element in α-Fe at annealing temperature for amorphous Fe-M-B after primary crystallization [69].

Figure 13 shows the relation between the grain size of the primary crystallization products in the Fe-M-B alloys and the solubility of each M element in α-Fe [71] at the annealing temperature. A small grain size of 10 - 15 nm is limited to the M element having a limited solubility less than 1 at%. This solubility limit is well below 7 at% and a large solute partitioning (C_{am} - C_{bcc}) is expected for M = Zr, Hf, Nb, Ta and W upon primary crystallization of the $Fe_{93-x}M_7B_x$ alloys. Hence, the grain growth as in Eq. (20) is retarded by this partitioning effect. Contrarily, more than 7 at% of V or Cr can be dissolved in α-Fe and thus, these two elements have little contribution to the retardation effect due solute partitioning. Naturally, the large solute partitioning (C_{am} - C_{bcc}) does not have much impact on the growth kinetics unless the diffusivity of the solute atoms is sufficiently lower than that of Fe. The diffusion coefficients in an amorphous $Fe_{91}Zr_9$ alloy have been measured by Kronmüller *et al.* [72], who used radioactive tracers and found that the diffusion coefficient of Zr is smaller by three orders of magnitude than that of Fe. Hence, the nanostructural formation in the $Fe_{93-x}M_7B_x$ (M = Zr, Hf, Nb, Ta and W) alloys is well understood by the small diffusivity of these M elements which are largely partitioned between the bcc-Fe and the residual amorphous phases.

As we discussed in the previous section, Cu addition is quite effective in preventing the formation of compounds in B-rich Fe-Zr-B alloys. This approach is also effective for the $Fe_{93-x}M_7B_x$ (M = Ti, Ta and W) alloys. The formation of the tetragonal-Fe_3B phase upon primary crystallization in these amorphous alloys is suppressed completely by an addition of 1 at% Cu. As a result, $\mu_e > 10^4$ at 1 kHz has been confirmed for nanocrystalline $Fe_{82}Ti_7B_{10}Cu_1$ and $Fe_{82}Ta_7B_{10}Cu_1$ [7].

5. APPLICATIONS

The primary function of magnetic core materials, that is magnetic flux multiplication, requires both high saturation magnetization and high permeability. Both of these are important for miniaturization. The desire for miniaturization also results in steadily increasing the operation frequencies which, of course, require low eddy-current losses. Similar to amorphous metals, their production-inherent low thickness of about 20 μm and their high residual resistivity of about 100 μΩcm minimize eddy currents in nanocrystalline ribbons and make them attractive up to frequencies of 100 kHz or even more.

The lowest coercivities ($H_c < 1$ A/m) and highest permeabilities ($\mu_i \sim 10^5$) are still found around the originally proposed compositions *i.e.* $Fe_{\sim 74}Si_{13-16}B_{6-9}Nb_3Cu_1$. They are comparable with the excellent properties possessed by established materials such as permalloys or Co-base amorphous alloys. The advantages, however, are a higher saturation magnetization of 1.2-1.3 T and a significantly better thermal stability of the soft-magnetic properties. The combination of high saturation magnetization and high permeability (see Fig. 1) together with good high frequency behaviour, low losses and the good thermal stability allows reductions in the size and weight of magnetic components used in, for example, switched mode power supplies or telecommunications.

Apart from its technical performance the material is based on the inexpensive raw materials iron and silicon. The amorphous precursor material for the Fe-Si-B-Nb-Cu alloys, furthermore, is easily accessible by rapid solidification from the melt – a well established technique for large scale production of amorphous metals.

The combination of the above factors has rendered the nanocrystalline solution competitive, not only with amorphous Co-based alloys, but also with classical crystalline alloys and ferrites. The consequence is a steadily increasing level of applications in magnetic cores for ground fault interrupters, common mode chokes and high frequency transformers. Fig. 14 shows some typical examples. The worldwide production rate meanwhile approaches an estimated 1000 tons/year, and the trend is increasing. The only drawback of the nanocrystalline material appears to be the embrittlement that occurs upon crystallization, which requires final shape annealing and, thus, restricts application mainly to toroidally wound cores.

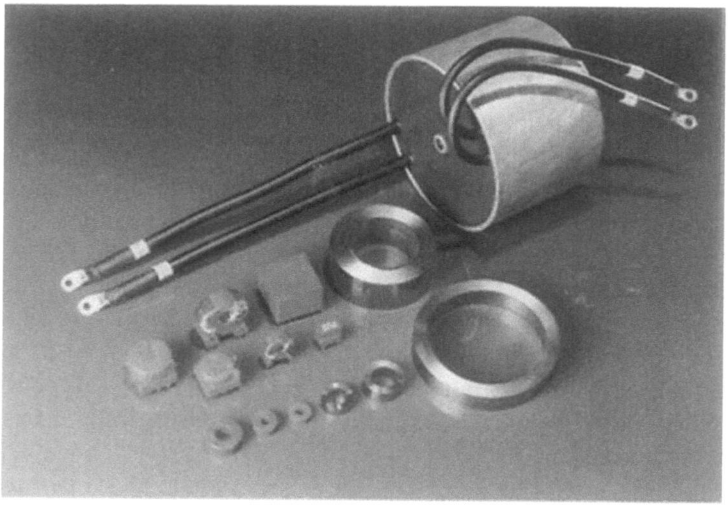

Figure 14. Toroidal wound cores and components of nanocrystalline VITROPERM 800F.

Nanocrystalline $Fe_{\sim 84\text{-}91}(Cu_1)\text{-}(Zr,Nb)_{\sim 7}B_{2\text{-}9}$ alloys appear even more interesting due to their still higher saturation magnetization of up to 1.7 Tesla. However, their application so far is limited to smaller quantities. One of the major reasons is that they require a much more sophisticated production technology due to their limited glass forming ability and/or the oxygen reactivity of the Zr addition.

The readers interested in the examples of device applications of nanocrystalline soft magnetic materials are referred to the articles by Naito *et al.* [73] for the Fe-*M*-B alloys and Petzold [45] for the Fe-Si-B-Nb-Cu alloys.

Acknowledgement

K. Suzuki is grateful to the Australian Research Council for its financial support.

References

[1] Y. Yoshizawa, S. Oguma, and K. Yamauchi, J. Appl. Phys. **64**, 6044 (1988).
[2] G. Herzer, IEEE Trans. Magn. MAG **25**, 3327 (1989)
[3] K. Suzuki, N. Kataoka, A. Inoue, A. Makino, and T. Masumoto, Mater. Trans. JIM **31**, 743 (1990).
[4] Y. Yoshizawa and K. Yamauchi, IEEE Trans. Magn. MAG **25**, 3324 (1989).
[5] Y. Yoshizawa and K. Yamauchi, Mater. Sci. Eng. A **133**, 176 (1991).
[6] Y. Yoshizawa, Scripta Mater. **44**, 1321 (2001).
[7] K. Suzuki, A. Makino, A. Inoue, and T. Masumoto, J. Appl. Phys. **70**, 6232 (1991).
[8] K. Suzuki, A. Makino, A. Inoue, and T. Masumoto, J. Appl. Phys. **74**, 3316 (1993).
[9] K. Suzuki, in "Properties and Applications of Nanocrystalline Alloys from Amorphous Precursors, NATO Science Series II: Mathematics, Physics and Chemistry, Vol. 184", Eds. B. Idzikowski, P. Švec, M. Miglierini, Kluwer Academic, Dordrecht 2005, p. 1.
[10] G. Herzer, in "Handbook of Magnetic Materials, Vol. 10, Ch. 3", Ed. K. H. J. Buschow, Elsevier Science, Amsterdam 1997, p. 415.
[11] G. Herzer, IEEE Trans. Magn. MAG **26**, 1397 (1990).
[12] G. Herzer, Mater. Sci. Eng. A **133**, 1 (1991).
[13] G. Herzer, Scr. Metall. Mater. **33**, 1741 (1995).
[14] K. Hono, Progress in Mater. Sci. **47**, 621 (2002).
[15] M. E. McHenry, M. A. Willard, D. E. Laughlin, Progress in Mater. Sci. **44**, 291 (1999).
[16] H. W. Fuller and M. H. Hale, J. Appl. Phys. **31**, 238 (1960).
[17] H. Rother, Z. Phys. **168**, 42 (1962).
[18] H. Hoffmann, J. Appy. Phys. **35**, 1790 (1964).
[19] H. Hoffmann, IEEE Trans. Magn. MAG **9**, 17 (1973).
[20] N. Kumasaka, N. Saito, Y. Shiroishi, K. Shiiki, H. Fujiwara, and M. Kudo, J. Appl. Phys. **55**, 2238 (1984).
[21] N. Terada, Y. Hoshi, M. Naoe, and S. Yamanaka, IEEE Trans. Magn. MAG **20**, 1451 (1984).
[22] M. Takahashi and T. Shimatsu, IEEE Trans. Magn. MAG **26**, 1485 (1990).
[23] R. Alben, J. J. Becker, and M. C. Chi, J. Appl. Phys. **49**, 1653 (1978).
[24] See for example, F. E. Luborsky, in "Amorphous Metallic Alloys, Ch. 1", Ed. F. E. Luborsky, Butterworths, London 1983, p. 1.

[25] T. Bitoh, A. Makino, T. Hatanai, and A. Inoue, J. Appl. Phys. **81**, 4634 (1997).
[26] K. Suzuki and J.M. Cadogan, Phys. Rev. B **58**, 2730 (1998).
[27] K. Suzuki, G. Herzer, and J. M. Cadigan, J. Magn. Magn. Mater. **177**, 949 (1998).
[28] G. Herzer, in "Properties and Applications of Nanocrystalline Alloys from Amorphous Precursors, NATO Science Series II: Mathematics, Physics and Chemistry, Vol. 184", Eds. B. Idzikowski, P. Švec, M. Miglierini, Kluwer Academic, Dordrecht 2005, p. 15.
[29] G. Herzer, J. Magn. Magn. Mater. in press (2005).
[30] R. Schäfer, A. Hubert, and G. Herzer, J. Appl. Phys. **69**, 5325 (1991).
[31] Y. Yoshizawa, Doctoral Thesis, Tohoku University, Sendai 1993, p. 160.
[32] M. Müller and N. Mattern, J. Magn. Magn. Mater. **136**, 79 (1994).
[33] Y. Fujii, H. Fujita, A. Seki, and T. Tomida, J. Appl. Phys. **70**, 6241 (1991).
[34] K. Suzuki, A. Makino, N. Kataoka, A. Inoue, and T. Masumoto, Mater. Trans. JIM **32**, 93 (1991).
[35] K. Suzuki, J. M. Cadogan, V. Sahajwalla, A. Inoue, and T. Masumoto, J. Appl. Phys. **79**, 5149 (1996).
[36] N. Hasegawa and M. Saito, J. Magn. Magn. Mater. **103**, 274 (1992).
[37] H. Hayashi, M. Hayakawa, W. Ishikawa, Y. Ochiai, H. Matsuda, Y. Iwasaki, and K. Aso, J. Appl. Phys. **61**, 3514 (1987).
[38] H. Watanabe, H. Saito, and M. Takahashi, J. Magn. Soc. Jpn. **17**, 191 (1993).
[39] Y. Yoshizawa and K. Yamauchi, J. Magn. Soc. Jpn. **13**, 231 (1989).
[40] K. Suzuki, A. Makino, A. Inoue, and T. Masumoto, J. Magn. Soc. Jpn. **18**, 800 (1994).
[41] S. Ito, K. Aso, Y. Makino, and S. Uedaira, Appl. Phys. Lett. **37**, 665 (1980).
[42] R. C. O'Handley, Solid State Commun. **21**, 1119 (1977).
[43] N. Ito and K. Suzuki, J. Appl. Phys. **97**, in press (2005).
[44] G. Herzer, Mater. Sci. Eng. A **181**, 876 (1994).
[45] J. Petzold, Scr. Mater. **48**, 895 (2003).
[46] Y. Yoshizawa, S. Fujii, D. H. Ping, M. Ohnuma, and K. Hono, Scr. Mater. **48**, 863 (2003).
[47] M. Müllaer, H. Grahl, N. Mattern, U. Kühn, and B. Schnell, J. Magn. Magn. Mater. **160**, 284 (1996).
[48] L. Kraus, K. Závěta, O. Heczko, P. Duhaj, G. Vlasák, and J. Schneider, J. Magn. Magn. Mater. **112**, 275 (1992).
[49] G. Herzer, IEEE Trans. Magn. MAG **30**, 4800 (1994).
[50] M. Ohnuma, K. Hono, T. Yanai, H. Fukunaga, and Y. Yoshizawa, Appl. Phys. Lett. **83**, 2859 (2003).
[51] A. Hernando, I. Navarro, and P. Gorria, Phys. Rev. B **51**, 3281 (1995).
[52] K. Suzuki, M. Kikuchi, A. Makino, A. Inoue, and T. Masumoto, Mater. Trans. JIM **32**, 961 (1991).
[53] M. A. Willard, D. E. Laughlin, M. E. McHenry, D. Thoma, K. Sickafus, J. O. Cross, and V. G. Harris, J. Appl. Phys., **84**, 6773 (1998).
[54] F. Johnson, H. Garmestani, S. Y. Chu, M. E. McHenry, and D. E. Laughlin, IEEE Trans. Magn. MAG **40**, 2697 (2004).
[55] I. Škorvánek, P. Švec, J. Marcin, J. Kováč, T. Krenický, and M. Deanko, Phys. Stat. Sol. A **196**, 217 (2003).

[56] U. Köster and U. Herold, in "Glassy Metals I. Topics in Applied Physics", Eds. by H.J. Guntherodt and H. Beck, Springer Verlag, Berlin 1981, p. 225.

[57] H. B. Aaron, D. Fainstein, and G. R. Kotler, J. Apply. Phys. **41**, 4404 (1970).

[58] U. Köster, U. Schünemann, M. Blank-Bewersdorff, S. Brauer, M. Sutton, and G.B. Stephenson, Mater. Sci. Eng. A **133**, 611 (1991).

[59] T. Pradell, D. Crespo, N. Clavaguera, and M. T. Clavaguera-Mora, J. Phys.: Condens. Matter **10**, 3833 (1998).

[60] K. Suzuki, A. Makino, A.P. Tsai, A. Inoue, and T. Masumoto, Mater. Sci. Eng. A **179**, 501 (1994).

[61] Y. Zhang, K. Hono, A. Inoue, A. Makino, and T. Sakurai, Acta Metal. Mater. **44**, 1497 (1996).

[62] K. Suzuki and J. M. Cadogan, J. Appl. Phys. **87**, 7097 (2000).

[63] Y. Yoshizawa and K. Yamauchi, Mater. Trans. JIM **31**, 307 (1990).

[64] K. Hono, K. Hiraga, Q. Wang, A. Inoue, and T. Sakurai, Acta Metall. Mater. **40**, 2137 (1992).

[65] Y. Zhang, K. Hono, A. Inoue, and T. Sakurai, Scr. Metall. Mater. **34**, 1705 (1996).

[66] T. Ohkubo, H. Kai, D. H. PingK. Hono, and Y. Hirotsu, Scr. Mater. **44**, 971 (2001).

[67] D. H. Ping, Y. Q. Wu, K. Hono, M. A. Willard, M. E. McHenry, and D. E. Laughlin, Scr. Mater. **45**, 781 (2001).

[68] M. Ohnuma, D. H. Ping, T. Abe, H. Onodera, K. Hono, and Y. Yoshizawa, J. Appl. Phys. **93**, 9186 (2003).

[69] K. Suzuki, A. Makino, A. Inoue, and T. Masumoto, Sci. Rep. RITU A **39**, 133 (1994).

[70] A. Makino, T. Bitoh, A. Inoue, and T. Masumoto, Scr. Mater. **48**, 869 (2003).

[71] "Binary Alloy Phase Diagrams", Ed. T. B. Massalski, American Society for Metals, Ohio 1986.

[72] H. Kronmüller, W. Frank, and A. Hörner, Mater. Sci. Eng. A **133**, 410 (1991).

[73] Y. Naitoh, T. Bitoh, T. Hatanai, A. Makino, A. Inoue, and T. Masumoto, Nanostructured Mater. **8**, 987 (1997).

Chapter 14

NANOSTRUCTURES FOR SPIN ELECTRONICS

P. P. Freitas, H. Ferreira, R. Ferreira and S. Cardoso
INESC Microsystems and Nanotechnologies, Rua Alves Redol 9 and Physics Department, Instituto Superior Tecnico, Avenida Rovisco Pais, 1000 Lisbon, Portugal

Sebastiaan van Dijken
SFI Trinity Nanoscience Laboratory
Physics Department, Trinity College
Dublin 2, Ireland

John Gregg
Clarendon Laboratory
University of Oxford
Parks Road OX1 3PU, UK

Abstract This chapter covers recent advances in spin electronic devices. Several topics will be addressed, starting with two terminal, metallic devices, from read head stacks for advanced magnetic data storage, where current in plane spin valve technology is reaching its limits in device performance (at 100-120 Gbit/in^2) and may be soon replaced by tunnel junction readers, and later by current perpendicular to the plane spin valves, to MRAMs, where new switching mechanisms will be discussed including spin transfer and thermally assisted switching, and finally field sensors, including recent applications to magnetoresistive biochips. New hybrid three terminal semiconductor-ferromagnetic devices are then discussed where polarized electrons are injected from a ferromagnet into a semiconductor. Examples are hot-electron devices such as the spin-valve transistor with a metallic GMR base, and magnetic tunnel transistors and devices where spin dependent transport in a semiconductor is analyzed, such as spin LEDs.

This chapter reviews several important subfields of spintronics. It is divided into four sections: read heads and magnetic data storage (including tunnel junctions), magnetic random-access memories, spintronic biosensors, and spin transistors. Throughout the chapter, the following abbreviations are used: MR (magnetoresistance); GMR (giant magnetoresistance); TMR (tunnel magnetoresistance); SV (spin-valve); MTJ (magnetic tunnel junction); CIP (current in plane); CPP (current perpendicular to the plane); AF (antiferromagnet); SAF (synthetic antiferromagnet); SF (synthetic free); NOL (nano-oxide layer); RA (resistance-area product); MRAM (magnetic random access memory); HD (hard-disk); and SNR (signal-to-noise ratio)

1. READ HEADS AND MAGNETIC DATA STORAGE

Current state-of-the-art disk drives (late 2004), using dual inductive-write, magnetoresistive read (MR) integrated heads reach areal densities of 100 Gbit/in^2, and are capable of putting a full 100GB in one platter (a 2 hour film occupies 10GB). Figure 1 represents schematically the integrated head. Notice the common top read shield and bottom write pole. At densities near 100 Gbit/in^2, the linear bit size is about 40 nm, and the width of the written tracks is 100 nm. At 1 Tbit/in^2 recording, bit size is expected to shrink to 25 nm x 25 nm. The written track width is limited by lithography tools and the bit length by the shield to shield spacing. Areal density in commercial drives increased steadily at a rate of nearly 100% per year from 1996 to 2001, but this annual growth has diminished in the past 3 to 4 years.

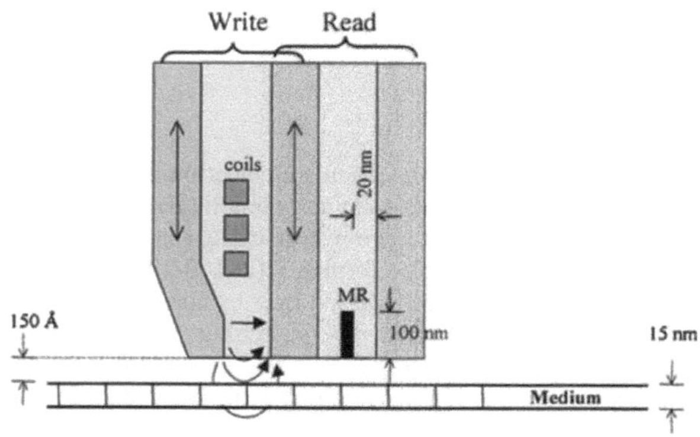

Figure 1. Integrated head with an inductive writer and a shielded magnetoresistive sensor used in present longitudinal recording.

The recent slow-down in areal density arises from several factors: on the writing side, head-media spacing can no longer be scaled down, and concerning write pole materials, the maximum head moment is limited to about 2.4 T (top of the Slater-Pauling curve) and this in turn limits the available writing field, and therefore the allowed media coercivity required to stabilize smaller bits. Perpendicular recording and the use of a soft underlayer in the recording media will in this case come to help by increasing the available head field. In this case the full deep gap field is available at the storage layer. On the reader sensitivity side, current in plane (CIP) spin valves can no longer be pushed to higher outputs without compromising stack integrity. Also the CIP geometry requires gap oxides which take un-necessary space, limiting bit length reduction. Readers with the current perpendicular to the plane (CPP) make the half-gap insulators obsolete and offer an extension of reader technology to higher densities.

1.1. Current In Plane (CIP) Spin Valves

In this section, state-of-the-art CIP spin valves, and new CPP spin valve and magnetic tunnel junction (MTJ) read heads will be discussed. Spin valves were introduced in 1991 [1], first sensors designed and tested in 1993-1994 [2, 3], and first head prototypes presented in 1994 [4]. A good review of the spin valve mechanism is given in reference [5]. Spin valves are based on the giant magnetoresistance effect [6], and were designed to replace anisotropic magnetoresistive read sensors. The physical mechanism responsible for the giant magnetoresistance effect is depicted in the inset of Fig. 2.

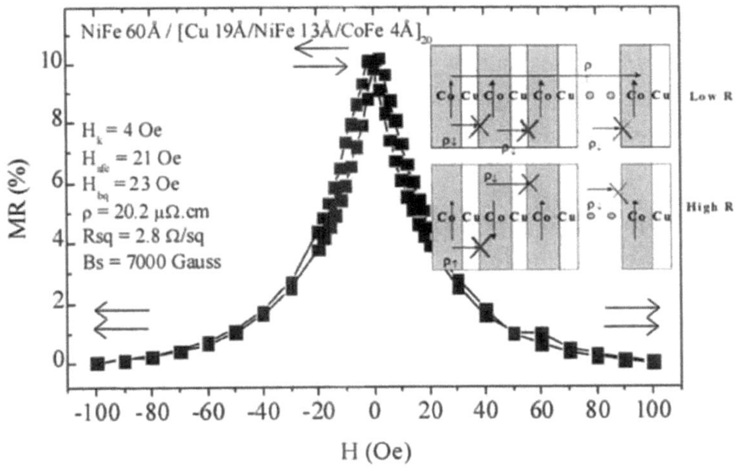

Figure 2. Asymmetric spin dependent scattering at CoFe/Cu and NiFe/Cu interfaces leads to the giant magnetoresistance effect observed in a $(CoFe/Cu/NiFe/Cu)_{x20}$ multilayer stack (S. Cardoso, INESC MN).

In 3d metals conduction is mainly carried by 4s electrons, but due to the large density of states of the 3d bands at the Fermi level, both s-s and s-d scattering events are relevant. In each spin sub-band the resistivity has therefore two terms, ($\rho_\uparrow = \rho ss_\uparrow + \rho sd_\uparrow$; $\rho_\downarrow = \rho ss_\downarrow + \rho sd_\downarrow$). These two channels now conduct in parallel ($1/\rho = 1/\rho_\uparrow + 1/\rho_\downarrow$). The GMR effect arises from the asymmetry in the spin dependent scattering at the non-magnetic/magnetic interfaces for spin up and spin down electrons. When the magnetic layers have parallel magnetization directions, spin up conduction electrons will be weakly scattered, and spin down electrons will be strongly scattered. The spin up channel will short the current leading to a low resistance state. When the magnetic layers have antiparallel magnetization directions, both spin up and spin down electrons will alternatively be strongly and weakly scattered, leading to a high resistance state.

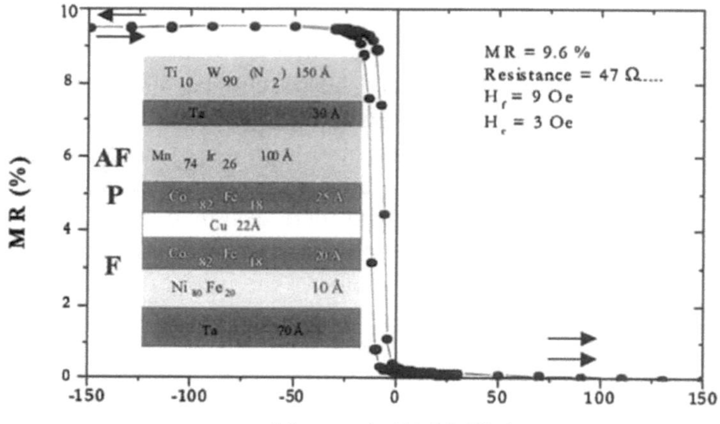

Figure 3. Magnetoresistance *vs.* magnetic field for a top-pinned spin valve coupon sample. (P) pinned, (F) free, and (AF) antiferromagnetic layers (A. Guedes, INESC MN).

Contrary to GMR multilayers, where the magnetization direction depends on the non-magnetic spacer thickness leading to oscillatory coupling [6], in spin valves, one layer is engineered to have its magnetization pinned by exchange to an antiferromagnet [7], while the other is free to rotate. A typical spin valve structure consists then of two ferromagnetic layers, separated by a Cu spacer (Fig. 3). The buffer layer (Ta in this case) is chosen in order to provide a <111> texture, and control grain size to 10 nm leading to soft free layer properties. With Ta buffers, typical MR values up to 10% are obtained, (Fig. 3, top-pinned, MnIr based spin valve, INESC-MN). The magnetoresistance dependence on the different layer thicknesses has been modeled in the semi-classical Boltzmann equation model [8]. MR increases strongly by minimizing Cu thickness down to the continuity level of the Cu layer (15 to 18 Å, on Ta buffers). However decrease in Cu thickness leads to increase in interlayer ferromagnetic Néel coupling so a compromise is usually found near 20 to 22 Å Cu. Within the Boltzmann equation model, the MR dependence on ferromagnetic layer thickness depends on the interfacial transmission coefficients at the Cu/F interfaces, and on the spin dependent bulk mean free paths in the ferromagnetic layer. Depending on ferromagnetic material ($Co_{90}Fe_{10}$, $Ni_{81}Fe_{19}$, Fe), maximum MR occurs for a ferromagnetic layer thickness varying from 60 Å down to 35 Å [5].

A strong magnetoresistance enhancement in spin valves was found by Parkin when introducing Co or $Co_{90}Fe_{10}$ at the Cu/NiFe interface [9]. Although controversy aroused initially on the cause of this enhancement (whether caused by increase in interfacial transmission coefficient asymmetry or caused by increase in asymmetry of bulk spin dependent scattering), it now appears that the enhancement essentially arises from a better interface structure (sharper interface, less intermixing, due to the immiscibility of Co or $Co_{90}Fe_{10}$ in Cu, contrary to the miscible NiFe/Cu interface).

Further improvements on the MR ratio of CIP spin valves were obtained by changing from Ta buffers into $(Ni_{81}Fe_{19})_{60}Cr_{40}$ buffers [10].The use of NiFeCr buffers dates back to the anisotropic magnetoresistance sensors days, where it was shown that NiFeCr can be made smoother than Ta (about 2 Å rms) achieving a higher degree of specularity at the NiFeCr/NiFe interface and leading to larger grain size, and therefore reducing grain boundary scattering and improving AMR [5]. The arguments that led to larger AMR also lead to larger MR in the NiFeCr based spin valve structures. MR values up to 15% were reported by different groups [10].

To further maximize the MR ratio, CIP spin valves can be made to incorporate Nano Oxide Layers (NOLs) to further confine the electron path within the active F/Cu/F layers (Fig. 4), and therefore reducing shunting effects and increasing specular reflection [11-13]. NOL layers are introduced at the free layer /cap interface and in the mid of the pinned layer. Increasing specularity decreases total spin valve resistivity, enhancing the MR ratio. The method of fabrication of the NOL layers, their optimum thickness, and their thermal stability were a subject of detailed study [14, 15]. Both magnetic ($CoFeO_x$) and non-magnetic (TaO_x, CuO_x) NOLs have been studied. The incorporation of the oxide layers in an optimized spin valve stack requires care, since free layer coercivity tends to increase and pinned layer exchange may be decreased. Figure 4 shows enhanced magnetoresistance (about 14.5%) in specular spin valves with two NOLs grown on Ta buffers [13]. Optimized specular spin valves with NiFeCr buffers and dual NOLs have reached MR values of 18-20% [16]. Further increase in MR can be also obtained by increasing the resistivity of the shunting layers (the antiferromagnet, the buffer). The use of oxide antiferromagnets (NiO) proved effective in providing high specularity at the CoFe/NiO interface and reduced shunting, but the exchange field was not large enough for head applications [17].

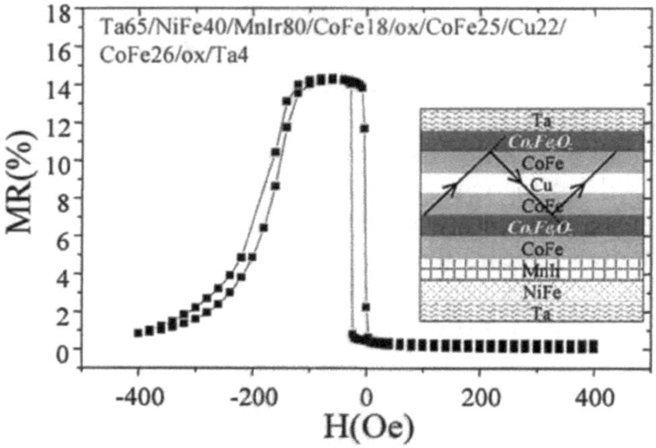

Figure 4. Magnetoresistance *vs.* applied field for a specular spin valve coupon sample with two nano-oxide layers (ref. 13).

For head applications, and apart from large MR values, the exchange field created at the pinned layer / exchange layer interface is important. The exchange energy should be large (> 0.2 erg/cm^2) such that the exchange field prevails against demagnetizing fields at head level. The blocking temperature (temperature where the exchange field vanishes) should exceed 300°C, to prevent accidental de-pinning of the pinned layer during head fabrication or head life. Also the corrosion resistance of this exchange layer should not be worse than that of $Ni_{80}Fe_{20}$ used as reference [18]. Another factor to take into account is the effective coupling field between the free and pinned layers [19], which should not exceed about 10Oe to allow for proper biasing (see the transfer curve in Fig. 3, with an effective coupling field of 9 Oe). This coupling arises from the competition between ferromagnetic Néel coupling [20] (caused by corrugated interface roughness), indirect RKKY-like oscillatory exchange coupling across the Cu spacer [21], and the coupled demagnetizing fields of both layers in patterned sensors [22].

Concerning exchange fields, most heads use synthetic antiferromagnets to increase exchange and improve the temperature dependence of the exchange field. The synthetic antiferromagnet (SAF) consists of two Co or CoFe layers with similar thickness strongly antiferromagnetically coupled through 0.5 to 0.7 nm Ru [23, 24]. To avoid a spin flop transition under an external transverse field [25], a conventional exchange layer must be coupled to one of the ferromagnetic layers. Although this strong AF coupling has a weak temperature dependence [26], this is not so for the conventional AF that must still have an optimum blocking temperature, but no longer large exchange. Figure 5 shows the magnetization *vs.* field loop for a

MnIr based synthetic antiferromagnet spin valve [27]. These SAF structures have another advantage. Since the effective moment of the pinned layer is low ($M_{eff} = M_1 - M_2$, where M_i is the moment of each ferromagnetic layer), its contribution to the demagnetizing and coupling fields acting on the free layer is much weaker than in conventional spin valves. Similar SAF structures can be used as synthetic free layers (SF), in order to reduce the effective magnetic thickness and moment of the free layer [28, 29].

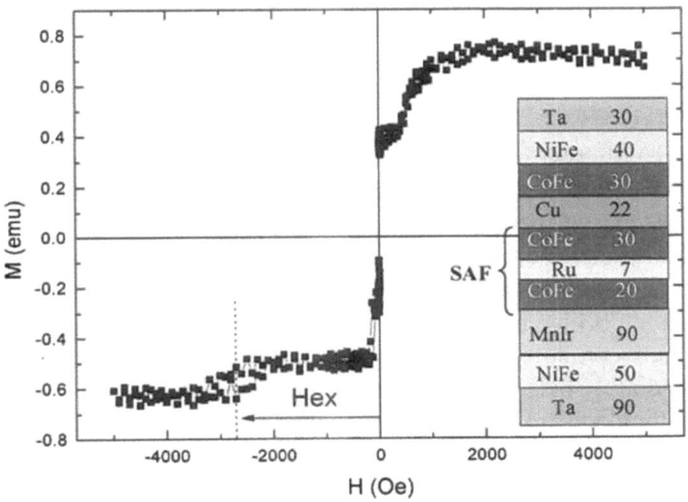

Figure 5. Exchange bias field Hex in a SAF bottom pinned structure (from reference 27).

Depending on applications, free and pinned layer easy axis can be set either parallel or orthogonal, by controlling the magnetic field direction during deposition. For sensor applications the spin valve output must be linearized. This can easily be achieved by inducing a transverse magnetization direction in the pinned layer, while the free layer, in its quiescent state is in the longitudinal direction. Figure 6(a) shows an unshielded spin valve sensor with pinned and free layers at 90° (sensor or head applications) [2, 22].

Equation (1) gives the unshielded spin valve sensor output voltage.

$$\Delta V = \tfrac{1}{2}\,(\Delta R/R)\ R_{sq}\ I\ \langle\cos(\Theta_f - \Theta_p)\rangle\ W/h \qquad (1)$$

Here, $\Delta R/R$ is the maximum MR signal of the spin valve sensor (reaching today 15 to 20% in specular spin valves), R_{sq} is the sensor square resistance (15 to 20 Ω/sq), W the track width of the read element (distance between leads), h is the sensor height, I the sense current, Θ_f is the angle between the free layer magnetization and the longitudinal direction, and Θ_p the angle between the pinned layer magnetization and the longitudinal direction. The average < ...> is taken over the active area of the sensor (between leads). For linearized output ($\Theta_P = \pi/2$). For the simple case where W>>h, and assuming uniform magnetization for the free and pinned layers, $<\sin \Theta_f> = (H_a + H_{bias}) / (H_{keff})$, where $H_{keff} = H_k + H_{demag}$. H_{keff} is the effective anisotropy field, incorporating shape anisotropy field (H_{demag}) and uniaxial anisotropy field (H_k). H_{bias} is the current bias field used to center the transfer curve response to an external signal field [22]. Eq.(1) can then be rewritten as:

$$\Delta V = (\Delta R/R)\ R_{sq}\ I\ (W/h)\ (H_a + H_{bias})/2H_{keff} \quad (2)$$

Fig. 6(b) shows the schematic transfer curve for an unpatterned spin valve coupon sample, and the linearized transfer curve for the underbiased, and properly biased (centered) patterned sensor [22]. Figure 6(c) shows the experimental transfer curve for an unshielded spin valve sensor (W = 6 μm, h = 2 μm). Unshielded sensors are used in a variety of applications, normally in a Wheastone bridge arrangement. For read head applications, the sensor is placed in between soft magnetic shields to increase response to the signal fields immediately below the magnetoresistive sensor.

For a shielded spin valve sensor used as a CIP read head, as in Fig. 6(d), the head output is given by Eq.(1) but now with $<\cos(\Theta_f - \Theta_p)> = E \cdot \phi_{ABS} / tw\mu_o M_s$,

$$\Delta V_{0\text{-}p} = (\Delta R/R)\ R_{sq} IWE\phi_{ABS}/2tW\mu_o M_s h \quad (3)$$

in SI units [30]. Here, ϕ_{ABS} is the media flux entering the shielded sensor at the head air bearing surface. The media flux leakage to the shields is described by the head efficiency $E = \tanh(h/2l_c)/(h/l_c)$, with l_c, the flux propagation length defined as, $l_c = (t\mu g_R/2)^{1/2}$, with μ the relative free layer permeability, M_s the free layer saturation magnetization, g_R the shield to sensor separation (read half gap), and t the free layer thickness [31]. As can be seen from Eq. (3), head output depends critically on head geometry, spin valve MR signal, and media parameters. Table I shows spin valve read head parameters for a 106 Gbit/in^2 recording demonstration [32]. In this table G_R corresponds to the read gap length, $(2g_R+t_{MR})$, where t_{MR} is the read sensor thickness.

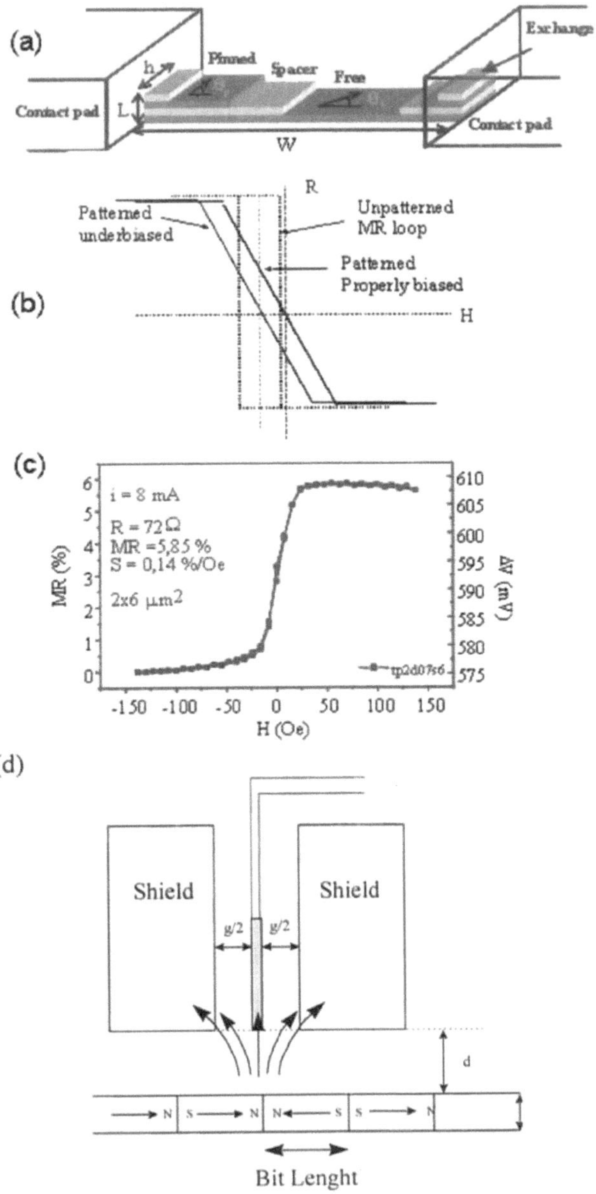

Figure 6. Spin-valve sensors: (a) Unshielded spin valve sensor schematics, (b) linearized transfer curve for the underbiased, and properly biased sensor, (c) experimental transfer curve for an unshielded spin valve sensor (H.Ferreira, INESC MN), and (d) shielded spin valve sensor used as a read head.

CIP spin valve sensors have reached a practical limit near 100 Gbit/in^2 since MR ratio has been essentially unchanged in the past 3 to 4 years, and further reductions in read gap (below 20 nm) are not possible in the sensor in gap geometry. Two alternatives appear for next generation read heads. CPP reads heads, either CPP spin valves or magnetic tunnel junctions. Both offer reduced shield to shield spacing (the current flows directly from shield to sensor) and therefore can immediately bring down bit length, to the limit of the sensor thickness (30 to 40 nm). Heat dissipation is also improved and larger current densities can cross the CPP device improving signal (CPP spin valves). The following two sections deal with the CPP head geometry.

Table 1. Read head parameters used for a 106 Gb/in^2 demonstration. Here w corresponds to the active read track width, h is the sensor height, G the shield to shield separation.

density (Gb/in^2)	w (nm)	h (nm)	G_R (nm)	V_{pp} (mV/μm)	ΔR/R (%)	Sensor type
106	124	150	70	9.3	12.4	SV

1.2. Low RA Magnetic Tunnel Junctions

Recent developments in materials technology have brought magnetic tunnel junction resistance down to few Ωμm^2 at sizable TMR (10 to 20%), opening a window of application for read heads in the 100-300 Gbit/in^2 area. In the spin-dependent tunneling effect, electrons tunnel across an insulating barrier between two ferromagnetic electrodes [33]. Figure 7 shows the simplified band structure for the electrons involved in the tunneling process. The tunneling can be either incoherent (for example across an amorphous AlOx barrier) or coherent (across a crystalline MgO barrier). For incoherent scattering (Julliere's model [34]), the tunnel magnetoresistance of such a junction is proportional to the product of both electrode polarizations P.

$$TMR = 2P_1P_2/(1 - P_1P_2) \tag{4}$$

where

$$P_i = (D_\uparrow(\varepsilon_F) - D_\downarrow(\varepsilon_F)) / (D_\uparrow(\varepsilon_F) + D_\downarrow(\varepsilon_F)) \tag{5}$$

and where $D(\varepsilon_F)$ is the interfacial density of states at the Fermi level for the particular spin up or spin down bands.

Figure 7. Schematic band structure for electrons tunneling across the tunneling barrier, in the parallel state (left) and antiparallel state (right).

Equations (4) and (5) lead to a low resistance state when the two electrode magnetizations are in the parallel state, and a high resistance state when they are in the antiparallel state. A typical magnetic tunnel junction has a structure similar to a spin valve, but with Cu replaced by the insulating barrier, and the current flowing from one electrode, across the barrier, into the other electrode. Figure 8 shows a 40% resistance change obtained upon free layer reversal, for a medium resistance, top-pinned, AlO_x tunnel barrier (Ta 50 Å /NiFe 70 Å /CoFe 20 Å/AlOx 15 Å /CoFe 25 Å /MnIr 100 Å /Ta 50 Å).

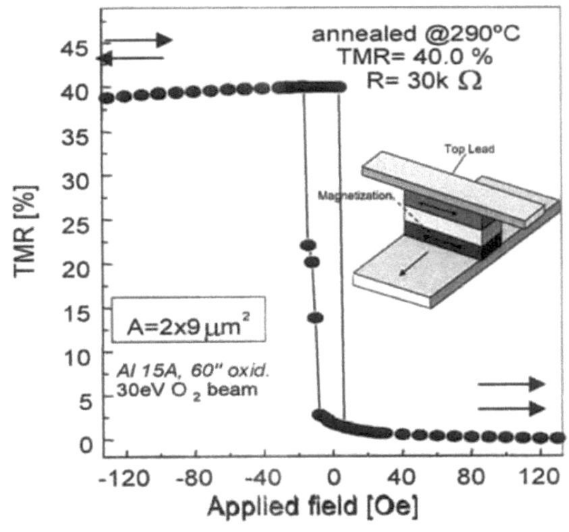

Figure 8. Minor-loop tunnel magnetoresistance (free layer reversal) for a 2 x 9 μm^2 tunnel junction [37].

The interfacial spin polarization involved in the tunneling process depends both on the ferromagnetic electrode and on the oxide barrier. Although known since the mid sixties, only in the late 90s have significant room temperature TMR signals been observed [34, 35-39] (TMR up to 70% with amorphous AlO_x barriers). In 2004, TMR over 200% was reported for coherent tunneling in crystalline MgO barriers [40, 41] In most of these studies, transition metal ferromagnetic electrodes are used (CoFe, CoFeB, Fe, NiFe). These results opened a realm of practical applications, among which, two of the most important are non-volatile magnetic tunnel junction random access memories (MRAM) and spin tunnel read heads.

In this section, low resistance tunnel junctions prepared by natural oxidation [42-45] or remote plasma oxidation techniques [46] will be described, since up to now they have provided the lowest junction resistances, required to supplant spin valves for read head applications above 100-200 Gbit/in^2. At the heart of tunnel junction fabrication lays the barrier fabrication. For read head applications, in the 100-300 Gbit/in^2 region, junction resistance must be around 1 $\Omega\mu m^2$ [47-51]. Figure 9(a) shows a TEM micrograph for a 1 nm thick AlHfOx low RA barrier [52].

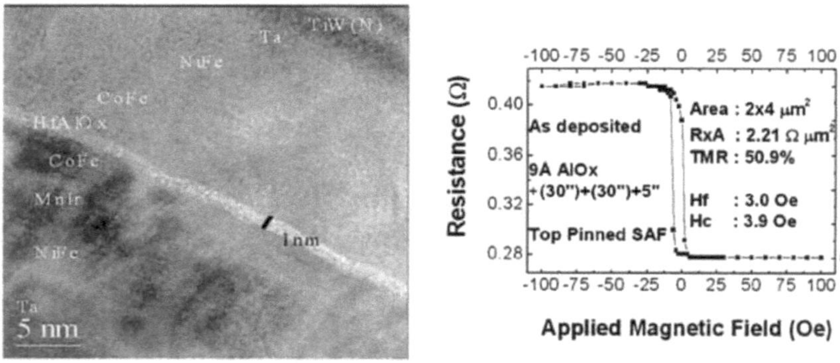

Figure 9. Low resistance tunnel junctions: (a) TEM micrograph of a low resistance ultrathin (1 nm) $AlHfO_x$ barrier (1 $\Omega\mu m^2$) and (b) transfer curve of a low resistance MTJ (R. Ferreira, INESC MN).

Figure 9(b) shows the transfer curve of a low resistance, high TMR magnetic tunnel junction prepared by remote plasma oxidation [46]. Figure 10 summarizes TMR *vs.* RA for low RA junctions prepared at INESC-MN by natural and remote plasma oxidation [46]. The shaded region corresponds to results obtained by natural oxidation in various laboratories. Both techniques lead to RA values close to 1 $\Omega\mu m^2$ with TMR ranging from 15 to 25%. For naturally oxidized samples ultrathin Al layers are used (t_{Al} = 4.5 to 6 Å) while for remote plasma oxidation thicker Al layers were used (t_{Al} = 8 or 9 Å). In the latter case the barrier is left underoxidized. Most of these low resistance barriers show defects either at the barrier (hotspots, pinholes and/or stoichiometry variations) leading to lower breakdown voltages (0.3 to

0.6 V) and gradual breakdown processes [53]. The ability to produce low RA barriers reliably is a key for their application as read heads in hard disk drives.

Figure 11(a) illustrates how the low RA tunnel junction is incorporated in a shielded read head. Figure 11(b) shows the tunnel junction head geometry for prototypes developed at INESC-MN [51]. In this CPP design, the MTJ is at the ABS, and contacts are made through the shields. Lateral permanent magnet hard bias is used. Notice the small lateral insulating layer preventing the permanent magnet of shorting the tunnel junction. Critical for tunnel head fabrication is controlling the lapping process into the read element.

Although low RA tunnel junctions promise relatively larger signal output when compared with CIP spin valves, they are intrinsically noisier [54]. Above the $1/f$ knee, the non-magnetic thermal background in spin valves (free layer saturated) is characterized by the Johnson noise $V^2_{Johnson} = 4k_BTR$ (noise power in V^2/Hz), while the non-magnetic thermal background in tunnel junctions at moderate bias (free layer saturated, $V_b >> 2k_BT/q \cong$ 50mV) is dominated by shot noise $V^2_{shot} = 2qV_bR$.

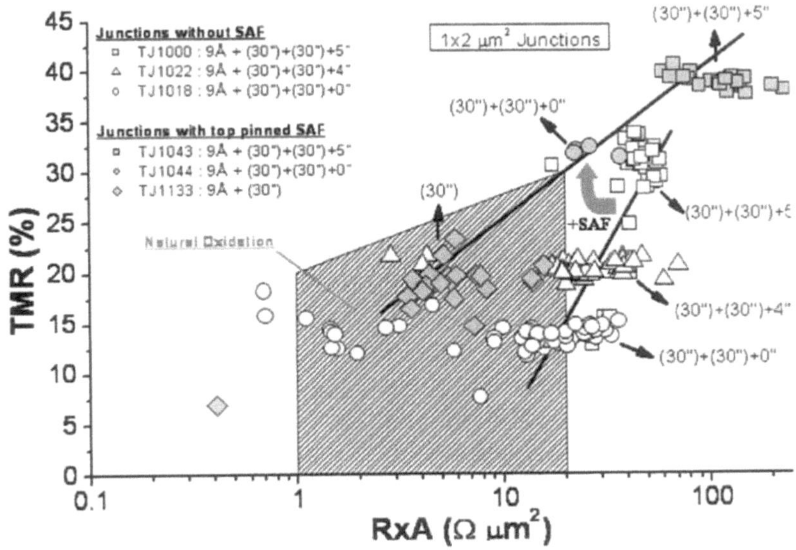

Figure 10. TMR and RxA for junctions patterned with 1×2 μm^2 area, with oxidation times between (30'') and (30'') + (30'') + 5'', and with a top pinned SAF (X = Co-Fe-B 4 / Ru 0.6 / Co-Fe-B 4) or a top pinned ferromagnet (X = Co-Fe-B 4). The grey region corresponds to the best reported results on samples produced with natural oxidation [46].

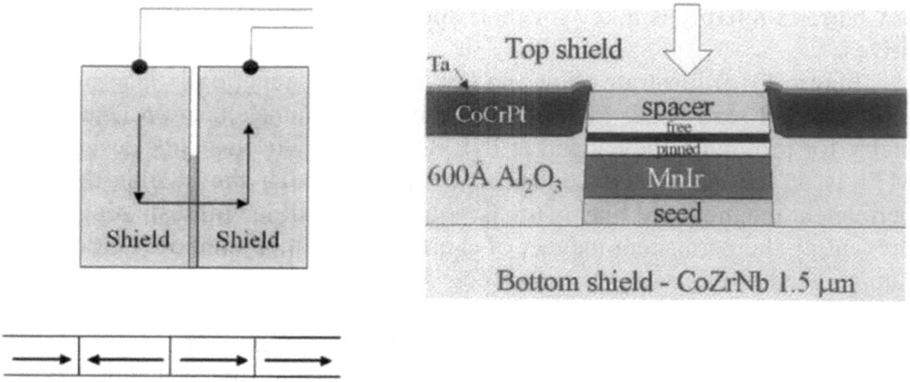

Figure 11. Tunnel junction head geometry developed at INESC-MN [51].

The signal to noise ratios (*SNR*) of spin valve sensors and MTJ sensors can be estimated from:

$$SNR = [(\Delta R/R)\, V_b/V_n]^2 \tag{6}$$

For MTJs to supplant spin valves in SNR, $SNR_{MTJ} > SNR_{SV}$ and the signal output satisfy:

$$\frac{(V_b\, \Delta R/R)_{MTJ}}{(V_b\, \Delta R/R)_{SV}} > \sqrt{\frac{V_b q R_{MTJ}}{2 k_B T R_{SV}}} \tag{7}$$

Taking sensor dimensions typical of 100 Gbit/in^2 applications (W = 100 nm, h = 120 nm, and a *RA* of 2 $\Omega\mu m^2$ for the MTJ and an R_{sq} of 20 Ω/sq for the *SV*) and for a bias voltage of 100 mV, $R_{MTJ} = 167\ \Omega$, and $R_{SV} = 17\ \Omega$, and the *MR* ratio for the MTJ must be 4.3 higher than the spin valve MR in order for MTJ based heads to supplant spin valve based heads.

Apart from the Johnson noise and shot noise contributions, and above the 1/f knee, both types of sensors (CIP SV and MTJs) will show magnetic noise coming from thermally induced magnetization fluctuations. This has a white component at medium frequencies (above the $1/f$ knee, and below magnetic resonance) [55], and a noise peak at the free layer magnetic resonance frequency [56]. At present recording densities and for a disk spinning at 15000 rpm the read head will see flux changes at frequencies around 500 MHz. In present head designs and materials, magnetic resonance noise starts to appear around 3-6 GHz [56] and becomes an important part in head design as sensor dimensions scale down [55, 57]. Controlling and reducing the damping parameter α involved in the free layer magnetization precession may play a key role in minimizing magnetic noise [57].

1.3. Current Perpendicular to the Plane (CPP) Spin Valves

The current perpendicular to the plane geometry, using either spin valve or magnetic tunnel junction sensors, leads to device resistance increase as sensor cross-sectional size is decreased. CPP spin valves become competitive as sensor dimensions go below 100 nm, as device resistance increases. As the sensor is connected directly to the metallic shields, thermal dissipation is improved compared with CIP devices, and larger currents can be flown through the device increasing signal. The CPP-GMR geometry also shows higher intrinsic MR ratio than the CIP geometry [58, 59]. A theoretical model describing CPP transport in GMR multilayers was introduced by T. Valet and A. Fert [60], and several groups analyzed the behavior of CPP spin valves and CPP GMR multilayers in order to determine intrinsic parameters characterizing perpendicular transport [59].The major problem concerning device implementations of CPP spin valves is the series resistance from the contact leads compared with the intrinsic small device resistance. Recently, different structures, such as single and dual spin valves, spin valves with multiple magnetic/non-magnetic interfaces in the free layers, and structures where a Cu-oxide nanolayer is inserted in the Cu spacer to achieve current confinement have been investigated in order to maximize the resistance change (ΔRA) and increase device resistance (RA). Resistance-area (RA) products at the level of about 81 $m\Omega\mu m^2$ in single spin valves, and 150 $m\Omega\times\mu m^2$ in dual spin valves, and resistance change-area products (ΔRA) of 0.35 $m\Omega\times\mu m^2$ (single spin valve) and about 2.5 $m\Omega\mu m^2$ (dual spin valve) were reported [61].

For the structures with a CuO_x discontinuous insulating layer introduced in the spacer Cu layer, $RA = 2.2\ \Omega\mu m^2$ and $\Delta RA = 9\ m\Omega\mu m^2$ were reported [61, 62]. More recently, use of magnetic Co-Fe-B NOLs in the Cu spacer led to RA values close to 1 $\Omega\mu m^2$ and $\Delta RA = 50\ m\Omega\mu m^2$ [62]. The latter structures are seen as current confined CPP structures where the current flows inhomogeneously through pinholes in the partially oxidized NOL.

Fig. 12 shows the schematics of dual CCP-CPP spin valves developed at INESC-MN [63], where two oxide layers are present on the stack below and above the active layers. The spin valve structure is Al 800 Å /**TaO_x**/Ta 70 Å/NiFe 40 Å/MnIr 80 Å/CoFe 40 Å/Cu 50 Å/CoFe 20 Å/NiFe 30 Å/CoFe 20 Å/Cu 50 Å/CoFe 40 Å/MnIr 80 Å/Ta 30 Å /**TaO_x**/TiW(N) 150 Å. The two TaO_x layers result from vacuum breaks during sample processing and can be easily seen in the TEM micrographs in Fig. 12 where a high angle annular dark field STEM image for the dual spin valve is shown. The TaOx layer at the bottom Ta/vacuum break/Ta interface can be clearly seen as a dark line. The narrower TaO_x layer at the top Ta/vacuum break/TiW(N) interface is more difficult to see but is also discernible as a narrow dark line. Figure 13 gives the measured resistance, R_m, and resistance change, ΔR (defined as $R_{antiparallel}$ - $R_{parallel}$) for the dual spin-valve structure with NOLs, versus 1/A, where A is the sensor cross-sectional area. The sense current was kept small (less than 5 mA) to minimize sense current Oe fields and heating effects. The measured resistance should follow $R = \rho t/A + R_0$, with R_0 the contact .resistance, and the measured $\Delta R = (\rho_{parallel} - \rho_{antiparallel})t/A$ should be linear in $1/A$. Both R and ΔR are proportional to $1/A$ as expected. From a fit to the data, $RA = 514$ m$\Omega\mu$m^2, and $\Delta R = 3.9$ m$\Omega\mu$m^2 respectively.

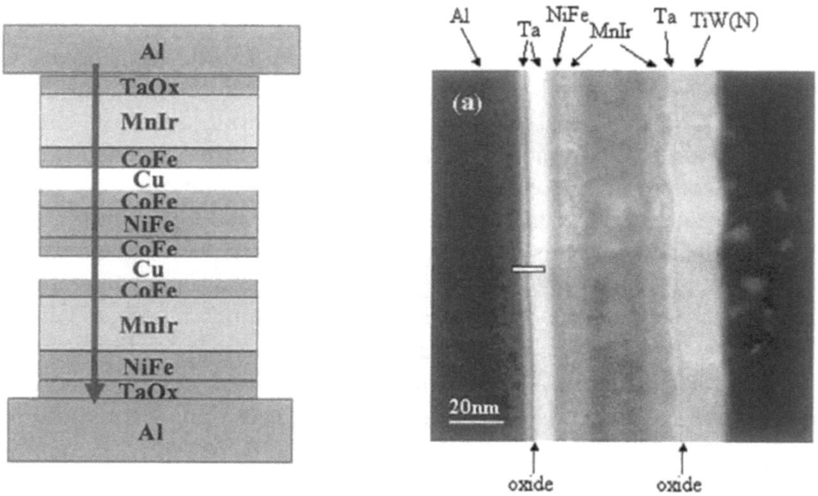

Figure 12. Dual CPP spin valve with two oxide layers at the top and bottom electrodes to confine the current path

The measured RA value includes contributions from bulk magnetic and non-magnetic layers as well as interfacial resistances. On the contrary, the resistance change-area product ΔRA originates only from spin dependent scattering. The CPP MR ratio is 2.02%. The measured ΔRA in the dual spin valve is about 4 times larger than that of the single spin valve. The enhancement of ΔRA in the dual spin valve is related to the increase in interfacial scattering from the increase in the number of interfaces and the bulk scattering in magnetic layers. When compared with similar dual structures prepared without NOLs, these CPP structures show both higher RA and ΔRA, showing that the NOLs are acting not only as series resistance (increase in RA) but also confining the current path and leading to higher ΔRA.

Despite these improved results, these two NOLs are not the best geometry for CCP-CPP spin valves, since pinholes existing on the top NOL do not mach topologically pinholes occurring on the bottom NOL, and the current paths are therefore not well defined. The best geometry for CCP-CPP spin valves seems to be that incorporating a discontinuous NOL within the Cu spacer. Latest results on CCP CPP devices start to show higher MR ratios (from 5 to 10%), relying on partial oxidation of a Al-Cu alloy layer within the Cu spacer, giving rise to the formation of Cu nanopillars confining the current and leading to high MR ratios [64].

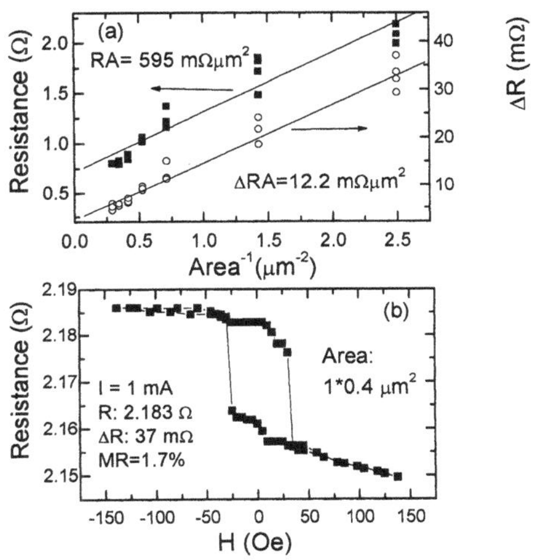

Figure 13. Spin-valve properties: (a) measured resistance R and resistance change ΔR versus the inverse of element size, 1/area for a dual spin valve.with NOLs inserted near the bottom and the top electrodes (b) MR transfer curve for a 0.4 μm^2 element (Haohua Li, PhD thesis, INESC-MN).

When large current densities (10+7 A/cm2) cross CPP spin valves, spin transfer effects start to be observed. When a spin polarized DC current flows across the CPP device, both well defined spin wave modes, and chaotic spin wave generation has been observed as a function of the injected DC current [65]. These spin-wave modes lead to excess magnetic noise. These effects were first predicted by Berger [66] and Slonczewski [67], and have been now experimentally observed by several groups. At low frequencies, spin transfer effects also affect the CPP device transfer curve [68], and can be used to induce free layer switching in magnetoresistive devices.

In comparison with magnetic tunnel junctions, CPP spin valves have the intrinsic advantage of having much lower resistances than MTJs, therefore being advantageous for scaling to lower dimensions. The major disadvantage is their lower signal due to series resistance contribution. Beyond 200 to 300

Gbit/in^2, tunnel heads will probably appear first as replacement for CIP spin valves. The ongoing studies of lower energy gap low resistance MgO barriers might extend the range of use of this technology. CPP spin valves or hybrid devices including discontinuous barriers at the Cu spacer, seem intrinsically better suited for scaling down to 25 to 30 nm minimum reader features. The technological problems arising with the production of these devices (ultrathin barriers, controllable pinhole fabrication) will in the end determine which technology will be used.

2. MAGNETIC RANDOM ACCESS MEMORIES

There is a strong interest in non-volatile memory devices based on magnetic materials, Magnetic Random Access Memories (MRAMs), due to their non-volatile characteristic, radiation hardness, non-destructive read-out, low-voltage, and very large (> 10^{15}) read-write cycle capability [69]. MRAMs can be almost as fast Static Random Access Memories (SRAMs), and almost as small as Dynamic Random Access Memories (DRAM) in cell size. First MRAM prototypes used minimum MTJ cell features not smaller than 0.25μm. Scaling MRAMs below 100 nm features, raises several technological issues regarding cell micromagnetics, free layer switching and power consumption. A recent review comparing alternative types of non-volatile memories can be found in ref. [70].

2.1. Field-Induced Magnetization Switching

2.1.1. Orthogonal line architecture

First generation MRAM devices as developed by IBM and Motorola used 1 magnetic tunnel junction and 1 transistor per base memory cell (see Fig. 14). The MTJ is gown in a back end of the line process, after the CMOS processing is complete, on top of a chemically mechanically polished (< 5Å rms) Cu bottom electrode. Memory writing is achieved by passing simultaneous short (< 5 ns) current pulses (few mA) on orthogonal word lines, creating simultaneous easy and hard axis fields that will switch the free layer. Memory read-out is obtained by passing a low (100 μA) read current through the MTJ and checking its resistance.

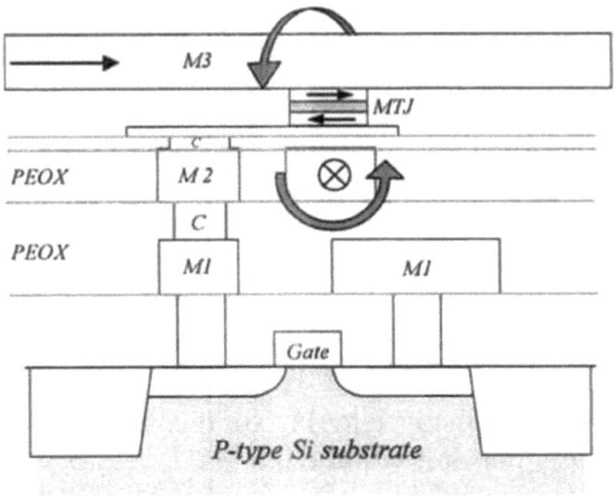

Figure 14. Memory cell with one MTJ / one transistor. The arrows indicate the easy axis and hard axis fields created during the write mode operation. Notice that the bottom word line is separated from the MTJ/bottom electrode by a thin insulating layer.

State-of-the-art MTJs for MRAM applications use typically bottom-pinned SAF MTJs with RA products from several hundred $\Omega\mu m^2$ to few $k\Omega\mu m^2$ (t_{Al} = 9-11 Å) and TMR in excess than 50%. Controlling Néel coupling (< 5 Oe) through ultrasmooth bottom electrodes (a-CoFeB for example), and achieving defect-free high quality barriers lead to high breakdown voltages (V_{br} > 1.5 V) and high $V_{1/2}$ values ($V_{1/2}$ > 700 mV). V ½ is defined as the bias voltage at which the tunnel junction TMR decreases by ½ of its zero bias value.

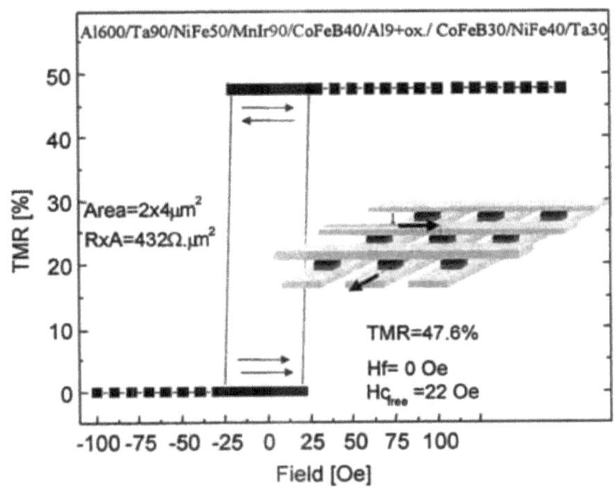

Figure 15. TMR minor loop for a MTJ with Co-Fe-B electrodes. Note the perfectly balanced (0 Oe) effective coupling between free and pinned layers. The inset shows a crossed wire MTJ only memory.

Figure 15 shows a bottom-pinned MTJ with Co-Fe-B electrodes. In this case, the low Néel coupling measured in an unpatterned portion of the sample (H_f = 6 Oe) is balanced by the magnetostatic coupling in the patterned junction, leading to a null effective coupling. The inset shows a crossed line memory architecture where MTJs only are placed at each intersection. This architecture allows 3D stacking but is slower in read-out due to the parallel arrangement of all junctions.

Free-layer switching during the write process can be basically modeled in the Stoner-Wohlfarth coherent-rotation model (Ch. 4), which yields an astroid switching curve [71]. Figure 16(a) shows an ellipsoidal cell, with magnetization **M** and applied field **H**. In the Stoner-Wohlfarth approximation, the cell energy per unit volume is

$$E/V = -H M_s \cos(\theta - \phi) + K_u \sin^2\theta + \tfrac{1}{2} M_s^2 (N_y - N_x) \sin^2\theta$$

$$= - H M_s \cos(\theta - \phi) + K_{eff} \sin^2\theta \qquad (8a)$$

where

$$K_{eff} = K_u + \tfrac{1}{2} M_s^2 (N_y - N_x) \qquad (8b)$$

Here K_u is the switching layer uniaxial anisotropy and N_y, N_x the demagnetizing factors along the hard axis (small axis of ellipsoid) and easy axis direction(long axis of ellipsoid).

The switching field H_s is obtained from $H_s = 2K_{eff}/M_s$. Figure 16(b) clearly shows the advantage of the use of simultaneous easy axis and hard axis write fields: the switching field is significantly reduced when compared with easy axis only or hard axis only fields. From Eq. 8 one can determine the energy barrier between switched (bit 1) and non-switched cells (bit 0).

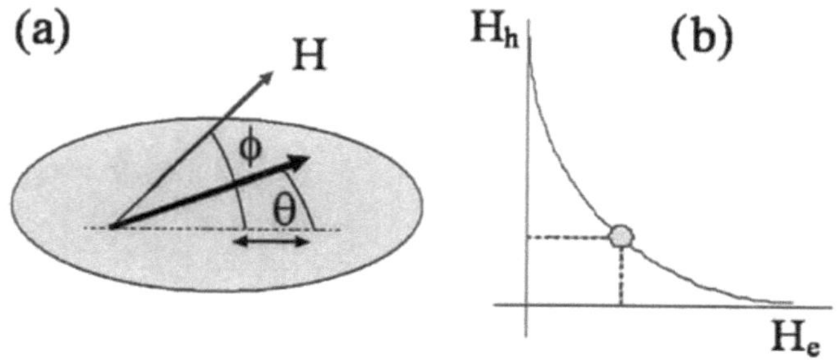

Figure 16. Stoner-Wohlfarth predictions: (a) schematic ellipsoidal cell used to calculate switching fields and the switching astroid in the Stoner-Wohlfarth model and (b) first quadrant of the switching asteroid, showing the switching boundary line.

For a field applied along the easy axis, the energy barrier

$$\Delta E = K_{\mathrm{eff}} V \left(1 - \frac{HM_s}{2K_{\mathrm{eff}}}\right)^2 \tag{9}$$

The energy diagram corresponding to the two stable configurations in the absence of an applied field, or for fields applied parallel to the easy axis is shown in Fig. 17.

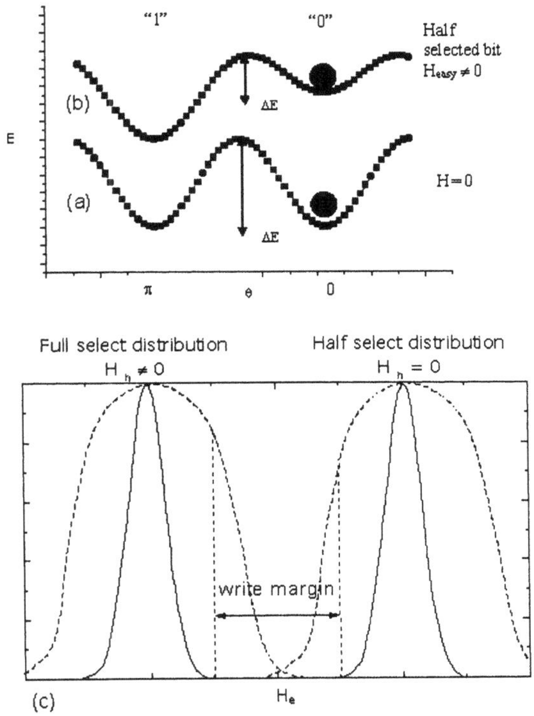

Figure 17. Energy barriers and switching in the Stoner-Wohlfarth model: (a) barrier between state "0" and state "1" when no field is applied, (b) reduced energy barrier under half-select conditions: only easy axis fields applied, and (c) switching field distribution for fully selected bits and for half-selected bits. For the full lines, the separation from the center of the two distributions is about 10 σ. In this situation half-select write errors do not occur. The dashed lines represent broadened switching field distributions. In this case the half-select and the full selected distributions start to overlap.

Under half-select conditions (field on), the energy barrier decreases. The existence of this energy barrier leads to stability criteria for the MRAM bits.

In order to prevent accidental thermal switching of a cell, it is required that $KV > 60\ k_BT$ [72]. Due to statistical distribution in the switching fields and thermal fluctuations becoming important as the magnetic volume decreases, some of the half-selected bits will eventually switch leading to write errors. Figure 18 represents schematically switching field distributions for selected (I_{hard} , $I_{easy} \neq 0$) and half-selected bits ($I_{easy} \neq 0$). The two distributions cannot overlap. As dimensions diminish the width of the switching field distribution increases, switching field dependence on bit shape becomes increasingly important, and minor lithography defects can produce unacceptable switching field variations. Due to this distribution, the switching field must be chosen so that even bits at the high end of the distribution will switch. However, at these fields, and with the help of thermal fluctuations, unselected bits at the low end of the switching distribution may now unintentionally switch. For standard free layers, it is therefore difficult to obtain tight switching windows, at moderate switching currents to avoid half-select disturbs.

Another point to consider is the dynamic nature MRAM operation. MRAMs are fast devices where switching speed is limited by the spin precession frequency (few GHz). For write pulses down to 10 to 20 ns, switching astroids are essentially similar to those obtained quasi-statically. Critical for reliable ultrafast switching is the single mode-single domain switching behavior of the cell. In a beautiful set of experiments, Schumacher *et al.* have shown that full bit reversal can be observed by tightly controlling the pulse width down to few hundreds of ps (half period) [73]. The magnetization precession around the effective field can then be stopped after half precession, avoiding the inevitable magnetization ringing for longer write pulse widths [74].

For the first generation of MRAM device prototypes [75, 76], with bit cells with minimum features from 0.25 to 0.6 μm, switching window control was achieved by understanding the micromagnetic behavior of the MTJ cell [77]. Shape anisotropy plays an important role in defining the switching field, and the formation of edge sates must be avoided to prevent increase in switching fields. Elliptical cells avoid edge states and a 2:1 aspect ratio was used in the first demonstrations. Power consumption (write current) was minimized by cladding the word lines with NiFe (see in Fig. 14, the cladding around the bottom word line), therefore increasing the write field by factors of 2 to 3 [78], minimizing the separation between the word line and the free layer (to around 400 to 500 Å), and reducing the free layer moment (reducing its thickness, and therefore reducing the switching field).

New types of MRAM devices are now being fabricated in order to allow easier scalability to features below 100 nm. The major problems that are faced are achieving the required write fields with reasonable power consumption, minimizing write disturb, narrowing switching field distribution, and maintaining bit stability. Three new types on MRAMs will be briefly discussed. The toggle switch MRAM, MRAMs using a thermally assisted writing process and an exchange biased storage layer, and MRAM s where storage layer switching is based on the spin transfer mechanism.

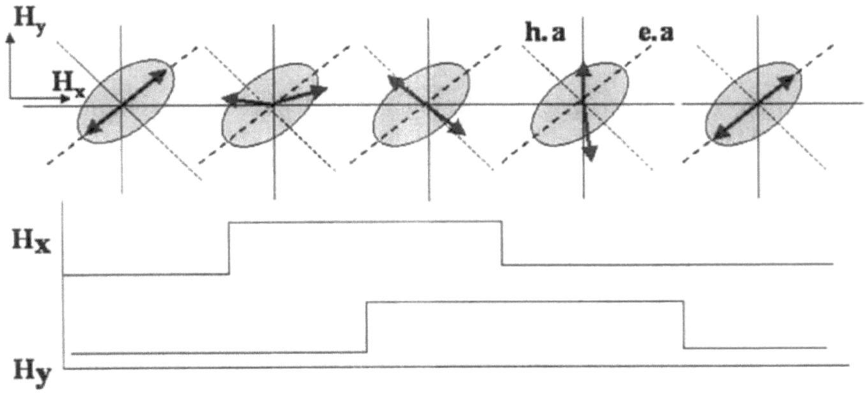

Figure 18. Write operation for toggle mode switching: the two magnetizations in the synthetic free layer toggle successively under the orthogonal applied fields leading to full switching. Under half-select conditions switching does not occur.

2.1.2. The toggle switch MRAM cell

In order to circumvent the write selectivity problem, researchers at Motorola introduced the toggle switch memory [79] as a direct evolution of the orthogonal line field-induced switching scheme. A balanced synthetic free layer [18, 80] with an easy axis at 45 degrees to the write fields created by the orthogonal word lines is used [81]. Figure 18 represents the pulse sequence used to switch the bit. The magnetizations of both ferromagnetic layers are also represented. The SF structure will flop at a critical field H_{flop}(rotate the magnetizations maintaining antiparallel alignment, such that the magnetizations remain orthogonal to the applied field) and scissor in order to minimize magnetic energy. Only when the two fields are applied simultaneously (resulting field applied along the bit easy axis) and if this field is above the bit flop field, will the magnetizations align at 90 degrees to their original orientation. The switching operation now continues by removing first the H_x field, and then the H_y field, leading to complete SF reversal. Notice that only the bottom magnetic layer of the SAF is effective in determining the junction resistance. The major advantage of this free layer geometry is that if only one line is operated (half selection), the magnetization of the half-selected bits will always go to step I, but will never cross over the SF flop field required to reach step II. By reducing half-select disturbs, this technique allows better cell scalability. The toggle write mode was applied by Motorola in the recent 4Mb MRAM demonstration [82].

The use of synthetic free layers can also increase bit stability [80, 83] (increase barrier height) for similar switching fields [84]. Defining the effective magnetization for the free layer as:

$$M_{\mathrm{eff}} = M_1 - M_2 \tag{10}$$

According to Eq.9, the barrier height is increased since the magnetic volume increases. The switching field becomes:

$$H_s = (2K_u/M_{\mathrm{eff}}) + M_{\mathrm{eff}}\,(N_y\text{-}N_x) \tag{11}$$

The conclusion from Eq. 11 is that SFs can be made with controllable switching field: the first term increases Hs due to the reduction of the net moment for the same intrinsic uniaxial anisotropy, but the second term, decreases the switching field from the decrease in shape anisotropy. Considering now the bit stability, since the magnetic volume is bigger, the energy barrier is increased for similar switching fields. The second advantage in using a synthetic free layer is that this structure allows the magnetic flux to close creating less stray field therefore reducing the magnetostatic coupling with close neighboring cells and layers. Despite these advantages, the toggle mode MRAM still requires reasonable currents for switching as the MRAM cell scales below 100 nm.

2.2. Thermally Assisted Switching

Solutions for the high current fields required for switching sub 100 nm cells include lowering the required switching fields, for example by thermally-assisted magnetization switching (TAS). Either Curie point writing or blocking temperature writing, can be used. In a first approach, junction heating has been proposed by passing current through the sense and /or word lines [85-87]. A more elegant approach uses the tunnel current through the junction itself for heating. In this approach, a transistor is used in series with the MTJ cell. The cell structure is similar to that in Fig. 14, but the cell transistor must cope with the full heat/write current. A single write line is required to create an easy axis reversal field. The typical cell structure is shown in Fig. 19. For writing, the weakly pinned storage layer in the MRAM is heated above its blocking temperature, and cooled under a smaller easy axis field [88-90]. This exchange-biased storage layer also stabilizes the bit against unintentional thermal write disturbs. Circular cells are now possible (reducing switching fields) since the exchange layer will stabilize the magnetization configuration down to the smallest dimensions. With this approach, the selectivity problem is greatly reduced (no disturb problems).

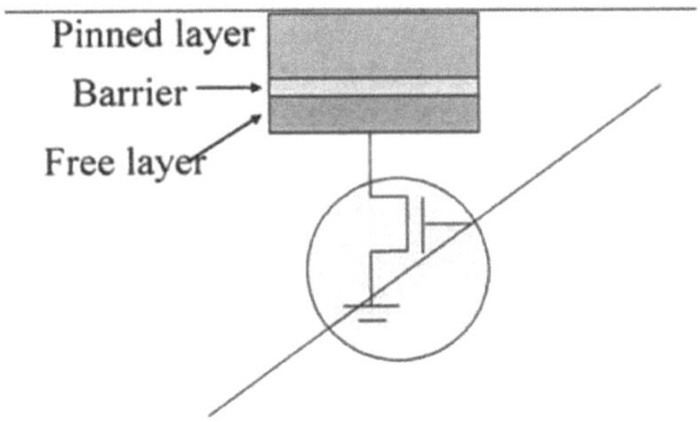

Figure 19. MRAM cell for thermally assisted switching with one MTJ in series with a transistor. In this architecture the transistor must cope with the full write/heat current, but a single word line is required.

In order to accommodate the required write current, this requires using medium resistance junctions (50 $\Omega\mu m^2$ to few hundred $\Omega\mu m^2$) with high breakdown voltage ($V_b > 1.5$ V). For a first proof of concept, single barrier and double barrier structures were studied [88] Fig. 20a) shows schematically the double barrier used in this study. In this case, the MRAM cell status had to be always interrogated since the information was thermo-magnetically stored on the storage layer.

The orientation of this storage layer was then determined by cycling the free layer. The reference storage layer was pinned by a thin MnIr layer leading to a blocking temperature of 120 °C. Single barrier junctions (about 40 $\Omega\mu m^2$) with a bottom pinned (MnIr 60 Å /CoFe 30 Å) storage layer structure were found to require too high current densities to achieve the required heating (44 mA/μm^2 at 40 ns pulse width). In order to increase Joule heating, and improve barrier breakdown and bias voltage dependence, double barrier tunnel junctions were used (~300 $\Omega\mu m^2$), where the weakly pinned layer (CoFe 30 Å /MnIr 60 Å /CoFe 30 Å) is sandwiched between two AlO_x barriers. The double barrier has a common pinned structure and two free layers (information is stored in the pinned layer). The AlO_x barriers were oxidized by a two-step natural oxidation (each step, 1 Torr for 5 minutes). Figure 20 (b), shows the minor TMR loops of double barriers junctions (solid squares, prior to pulse heating; open triangles, after heating pulse and reversing the pinned layer in a 60 Oe easy axis external field). The junction area is 1 μm^2. TMR of 25% and a resistance × area product of 300 $\Omega\mu m^2$ were obtained. The amplitude and width of the current pulse needed to heat up the double barrier junctions is much smaller when compared to that of the single barrier junction. A 10ns current pulse of 9mA is enough to heat the double barriers junctions up to blocking temperature (without current in word line). The open circles correspond to a pinned layer direction at 90 degrees with the free layers easy axis. This pinned layer orientation is obtained by cooling under a transverse hard axis field (60 Oe) created by a 30 mA current in the world line above the top electrode. For these double barriers junctions with *RA* ranging from 498~280 $\Omega\mu m^2$ the required critical currents are of 7.5 to 9 mA/μm^2 (about 20 mW/μm^2) for 10 ns pulses.

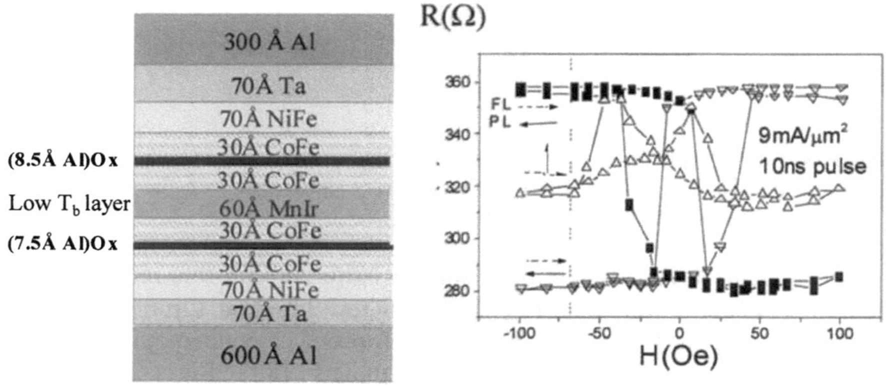

Figure 20. Example of double barrier magnetic tunnel junctions used in thermally assisted switching: double free layer and weakly pinned storage layer (left) and minor TMR loops of double barriers junctions (right). Solid squares: prior to pulse heating, open triangles: after heating pulse and reversing the pinned layer in a 60 Oe easy axis external field. The junction area is 1 μm^2, and a TMR of 25% and a resistance × area product of 300 $\Omega\mu m^2$ are obtained.

A more efficient TAS MRAM cell will use a weakly pinned low blocking temperature storage layer(thin MnIr, low Tb), and a strongly pinned, high blocking temperature, reference layer (PtMn, thick MnIr, high Tb) [89, 90]. Figure 21 shows schematically a double barrier structure using this architecture. The blocking temperature difference in the two antiferro-magnets should exceed 100 °C to prevent reference layer disturb during storage layer writing. Assuming the blocking temperature of AF1 to be around 160 °C to prevent accidental thermal depinning (thin MnIr), AF2 must have a blocking temperature in excess than 250 °C (PtMn, NiMn). In a first experimental demonstration, using single MTJ cells with exchange biased storage layers (Ta 30 Å/ NiFe 70 Å/ IrMn 60 Å/ NiFe 30 Å/ AlOx 6 Å/ CoFe 35 Å/ Ru 7 Å/ CoFe 25 Å/ IrMn 250 Å/ Ru 30 Å/ TiW(N) 150 Å), critical write powers increasing from 20 mW/μm^2 to 80 mW/μm^2 were observed as cell size decreased from 1 μm^2 to 0.1 μm^2. As the junction area is decreased, it will be no longer valid to assume that heat flows mostly perpendicular to the junction top and bottom surfaces. While the assumption might be true at the centre of the junction, the horizontal component of the heat flow will increase as one approaches the junction edges, since the leads are larger than the junction area. As a consequence the power density required to achieve the same ΔT will not remain constant, but will increase as the junction area is reduced. However, the total absolute power required to elevate the junction temperature by ΔT will continue to decrease with the junction area. Dynamic thermo-magnetic writing at 5-10ns was also demonstrated [89, 90].

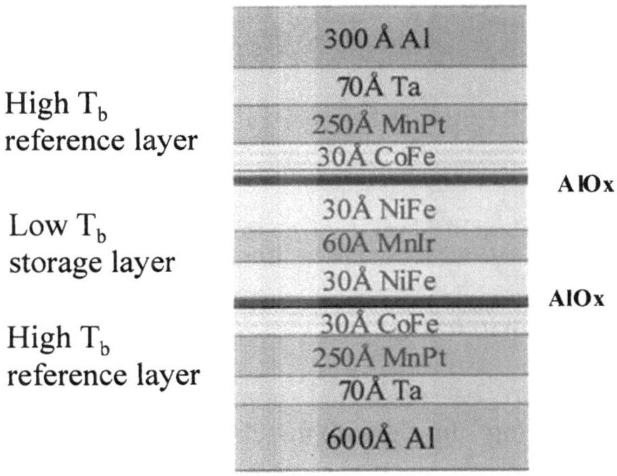

Figure 21. TAS MRAM cell using a weakly pinned low blocking temperature storage layer (thin MnIr, low Tb), and a strongly pinned, high blocking temperature, reference layer (PtMn, thick MnIr, high Tb) [89, 90].

Figure 22 shows experimental data on a 0.3 × 0.2 μm^2 single barrier junction. Selective writing of the memory cell is achieved, applying a current heating pulse (2 V, 20 ns) and a simultaneous +/-200 Oe magnetic field value. Each time a heating pulse is applied to the junction, the temperature in the storage layer exceeds the Tb of the 60A MnIr AF and the magnetization can be switched. This cycling sequence shows that the memory is correctly written as the resistance jumps each time between two well defined resistance levels corresponding to the AP or P states.

This measurement highlights two important features of the thermo-magnetic write scheme. First, excellent selectivity as only heated junctions can be magnetically written and second, that the temperature-rise time can be only a few nanoseconds. The use of extra heating layers in the junction stack is being considered, in order to enable scaling the TAS MRAM approach below 100 nm features.

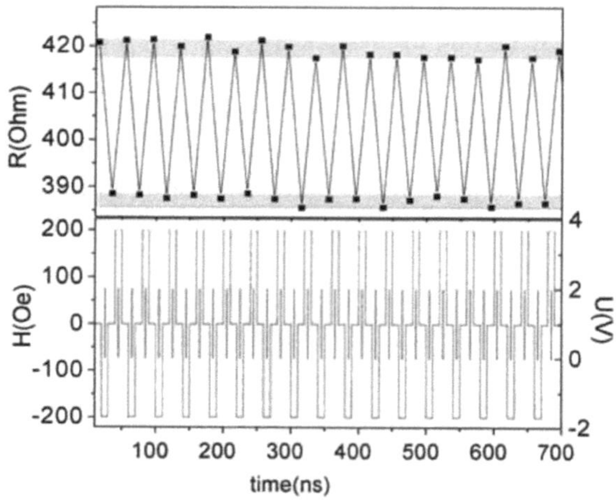

Figure 22. (Top) measured zero field values of the junction resistance and (bottom) sequence of applied magnetic fields and the position of the 20 ns heating pulses of 2V.

2.3. Spin Transfer Switching

The last approach to be discussed comes from direct spin transfer magnetization switching. In this approach, a polarized electron current crosses the bottom polarizer electrode, crosses the barrier and enters the free layer. This method induces full magnetization switching if the polarized current exceeds a critical value J_{cr}, where $J_{cr} \cong q\, \alpha\, M_s\, t\, H_{eff}/h$ [66, 67]. Here α is the damping parameter for the precessing spins, q the electron charge, M_s and t the saturation magnetization and thickness of the switching free layer, and H_{eff} the effective field in the free layer. This approach solves the selectivity problem as the TAS approach, but requires at the moment too high write currents to be competitive ($J_{cr} > 10^6 A/cm^2$) [91-93]. Furthermore, the high current densities required for the spin transfer effect imply the use of low resistance MTJs (as used for read heads) in high current regimes, leading to barrier degradation and or charge migration in or out of the barrier [94-96]. First results on current-induced magnetization or resistance change in TMJ cells led to the discovery of several effects related to the inhomogeneous nature of the tunnel current. Depending on barrier microstructure, both spin transfer [91-93], electromigration [94-96], or joule heating effects (see previous section) can be observed when high current densities cross the ultra-thin tunnel barrier. Figure 23 shows one example of current induced magnetization switching (spin transfer driven, Jianguo Wang, INESC MN).

In this case the switching currents depend strongly on an external applied field. Figure 23(a) shows that the *R vs. H* loop and *R vs. I* loops. In this case, positive currents favor the parallel alignment, corresponding to electrons being polarized in the reference layer and flowing into the free layer. The *R vs. I* loop is traced under a constant – 20 Oe field (Néel coupling) used to center the free layer switching loop. Figure 23(c) shows the effect of an external field on the *R vs. I* loop. As the negative applied field increases, the negative critical current decreases. For $H_{ex} < -70$ Oe, the free layer is already on the antiparallel state.

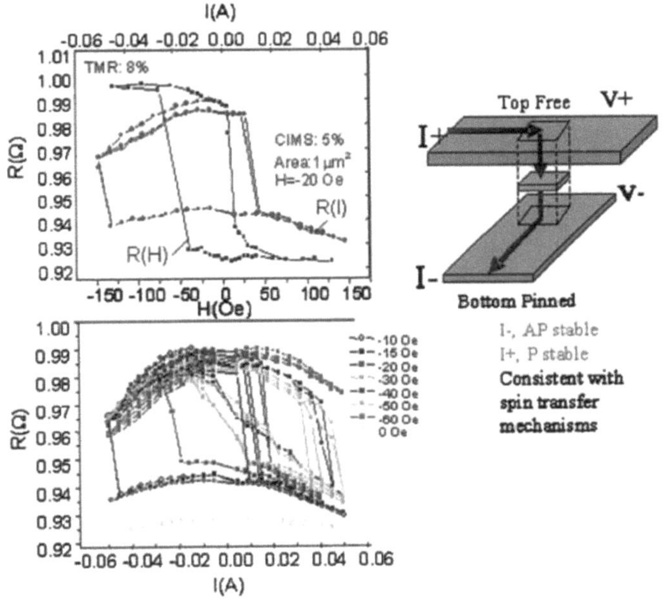

Figure 23. Current induced magnetization switching (spin transfer driven): (a) resistance *vs.* field and resistance *vs.* current loops (top left), current direction definition with respect to reference and free layers (right), strong dependence of the switching currents on an external applied field (bottom left). Typical critical currents are 3 to 5 MA/cm^2.

Depending on barrier structure, a different physical effect can be observed when passing large current densities through thin insulating barriers. Figure 24 shows current induced resistance change (electromigration-driven) in thin tunnel junctions fabricated at INESC MN. Although the *R vs. I* loops are similar to those obtained by spin transfer switching, in this case, the critical switching currents on the *R vs. I* cycles do not depend on the external applied field(see Fig. 24c). This is a direct proof that a non-magnetic effect is at the origin of these current-induced resistance changes. Electromigration driven movement of electric charges/defects at the barrier/electrode interface or within the barrier are thought to be at the origin of the observed effects [95, 96].

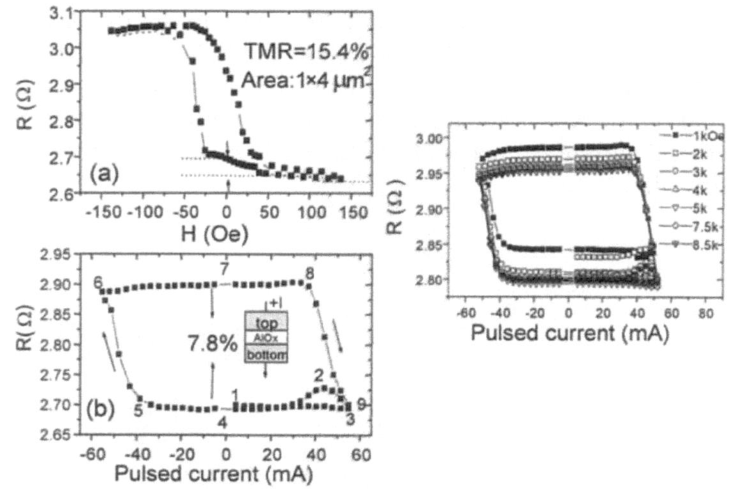

Figure 24. Current induced resistance change (electromigration driven) in thin tunnel junctions. Notice that the switching currents on the *R vs. I* cycles do not depend on the external applied field.

In conclusion, new MRAM architectures are being investigated allowing scaling below 100 nm minimum features, while the orthogonal line field induced switching devices are nearing industrial production (4MB and 16MB demonstrators made by Motorola and IBM/Infineon).

3. SPINTRONIC BIOSENSORS

A new area of application of spintronic nanostructures concerns biomedicine and biotechnology (Ch. 15). Biomolecular recognition has been playing an ever important role in health care, pharmaceutical industry, environmental analysis and broad biotechnological applications. In particular, a great deal of effort is being placed in developing high-performance and low-cost tools for the detection of DNA-DNA hybridization in genetic disease diagnostics, mutation detection or gene expression quantification, and for the detection of antibody-antigen interaction in microorganism identification and biological warfare agent screening.

The idea behind a spintronic biochip or biosensor is to replace traditionally used fluorescent markers by magnetic labels. Instead of detecting biomolecular recognition using expensive optical or laser-based fluorescence scanner systems, the detection is made by a spintronic element, such as a magnetoresistive sensor, that senses the magnetic labels stray field and provides a straightforward electronic signal at low-cost. In addition, since biological samples are usually non-magnetic, the background is greatly diminished when compared to fluorescence methods. Other advantages of the spintronic biochip are the fast response, high sensitivity and ease of integration and automation, making it competitive in the analysis of few (100's to 1000's) of biological analytes (e.g. screening for mutations in genetic diagnosis).

A typical spintronic biochip consists of an array of sensing elements (such as magnetoresistive sensors); an array of probes (biomolecules of known identity such as gene specific oligonucleotides or antibodies) that are immobilized onto the surface of the sensors (through microspotting, or electrical or magnetic arraying); an hybridization chamber (normally a microfluidic channel arrangement); and an optional target arraying mechanism (electric fields for charged molecules such as DNA or magnetic field generating lines for magnetically labeled targets, see Fig. 25).

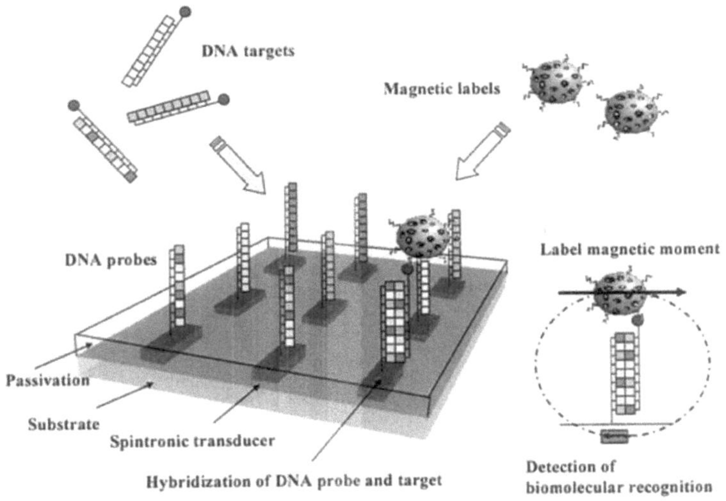

Figure 25. Schematic of a spintronic biochip which is composed of an array of spintronic transducers, an array of probe biomolecules immobilized onto the surface of the sensors (in these case single-strand DNA molecules are represented), solutions with target biomolecules (DNA strands) and magnetic labels that bind to the surface through biomolecular recognition (DNA hybridization). At the side: the detection of biomolecular recognition is achieved through the sensing of a magnetic stray field created by the magnetic label using a spintronic transducer.

The targets (biomolecules to identify in a sample such as a DNA strand complementary to the immobilized DNA probe, or antigens complementary to the immobilized antibodies) are incubated with the chip for biomolecular recognition to occur. They can be already magnetically labeled before or be labeled after the recognition step. Magnetic labels usually are superparamagnetic or non-remanent ferromagnetic in nature, with nano- or micrometer dimensions, and can attach to the target biomolecules. Under a magnetic applied field these particles or beads acquire a moment and their fringe field can induce a change in resistance of the spintronic sensor, enabling biomolecular recognition detection.

Magnetoresistive-based biochips were first introduced in 1998 by the Naval Research Laboratory (NRL) [97] and since then an increasing number of research laboratories and companies has been developing such systems. Magnetic label detection has been accomplished by using different types of integrated magnetoresistive sensors, such as GMR stripes [98, 99], meander GMR structures [100] and GMR spirals [101, 102]; linear [103-106], serpentine [107] and U-shaped [108] spin-valves; AMR rings [109]; planar Hall effect crosses [110, 111]; and magnetic tunnel junctions [112]. These labels also enable the use of magnetic fields for stringency control [97] and for on-chip manipulation and transport integrated with the detection [103, 104, 108, 113]. Recent reviews on the different magnetic biosensing platforms can be found on references [100, 114, 115].

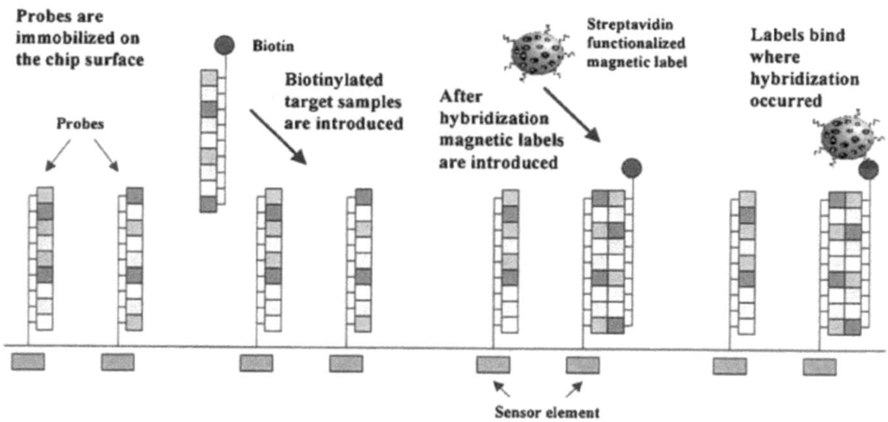

Figure 26. Schematic of the post hybridization detection method. A DNA target solution labeled with biotin is first incubated with the DNA probe functionalized chip. Targets diffuse passively from the solution to the surface where they hybridized with the probes if complementary. A solution containing streptavidin-functionalized magnetic labels is then incubated with the chip. Labels bind through the biotin-streptavidin interaction to where hybridization occurred. DNA hybridization is detected with spintronic transducers.

The biological systems studied so far concern mainly to the detection of DNA-DNA hybridization. Two methodologies have been approached, post-hybridization detection and magnetic field assisted hybridization and simultaneous detection. The first method relates to the incubation of streptavidin-functionalized magnetic particles with the chip. Labels bind to the surface where previously hybridization with a biotinylated target has occurred, through the strong biotin-streptavidin interaction. Hybridization in this case takes long time (3 to typically 12 hours) to happen as targets need to diffuse in solution to find the complementary match (see schematic for the method in Fig. 26). This approach has been followed by the groups at NRL [97-100], INESC [114, 115], and the University of Bielefeld [102], but is unsuitable for applications where a fast response is required or desired such as biological warfare agent or pathogen microorganism detection, or a clinical diagnostics during a medical doctor's appointment.

The second method has been followed by INESC and relates to the incubation of magnetically-labeled target DNA. In this case particles or beads are transported on-chip and focused at probe functionalized sensing sites using tapered current line designs [103, 104] or U-shaped current line [116]. A variety of other on-chip magnetic particle transport mechanisms have been developed and can, in principle, be integrated with spintronic sensors [117-120]. The proximity of target and probe increases the rate at which hybridization of complementary DNA strands. Hybridization times of 5 to 30 min where observed and detected [108, 113] which is a factor of at least 6 to 150 times faster than with the previous approach (see schematic for the method in Fig. 27).

In order to meet specific biological and detection requirements the spintronic transducers can be tailored in size and shape. In the post-hybridization detection approach the biological sensitivity of the assay (capability of a biomolecule in solution to recognize its complementary biomolecule immobilized to the sensor surface) increases with the functionalized area and so the sensor should then be of the same size as that area for maximum sensor biological sensitivity (capability of detecting the bound biomolecule using the spintronic transducer). For instance for DNA spots of typical sizes of 50 to 100 μm in diameter, GMR serpentines [100] and spirals [101, 102] have been designed and fabricated. For these geometries though, only roughly half of the area occupied is sensing material, and so the corresponding biological sensitivity is approximately half the size of the DNA spot. The transducer can also be made such that the full sensor area is biologically sensitive such as in planar Hall sensors [110, 111].

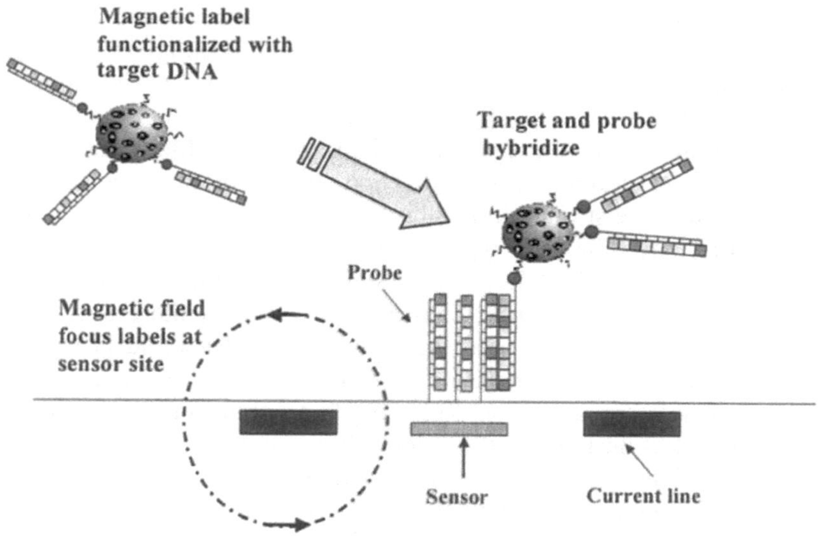

Figure 27. Schematic of the magnetic field assisted hybridization and simultaneous detection method. A solution of magnetic-labeled DNA target is first incubated with the DNA probe functionalized chip. Targets are focused to sensing areas using magnetic field gradients created by on-chip current lines and hybridize with the probes if complementary. DNA hybridization is detected with spintronic transducers

The increase of the sensor size comes with a decrease in sensor label sensitivity, that is, the sensor response for a single magnetic label is smaller. On the other hand, the sensor dynamic range is larger as the sensor is able to detect a larger number of particles. In order to be sensitive to single labels [106] the sensor size can be reduced to match the label size [121], but in this case the assay biological sensitivity is smaller.

In the magnetic field assisted hybridization and detection approach some of these sensitivity issues are circumvented. Since targets are controllably transported and focused onto sensing regions the biologically functionalized area can be made smaller and matched to the sensing area. This way both the biological sensitivity of the assay and the sensor are increased. Linear $2 \times 6\ \mu m^2$ [103] and U-shaped $2.5 \times 80\ \mu m^2$ [108] spin-valve sensors were designed for respective dynamic ranges of 1 to 200 and 1 to 3200 magnetic particles of 250 nm in diameter (see Fig. 28).

In addition, in terms of biological sensitivity of the sensor, since it can be designed and fabricated to detect a single magnetic label, it can in principle be used to detect single biomolecular recognition events (such as a single DNA hybridization) [115], making spintronic transducers a potentially valuable tool in molecular biology studies.

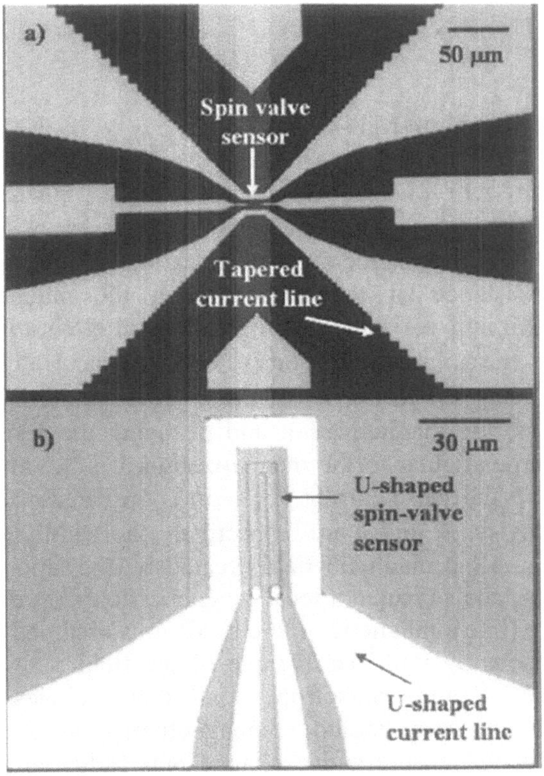

Figure 28. Photographs of spintronic biochip architectures from INESC for use in magnetic field assisted hybridization detection experiments. (a) biochip using linear spin-valve sensors ($2 \times 6\ \mu m^2$) and tapered current lines, and (b) biochip using u-shaped spin-valve sensors ($2.5 \times 80\ \mu m^2$) and u-shaped current lines

These sensors usually work in the linear regime of their transfer curves such that particle quantification is straightforward from the output signals. The linearization (and the linear range) of the sensor response and can be obtained by sensor fabrication, geometry (such as demagnetization fields in spin-valves) or/and by external or on-chip magnetic bias fields. Electrical responses (voltage drop ΔV_s) from the mainly used spintronic transducer in biological applications are shown below [114].

GMR:

$$\Delta V_s = -(\Delta R/R)_s I R_{sq} W < |H_{labels}| > / h H_k \tag{12.a}$$

Spin-valves (SV):

$$\Delta V_s = -\tfrac{1}{2} (\Delta R/R)_s I R_{sq} W < H_{labels} > / h H_k \tag{12.b}$$

Planar Hall (PH):

$$\Delta V_s = -\Delta R I \langle H_{labels} \rangle / H_k \quad (12.c)$$

Magnetic Tunnel Junction (MTJ):

$$\Delta V_s = -\tfrac{1}{2}(\Delta R/R)_s I R_{sq} RA \langle H_{labels} \rangle / WhH_k \quad (12.d)$$

Here $(\Delta R/R)_S$ is the magnetoresistance ratio, defined as the difference between the resistance of the sensor when the magnetic layers are antiparallel (maximum resistance state) and parallel (minimum resistance state) divided by the minimum resistance. In the planar Hall sensor $\Delta R = (\rho_{\|} - \rho_{\perp})/t$, with $\rho_{\|}$ being the resistivity of the magnetic material with the sense current parallel to the magnetization and $\rho_{\perp}$ being the resistivity with the sense current perpendicular to the magnetization, t is the magnetic material film thickness. I is the sense current, $R_{sq} = \rho/t$ is the sensor square resistance, with ρ the resistivity. *W* and *h* are respectively the width and the height of the sensor. H_{labels} is the magnetic field created by the labels at the sensing layers. Although, labels created a local magnetic field that decreases rapidly with the distance (like a magnetic dipole field), it is assumed that the sensing layer rotates coherently with the average of the field $\langle\ \rangle$ on the sensing layer. H_k is the effective anisotropy field that includes the crystalline anisotropy field and may include demagnetizing fields such as in the fabricated spin-valve sensors and exchange bias fields as in the exchange bias planar Hall sensors.

These different sensor types have distinct maximum outputs (magneto-resistive change) which impose constraints on the applicability and design of the biosensor system. It is desirable to have a maximum signal per label, and in that respect MTJs seem to have an advantage, followed by SVs, GMR, and PH and AMR sensors. Nevertheless, the figure of merit for magnetic label and thus biomolecular recognition is not so much the amplitude of the signal but the signal-to-noise ratio (SNR). All sensors show Johnson noise given by $V^2_{Johnson} = 4k_BTR$, where k_B is the Boltzmann constant, T is the absolute temperature, R is the sensor resistance and Δf is the frequency bandwidth of the measurement. As seen from the expression the highest the sensor resistance the highest the thermal noise. Nevertheless the dominant source of noise for low frequency measurements (DC-to few 100's of Hz as used so far) is $1/f$ magnetic noise given by $V^2_{1/f} = (\gamma/N_c) R^2 I^2_{rms} (1/f)$, where γ is a phenomenological constant (Hooge's constant), N_c is the number of noise carriers in the sensor, I is the sensing current and f is the measuring frequency [122]. For MTJ's the $1/f$ noise is given by $V^2_{1/f} = (\alpha_f/A) R^2 I^2_{rms}$ $(1/f)$, where α_f is a phenomenological parameter and A is the junction area

[126]. In addition MTJs show shot noise due to tunneling electrons and given by $V^2_{shot} = 2qV_bR$, where q is the electron charge and V_b is the voltage bias across the junction. For low frequency measurements the planar Hall sensor shows the lowest $1/f$ noise and highest SNR, followed by SV, GMR, low RA MTJ and AMR rings [102]. In particular, SVs show lower $1/f$ noise than MTJs and in order for MTJ SNR to be comparable to SVs', a low *RA* of about 1Ωμm² with large TMR, as mentioned previously. To obtain the maximum SNR possible, sensors must operate above the $1/f$ knee in the thermal noise regime, which happens at frequencies of 100's of kHz for spin valves, but above 100 MHz for tunnel junctions (strongly dependent on junction resistance). High frequency measurements can in principle be used for the detection of smaller size magnetic labels (about 10 to 20 nm in diameter) of very small magnetic moments, which can be bound to single biomolecules, providing the ultimate biological sensitivity of the sensor. Another important reason for using small labels is that being comparable in size to the molecules they are labeling the steric hindrance of biomolecular recognition is greatly diminished when compared when using larger sized micrometer labels.

Spintronic biochip platforms have been used for the detection of biomolecular recognition in binding models such as biotin-streptavidin [124, 125], immunoglobulinG – Protein A, and DNA-cDNA (cf. cystic fibrosis) [114, 115], and in the development of applications in the detection of biological warfare agents [98, 99] and most recently in the detection of cells from pathogenic microorganisms [126]. The two chip architectures developed at INESC for magnetic field assisted hybridization and detection of cystic fibrosis related DNA targets were shown in Fig. 28. Using the platform of figure 28a that features tapered current line designs and small 2 × 6 μm² spin-valve sensors, the hybridization of magnetic labeled PCR amplified (Polymerase Chain Reaction) DNA targets to surface immobilized complementary DNA probes was detected in few minutes. The DNA probe, a single strand of 50 nucleotides (50-mer), spans a region of the Cystic Fibrosis Conductance Regulator (CFTR) gene that contains the most common mutation F508del, a single aminoacid deletion at phenylalaline residue 508 in the CFTR protein. Real-time DC detection signals were obtained using 8 mA sense current, 20 mA currents through the tapered lines for focusing the labels (250 nm in diameter) and an external 15 Oe in-plane field for magnetizing the labels and biasing the sensor [124, 125]. Figure 29 shows a result obtained by interrogating the chip with complementary DNA targets. After addition, of the magnetically-labeled biomolecules, current was sent through the current lines for 3 min attracting the particles to the sensing region, after which particles were left to settle on to the sensor for another 3 minutes to further promote hybridization. The chip was then washed to remove unspecifically and weakly bound labels. A residual signal of about 1mV was obtained due to hybridization. This signal corresponds to about 50 nanoparticles bound to the surface. When using a non-complementary target the signal returned to the baseline indicating that no hybridization occurred. These small size (2 × 6 μm²) sensors have a small dynamic range of up to 200 nanoparticles of 250 nm in diameter, but provide a higher signal per particle.

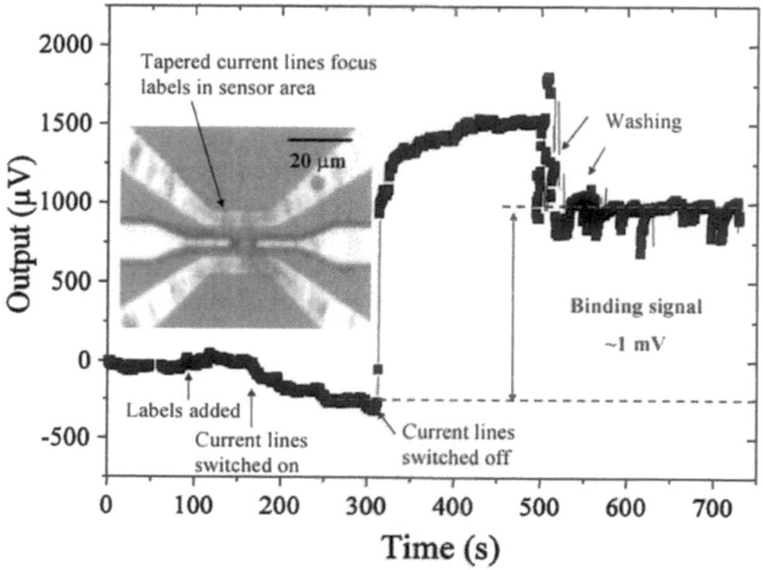

Figure 29. Real-time data obtained for a magnetic field assisted hybridization experiment using the architecture of fig. 27a). A 50-mer CFTR-related DNA probe immobilized on the chip surface was hybridized with complementary DNA targets labeled with 250 nm magnetic particles. A sense current of 8 mA was used together with currents of 20 mA through the tapered current lines and a 15 Oe external bias field. Inset: photograph showing magnetically-labeled DNA target focusing at the sensing area using tapered current lines.

This sensor size may thus be better suited for biological assays where a yes-or-no answer is required or only a small degree of signal quantification is needed. A potential application is the detection of mutations in DNA that occur in one or few nucleotides only, the so called SNPs (Single Nucleotide Polymorphisms). In this case it is then necessary to distinguish accurately between almost identical DNA strands and it only matters if the particular mutation being probed is present or not.

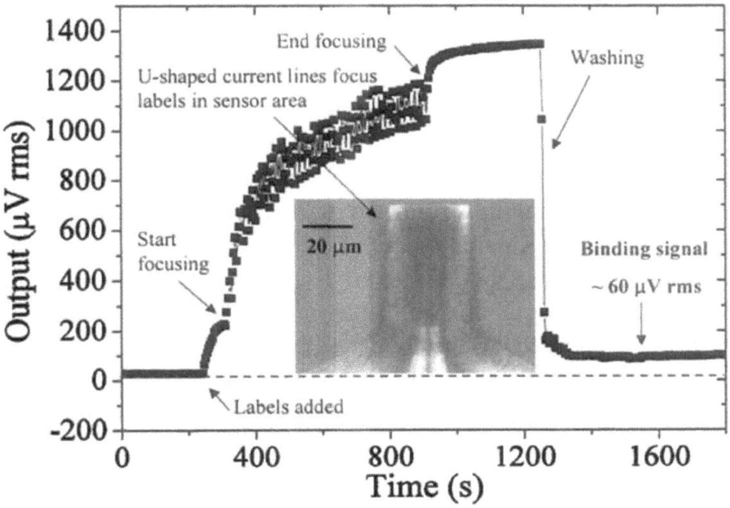

Figure 30. Real-time data obtained for a magnetic field assisted hybridization experiment using the architecture of Fig. 21(b). A 50-mer CFTR-related DNA probe immobilized on the chip surface was hybridized with complementary DNA targets labeled with 250 nm magnetic particles. A sense current of 1 mA was used together with currents of 40 mA rms (at a frequency of 0.2 Hz) through the u-shaped current lines and an external magnetic field of 13.5 Oe rms (at a frequency of 30 Hz) with a 24 Oe DC component. Inset: photograph showing magnetically-labeled DNA target focusing at the sensing area using a u-shaped current line.

In the case where quantification in biological assays is needed, such as in gene expression (to assess if a gene is more or less expressed) and microbial cell counting, sensors with larger dynamic ranges are required. In that respect, larger ($2.5\times80\ \mu m^2$) u-shaped spin-valve sensors were designed and fabricated, together with u-shaped current lines for AC field focusing of magnetic labels (see Fig. 28b). This platform was used to detect the hybridization of cystic fibrosis related DNA targets. Single-stranded 50-mer probes and targets were designed to correspond to genes that have been found to be either up- or down-regulated (that is over or sub-expressed) in cystic fibrosis related cell lines *vs.* healthy cell lines by preliminary microarray analysis of gene expression [127]. Real-time AC detection signals were obtained using 1 mA sense currents, an external in-plane AC excitation field of 13.5 Oe rms at a frequency of 30 Hz together with a DC bias component of 24 Oe, and currents of 25 to 40 mA rms at a frequency of 0.2 Hz were sent through the u-shaped lines for focusing the labels. Figure 30 shows a result obtained by interrogating the chip with complementary

DNA targets. After addition, of the magnetically-labeled biomolecules, a current 40 mA rms at 0.2 Hz was applied to the U-shaped current line for 10 min attracting the particles to the sensing region. After turning of the current particles were left for 5 minutes to settle onto the surface to further promote hybridization between target and immobilized probe. The chip was then washed to remove unspecifically and weakly bound labels. A residual signal of about 60 μV rms was obtained due to hybridization. When using a non-complementary target signals returned to the baseline indicating that no hybridization occurred. Signals of about 10 μV rms were obtained for non-specific binding in these experiments. Recently statistical data on chips functionalized with single and multiple probes were obtained using this platform showing hybridization signals from 100 to 300 μV rms and non-complementary and background (no probe) signals of about 10 to 15μV rms, showing the potential of this system to be used for gene expression analysis where, in minimum, a two fold up and two fold down in expression of genes is required.

Spintronic transducers show a great potential for integration in novel biosensors and biochips. Their unique features, such as the magnetic field transduction mechanism with electronic readout, fast response, high sensitivity, scalability, automation and easy CMOS process integration, makes them versatile enough to be tailored to the desired biological applications, in areas from biomedicine and biotechnology to food and environmental analysis. As a result, interest from laboratories and companies, and research on this field of spintronics is continuously increasing.

4. SPIN TRANSISTORS

Much of the sophistication of modern electronics is enabled by active three terminal devices such as bipolar transistors, field-effect transistors, silicon controlled rectifiers, triacs, unijunction transistors to name some principal players. The lure of devices that resemble these but that have the additional feature that their characteristics are magnetically tunable needs little apology. True, such devices may be mimicked for simple tasks such as basic sensing by coupling giant-magnetoresistive elements with conventional semiconductor packages. However integrated devices in which electronic power gain and magnetic sensitivity are implemented in the same location have considerably more to offer. Moreover, an active device whose parameters are susceptible to magnetic modulation opens the way to higher speed signal handling and more sophisticated electronic processing such as magnetically driven signal mixing to give just one example.

Various schemes have emerged in the quest for a magnetically sensitive transistor and in this section we shall focus on two such approaches which are different in concept yet which turn out to have some surprising similarities. For example both devices operate by creating and propagating in the central part of the structure a non-equilibrium state whose relaxation determines the device characteristics: and in the analysis of both device types there ultimately emerges a trade-off between the conventional electron gain parameters and the magnetic sensitivity.

Chronologically the first device to appear was the spin-valve transistor (SVT) which later developed into the MTT, so we will first turn our attention to describing these seminal devices.

4.1. Hot Electron Spin Transistors

Hot electron spin transistors are hybrid metal/semiconductor devices that rely on spin-dependent transport of hot (nonthermalized) electrons rather than electrons near the Fermi level. The spin-valve transistor (SVT) was the first example of this new class of spintronic devices [128, 129]. It has a three-terminal structure consisting of a metallic spin-valve base that is sandwiched between two semiconductor substrates, serving as the emitter and the collector, respectively. The electrons in this device are transported perpendicular to the spin-valve layers at energies just above the collector Schottky barrier height.

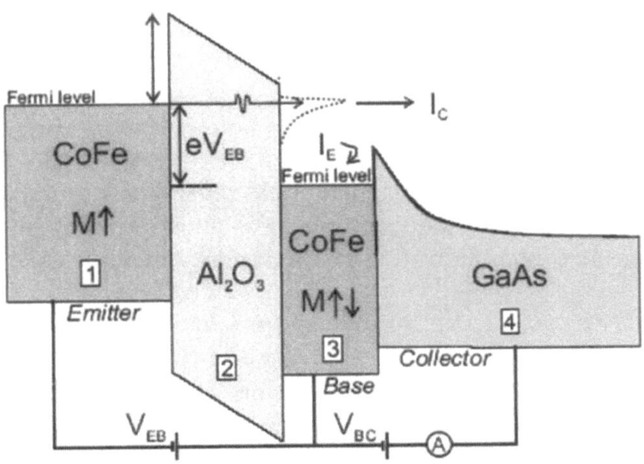

Figure 31. Schematic energy diagram of a MTT with a single ferromagnetic base layer. Region 1 is the emitter, region 2 is the Al_2O_3 tunnel barrier, and region 3 is the base. In this case a GaAs substrate acts as the semiconductor collector. From Ref. 157.

An alternative to the SVT is the magnetic tunnel transistor (MTT), in which a conventional tunnel barrier replaces the Schottky barrier emitter [130-134]. The hot electron energy in this device can be tuned by varying the bias voltage across the tunnel barrier (see Fig. 31). The MTT is therefore a powerful tool to study the energy dependence of electron transport in ferromagnetic films and across metal/semiconductor interfaces. There are two basic MTT designs. The first one consists of a ferromagnetic emitter and a single ferromagnetic base layer. Due to the thin base layer this structure can exhibit relatively large collector current outputs. The second design has a ferromagnetic spin-valve base instead of a single ferromagnetic base layer. Just like the SVT it has very high magnetic field sensitivity, which can be nearly two orders of magnitude larger than that observed in metallic multilayers and magnetic tunnel junctions (MTJs).

In SVT and MTT devices hot electrons are injected from the emitter into the base by applying an emitter/base bias voltage (V_{EB}). Scattering of these electrons in the base layers leads to a loss of energy and/or a change of momentum. Only those electrons that retain enough energy to overcome the Schottky barrier at the base/collector interface and are transmitted into one of the available conduction band states of the semiconductor are collected. In ferromagnetic *3d* transition metals and their alloys, hot electron scattering is much stronger for minority than for majority electrons. As a consequence, the collector current (I_C) of the spin transistor depends critically on the orientation of the magnetic moment of the two ferromagnetic layers. The change in collector current that occurs upon a switch from an antiparallel (AP) to a parallel (P) alignment of the magnetic moments is called the magnetocurrent (MC) defined as

$$MC = (I_{C,P} - I_{C,AP})/I_{C,AP}. \quad (13)$$

For a comparison of the magnetic field sensitivity of the SVT and MTT with that of other spintronic devices it is illustrative to define the hot-electron transmission polarization of a ferromagnetic layer as

$$P = (n_\uparrow - n_\downarrow)/(n_\uparrow + n_\downarrow) \quad (14)$$

where $n_\uparrow$ and $n_\downarrow$ are the number of majority and minority electrons after transmission through the ferromagnetic layer. Ignoring interface contributions and taking into account that the transmission of hot electrons through a metal film decays exponentially with film thickness t, the transmission polarization can be written as

$$P = (\exp(-t/\lambda_\uparrow) - \exp(-t/\lambda_\downarrow))/(\exp(-t/\lambda_\uparrow) + \exp(-t/\lambda_\downarrow)). \quad (15)$$

In this formula $\lambda_\uparrow$ and $\lambda_\downarrow$ are the spin-dependent hot electron attenuation lengths, which have been measured by various experimental techniques for different ferromagnetic materials [135-145]. The SVT and MTT have also successfully been used to measure $\lambda_\uparrow$ and $\lambda_\downarrow$ for *3d* transition metal alloys [146-148]. In these experiments the collector current is measured as a function of the ferromagnetic base layer thickness.

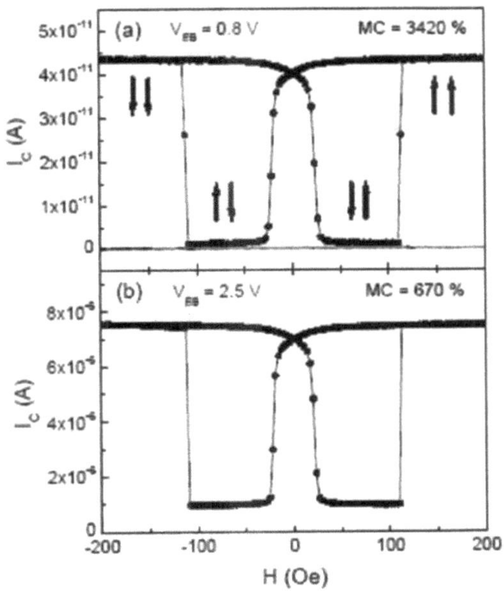

Figure 32. Magnetic field dependence of the MTT collector current at low (0.8 V) and high (2.5 V) emitter/base bias voltage. The MTT has a 5 nm $Ni_{81}Fe_{19}$/4 nm Cu/5 nm $Co_{70}Fe_{30}$ spin-valve base and a GaAs(001) collector. From Ref. 153

The results show that the attenuation length of minority electrons in NiFe and CoFe is only about 1 nm, while that of majority electrons is a factor 4 to 6 larger. Consequently, thin ferromagnetic films act as efficient spin filters for hot electron currents. For example, the calculated transmission polarization of a 4 nm thick ferromagnetic film with $\lambda_\downarrow$ = 1 nm and $\lambda_\uparrow$ = 5 nm amounts already 92%.

The magnetocurrent of SVT and MTT devices can be written in terms of the transmission polarization of the two ferromagnetic layers [149], as previously described in Eq.14,

$$MC = 2P_1P_2/(1 - P_1P_2). \tag{16}$$

This expression is equivalent to Jullière's formula [34], which expresses the tunneling magnetoresistance (TMR) of a MTJ in terms of the spin polarization of the two electrodes. The polarization of ferromagnetic 3*d* transition metal alloys is typically 45% [150], which according to Jullière's formula limits the TMR for incoherent tunneling to 50%. The transmission polarization of a FM layer, on the other hand, can be tuned by varying its thickness (see Eq. 15). In an SVT and MTT with a spin-valve base both ferromagnetic layers are utilized to spin-filter the hot electron current. The formula for the magnetocurrent (Eq. 16) predicts a giant MC of about 1100 % when $P_1 = P_2 = 92$ % is used. Very large effects of this order of magnitude have indeed been measured on hot electron spin transistors with a spin-valve base. SVTs with magnetocurrent effects up to 400% at room temperature and several hundred percent more at low temperature have been fabricated [151, 152] and for MTTs it has been shown that a careful optimization of the base layer growth can result in magnetocurrent values exceeding 3400%, Fig. 32(a) [148]. The magnetic field sensitivity of an MTT with a ferromagnetic emitter and one ferromagnetic base layer is considerably smaller. In this case, the spin-polarization of the current that is injected from the emitter into the base is limited to about 45%. In combination with efficient spin filtering in the ferromagnetic base, $P_2 = 92\%$ in this example, the predicted MC yields 140%. Measurements on MTTs with a single ferromagnetic base layer of Fe, NiFe or CoFe indicate that this value is an upper limit for the MC in this type of structures [129, 130, 132, 148, 153, 154].

While strong spin-dependent hot electron scattering in the ferromagnetic base layers is the reason for large magnetocurrent effects, it also leads to a relatively small collector current. The collector current of hot electron spin transistors can be written as

$$I_C = \alpha I_E, \tag{17}$$

where I_E is the emitter current and α is the transfer ratio of the device. The transfer ratio depends on hot electron scattering in the metal base layers, hot electron scattering at their interfaces, and hot electron collection at the metal/semiconductor interface. For MTTs it has been demonstrated that the transfer ratio can be enhanced over several orders of magnitude by increasing the bias voltage across the tunnel barrier emitter (see Fig. 33) [132, 148, 149, 153-155]. This dramatic dependence of α on V_{EB} is mainly due to a change in the number of available conduction band states in the semiconductor collector with varying hot electron energy [153, 154]. For GaAs collectors, for example, hot electron collection is limited to Γ conduction band states at energies that are only slightly larger than the collector Schottky barrier height. At elevated electron energy, the number of

Γ states increases rapidly and additional L and X conduction band states become available for electron collection. The transfer ratio of an MTT saturates at large emitter/base bias.

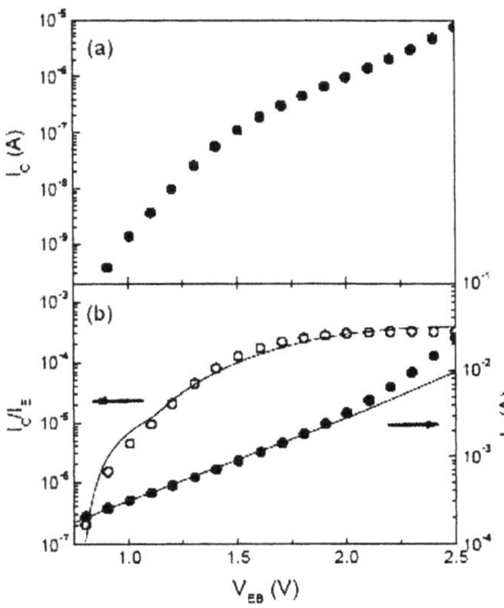

Figure 33. Emitter/base bias voltage dependence of (a) $I_{C,P}$ and (b) the transfer ratio (open circles) and I_E (solid circles) for an MTT with a 5 nm $Ni_{81}Fe_{19}$/4 nm Cu/5 nm $Co_{70}Fe_{30}$ spin-valve base and a GaAs(001) collector. The solid lines in (b) represent a model calculation of the transfer ratio and a Simmons' formula fit to I_C [149].

This saturation is due to a reduced energy dependence of the electron collection efficiency at the metal/semiconductor interface (at large electron energy almost the entire Brillouin zone has available conduction band states) and enhanced inelastic hot electron scattering in the metal base layers. Besides α, the emitter current also increases with tunnel barrier bias voltage. The combination of both effects results in collector currents in the μA range at large V_{EB} [132, 149].

Opposite to the collector current, the magnetocurrent of an MTT device is reduced at large emitter/base bias voltage. The decrease of the MC with electron energy is the result of spin-dependent inelastic hot electron scattering in the ferromagnetic base layers [153, 154]. Since minority electrons are scattered more than majority electrons, the energy distribution of minority electrons at the metal/semiconductor interface is broader than that of majority electrons. If the emitter/base bias voltage is slightly larger than the collector Schottky barrier height, only those electrons that do not lose any energy while traversing the MTT base layers can be collected. Since these are mainly majority electrons a very large MC is measured, i.e., the ferromagnetic base layers in combination with the Schottky barrier act as a nearly perfect filter to remove minority electrons. At larger emitter/base bias voltage, however, some of the inelastically scattered minority electrons retain enough energy to surmount the Schottky barrier. This increases the contribution of minority electrons to the collector current and hence decreases the MC. Although smaller than at low bias voltage, the MC that can be obtained at large V_{EB} is still extraordinary large (see Fig. 32(b)).

Contrary to the MTT the SVT operates at constant hot electron energy that is determined by the collector Schottky barrier height (~0.8 eV). As a result, the transfer ratio and the MC are only weakly dependent on the emitter/base bias voltage. From Eq. 17 it then follows that the SVT collector current increases almost linearly with the emitter current. It has recently been shown that collector currents exceeding 40 μA and large MC of around 400% can be obtained with SVT devices when large emitter currents are used [156].

For practical applications of the SVT and MTT a large collector current is pivotal. Apart from an increase in the emitter current, the collector current can be improved by a further optimization of the base layer structure. The transfer ratio of a hot electron transistor is limited by electron scattering in the metal base and scattering at the base/collector interface. Although a decrease of the base layer thickness will lead to a drastic increase of the collector current, it will be at the cost of magnetic field sensitivity. An improved metal/semiconductor interface, on the other hand, will enhance the collector current without a reduction of the MC. For many ferromagnetic metal/semiconductor combinations the formation of an alloy is very common and it generally leads to an interfacial layer in which hot electron scattering is intense. It has been demonstrated that the use of a non-magnetic seed layer can drastically improve the quality of the base/collector interface with larger collector output currents as result [157-159].

Besides its potential as magnetic field sensor (either magneto-electric or magneto-optic [160]) and its possible memory applications, the hot electron transistor can be used as a sensitive diagnostic tool to characterize spin-dependent scattering processes in ferromagnetic materials and at metal/semiconductor interfaces. Furthermore, the SVT and MTT are unique room-temperature sources of highly spin-polarized electron currents that can be utilized in spin-injection devices. The advantages of an SVT or MTT as spin injector over conventional ferromagnetic metal/semiconductor contacts are threefold. First of all, due to strong spin-filtering in the ferromagnetic base the hot electron current at the base/collector interface is nearly 100% spin-polarized. This is much larger than the polarization that is obtained from the Fermi level of ferromagnetic *3d* transition metals. Second, in a hot electron spin transistor the spin-polarization is generated within the ferromagnetic base and not at the ferromagnetic metal/semiconductor interface.

As a result, nearly perfect interfaces between the ferromagnetic material and the semiconductor are not a prerequisite for efficient spin injection. It is for example possible to insert a non-magnetic seed layer between the ferromagnetic base layer and the semiconductor collector. Since hot electrons retain their spin moment while traversing the thin non-magnetic layer this will not drastically reduce the spin polarization of the injected current. Finally, since electron injection is ballistic in SVT and MTT devices the spin injection efficiency is not fundamentally limited by a substantial conductivity mismatch between metals and semiconductors [161, 162]. The latter is the case in diffusive ferromagnetic metal/semiconductor contacts [163].

Jiang *et al.* [164] were the first to demonstrate efficient spin injection from a hot electron spin transistor source. In their experiment they used an MTT with a multiple quantum well light-emitting diode collector. In this structure the injection of hot electrons leads to the emission of photons through electron-hole recombination in the quantum wells. The polarization of the emitted light is correlated to the spin polarization of the injected electrons according to the optical selection rules [165, 166]. The measurements on an MTT with a 5 nm thick ferromagnetic base layer clearly show that the electroluminescence (EL) from the quantum wells is polarized. After background subtraction the polarization of EL amounts 10% and its polarity reveals the injection of a majority electron current. The modest values of spin-polarized current that are inferred from these first experiments are mainly due to spin relaxation effects in the quantum well detector. It is expected that a further optimization of the device structure will result in spintronic devices that operate with much higher spin-injection efficiencies.

4.2. The SPICE Transistor Design

The second spin transistor design that is addressed uses a spin injecting current emitter, hence its acronym SPICE [167, 168]. It uses semiconductors in its construction, like the SVT and MTT: unlike them, however, its operation also requires spin-dependent transport in the semiconductor itself and thereby arise a small family of additional materials problems which are discussed below.

The operation of the SPICE design is closely analogous to that of a conventional bipolar transistor comprising distinct emitter, base, and collector stages but with some differences that are forced by materials considerations. The central base layer is a doped semiconductor and in it is established a non-equilibrium minority carrier population by injection from the emitter. Additionally these injected carriers are partly spin polarized and when they are harvested by the spin-dependent collector, the collection efficiency is dependent on the relative orientations of the emitter and collector orientations. The feasibility of these devices depend on several issues, respectively the characteristic length-scale for spin transport in semiconductors , the efficient injection of spins into a semiconductor, and the achieved spin polarization in the semiconductor.

The first of these issues was elegantly answered by an optical pumping experiment [169] in which spin polarized carriers were generated *in situ* in a GaAs semiconductor by optical pumping with a high intensity localized laser pulse and their behavior then analyzed by a low intensity probe beam. Small applied electric and magnetic fields respectively induced spatial drift and spin precession in the localized spin packet. The spin precession was picked up by observing the Faraday rotation of the scanned probe beam, by which measurement the temporal and spatial evolution of the packet could be mapped and the corresponding spin diffusion length inferred for the carriers. Spin diffusion lengths of order hundreds of microns were shown to be feasible, given the correct choice of material and doping type/level. These dimensions are more than adequate to avoid conflict with standard semiconductor fabrication methods.

This experiment used polarized optics to create and analyze the spin polarized carriers in the semiconductor. For spin-device purposes, the polarized carriers usually need to be injected into the semiconductor from an external source. That this is feasible with high efficiency has been proven by using diluted magnetic semiconductors, for example Mn-doped GaAs to spin-inject a light-emitting-diode (LED) structure [170-173]. The injected polarization was monitored by measuring the optical polarization of the emitted light which was shown to track the hysteresis loop of the magnetic semiconductor. Moreover, with rising temperature, the optical polarization diminished and traced out the magnetization curve $M(T)$ of the magnetic semiconductor thereby affording further neat confirmation of successful spin injection. The injection efficiency observed in such experiments can be of order many tens of percent.

The success of this class of experiment has led to much interest in novel types of magnetic semiconductor, and the most interesting recent discovery in this domain is that zinc oxide is such a candidate when doped with small percentages of such metals as cobalt, iron and manganese. It displays a rich phase diagram whose details have yet to be fully explored and its very major attraction is the ease with which it may be made in thin film form and this procedure integrated into device fabrication routines.

Pending further work on these new magnetic semiconductors, metallic ferromagnets are in principle, the most convenient spin polarized sources for spin device work. The obvious configuration of direct Ohmic contact between metal and semiconductor proved to have fundamental shortcomings. The conductivity mismatch between the two materials implies very indifferent spin injection efficiency [174, 175]. However it transpires that this difficulty is surmountable [176] by placing a tunnel barrier between the two components and spin injection efficiency is thus restored that is compatible with the spin polarization of the injecting ferromagnet.

In the light of these various experiments a clear choice of geometry for the SPICE transistor design uses tunnel injectors for both emitter and collector. Hot polarized carriers arrive from the emitter high into the base conduction band and thermalise to produce a non-equilibrium minority carrier excess that part diffuses, part drifts towards the base contact. These are in turn harvested in passing by the back biased tunnel collector, the collector spin selectivity deriving from the spin asymmetry of the ferromagnetic collector at the point in the band structure corresponding to the conduction band bottom.

Compared with a conventional bipolar transistor, the tunnel collector efficiency is low and this means the collector base current gain, is modest for such transistors. This current is determined not just by base recombination as is usual in standard transistors but also by the fraction of carriers that evade the collection process. Another unusual feature is the form of the output admittance, which is determined by the upper band structure of the ferromagnetic collector material and the effective tunnel barrier width. It is therefore non-negligible and is predictably a function of V_{ce}.

Possible choices of semiconductor material for the SPICE design include gallium arsenide, silicon, and germanium. In the work described here, silicon was chosen as a reflection of its commercial importance, however this choice brings with it materials compatibility issues and surface chemistry complications. Silicon has a native oxide SiO_2 that is typically 2 nm thick which makes an inefficient spin injection tunnel barrier.

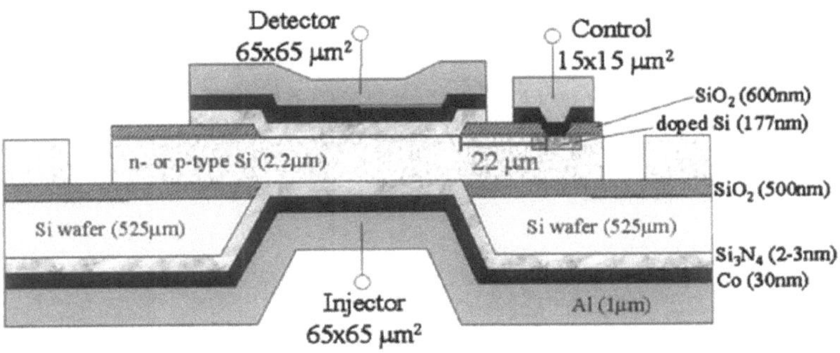

Figure 34. Example of a practical realization of the SPICE transistor design.

Growing a quality tunnel barrier on a silicon surface involves cleaning off this oxide and depositing a metal, such as aluminium for example, that is subsequently oxidized to form a thin insulating layer. Two potential difficulties lurk in the process: formation of a conducting metal silicide at the interface, of partial diffusion/penetration of the mobile metal atoms into the silicon surface via dislocations and channelling. Either is seriously damaging to spin injection efficiency since the situation has been recreated [174, 175] in which the spins arrive in the semiconductor via a direct metal semiconductor interface. These materials issues represent the biggest challenge to making a device of this type. Solutions are at hand and are being developed. For example it has been established that silicide formation and metallic penetration may be minimized by leaving a partial SiO_2 layer which acts as a barrier to the aluminum by which it is chemically reduced [177].

Maximizing spin injection efficiency is a necessary but insufficient condition for a magnetically sensitive transistor: the collector efficiency is also critical. As with the MTT it turns out that there is conflict of interest between the current gain and the magnetic sensitivity. A physically small collector gives a high degree of magnetic selectivity: however in doing this it collects only a fraction of the carriers and the current gain is poor. A large extended collector collects current carriers of both polarizations, hence affords good current gain and indifferent magnetic selectivity. The solution to this problem lies initially in careful modeling of the diffusion/drift characteristic of the base [178]. A more radical solution is introduction of a fourth screening electrode analogous to the screened grid in a vacuum tetrode or a dual-gate mosfet. With this fourth electrode clamped to AC ground, it is possible to restore reasonable values of current gain while maintaining magnetic selectivity. Several of the above aspects of SPICE transistor design have been incorporated into the practical realization shown in Fig. 34.

The spin transistor as represented in Fig. 34 is a vertical spin transistor. Figure 35 is a schematic picture of a lateral spin transistor as originally proposed by Datta and Das [179]. In this case, a iron emitter injects spins into a 2D electron gas. A Schottky gate can rotate the spin polarization by the Rashba effect, and another iron analyzer detects transmitted spin polarized current.

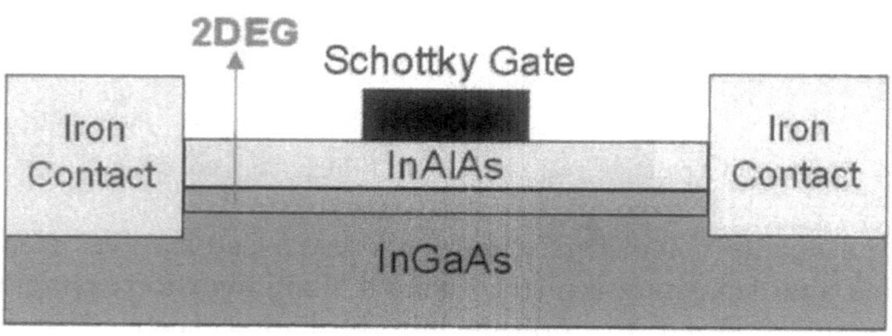

Figure 35. Datta and Das spin transistor.

Recently, a successful implementation of electrical spin injection and electrical detection using a vertical GaMnAs/AlAs/GaAs/AlAs/GaMnAs double tunnel junction was reported [180]. In this case, the emitter and analyzer are semiconductor ferromagnets. In this structure 40% TMR values are observed at 4 K, as the emitter and collector magnetizations change from a parallel to antiparallel orientations. Hybrid semiconductor/ferromagnetic devices are still at their infancy, but the technical details involved in injecting spin polarization from a metal into a semiconductor start now to be understood. Device applications are still in the drawing board.

5. CONCLUSION

In conclusion, this chapter covered various applications of spintronic devices, from advanced read heads for magnetic data storage, to hybrid semiconductor/ferromagnetic devices, including MRAMs and spintronic biochips. In all the different areas, rapid developments are occurring on almost a daily basis. Despite the length of the chapter areas such as magnetic logic and integrated and tunable high Q oscillators that are attracting recent attention could not be covered. We just hope that this chapter can provide the reader an insight on the richness of applications coming out from spintronic nanostructures.

References

[1] B. Dieny, V. S. Speriosu, S. S. Parkin, B. A. Gurney, D. R. Wilhoit, and D. Mauri, Phys. Rev. B **43**, 1297 (1991).

[2] D. E. Heim, R. E. Fontana, C. Tsang, V. S. Speriosu, B. A. Gurney, and M. L. Williams, IEEE Trans. Magn. **30**, 316 (1994).

[3] P. P. Freitas, J. L. Leal, L. V. Melo, N. J. Oliveira, L. Rodrigues, and A. T. Sousa, Appl. Phys. Lett. **65**, 493 (1994).

[4] C. Tsang, R. E. Fontana, T. Lin, D. E. Heim, V. S. Speriosu, B. A. Gurney, and M. L. Williams, IEEE Trans. Magn. **30**, 3801 (1994).

[5] B. Dieny, "Spin Valves", in Magnetoelectronics, Ed. M. Johnson, Elsevier Academic Press, 2004.

[6] M. Baibich, J. M. Broto, A. Fert, Van Dau Nguyen, F. Petroff, P. Etienne, G. Creuzet, A. Friederich, and J. Chazelas, Phys. Rev. Lett. **61**, 2472 (1988).

[7] M. Lederman, IEEE Trans. Magn. **35**, 794 (1999).

[8] B. Dieny, J. Phys. Condens. Matter. **4**, 8009 (1992).

[9] S. A. Parkin, Phys. Rev. Lett. **71**,1641(1993).

[10] Ch.-L. Lee, A. J. Devasahayam, Chih-Ching Hu, Yingbo Zhang, and Ming Mao, J. C. S. Kools, K. Rook, IEEE Trans. Magn. **40**, 2209 (2004).

[11] Y. Kamiguchi, H. Yuasa, H. Fukuzawa, K. Koui, H. Iwasaki, and M.Sahashi, Digest DB-01, presented at INTERMAG 1999, Korea.
[12] H. Sakakima, M. Satomi, Y. Sugita, and Y. Kawawake, J. Magn. Magn. Mater. **210**, L20 (2000).
[13] A. Veloso, P. P. Freitas, P.Wei, N. P. Barradas, J. C. Soares, B. Almeida, J. B. Sousa, Appl. Phys. Lett. **77**, 1020 (2000).
[14] S. Cardoso, Zhongzi Zhang, Haohua Li, R. Ferreira, P. P. Freitas, Peng Wei, J. C. Soares, E. Snoeck, X. Battle, IEEE Trans. Magn. **38**, 2755 (2002).
[15] Y. M. Huai, Z. T. Diao, and J. Zhang, IEEE Trans.Magn. **38**, 20 (2002).
[16] N. Hasegawa, F. Koile, K. Ikarashi, M. Ishizone, M. Lawamura, Y. Nakazawa, and A. Takahashi, J. Appl. Phys. **91**, 8774 (2002).
[17] W. F. Egelhoff, T. Ha, R. D. K. Misra, Y. Kadmon, J. Nir, J. C Powell, M. D. Stiles, R McMichael, C. L. Lin, J. M. Silversten, J. H. Judy, K. Takano, A. E. Berkowitz, T. C. Anthony, and J. A. Brugg, J. Appl. Phys. **78**, 273 (1995).
[18] A. Veloso, N. J. Oliveira, and P. P. Freitas, IEEE Trans. Magn. **34**, 2343 (1998).
[19] J. C. S. Kools, Th. G. S. M. Rijks, A. E. M. de Veirman, and R. Coehoorn, IEEE Trans. Magn. **31**, 3918 (1995).
[20] L. Néel, C. R. Acad. Sci. **255**, 1545 (1962).
[21] P. Bruno and C. Chappert, Phys. Rev. Lett. **67**, 1602 (1991)
[22] G. B. Albuquerque and P. P. Freitas, Physica **B233**, 294 (1997).
[23] K. R. Coffey, B. A. Gurney, D. E. Heim, H. Lefakis, D. Mauri, V. S. Speriosu, and D. R. Wilhoit, U.S.Patent No. 5 583 725, 10 December 1996.
[24] H. van den Berg, W. Clemens, G. Gieres, G. Rupp, W. Schelter, and M.Vieth, IEEE Trans. Magn. **32**, 4624 (1996).
[25] J. G. Zhu, IEEE Trans. Magn. **35**, 655 (1999).
[26] J. L. Leal and M. Kryder, IEEE Trans. Magn. **35**, 800 (1999).
[27] M. Rickart, A. Guedes, B. Negulescu, P. P. Freitas, J. Ventura, J. B. Sousa, P. Diaz, M. MacKenzie, and J. N. Chapman, Eur. Phys. Journ. B, in press.
[28] A. Veloso and P. P. Freitas, J. Appl. Phys. **87**, 5744 (2000).
[29] A. Veloso, R. H. Dee, and P. P. Freitas, IEEE Trans. Magn. **38**, 1928 (2002).
[30] H. N. Bertram, IEEE Trans. Magn. **31**, 2573 (1995).
[31] C. D. Mee and E. D. Daniel, in "Magnetic Recording Technology", McGraw-Hill, New York 1995, pp.6.45-6.48.
[32] J. Hong, J. Hashimoto, M. Yamagishi, K. Noma, and H. Kanai, IEEE Trans. Magn. **38**, 15 (2002).
[33] J. Moodera, J. Nassar, and G. Mathon, Annu. Rev. Mater. Sci. **29**, 381(1999).
[34] M. Julliere, Phys.Lett. **54A**, 225 (1975).
[35] S. S. Parkin, K. P. Roche, M. G. Sammant, P. M. Rice, R. B. Beyers, R. E. Scheurlein, E. J. O'Sullivan, S. L. Brown, J. Bucchiganno, D. W. Abraham, Y. Lu, M. Rooks, P. L. Trouiloud, R. A. Wanner, and W. J. Gallagher, J. Appl. Phys. **85**, 5828 (1999).
[36] R. C. Sousa, J. J. Sun, V. Soares, P. P. Freitas, A. Kling, M. F. daSilva, and J. C. Soares, Appl. Phys. Lett. **73**, 3288 (1998).
[37] S. Cardoso, V. Gehanno, R. Ferreira, and P. P. Freitas, IEEE Trans. Magn. **35**, 2952 (1999).

[38] Z.-Zh. Zhang, S. Cardoso, P. P. Freitas, X. Battle, P. Wei, N. Barradas, and J. C. Soares, J. Appl. Phys. **89**, 6665 (2001).

[39] D. Wang, C. Nordman, J. M. Daughton, Z. Qian, and J. Fink, IEEE Trans. Magn. **40**, 2269 (2004).

[40] S. Yuasa, T. Nagahama, A. Fukushima, Y. Suzuki, and K. Ando, Nature Mater. **3**, 868 (2004).

[41] S. S. Parkin, C. Kaiser, A. Panchula, P. M. Rice, B. Hughes, M. Samant, and S.-H. Tang, Nature Mater. **3**, 862 (2004).

[42] J.-G. Wang, P. P. Freitas, and E. Snoeck, Appl. Phys. Lett. **79**, 4553 (2001).

[43] Zh.-G. Zhang , P. P. Freitas, A. R. Ramos, N. P. Barradas, and J. C. Soares, Appl. Phys. Lett. **79**, 2219 (2001).

[44] Zh.-G. Zhang, P. P. Freitas, A. R. Ramos, N. P. Barradas, and J. C. Soares, J. Appl. Phys. **91**, 8786 (2002).

[45] J.-G. Wang, P. P. Freitas, E. Snoeck, X. Battle, and J. Quadra, IEEE Trans. Magn. **38**, 2703 (2002).

[46] R. Ferreira, P. P. Freitas, M. MacKenzie, and J. Chapman, J. Appl. Phys. 97, 10C903 (2005).

[47] J. J. Ruigrok, R. Coehoorn, S. R. Cumpson, and H. van Kesteren, J. Appl. Phys. **87**, 5398 (2000).

[48] J. J. Sun, K. Shimazawa, N. Kasahara, K. Sato, S. Saruki, T. Kagami, O. Redon, S. Araki, H. Morita, and N. Marsuzaki, Appl. Phys. Lett. **76**, 2424 (2000).

[49] P. P. Freitas, S. Cardoso, R. Sousa, Wanjun Ku, R. Ferreira, V. Chu, J. P. Conde, IEEE Trans. Magn. **36**, 2796 (2000).

[50] S. Araki, K. Sato, T. Kagami, S. Saruki, T. Uesugi, N. Kasahara, T. Kuwashima, N. Ohta, Jijun Sun, K. Nagai, Sh.-X. Li, N. Hashisuka, H. Hatate, T. Kagonati, N. Takahasi, K. Ueda, and M. Matsuzaki, IEEE Trans. Magn. **38**, 72 (2002).

[51] S.-N. Mao, J. Nowak, D. Song, P. Kolbo, Lei Wang, E. Linville, D. Saunders, E. Murdock, and P. Ryan , IEEE Trans.Magn. **38**, 78 (2002).

[52] M. Mackenzie, J. N. Chapman, H.-H. Li, R. Ferreira, P. P. Freitas, Eur. Phys. Journ. B, in press.

[53] B. Oliver, G. Tuttle, Q. He, X.-F. Tang, and J. Nowak, J. Appl. Phys. **95**, 1315 (2004).

[54] K. B. Klaassen, X.-Zh. Xing, and J. C. L. van Peppen, IEEE Trans.Magn. **40**, 195 (2004).

[55] N. Smith and P. Arnett, Appl.Phys.Lett. **78**, 1448 (2001).

[56] J. C. Jury, K. B. Klaassen, J. C. L. van Peppen, and S. X. Wang. IEEE Trans. Magn. **38**, 3545 (2002).

[57] M. Mallary, A. Torabi, and M. Benakli, IEEE Trans.Magn. **38**, 1719 (2002).

[58] A. Vedyayev, M. Chshiev, N. Ryzhanova, B. Dieny, C. Cowache, F. Brouers, J. Magn. Magn. Mater. **172**, 53 (1997).

[59] W. P. Pratt, Jr., S.-F. Lee, J. M. Slaughter, R. Loloee, P. A. Schroeder, and J. Bass, Phys. Rev. Lett. **66**, 3060 (1991).

[60] T. Valet and A. Fert, Phys. Rev. B **48**, 7099 (1993).

[61] A. Tanaka, Y. Shimizu, Y. Seyama, K. Nagasaka, R. Kondo, H. Oshima, Sh. Eguchi, and Hitoshi Kanai, IEEE Trans. Magn. **38**, 84 (2002).

[62] H. Oshima, K. Nagasaka, Y. Seyama, A. Jogo, Y. Shimizu, A. Tanaka, and Y. Miura, IEEE Trans. Magn. **39**, 2377 (2003).

[63] H.-H. Li, R. Ferreira, P. P. Freitas, M. MacKenzie, J. N. Chapman, PhD Thesis, Instituto Superior Tecnico, Lisbon, 2005.

[64] H. Fukuzawa, H. Yuasa, S. Hashimoto, K. Koi, H. Iwasaki, M. Takagashi, Y. Tanaka, and M. Sahashi, IEEE Trans. Magn. **40**, 2236 (2004).

[65] X.-Ch. Zhu, J.-G..Zhu, and R. M. White , J. Appl. Phys. **95**, 6630 (2004).

[66] L. Berger, Phys. Rev. B **54**, 9353 (1996)

[67] J. C. Slonczewski, J.Magn.Magn.Mater. **159**, L1 (1996)

[68] K. J. Lee, Y. Liu, A. Deac, M. Li, J. W. Chang, S. Liao, K. Ju, O. Redon, J. P. Nozieres, and B. Dieny, IEEE Trans. .Magn., in press.

[69] G. Grynkewich, J. Akerman, P. Brown, B. Butcher, R. W. Dave, M. DeHerrera, M. Durlam, B. N. Engel, J. Janesky, S. Pietambaram, N. D. Rizzo, J. M. Slaughter, K. Smith, J. J. Sun, and S. Tehrani, MRS Bulletin **29**, 818 (2004).

[70] H. Goronkin and Y. Yang, MRS Bulletin **29**, 805 (2004).

[71] E. C. Stoner and E. P. Wohlfarth, Phil. Trans. R. Soc. Lond. **A240**, 599 (1948).

[72] J. Ackerman, M. DeHerrera, M. Durlam, B. Engel, J. Janesky, F. Mancoff, J. Slaughter, and S. Tehrani, "Magnetoelectronics" in Magnetic Tunnel Junction Based Magnetoresistive Random Access Memory", Ed. M. Johnson, Elsevier Academic Press, 2004.

[73] H. W. Schumacher, C. Chappert, R. C. Sousa, and P. P. Freitas, Appl. Phys. Lett. **83**, 2205 (2003).

[74] H. W. Schumacher, C. Chappert, P. Crozat, R. C. Sousa, and P. P. Freitas, Appl. Phys. Lett. **80**, 3781 (2002).

[75] M. Durlam, P. Naji, A. Omair, M. DeHerrera, J. Calder, J. M. Slaughter, B. Engel, N. Rizzo, G. Grynkewich, B. Butcher, C. Tracy, K. Smith, K. Kyler, J. Ren, J. Molla, B. Feil, R. Williams, and S. Tehrani, IEEE 2002 Symp. VLSI Circuits **12.4**, 158 (2002).

[76] W. J. Gallagher, S. S. Parkin, Y. Lu, X. P. Bian, A. Marley, R. A Altman, S. A. Rishton, K. P. Roche, C. Jahnes, T. M. Shaw, and G. Xiao, J. Appl. Phys. **81**, 3741 (1997).

[77] J. Shi, S. Tehrani, M. R. Scheinfein, Appl. Phys. Lett. **76**, 2588 (2000).

[78] E. Chen, S. Tehrani, M. Durlam, and T. Zhu, US Patent 5659499, Aug.1997.

[79] L. Savtcenko, B. N.Engel, N. D. Rizzo, M. F. DeHerrera, and J. Janesky, U.S. Patent No. 6,545,906 (April 8, 2003).

[80] R. C. Sousa, Zh.-Zh. Zhang, and P. P. Freitas, J. Appl. Phys. **91**, 7700 (2002).

[81] J.-G. Zhu, IEEE Trans.Magn. **35**, 655 (1999).

[82] B. N. Engel, J. Akerman, B. Butcher, R. W. Dave, M. DeHerrera, M. Durlam, G. Grinkewich, J. Janesky, S. V. Pietambaram, N. D. Rizzo, J. M. Slaughter, K. Smith, J. J. Sun, and S. Tehrani, IEEE Trans.Magn. **41**, 132 (2005).

[83] J. Janesky, N. D. Rizzo, B. N. Engel, and S. Tehrani, Appl. Phys. Lett. **85**, 2289 (2004).

[84] K. Inomata, N. Koike, T. Nozaki, S. Abe, and N. Tezuka, Appl. Phys. Lett. **82**, 2667 (2003).

[85] R. S. Beech, J. A. Anderson, A. V. Pohm, and J. M. Daughton, J. Appl. Phys. **87**, 6403 (2000).

[86] J. M. Daughton and A. V. Pohm, US Patent 6,535,416 B1 (2003).
[87] J. M. Daughton and A. V. Pohm, J. Appl. Phys. **93**, 7304 (2003).
[88] J.-G. Wang and P. P. Freitas, Appl. Phys. Lett. **84**, 945 (2004).
[89] R. C. Sousa, I. L. Prejbeanu, D. Stanescu, B. Rodmacq, O. Redon, and B. Dieny , J.-G. Wang, and P. P. Freitas , J.Appl.Phys. **95**, 6783 (2004).
[90] M. Kerekes, R. C. Sousa, I. L. Prejbeanu, O. Redon, U. Ebels, C. Baraduc, B. Dieny, J.-P. Nozières, P. P. Freitas, P. Xavier, J. Appl. Phys. **97**, 10P501 (2005).
[91] Y. Huai, F. Albert, P. Nguyen, M. Pakala, and T. Valet, Appl. Phys. Lett. **84**, 3118 (2004).
[92] G. D. Fuchs, N. C. Embley, I. N. Krivorotov, M. P. Bragança, E. M. Ryan, S. I. Kiselev, J. C. Sankey, D. C. Ralph, R. A. Buhrmam, and J. A. Katine, Appl. Phys. Lett. **85**, 1205 (2004).
[93] J.-G. Wang, C. Boubeta, and P. P.Freitas, unpublished
[94] Yaowen Liu, Zongzhi Zhang, Jianguo Wang, P. P. Freitas, J. L. Martins, J. Appl. Phys. **93**, 8385 (2003).
[95] A. Deac, O. Redon, R. C. Sousa, B. Dieny, and J. P. Nozières , Z. Zhang, Y. Liu, and P. P. Freitas , J. Appl. Phys. **95**, 6792 (2004).
[96] Y.-W. Liu, Zh,-Zh. Zhang, and P.P.Freitas, IEEE Trans. Magn., **39**, 2833 (2003).
[97] D. R. Baselt, G. U. Lee, M. Natesan, S. W. Metzger, P. E. Sheehan, and R. J. Colton, Biosens. Bioelectron **13,** 731 (1998).
[98] R. L. Edelstein, C. R. Tamanaha, P. E. Sheehan, M. M. Miller, D. R. Baselt, L. J. Whitman, and R. J. Colton, Biosens. Bioelectron. **14**, 805 (2000).
[99] M. M. Miller, P. E. Sheehan, R. L. Edelstein, C. R. Tamanaha, L. Zhong, S. Bounnak, L. J. Whitman, and R. J. Colton, J. Magn. Magn. Mater., **225,** 138 (2001).
[100] J. C. Rife, M. M. Miller, P. E. Sheehan, C. R. Tamanaha, M. Tondra, and L. J. Whitman, Sens. Actuators A **107**, 209 (2003).
[101] J. Schotter, P. B. Kamp, A. Becker, A. Puhler, D. Brinkmann, W. Schepper, H. Bruckl, and G. Reiss, IEEE Trans. Magn. **38,** 3365 (2002).
[102] J. Schotter, P. B. Kamp, A. Becker, A. Puhler, G. Reiss, and H. Bruckl, Biosens. Bioelectron. **19,** 1149 (2004).
[103] D. L. Graham, H. Ferreira, J. Bernardo, P. P. Freitas, and J. M. S. Cabral, J. Appl. Phys. **91,** 7786 (2002).
[104] L. Lagae, R. Wirix-Speetjens, J. Das, D. Graham, H. Ferreira, P. P. Freitas, G. Borghs, and J. D. Boeck, J. Appl. Phys. **91,** 7445 (2002).
[105] G. Li, V. Joshi, R. L. White, S. X. Wang, J. T. Kemp, C. Webb, R. W. Davis, and S. Sun, J. Appl. Phys. **93,** 7557 (2003).
[106] G. Li, S. X. Wang, and S. Sun, IEEE Trans. Magn. **40,** 3000 (2004).
[107] A. Anguelouch, D. H. Reich, C. L. Chien, and M. Tondra, IEEE Trans. Magn. **40,** 2997 (2004).
[108] H. A. Ferreira, D. L. Graham, N. Feliciano, L. A. Clarke, M. D. Amaral, and P. P. Freitas, IEEE Trans. Magn. (in press).
[109] M. M. Miller, G. A. Prinz, S. F. Cheng, and S. Bounnak, Appl. Phys. Lett. **81,** 2211 (2002).
[110] L. Ejsing, M. F. Hansen, A. K. Menon, H. A. Ferreira, D. L. Graham, and P. P. Freitas, Appl. Phys. Lett. **84,** 4729 (2004).

[111] L. Ejsing, M. F. Hansen, A. K. Menon, H. A. Ferreira, D. L. Graham, and P. P. Freitas, J. Magn. Magn. Mater. 293, 677 (2005).
[112] W. Shen, X. Liu, D. Mazumdar, and G. Xiao, in Proceedings, 49th Annual Conference on Magnetism and Magnetic Materials (Jacksonville, Florida, USA, 2004).
[113] D. L. Graham, H. A. Ferreira, N. Feliciano, P. P. Freitas, L. A. Clarke, and M. D. Amaral, Sens. Actuators B (in press).
[114] P. P. Freitas, H. A. Ferreira, D. L. Graham, L. A. Clarke, M. D. Amaral, V. Martins, L. Fonseca, and J. S. Cabral, "Magnetoresistive DNA Chips", in "Magnetoelectronics", Ed. M. Johnson, Academic Press, New York 2004, p. 331-373.
[115] D. L. Graham, H. A. Ferreira, and P. P. Freitas, Trends Biotechnol. **22,** 455 (2004).
[116] D. L. Graham, H. A. Ferreira, N. Feliciano, P. P. Freitas, and P. Galvin, patent pending.
[117] T. Deng, G. M. Whitesides, M. Radhakrishnan, G. Zabow, and M. Prentiss, Appl. Phys. Lett. **78,** 1775 (2001).
[118] C. S. Lee, H. Lee, and R. M. Westervelt, Appl. Phys. Lett. **79,** 3308 (2001).
[119] R. Wirix-Speetjens and J. D. Boeck, IEEE Trans. Magn. **40,** 1944 (2004).
[120] M. Brzeska, M. Panhorst, P. B. Kamp, J. Schotter, G. Reiss, A. Puhler, A. Becker, and H. Bruckl, J. Biotechnol. **112,** 25 (2004).
[121] M. Tondra, M. Porter, and R. J. Lipert, J. Vac. Sci. Technol. **18,** 1125 (2000).
[122] B. Raquet, in "Spin Electronics", Eds.M. Ziese and M. J. Thornton, Springer, Berlin 2001, p. 232.
[123] E. R. Nowak, M. B. Weissman, and S. S. P. Parkin, Appl. Phys. Lett. **74**, 600 (1999).
[124] D. L. Graham, H. A. Ferreira, P. P. Freitas, and J. M. S. Cabral, Biosens. Bioelectron. **18,** 483 (2003).
[125] H. A. Ferreira, D. L. Graham, P. P. Freitas, and J. M. S. Cabral, J. Appl. Phys. **93,** 7281 (2003).
[126] V. Martins, L. P. Fonseca, H. A. Ferreira, D. L. Graham, P. P. Freitas, and J. M. S. Cabral, (unpublished).
[127] A. Clarke, C. Braz, and M. D. Amaral, Pediatric Pulmonol. **38**, 219 (2004).
[128] D. J. Monsma, J. C. Lodder, T. J. A. Popma, and B. Dieny, Phys. Rev. Lett. **74,** 5260 (1995).
[129] D. J. Monsma, R. Vlutters, and J. C. Lodder, Science **281**, 407 (1998).
[130] K. Mizushima, T. Kinno, T. Yamauchi, and K. Tanaka, IEEE Trans. Magn. **33**, 3500 (1997).
[131] K. Mizushima, T. Kinno, K. Tanaka, and T. Yamauchi, Phys. Rev. B **58**, 4660 (1998).
[132] R. Sato and K. Mizushima, Appl. Phys. Lett. **79**, 1157 (2001).
[133] S. van Dijken, X. Jiang, and S. S. P. Parkin, Appl. Phys. Lett. **80**, 3364 (2002).
[134] D. P. Pappas, K. P. Kämper, B. P. Miller, H. Hopster, D. E. Fowler, C. R. Brundle, A C. Luntz, and Z. X. Shen, Phys. Rev. Lett. **66**, 504 (1991).
[135] G.Schönhense and H. C. Siegmann, Ann. Phys. (Leipzig) 2, 465 (1993).
[136] M. Getzlaff, J. Bansmann, and G. Schönhense, Solid State Commun. **87**, 467 (1993).
[137] J. C. Gröbli, D. Oberli, and F. Meier, Phys. Rev. B **52,** R13095 (1995).
[138] E. Vescovo, C. Carbone, U. Alkemper, O. Rader, T. Kachel, W. Gudat, and W. Eberhardt, Phys. Rev. B **52,** 13497 (1995).

[139] H. J. Drouhin, A. J. v. d. Sluijs, Y. Lassailly, and G. Lampel, J. Appl. Phys. **79**, 4734 (1996).
[140] M. Aeschlimann, M. Bauer, S. Pawlik, W. Weber, R. Burgermeister, D. Oberli, and H. C. Siegmann, Phys. Rev. Lett. **79**, 5158 (1997).
[141] A. Filipe, H.-J. Drouhin, G. Lampel, Y. Lassailly, J. Nagle, J. Peretti, V. I. Safarov, and A. Schuhl, Phys. Rev. Lett. **80**, 2425 (1998).
[142] D. Oberli, R. Burgermeister, S. Riesen, W. Weber, and H. C. Siegmann, Phys. Rev. Lett. **81**, 4228 (1998).
[143] R. Knorren, K. H. Bennemann, R. Burgermeister, and M. Aeschlimann, Phys. Rev. B **61**, 9427 (2000).
[144] W. H. Rippard and R. A. Buhrman, Phys. Rev. Lett. **84**, 971 (2000).
[145] W. Weber, S. Riesen, and H. C. Siegmann, Science **291**, 1015 (2001).
[146] R. Vlutters, R. Jansen, O. M. J. van 't Erve, S. D. Kim, and J. C. Lodder, J. Appl. Phys. **89**, 7305 (2001).
[147] R. Vlutters, O. M. J. van 't Erve, S. D. Kim, R. Jansen, and J. C. Lodder, Phys. Rev. Lett. **88**, 027202 (2002).
[148] S. van Dijken, X. Jiang, and S. S. P. Parkin, Phys. Rev. B **66**, 094417 (2002).
[149] S. van Dijken, X. Jiang, and S. S. P. Parkin, Appl. Phys. Lett. **83**, 951 (2003).
[150] D. J. Monsma and S. S. P. Parkin, Appl. Phys. Lett. **77**, 720 (2000).
[151] P. S. Anil Kumar, R. Jansen, O. M. J. van 't Erve, R. Vlutters, P. de Haan, and J. C. Lodder, J. Magn. Magn. Mater. **214**, L1 (2000).
[152] R. Jansen, P. S. Anil Kumar, O. M. J. van 't Erve, R. Vlutters, P. de Haan, and J. C. Lodder, Phys. Rev. Lett. **85**, 3277 (2000).
[153] S. van Dijken, X. Jiang, and S. S. P. Parkin, Phys. Rev. Lett. **90**, 197203 (2003).
[154] X. Jiang, S. van Dijken, and S. S. P. Parkin, Phys. Rev. B **69**, 014413 (2004).
[155] S. van Dijken, X. Jiang, and S. S. P. Parkin, Appl. Phys. Lett. **82**, 775 (2003).
[156] R. Jansen, H. Gokcan, O. M. J. van 't Erve, F. M. Postma, and J. C. Lodder, J. Appl. Phys. Lett. **95**, 6927 (2004).
[157] O. M. J. van 't Erve, R. Vlutters, P. S. A. Kumar, S. D. Kim, F. M. Postma, R. Jansen, and J. C. Lodder, Appl. Phys. Lett. **80**, 3787 (2002).
[158] O. M. J. van 't Erve, Ph.D. Thesis, University of Twente, The Netherlands (2002).
[159] S. van Dijken, X. Jinag, and S. S. P. Parkin, submitted to J. Appl. Phys.
[160] I. Appelbaum, K. J. Russell, D. J. Monsma, V. Narayanamurti, C. M. Marcus, M. P. Hanson, and A. C. Gossard, Appl. Phys. Lett. **83**, 4571 (2003).
[161] V. Ya. Kravchenko and E. I. Rashba, Phys. Rev. B **67**, 121310 (2003).
[162] E. I. Rashba, Phys. Rev. B **68**, 241310 (2003).
[163] G. Schmidt, D. Ferrand, L. W. Molenkamp, A. T. Filip, and B. J. van Wees, Phys. Rev. B **62**, R4790 (2000).
[164] X. Jiang, R. Wang, S. van Dijken, R. Shelby, R. Macfarlane, G. S. Solomon, J. Harris, and S. S. P. Parkin, Phys. Rev. Lett. **90**, 256603 (2003).
[165] F. Meier and B. P. Zakharchenya, "Optical Orientation", North-Holland, New York 1984.
[166] C. Weisbuch and B. Vinter, "Quantum Semiconductor Structures: Fundamentals and Applications", Academic, New York 1991.

[167] J. F. Gregg, W. D. Allen, N. Viart, R. Kirschman, C. Sirisathitkul, J-P. Schille, M. Gester, S. M. Thompson, P. Sparks, V. da Costa, K. Ounadjela, M. Skvarla, J. Magn. Magn. Mater. **175**, 1 (1997).

[168] J. F. Gregg and P. D. Sparks, British Patent Application No 9608716.8

[169] J. M. Kikkawa and D. D. Awschalom, Nature **397**, 139 (1999)

[170] M. Oestreich, J. Hubner, D. Hagele, P. J. Klar, W. Heimbrodt, W. W. Ruhle, D. E. Ashenford, and B. Lunn, Appl. Phys. Lett. **74** , 1251 (1999).

[171] R. Fiederling, M. Keim, G. Reuscher, W. Ossau, G. Schmidt, A. Waag, and L. W. Molenkamp, Nature **402**, 787 (1999).

[172] Y. Ohno, D. K. Young, B. Beschoten, F. Matsukura, H. Ohno, and D. D. Awschalom, Nature **402**, 790 (1999).

[173] B. T. Jonker, Y. D. Park, B. R. Bennett, H. D. Cheong, G. Kioseoglou, and A. Petrou, Phys. Rev. B **62**, 8180 (2000).

[174] G. Schmidt D. Ferrand, L. W. Molenkamp, A. T. Filip, and B. J. van Wees, Phys. Rev. B **62**, 4790 (2000).

[175] R. P. Borges, C. Dennis, J. F. Gregg, E. Jouguelet, K. Ounadjela, I. Petej, S. M. Thompson, and M. J. Thornton , J. Phys. D **35**, 186 (2002).

[176] E. I. Rashba, Phys. Rev. B **62**, R16267 (2000).

[177] C. L Dennis *et al.* Proc of JEMS '04, JMMM

[178] C. L. Dennis *et al.* J. Phys. D: Appl. Phys. **36** ,81 (2003).

[179] S. Datta and B. Das, Appl.Phys.Lett. **56**, 665 (1990).

[180] R. Mattana, J.-M. George, H. Jaffres, F. Nguyen van Dau, A. Fert, B. Lepine, A. Guivarc'h, and G. Jezequel, Phys.Rev.Lett. **90**, 166601 (2003).

Chapter 15

NANOBIOMAGNETICS

Diandra L. Leslie-Pelecky
Department of Physics and Astronomy
Center for Materials Research & Analysis
University of Nebraska
Lincoln, NE 68588-0111, USA

V. Labhasetwar
Department of Pharmaceutical Sciences
College of Pharmacy
986025 Nebraska Medical Center
Omaha, NE 68198-6025, USA

R. H. Kraus, Jr.
Biophysics Group
Los Alamos National Laboratory
Los Alamos, NM 87545, USA

Abstract The application of nanomagnetic materials to biological systems has produced significant advances in research, diagnosis, and treatment of numerous pathologies. This chapter summarizes the major applications of magnetic materials: magnetic targeting, drug and gene delivery, magnetic separation, the use of magnetic beads in manipulating single molecules, as contrast agents in magnetic resonance imaging, and for hyperthermia. Biocompatibility requirements for magnetic materials used in these applications are reviewed.

1. INTRODUCTION

'Nanobiomagnetism' is the intersection of nanomagnetism and medicine that focuses on biological systems and/or processes. Magnetism is an inherent facet of life, from iron in blood to the ability of magnetotactic bacteria, birds, honeybees and other creatures to navigate by the Earth's magnetic field. Iron plays a critical part in many aspects of human neurophysiology. Naturally occurring iron in the body usually is stored within ferritin, which are 12-nm hollow spherical shells that each can hold up to 2500 iron atoms in the form of mineralized ferrihydrite. Anomalous amounts of iron – possibly in nanoscale form – are associated with many neurodegenerative disorders such as Alzheimer's, Parkinson's, and Huntington's diseases [1].

The ability of magnets to act on objects at a distance makes them valuable medical tools. A 1624 report described the extraction of an iron splinter from an eye using a magnet [2]. Safety pins, bullets and grenade splinters were removed using magnets [3-5]. Grazing cows are fed magnets to prevent sharp metallic objects they eat from damaging the intestines. The invention of stronger, smaller permanent magnets made possible more delicate applications, such as temporarily fixing prosthesis in dentistry, guiding catheters through the body, and navigating within the brain [6-8].

Nanoscale materials have a special relevance to biomedical applications due to their size compatibility with cells (10-100 μm), viruses (20-450 nm), proteins (5-50 nm) and genes (2 nm wide by 10-100 nm long). Nanoparticles are small enough to move inside the body without disrupting normal functions, and can access spaces inaccessible by other means. Cells react to the topography of their environment on size scales as small as 5 nm – up to 1000 times smaller than their own size [9, 10]. Changes in response to topography literally can induce growth or death. Nanostructured materials allow study of these critical processes on a single-cell level [11].

2. MATERIALS

Magnetic biomaterials have different constraints than materials used for other applications. *In vivo* (in the body) applications require strict biocompatibility. *In vitro* (outside of the body) applications have less strict requirements, but techniques involving living cells still must consider the effect of the materials on the sample under study. In addition to biocompatibility, materials must be capable of being functionalized with one or more molecules, must retain their magnetic properties for a reasonable period of time in aqueous media with varying pH, must not be cleared too quickly from the bloodstream, and must form stable, non-aggregating dispersions [12, 13].

2.1. Biotoxicity

Cells can be killed by external agents or can be induced to 'commit suicide' via a pattern of events called programmed cell death or *apoptosis*. Injuries from external agents include mechanical damage and exposure to toxic chemicals (e.g. chemotherapy). Substances toxic to cells are called cytotoxic, and different types of cells can have different responses to the same material.

The vast majority of research in this area involves iron-oxide particles, as they are highly biocompatible, cheap, can be made in a variety of ways and sizes, and can be made as superparamagnets or ferrimagnets. Iron oxides are metabolized into elemental iron and oxygen by hydrolytic enzymes in the body. The iron joins the normal body stores and the body compensates by taking up less iron from the stomach. Intravenous injections of up to 250 mg iron/kg body weight does not produce chronic or acute hepatotoxicity in rats [14], while 1-3 mg/kg has been used clinically in humans [15].

The disadvantage of iron oxides, however, is their low magnetic susceptibilities. Iron nanoparticles offer an order of magnitude greater susceptibility than iron oxides at room temperature, but are easily oxidized and are not as biocompatible as their oxides. Fe-C composite particles made by mechanical milling, chemical reduction, or plasmochemical recondensation have been used in clinical trials for hyperthermia and drug delivery [16-18]. These particles are biocompatible and the carbon in the particles may assist in reversing drug-induced toxicities via physical adsorption. The need for monodisperse particles for magnetic recording has stimulated a many chemical methods that produce a broad variety of element and alloy nanoparticles, including Co, Ni, Cu, FePt, CoPt, Fe, $CoFe_2O_4$, $MnFe_2O_4$, $SmCo_5$ and even core-shell particles such as $FePt/Fe_3O_4$ [13, 19-27]. Physical deposition techniques such as inert gas condensation [28-31] and laser ablation [32-35] also are used.

2.2. Coatings

Coatings can improve oxidation resistance, colloidal stability, the ability to functionalize, phagocyte resistance, mechanical stability, and biocompatibility. Unfortunately, coating a magnetic nanoparticle with a biocompatible material does not necessarily render the nanoparticle biocompatible [36]. The chemistry for functionalizing gold is well established and gold coating offers corrosion resistance; however, biotoxicity issues may need to be resolved [37, 38]. Polysaccharide coatings such as dextran, starch, and chitosan are biocompatible and offer a range of functionalization options [39-41]; however, they can be structurally weak and can be dissolved by highly acidic environments. Silicon-based coatings are used to protect particles from lysomal enzymatic digestion, and improve mechanical properties and chemical stability [42-45]. Silica-coating can improve chemical stability, but a porous coating may allow the contents inside to be dissolved or oxidized [46, 47].

Many polymers are biocompatible and may be used as coatings for metallic or ceramic particles, or can serve as hosts by either capturing nanoparticles inside a larger polymer particle or attaching nanoparticles to their surfaces. Polyethylene Glycol (PEG) and related polymers covalently bond to surfaces or are adsorbed on magnetic nanoparticles and can prolong the circulation time in the bloodstream [39, 40].

2.3. Clearance Time

Particles introduced into the bloodstream are covered rapidly by components of the circulation, such as plasma proteins, in a process called *opsonization*. Opsonization makes the particles recognizable to the body's major defense system, the reticuloendothelial system (RES). The RES comprises a diffuse system of phagocytic cells (which engulf inert material) that are primarily associated with the connective tissues in the liver, spleen, and lymph nodes. Macrophage (Kupffer) cells in the liver and macrophages of the spleen and circulation are important in removing particles identified by opsonization. A significant fraction of nanoparticles can be cleared from the circulation system in as little as 15 minutes [48, 49].

The clearance rate is dependent on size, charge, surface hydrophobicity and the number and nature of functional groups on the surface [48, 50]. These variables are interdependent, making understanding the role of each one independently challenging. Some of these variables also may affect the magnetic properties. For example, smaller particles more easily evade the RES; however, the smaller size usually results in a smaller moment.

Anionic particles with negative surface charge have a high affinity for the cell membrane and are typically taken up by the endocytic process [51]. Cationic magnetite particles show significantly lower cell-survival rates but their toxicity depends highly on the magnitude of the surface charge. More highly cationic particles tend to be more toxic [52]. Hydrophilic surfaces such as dextran, polyethylene glycol, polyethylene oxide, poloxamers, polysorbates and polyoxamines provide a dynamic 'cloud' of hydrophilic and neutral chains at the particle surface that repel plasma proteins and prevent rapid removal of particles from circulation [53-55].

Related to clearance time is the manner in which the nanoparticles attach to cells. Nanoparticles can be internalized or remain adhered to the surface. The mechanism is determined by the surface charge, adhesion properties, and chemical functionality of the cells with respect to the nanoparticle.

2.4. Magnetic Fluids

The delivery of nanoparticles in the human body usually requires suspending the nanoparticles in a water-based fluid. *In-vitro* applications also usually require an aqueous environment. Magnetic nanoparticles must remain suspended in fluid (or be easily re-dispersed when needed) and cannot form aggregates due to van der Waals or magnetic interactions.

Magnetic particles in a solution undergo two types of relaxation: Brownian relaxation, in which the entire particle rotates, and Néel relaxation, in which the moment rotates while the particle remains still. The Brownian relaxation time τ_B is

$$\tau_B = \frac{3\eta V_H}{kT} \tag{1}$$

where V_H is the hydrodynamic volume of the particle, k is the Boltzmann constant, T is the temperature, and η is the dynamic viscosity of the carrier liquid. The hydrodynamic diameter characterizes how a particle moves through the fluid in which it is suspended and may be different than the magnetic size due to agglomeration, coating, or interactions between the fluid and the nanoparticle surface [56, 57].

The Néel relaxation time τ_N is defined by [58, 59]

$$\tau_N = \tau_o \exp\left(\Delta E / kT\right) \tag{2}$$

where ΔE is the energy barrier over which the magnetization must reverse, and $\tau_o \sim 10^{-9}$ s. ΔE usually is determined by the product of the anisotropy K and the magnetic volume V: $\Delta E=KV$. Particles with relaxation times faster than 100 s are called superparamagnetic. Since ΔE depends on the nanoparticle size as the radius cubed, τ_N can range easily over 9-10 orders of magnitude. The hydrodynamic and magnetic diameters of the particle can be determined by the peak frequency of the ac susceptibility [57].

The relaxation rates are affected differently by the immobilization of the nanoparticles. The Brownian relaxation time can be changed by changing the viscosity of the carrier while the Néel relaxation should be independent of the carrier liquid. The relaxation rates also are affected differently by phenomena such as immobilization of a nanoparticle due to endocytosis.

3. TARGETING

Some materials are taken up easily by all types of cells, while others are preferentially taken up by specific types of cells. A disadvantage to systemic treatment is that healthy cells can be negatively affected, as in chemotherapy. Often, a specific type of molecule needs to be separated for further study, so the ability to differentiate between different types of cells is very important. Nanoparticles can be localized by physical, chemical, and/or magnetic targeting.

Physical or passive targeting uses surface features of nanoparticles, such as hydrophobicity, charge or pH, to induce reactions that cause the nanoparticles to stick to or enter the cells [60]. This mechanism is highly non-specific, but may be used and to take advantage of the natural clearance of the RES if the targeted organs are within the RES [61]. Nanoparticles can be targeted to certain pathologies such as tumor or inflamed tissue because of leaky vasculature. This phenomenon is known as the enhanced permeation retention (EPR) effect [62].

Chemical targeting uses functionalization of particles to increase the specificity of binding [63]. Functionalization is the physical or chemical association of ligands – targeting agents, therapeutics, surfactants, etc. – with a magnetic nanoparticle. Functional groups may be incorporated using covalent or non-covalent bonding, and/or physical adsorption. Binding to a receptor of interest is called specific binding, while binding to that *and* other sites is called non-specific binding. Antibodies, for example, specifically bind to their antigen, providing an effective means of tagging. Molecules used for targeting include proteins, oligonucleotides, antibodies and their fragments, lectins, hormones, charged molecules, nucleic acids, peptides, and receptor ligands.

Magnetic targeting is used when a therapy has limited ability to be chemically targeted to specific types of cells or tissues due to high non-specific binding. The force felt by a magnetic moment of value $\vec{m}$ in a gradient field is

$$\vec{F}_m = (\vec{m} \cdot \vec{\nabla})\vec{B} \tag{3}$$

The magnetic targeting force must compete with the force due to linear blood-flow rates of about 0.05 cm/s in capillaries to 10 cm/s in arteries and 50 cm/s in the aorta. Iron-oxide nanoparticles require flux densities at the target site on the order of 0.1 to 1.0 T with field gradients ranging from 8 T/m (femoral arteries) to over 100 T/m for carotid arteries [39, 64].

The accuracy of magnetic targeting also is dependent on the depth of the target tissue within the body: organs such as the liver and the lungs are harder to target than organs closer to the surface or in the extremities. The primary challenges are producing a focused field of sufficient magnitude and gradient, and fabricating nanoparticles with sufficiently high moment [65]. As an example, magnetic targeting was useful for treating subcutaneous mouse lymphoma models, but was not effective in intraspinal gliobastoma models [64]. Similarly, permanent magnets are more effective on surface tumors and tumors in extremities, while electromagnets or superconducting magnets (such as in clinical MRI machines) can be used in locations requiring higher gradients.

Magnetic targeting has been shown to reduce significantly the movement of particles to undesired organs and tissues during the time the magnetic field is applied [64]. Micron-sized FeC nanoparticles have been targeted successfully to various organs, including the liver, lungs, and the brain [66, 67]. A complication is that, in some cases, the particles are retained only while the magnetic field is applied. Some studies indicate that the magnetic field promotes extravasation (movement of nanoparticles out of blood vessels) and the particles remain in the target area after removal of the field; however, other studies show that nanoparticles migrate after the targeting field is removed [66, 68-70].

One relatively new approach proposed for magnetic targeting is 'ferromagnetic seeding' [71-73]. Nano-sized ferromagnetic objects ('dockers') could be inserted by catheter. The docker would reinforce the magnetic gradient of the external magnetic field, decreasing the need for externally applied high-gradient fields. A similar idea has been suggested for stents in treatment of vascular disease [74-76].

4. MAGNETIC SEPARATION

The detection of specific molecules is critical for diagnosis, treatment, and prevention of disease. The development of fast, handheld analysis units capable of detecting multiple species is made more urgent by fears of biological and chemical terrorism. Magnetic separation has been applied to everything from separating tin from stainless steel at recycling centers to separating pure natural diamonds from diamonds with inclusions of other (magnetic) minerals.

Few cells are naturally magnetic enough to be separated due using their own inherent magnetism, so the cells must be attached to a magnetic nano- or micro-particle with a detectable magnetic moment. Magnetic cell sorting first was proposed using surface markers for cell receptors [77]. Magnetic cell separation allows separation of target cells directly from blood, bone marrow and other fluids in short times due to the fast reaction kinetics. The limiting factor for magnetic separation is identifying a linking molecule with high specificity for the desired cell.

Magnetic sorting may be accomplished with micron-sized or nanometer-sized particles [78]. Smaller nanoparticles produce suspensions that are stable against sedimentation due to gravity or an applied magnetic field, while larger particles can be used to take advantage of sedimentation as part of the separation process [79].

Applications include purging malignant cells from autologous stem cell products [80], water purification [81], minimizing and recycling nuclear waste [82, 83] and recovering heavy metals [84]. Blood purification using magnetic carriers has been used to treat autoimmune and inflammatory diseases, including myasthenia gravis, lupus and Guillain-Barré syndrome [85]. Gram-negative bacteria, such as E. coli can be detected at concentrations of 15 cfu/mL, while gram-positive bacteria such as vancomycin-resistant enterococci can be detected at even lower concentrations [86, 87]. T4 and T8 cells in HIV-infected patients have been isolated using magnetic separation, thus allowing study of the effect of different drugs on specific types of cells [88]. Isolation of rare cell populations such as endothelial cells in blood down to 10 cells/ml has been accomplished [89].

Magnetic separation consists of three parts: tagging or labeling the desired cells with a magnetic marker as described earlier, separating magnetically labeled cells from unlabeled cells, and measuring the magnetic properties to quantify the number of cells present.

4.1. Separation

Separation may be done in batch or flow configurations, depending on the specific application. In batch processing, the magnetic beads and the analyte material are mixed. The reaction kinetics determine the amount of time necessary to wait for a sufficient amount of binding to occur. A magnet is used to separate the magnetically targeted cells from the non-targeted cells, as shown in Fig. 1. The most commonly used (and commercially available) materials for cell separations are micron-sized polymer beads into which a magnetic material – usually maghemite – has been embedded.

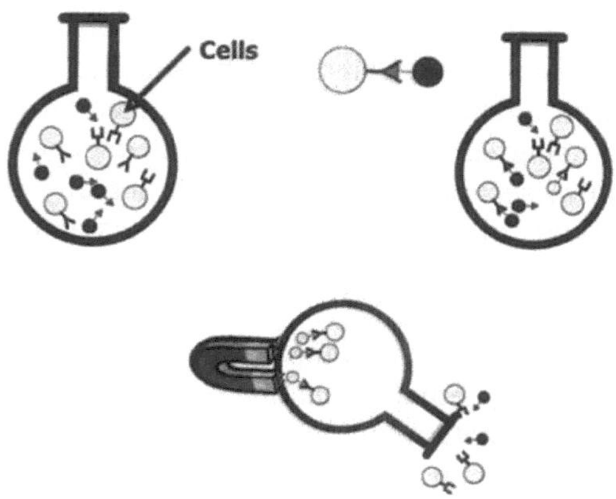

Figure 1. A schematic illustration of the magnetic cell separation process.

Fluid-flow techniques allow continuous processing and are advancing rapidly due to microfabrication capabilities [89, 90]. A permanent magnet can be used to either deflect or collect magnetically labeled particles. Several flow channels may be used in parallel to increase throughput. The magnetic force acting on a magnetic carrier is given by [91]:

$$F_b = \frac{1}{2\mu_o} \Delta\chi V_H \nabla B^2 \tag{4}$$

where F_b is the force on a single magnetic carrier, ∇B^2 is the magnetic energy gradient, V_H is the hydrodynamic volume of the magnetic carrier, and $\Delta\chi$ is the difference in magnetic susceptibility between the carrier and the suspending medium. This deflective force competes with the drag force F_d of the fluid on the particle.

$$F_d = 3\pi v_m D_H \eta \tag{5}$$

where D_H is the hydrodynamic diameter of the magnetic carrier, v_m is the velocity of the magnetic carrier and η is the fluid viscosity.

4.2. Detection

Magnetic sorting techniques have high potential for real-time detection and monitoring of bacterial, viral and other pathogenic contamination [92]. Integrated structures utilizing nanolithography can perform sorting and quantitative analysis in a single device. Magnetic transducers have low interference, low background signal, do not require sample pre-treatment, and can be small enough to be portable. Magnetoresistive techniques have an advantage over techniques that use, for example, MFM or AFM tips to manipulate magnetic beads attached to molecules, in that they are much faster and have the potential to detect more than one molecule at a time.

Spin-valve and other magnetoresistive devices detect the stray field from a magnetic micro- or nanobead, as illustrated by Fig. 2. Lithographically fabricated microcircuits [93-96] may be used to manipulate the magnetic particles. Detection limits in the 10^2 nM can be achieved, and detection of single particles is theoretically possible.

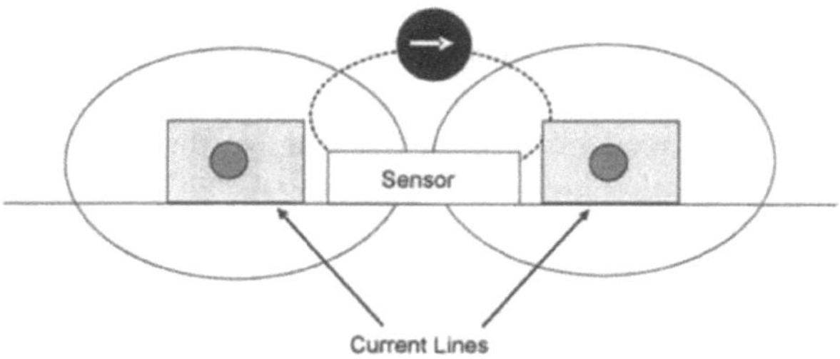

Figure 2. Sensing the stray field of a magnetic bead using a magnetoresistive sensor [93].

Detection of multiple species on a single chip is possible by fixing a probe molecule (often DNA) to a polymer layer covering the sensor, as shown in Fig. 3. The analyte DNA is a single strand complementary to the probe DNA and is labeled (often with biotin). Magnetic microspheres functionalized with streptavidin (which attaches to biotin) are then introduced; the microspheres bind to the biotin, which is present only on the successfully trapped DNA.

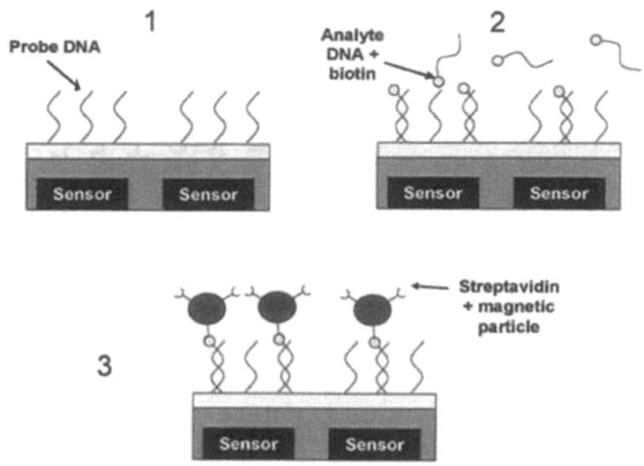

Figure 3. Process of detecting DNA via a magnetoresistive device. (After [97]) The probe DNA is fixed to a polymer layer on the chip (1), the analyte DNA tagged with a molecule such as biotin in introduced and allowed to bind (2). The excess analyte is removed and a molecule that binds to the biotin (streptavidin) is attached to the magnetic label and introduced (3).

The signal measured by sensor can be used to quantify the amount of analyte present. The response to the sensor is determined by the in-plane component of the stray fields induced by the magnetized microspheres. Concentrations as low as 3.2 pg/ml have been detected [98]. The introduction of tunneling magnetoresistance (TMR) sensing elements and smaller magnetic markers will increase the sensitivity of the method.

The Bead Array Counter (BARC) was one of the first sensors to measure DNA-DNA, antibody-antigen and ligand-receptor pairs at the level of single molecules [99-101]. Adapting technology developed for MRAM allows the possibility of testing for a large number of different molecules with a single chip because different sensors can share circuitry.

A primary challenge in 'lab on a chip' devices is to integrate the fluidics and the sensing devices on a scale that allows a handheld device [102]. A microfabricated chip comprising a compact electromagnet, a GMR sensor and a microfluidic flow cell that can simultaneously detect eight different analytes and is small enough to sit on a table top already has been developed [103]. More recently, a planar Hall effect magnetic sensor using exchange-biased Permalloy has been shown capable of detecting a single 2-micron magnetic bead, which corresponds to a 300 nV signal at 10 mA and 15 Oe applied field [104]. The advantage of planar Hall sensors is that the entire active surface is used for bead detection, in contrast to GMR or spin-valve sensors. A higher signal-to-noise ratio is expected, and the lower noise of the planar Hall sensors is expected to make detection of single nanobeads possible [105]. Magnetic tunnel junctions and anisotropic magnetoresistive rings are also possible candidates.

An alternative detection technique for biological molecules is based on changes in the Brownian relaxation due to binding [106, 107]. The relaxation frequency of a nanoparticle changes when it binds to another molecule, as the binding increases its hydrodynamic size. The shift is proportional to the hydrodynamic size of the nanoparticle, allowing discrimination between target molecules with different sizes (although molecules with different functionalities, but similar sizes cannot be independently detected). An advantage of this technique is a signal is present both before and after binding, which allows for reliability checks.

Superconducting Interference Device (SQUID) sensors are more sensitive than GMR devices, but require low temperatures and magnetic shielding. RF Squids have been used to measure magnetic markers to which monoclonal antibodies have been attached [108]. SQUID sensors also can be used to detect changes in relaxation due to binding or changes in local environment. Magnetic nanoparticles immobilized *in vivo*, homogeneously distributed over a total volume of 0.1 ml, could be detected at a limit of 0.3 nmol using a SQUID gradiometer [109, 110].

5. MAGNETIC TWEEZERS

Magnetic beads can provide a 'handle' that can be used to manipulate molecules and investigate intermolecular interactions [111]. DNA strands can be attached to a glass surface on one end and a micron-sized magnetic bead on the other. Magnetic fields can be used to apply linear or torsional force to the DNA, allowing investigation of the extension and elasticity of the molecule using forces from fN to a few tens of pN [112-115].

Mechanical manipulation is interesting because cell morphology regulates many functions, including cell growth, proliferation, protein synthesis and gene expression; however, there is much to be understood about how mechanical signals are translated into biological processes. Magnetic twisting cytometry can be used to apply controlled mechanical stresses directly to specific cell-surface receptors via ligand-coated particles. Cellular mechanical properties such as cell stiffness and viscosity can be measured in this manner [116-118]. The reliability of these measurements, however, can be affected by how the magnetic bead binds to the cell. Although nanoparticles may localize at a cell-membrane receptor and remain external to the cell, most nanoparticles are taken into the cell via receptor-mediated endocytosis, which can produce an inhomogeneous cell shape. Applied forces and torques may not be as evenly distributed across these regions and a shear stress may result. The internalization of magnetic nanoparticles by cells can, however, allow probes of the cell interior. Anionic ferromagnetic nanoparticles can be internalized by endosomes via endocytosis, producing a superparamagnetic endosome. The magnetized endosomes elongate in the field, allowing study of the rheological properties of the cell interior [119, 120].

6. DRUG AND GENE DELIVERY

Controlled drug delivery has the potential to improve drug efficacy, as well as patient convenience and compliance [121, 122]. Although overall drug dosage can be reduced by 50-80%, dosage at the target site is increased and systemic uptake is decreased. Local drug delivery also reduces the patient-to-patient pharmacokinetic variability inherent in oral and intravenous applications. Protecting the drugs until they reach their target area increases the usability of drugs that have a short half-life in the body. Allowing the release of a drug over a prolonged period of time maximizes the effect of drugs such as chemotherapeutics that are effective only during a specific part of a cell's life cycle.

Nanoparticles were first used for drug delivery starting around 1970, when they were developed as carriers for vaccines and anticancer drugs, and are now widely used [123-125]. The ideal nanoparticle for drug delivery must be able to efficiently incorporate a reasonably high weight fraction (loading) of the drug, must form a stable suspension in an aqueous medium, must be biocompatible and biodegradable, and must not be cleared too rapidly from the bloodstream. In addition, nanoparticles should be able to be made in a range of sizes, but with uniform size distribution, and should be able to be further functionalized. The main attraction of magnetic nanoparticles from the perspective of drug delivery is the ability to use the magnetic properties to either limit the drug to a particular region using magnetic targeting, or to release the drug remotely.

Drug may be encapsulated in, conjugated to, or adsorbed onto the surface of a nanoparticle. The drug may be released via degradation of the carrier particle or may be triggered by heat or pH. Many drug delivery systems use polylactic-co-glycolic acid (PLGA), polycaprolactone (PCL) and their copolymers, as these are biodegradeable, FDA-approved materials and the degradation rate can be controlled by the particle formulation [126]. Careful choice of surfactants can allow hydrophobic drugs to be transported throughout the body [127].

Ionically bound pharmaceuticals have the advantage that the active-low-molecular weight substances can desorb from the carriers after a defined time span and diffuse from the vascular wall into the tissue. The diffusion through the vascular wall can significantly change the desorption kinetics of the pharmaceutical [128]. Epirubicin chemoadsorptively bound to a polymer-coated particle can desorb according to physiological environment (pH, osmolarity, and temperature) [129]. The half-life of the drug desorption can be fixed to be approximately the same as the desired time for magnetic field targeting.

6.1. Chemotherapy

Cancer is a leading cause of death worldwide. Chemotherapy, a common treatment, is non-selective and causes significant negative side effects. Chemotherapy dosages are calculated primarily by the individual tolerance levels of a particular patient, which means that physically weaker patients are not able to receive adequate doses for successful treatment [122]. Targeting the drugs and preventing release until they reach the tumor decreases the damage to normal cells and increases the dosage at the tumor.

The first evidence for the utility of nanospheres in cancer treatment was demonstrated in the early 1980s. Widder, *et al.* showed significant remission of Yoishida sarcoma without drug toxicity in rats using magnetically targeted albumin microspheres with doxorubicin [69, 130, 131]. The majority of magnetic nanoparticles used for drug delivery are based on iron-oxide or iron-carbon combinations. Mitoxantrone, doxorubicin, mitomycin C, etoposide, paclitaxel, oxaliplatin and epirubicin have been bound to iron oxide or Fe-C fluids for magnetically targeted cancer treatment [16-18, 39, 67, 69, 131, 132]. MTC™ is a micron-sized Fe-C particle with 3-10 times greater susceptibility than the corresponding-sized Fe_3O_4 particles [133]. The high susceptibility is necessary for the goal of magnetically targeting hepatocellular (liver) carcinoma, which requires the particles to be captured at depths up to 14 cm. Despite the recent suspension of one U.S. Phase II/III liver-cancer trial that was using magnetically targeted Fe-C particles, other studies suggest high potential for treatment of cancer in humans.

6.2. Radionuclide Therapy

Radiotherapeutics attack cancer by causing radiation damage to DNA in cells. The requirements differ from those for drug delivery because the radiotherapeutic can act at a distance and does not have to separate from the delivery particle. Radiotherapeutics can be selected to provide action over a range of distances, from tens of nanometers to hundreds of microns. Three radiotherapy modalities can be identified. Brachytherapy, most often used with beta-emitters, uses tightly enclosed radioactive material that is brought in close proximity to the tumor. A second modality is intravenous injection so that the radiopharmaceutical binds to the outside of the tumor cells or is taken up by the cell and irradiates from within. The third approach uses a carrier loaded with the radiotherapeutic that is transported to the vicinity of the target cells, and then released.

Y-90 is a stable beta emitter, releasing less than 5% of the bound radioactivity within 3 weeks [134, 135]. The polymer decay rate can be matched to the 64.1-hour half-life, so that once the particles loose a significant amount of their radioactivity, they decay into lactic acid. This treatment could deliver up to 100 times greater dose of radiation to the tumor than conventional external beam radiation therapy and minimize the damage to healthy tissue [135, 136]. Squamous cell carcinoma in rabbits has been treated using 100-nm multidomain magnetic iron oxide/hydroxide particles onto which I^{123} has been ionically bonded [137, 138]. A permanent magnet is used for magnetic targeting and the biodistribution of the ferrofluid is detected using a gamma camera [138].

6.3. Magnetic Switches

Magnetic fields can be used to activate the release of a drug remotely. For example, liposomes can be made to encapsulate magnetic nanoparticles and drugs. An applied ac field causes the magnetic particles to heat, which in turn opens the lipid layer and releases the encapsulated drug [139, 140]. A similar effect can be used with thermoresponsive gels, which are chemically cross-linked polymer network characterized by pores, elasticity, and the ability to change volume when stimulated by temperature. Entrapping magnetic nanoparticles in the gel allows control of the pore size via the external ac magnetic field. Drug entrapped in the pores is released when the gel swells and the pores expand [141, 142].

Magnetically induced stress has been used to control drug release. Polymer spheres filled with magnetic nanoparticles and drug subjected to an oscillating magnetic field produced small stress-induced cracks in the polymer. The cracks allowed liquid to enter the spheres and carry out drug. Using magnetism as a 'release-on-demand' mechanism could be useful for insulin-dependent diabetics [143, 144].

Innovent, Inc. has developed a magnetic capsule made of two or more parts that are held together magnetically. Demagnetizing the magnetic capsule by applying alternate pulses of opposite magnetic polarity allows the capsule to open. The resulting capsule parts are small enough for the patient to eliminate. The capsule is smaller than ones now used for endoscopy and could be used for delivery of drugs to the gastrointestinal tract [145].

6.4. Gene Delivery

Transfection refers to the incorporation of exogenous DNA into a cell; however, the incorporation process is diffusion limited and thus slow. 'Magnetofection' uses magnetic targeting of polyethylenimine–coated magnetic nanoparticles to which gene vectors are electrostatically attached. The magnetic targeting overcomes the limitations of diffusion, allowing increases in uptake in the target tissue [146-148]. Research remains to be done on the influence of the externally applied magnetic field beyond merely attracting the nanoparticles to the tissue. Magnetic targeting also has been used to deliver stem cells [149].

7. MAGNETIC RESONANCE IMAGING

Magnetic resonance imaging (MRI) aids in diagnosis, research, and treatment of a wide variety of pathologies. The first attempts to use nuclear magnetic resonance for medical purposes were made by Odeblad and Lindström in 1955 [150], but it was not until 1973 that Lauterbur developed magnetic-field-gradient methods capable of generating images that could be used clinically [151]. Although hydrogen atoms are the usual targets, magnetic nuclei such as ^{13}C, ^{19}F, ^{23}Na and ^{31}P can be detected in biological tissue. MRI measures change in the magnetization of the protons in water molecules in a magnetic field after being subjected to a radio-frequency magnetic field pulse. Protons in different types of tissues relax differently, thus providing the contrast necessary to distinguish between different types of tissues.

One mm^3 of water contains 6.7×10^{19} H nuclei. The nuclear magnetic moment of the sample is proportional to the magnetic field strength and inversely proportional to temperature and an external magnetic field creates a surplus of nuclear spins aligned parallel to the field. The surplus in room-temperature water at a magnetic field of 1 T corresponds to a fraction of 3.2×10^{-6} of the protons. A radio-frequency field at a frequency near the Larmor frequency stimulates precession of the nuclear magnetization about the static field. The amplitude of the precessing nuclear magnetization decays with time until the original equilibrium state is recovered.

Two time constants describe the relaxation process: T_1 is the longitudinal relaxation time (a spin-lattice relaxation involving energy transmission to the surroundings) and T_2 is the transverse relaxation time (a spin-spin relaxation). An individual moment sees many other moments in thermal (Brownian) motion and thus experiences a continuously changing magnetic perturbation field. Spectral components of the field that correspond to the Larmor frequency induce longitudinal relaxation, while transverse relaxation is due to the frequency of collisions between the molecules [152]. The slope of the inverse relaxation time vs. concentration plot is called the relaxivity, with r_1 being the relaxivity corresponding to T_1 relaxation and r_2 being the relaxivity corresponding to the T_2 relaxation. Fat has a very efficient energy exchange and thus a short T_2, while water is less efficient and has a longer T_2. The white matter of the brain has a T_2 around 60-100 ms and a T_1 an order of magnitude larger.

Magnetic contrast agents work by changing the relaxivity of the water in the area near the contrast agent, thus improving differentiation between otherwise similar types of tissue. Individual atoms such as Gd and Fe decrease T_1 by orders of magnitude; however, they must be chelated for *in vivo* use to prevent toxicity. A disadvantage of these types of contrast agents is that they equilibrate rapidly throughout the interstitial space, making it difficult to identify specific regions with high precision [153]. Superparamagnetic nanoparticles, either coated metal particles or polymeric nanoparticles loaded with metal atoms, preferentially affect the *transverse* relaxation time T_2 as protons diffuse within the inhomogeneous magnetic field created by the magnetic clusters.

Most clinically used nanoparticle contrast agents are iron-oxide based and are grouped into two categories: Superparamagnetic iron oxides (SPIOs) are between 500 nm and 50 nm, and ultra-small SPIOs (USPIOs), which are smaller than 50 nm.

SPIOs are used for gastrointestinal tract, liver, and spleen imaging and take advantage of the natural clearance of the RES, the transit time through the gut or preferential uptake by specific cells. The most promising uses of SPIOs are to improve the sensitivity of detection and localization of primary and metastatic brain tumors, inflammation and ischemia (insufficient supply of blood to an organ) [154-156].

USPIOs can be used as 'blood pool agents' because their smaller size allows them to remain in the bloodstream for longer times. They can be used to assess perfusion (the passage of blood into an organ or tissue) in areas of ischemia and provide information about capillary permeability. They also can be used to study the extent of tumor neovascularity and associated permeability changes [153].

Nanoparticle T_1 contrast agents are being studied as well. Gd-chelates can be surrounded by a polymer with a high affinity for the metal and again by a porous hydrophobic polymer shell that modulates access to the core [157]. The coatings increase the circulation time, which has been the primary limitation of Gd chelates. When administered intravenously, the nanoparticles remain in the intravascular space, and thus provide excellent visualization of the vasculature.

Natural targeting can be used by incorporating magnetic nanoparticles within the lipid phase of liposomes. The relaxivities of magnetoliposomes vary enormously, depending on size and surface characteristics, and on whether the magnetoliposomes are free or bound [139]. Changes in the relaxivity can indicate bound vs. unbound magnetoliposomes and even changes in the ratio r_1/r_2 also can be used for contrast.

Magnetic contrast agents can be targeted; however, targeting is usually chemical due to the magnetic nature of the resonance measurement. Restenosis, a complication of coronary angioplasty that involves the proliferation and migration of vascular smooth muscle cells, and arterial plaques can be detected by targeted Gd contrast agents [158]. The acoustic reflectivity increases when the nanoparticles are bound, thus allowing high contrast between the targeted tissue and the background. MRI also can be used to confirm the targeting of drug-delivery nanoparticles [159].

MRI resolution can be 20-25 μm, which means that real-time tracking of single cells is possible if a cell can be loaded with sufficient magnetic material [160]. These techniques are useful in understanding how cells migrate in response to diseases. Biocompatible magnetic nanosensors have been designed to detect molecular interactions in biological media via MRI. Changes in relaxation times due to the nanoparticles binding with the target molecule are detected by MRI contrast, thus allowing simultaneous study of the location and chemical functionality of specific types of cells. These magnetic nanosensors can detect specific mRNA, proteins, enzymatic activity, and pathogens with sensitivity in the low femtomole range (0.5 - 30 fmol) [161-163]. Other applications include imaging gene expression [164, 165].

Work is in progress on chip-scale integration for MRI of very small samples. Micromechanical cantilever oscillators allow for high sensitivity magnetic measurements [166]. Integrating DC and RF magnetic field sources into such a chip would allow for magnetic resonance measurements with significantly improve sensitivity.

8. HYPERTHERMIA

Cancer growth is slowed or stopped at temperatures in the range of 42-48 °C, while normal cells can tolerate even higher temperatures [167, 168]. Heat treatments can be characterized as hyperthermia, in which the temperature is limited to less than about 50 °C, and thermal ablation, which involves higher temperatures.

Hyperthermia induces almost reversible damage to cells and tissues; however, it can enhance radiation and chemotherapy injury of tumor cells [169]. The enhancement is attributed to heat-induced malfunction of the processes that ordinarily repair DNA. Hyperthermia also affects the activity of regulatory proteins, kinases and cyclins, which in turn alters cell growth and differentiation, and can induce apoptosis. Thermoablation produces necrosis, coagulation or carbonization that could be sufficient to eliminate the need for radio- or chemo-therapy [68, 170]. Thermoablation has been studied primarily for difficult-to-treat cancers (such as liver cancer) with limited treatment options, and for areas of the body far away from vital organs (e.g. breast cancer). Targeting of the magnetic nanoparticles to specific sites eliminates the systemic side effects that result from oncologic treatments (i.e. nausea or radiation pneumonitis); however, whole body hyperthermia produces non-selective damage and thermal ablation has a higher risk for collateral damage.

Hyperthermia and thermoablation have been accomplished using capacitive or inductive coupling of rf fields (10-100 MHz), microwaves (> 300 MHz), ultrasound, lasers or external heat [171-177]. Macroscopic metal implants of Cu and other high-conductivity metals have been used to induce eddy-current heating. The absorbed power per mass is called the specific absorption rate (SAR), which can be expressed as

$$SAR = \frac{\Delta Q}{\Delta t}\left(\frac{1}{m_f}\right) \tag{6}$$

where ΔQ is the energy converted into heat, Δt is the time over which the conversion occurs, and m_f is the magnetic material mass [170]. The SAR is determined by the "rate-of-temperature-rise" method [178]

$$SAR = c\frac{dT}{dt} \tag{7}$$

where c is the specific heat and dT/dt is the rate of temperature change. The density of absorbed power is related to the SAR by

$$P = SAR\frac{m_f}{V} \tag{8}$$

The coupling of an external RF magnetic field to magnetic particles in the body results transfer of energy to the tissue by: 1) eddy current heating, 2) hysteretic heating: heat generated when a magnetic material is forced around part or all of the hysteresis loop, 3) viscous heating: heat generated by the kinetic motion of a particle within a viscous fluid, and 4) magnetic resonance. The loss power of the magnetic particles should be as high as possible so as to allow the lowest possible dose [179].

Avoiding neuromuscular electrostimulation requires frequencies greater than 50 kHz, and the penetration depth limits the frequency to less than 5-10 MHz, effectively eliminating effects from magnetic resonance [180]. Typical values used with iron-oxide nanoparticles are magnetic fields of frequency in the range 50-500 kHz and amplitude 1-15 kA/m [170, 181]. Jordan suggests that materials should be investigated for use at frequencies near 100 kHz so as to optimize the SAR in the magnetic material in comparison to the potential SAR in tissues due to eddy currents [182]. For thermal ablation, a frequency of 400 KHz and 6.5 kA/m should be tolerable for the exposure of parts of the body with diameters of up to 15 cm, if short exposure times are used [170, 183]. In many nanoparticle systems, the SAR is best be described by

$$SAR = kf^n H^2 \tag{9}$$

where n ranges from 1.1 to 1.5, suggesting that there must be a frequency-dependent process of magnetic relaxation that accounts for the changes in the power n [180]. Theoretical predications for Rayleigh loops suggest a H^3 dependence.

Modeling magnetic hyperthermia is difficult due to the complex magnetization reversal mechanisms found in nanoparticles. Optimization of nanoparticle properties is important to limit the amount of material that must be introduced. The SAR achievable for a given combination of field, frequency, and type of particle usually must be determined experimentally. SAR values range from a few W/g to a few hundred W/g for optimized values (i.e. using the field and frequency that provides the best results for the particular system being studied) [167]. The SAR depends on many factors, including the effect of coating on surface spin dynamics, the effect of surface properties on Brownian relaxation, size, and crystallinity among other factors.

The SAR due to losses along a hysteresis loop is:

$$SAR_{hyst} = \frac{v\mu_0}{m_f} f \oint H\mathrm{d}M \tag{10}$$

Magnets with large-area loops are thus preferred for hysteresis hyperthermia; however, the size of the magnetic field at the location of the nanoparticles may be limited, making only minor loops accessible.

In macroscopic implants, shape anisotropy can be used to maximize the area of the loop; however, local heating ('hot spots') is a concern. In nanoparticles, overcoming the energy barrier to rotation of the magnetization (with the particle fixed) determines the properties of the loop. Ferromagnetic particles much larger than the superparamagnetic limit have no implicit frequency dependence to the hysteresis in the frequency range considered. The physical basis of the heating due to superparamagnetic particles shows that the frequency dependence is more important than for their ferromagnetic counterparts [184].

The use of iron oxides for hyperthermia of tumors was first proposed by Gilchrist *et al.* [185]. Magnetic nanoparticle hyperthermia has a potential advantage over radio- and chemo- therapies because there is no systemic buildup in organs, so larger doses are possible. The nanoparticles can be introduced into the body once, and then used for multiple treatments. Nanoparticles can be magnetically targeted, injected directly into the tumor in some cases, or injected into the vasculature supplying the tumor. Chemical and/or magnetic targeting help limit side effects [186]. Tumors derive their nourishment from the supplying vasculature, so blocking the blood supply via magnetic nanoparticle thermoablation has been attempted [187-189].

The first clinical human trials using magnetic hyperthermia were reported by Lübbe, *et al.* [70, 129, 137, 190] who used 100-nm starch-coated iron-oxide particles bound with epirubicin for treatment of advanced solid cancers. Jordan recently reported positive results from ongoing trials of advanced cancer patients who received magnetic nanoparticle hyperthermia in conjunction with conformal external beam radiation therapy [191]. The therapy was well tolerated by the patients and significant increases in the length and quality of life were observed.

The greatest challenge to the efficacy of hyperthermia using magnetic nanoparticle fluids is balancing the rate of thermal energy deposition with the mechanisms responsible for thermal dissipation. Modeling heating is complex due to the multiple factors affecting temperature change, including tissue density, amount of fat, and blood flow. Tumors in highly perfused organs, such as the kidney, lung and liver, are harder to heat due to the high blood flow [68, 170]. The dependence of temperature rise on distance is steep, and there are delays in heating and cooling after the field is applied or removed. Larger bones, such as the pelvis and skull shield tissues and produce inhomogeneous heating. Non-uniformity of tumors also poses a complication, as large tumors heat at a greater rate than small tumors due to the poorer tissue cooling and differences in heat conduction in the necrotic regions of large tumors [167, 192, 193].

One approach to controlling temperature is to use materials with a Curie temperature between 42 and 50 °C, as these materials automatically 'turn off' when the temperature becomes too high. Substituted ferrites such as $(Co_{1-x}Zn_x)Fe_2O_4$, manganates such as $La_{1-x}Me_xMnO_3$ [Me=Sr, Ba, Pb, Ag, Na] and substituted yttrium-iron garnet $Y_3Fe_{5-x}Al_xO_{12}$ are ideal candidates due to their stability against oxidation (relative to metals) and the ability to tune the Curie temperature by composition [56, 194, 195]. Ni-Cu and Ni-Pd alloys have been investigated, but biocompatibility issues must be addressed [196-200].

9. OTHER APPLICATIONS

Magnetic nanoparticles can be used simultaneously for more than one of the applications discussed; for example, MRI can be used to confirm magnetic nanoparticle distribution prior to using the same particles to administer hyperthermia [201]. A combination of fluorescence and magnetic-nano-particle-enhanced MRI was used for preoperative magnetic resonance imaging and as an intraoperative optical probe during surgery, allowing clearer delineation of brain tumors. These 'multimodal nanoparticles' may allow radiologists and neurosurgeons to see the same probe in the same cells and thus improve identification of tumor margins [202, 203]. A nanoparticle combining near-infrared fluorescent dye with magnetism allows the particles to be located by MRI while the fluorescent dye provides simultaneous information about the molecular environment about the nanoparticle [204].

Although most applications try to avoid agglomeration, the intentional formation of a blockage of magnetic nanoparticles to block blood supply to a tumor has been investigated [67, 70, 137]. Aggregation and selective uptake has been used to destroy cells via the application of pulsed magnetic fields and subsequent rupture of cells [205].

Retinal detachment is a major cause of vision loss in adults. The usual treatment is the scleral buckle, which is a silicone band sewn to the outside of the eye that compresses the wall of the eye inward to close the holes in the retina. Sterically stabilized 4-10 nm magnetic particles in a poly(di-methylsiloxane) biocompatible fluid can be held in place with an external magnetized scleral buckle, thus providing a stable internal blockage that encircles the entire eye periphery with a ring of silicone oil. No fluid contacts the central vitreous cavity or the lens, thus decreasing the chances of undesired contact and/or damage [56, 206].

Many neurodegenerative diseases indicate the disruption of normal iron homeostasis in the brain. Recent experimental work indicates that nanoscale magnetic biominerals (primarily magnetite and maghemite) may be associated with senile plaques and fan filaments found in brain tissue affected by these diseases. Understanding the role of iron in neurodegenerative disease could help understand the origin, diagnosis and treatment of these diseases [175].

10. CONCLUSION

The field of nanobiomagnetics is exceptionally broad, involving researchers from medicine, pharmaceutical science, chemistry, physics, biology, engineering, and materials science. The literature is similarly decentralized, making it impossible not to omit some contributions in this article. Recent reviews mentioned in this chapter provide additional information for the reader interested in more detail on a specific topic. Nanomagnetics researchers have much to contribute to this new and exciting field.

Acknowledgements

The authors appreciate the assistance of Shannon Fritz, Tapan Jain, Raymond Lemoine, Marco Morales, Hai Nguyen, Mick O'Shea, Shaina Remboldt, Dave Schmitter, Michelle Strand and Steve Wignall for their assistance.

References

[1] D. Hautot, Q. A. Pankhurst, N. Khan, and J. Dobson, Proc. Roy. Soc. B (London) **270**, S62 (2003).

[2] W. Andrä, in "Magnetism in Medicine: A Handbook", Eds W. Andrä and H. Nowak, Wiley, New York - Berlin 1998, p. 425.

[3] S. A. Douglas, S. Mirza, and F. W. Stafford, Int. J. Pediatr. Otorhi. **62**, 165 (2002).

[4] E. H. Frei, J. Appl. Phys. **40** (1969).

[5] F. E. Luborsky, B. J. Drummond, and A. Q. Penta, AJR Am. J. Roentgenol. **92**, 1021 (1964).

[6] U. Häfeli, in "Scientific and Clinical Applications of Magnetic Carriers", Eds U. Häfeli, W. Schütt, J. Teller, and M. Zborowski, Plenum Press, New York 1997, p. 1.

[7] G. T. Gillies, R. C. Ritter, W. C. Broaddus, M. S. Grady, M. A. Howard, and R. G. McNeil, Rev. Sci. Instrum. **65**, 533 (1994).
[8] H. Tillander, Acta Radiol. **45**, 21 (1956).
[9] A. S. Curtis and M. Varde, J. Natl. Cancer Inst. **33**, 15 (1964).
[10] A. Curtis and C. Wilkinson, Trends Biotechnol. **19**, 97 (2001).
[11] C. S. Chen, M. Mrksich, S. Huang, G. M. Whitesides, and D. E. Ingber, Science **276**, 1425 (1997).
[12] B. Bonnemain, J. Drug. Target. **6**, 167 (1998).
[13] P. Tartaj, M. D. Morales, S. Veintemillas-Verdaguer, T. Gonzalez-Carreno, and C. J. Serna, J. Phys. D **36**, R182 (2003).
[14] B. R. Bacon, D. D. Stark, C. H. Park, S. Saini, E. V. Groman, P. F. Hahn, C. C. Compton, and J. T. Ferrucci Jr., J. Lab. Clin. Med. **110**, 164 (1987).
[15] A. Jordan, R. Scholz, P. Wust, H. Fahling, J. Krause, W. Wlodarczyk, B. Sander, T. Vogl, and R. Felix, Int. J. Hyperther. **13**, 587 (1997).
[16] A. A. Kuznetsov, A. R. Harutyunyun, E. K. Dobrinsky, V. I. Filippov, A. G. Malenkov, A. V. Vanin, and O. A. Kuznetsov, in "Scientific and Clinical Applications of Magnetic Carriers", Eds U. Häfeli *et al.*, Plenum Press, New York 1997, p. 379.
[17] A. A. Kuznetsov, V. I. Filippov, O. A. Kuznetsov, V. G. Gerlivanov, E. K. Dobrinsky, and S. I. J. M. M. M. Malashin, J. Magn. Magn. Mater. **194**, 22 (1999).
[18] J. Johnson, T. Kent, J. Koda, C. Peterson, S. Rudge, and G. Tapolsky, Eur. Cells. Mater. **3**, 12 (2002).
[19] S. H. Sun, H. Zeng, D. B. Robinson, S. Raoux, P. M. Rice, S. X. Wang, and G. X. Li, J. Am. Chem. Soc **126**, 273 (2004).
[20] S. H. Sun, C. B. Murray, D. Weller, L. Folks, and A. Moser, Science **287**, 1989 (2000).
[21] S. H. Sun, S. Anders, T. Thomson, J. E. E. Baglin, M. F. Toney, H. F. Hamann, C. B. Murray, and B. D. Terris, J. Phys. Chem. B **107**, 5419 (2003).
[22] X. X. Zhang, G. H. Wen, G. Xiao, and S. H. Sun, J. Magn. Magn. Mater. **261**, 21 (2003).
[23] C. B. Murray, S. H. Sun, H. Doyle, and T. Betley, MRS Bull. **26**, 985 (2001).
[24] H. Zeng, J. Li, Z. L. Wang, J. P. Liu, and S. H. Sun, Nano Lett. **4**, 187 (2004).
[25] H. W. Gu, B. Xu, J. C. Rao, R. K. Zheng, X. X. Zhang, K. K. Fung, and C. Y. C. Wong, J. Appl. Phys. **93**, 7589 (2003).
[26] T. Hyeon, S. S. Lee, J. Park, Y. Chung, and H. B. Na, J. Am. Chem. Soc **123**, 12798 (2001).
[27] G. Viau, V. F. Fievet, and F. Fievet, J. Mater. Chem. **6**, 1047 (1996).
[28] I. Nakatani, T. Furubayashi, T. Takahashi, and H. Hanaoka, J. Magn. Magn. Mater. **65**, 261 (1987).
[29] H. Yamamoto, T. Kanno, and I. Nakatani, J. Magn. Magn. Mater. **122**, 15 (1993).
[30] M. Wagener, B. Gunther, and E. Blums, J. Magn. Magn. Mater. **201**, 18 (1999).
[31] N. H. Hai, R. Lemoine, S. Rembolt, M. Strand, J. E. Shield, D. Schmitter, R. H. Kraus Jr., M. Espy, and D. L. Leslie-Pelecky, J. Magn. Magn. Mater. **293**, 75 (2005).
[32] B. L. Cushing, V. Golub, and C. J. O'Connor, J. Phys. Chem. Solids **65**, 825 (2004).
[33] S. R. Shinde, S. D. Kulkarni, A. G. Banpurkar, R. Nawathey-Dixit, S. K. Date, and S. B. Ogale, J. Appl. Phys. **88**, 1566 (2000).

[34] G. X. Chen, M. H. Hong, B. Lan, Z. B. Wang, Y. F. Lu, and T. C. Chong, Appl. Surf. Sci. **228**, 169 (2004).

[35] M. P. Morales, O. Bomati-Miguel, R. P. de Alejo, J. Ruiz-Cabello, S. Veintemillas-Verdaguer, and K. O'Grady, J. Magn. Magn. Mater. **266**, 102 (2003).

[36] Z. G. M. Lacava, R. B. Azevedo, L. M. Lacava, E. V. Martins, V. A. P. Garcia, C. A. Rebula, A. P. C. Lemos, M. H. Sousa, F. A. Tourinho, P. C. Morais, and M. F. Da Silva, J. Magn. Magn. Mater. **194**, 90 (1999).

[37] E. E. Carpenter, J. Magn. Magn. Mater. **225**, 17 (2001).

[38] M. Chen, S. Yamamuro, D. Farrell, and S. A. Majetich, J. Appl. Phys. **93**, 7551 (2003).

[39] C. Alexiou, W. Arnold, P. Hulin, R. J. Klein, H. Renz, F. G. Parak, C. Bergemann, and A. S. Lubbe, J. Magn. Magn. Mater. **225**, 187 (2001).

[40] C. Grüttner, S. Rudershausen, and J. Teller, J. Magn. Magn. Mater. **225**, 1 (2001).

[41] C. Grüttner, J. Teller, W. Schütt, F. Westphal, C. Schümichen, and B.-R. Paulke, in "Scientific and Clinical Applications of Magnetic Carriers", Eds U. Häfeli *et al.*, Plenum Press, New York 1997, p. 53.

[42] A. Jordan, R. Scholz, P. Wust, H. Schirra, T. Schiestel, H. Schmidt, and R. Felix, J. Magn. Magn. Mater. **194**, 185 (1999).

[43] G. A. van Ewijk, G. J. Vroege, and A. P. Philipse, J. Magn. Magn. Mater. **201**, 31 (1999).

[44] C. Grüttner and J. Teller, J. Magn. Magn. Mater. **194**, 8 (1999).

[45] J. Connolly, T. G. St. Pierre, M. Rutnakornpituk, and J. S. Riffle, Eur. Cells. Mater. **3**, 106 (2002).

[46] S. Mornet, F. Grasset, J. Portier, and E. Duguet, Eur. Cells. Mater. **3**, 110 (2002).

[47] H. H. Boennemann, W. J. Brijoux, R. Brinkmann, N. Matoussevitch, and N. Waldoefner, Abs. Pap. Am. Chem. Soc. **226**, U349 (2003).

[48] R. H. Müller, M. Lück, S. Harnisch, and K. Thode, in *Scientific and Clinical Applications of Magnetic Carriers*, Eds U. Häfeli *et al.*, Plenum Press, New York1997, p. 135.

[49] J. Kreuter, Eur. J. Drug Metab. Ph. **19**, 253 (1994).

[50] R. Gref, Y. Minamitake, M. T. Peracchia, V. Trubetskoy, V. Torchilin, and R. Langer, Science **263**, 1600 (1994).

[51] C. Wilhelm, C. Billotey, J. Roger, J. N. Pons, J. C. Bacri, and F. Gazeau, Biomaterials **24**, 1001 (2003).

[52] I. Hilger, S. Fruhauf, W. Linss, R. Hiergeist, W. Andra, R. Hergt, and W. A. Kaiser, J. Magn. Magn. Mater. **261**, 7 (2003).

[53] I. Brigger, C. Dubernet, and P. Couvreur, Adv. Drug Deliver. Rev. **54**, 631 (2002).

[54] U. Gaur, S. K. Sahoo, T. K. De, P. C. Ghosh, A. Maitra, and P. K. Ghosh, Int. J. Pharm. **202**, 1 (2000).

[55] L. M. Lacava, Z. G. M. Lacava, M. F. Da Silva, O. Silva, S. B. Chaves, R. B. Azevedo, F. Pelegrini, C. Gansau, N. Buske, D. Sabolovic, and P. C. Morais, Biophys. J. **80**, 2483 (2001).

[56] F. Grasset, S. Mornet, A. Demourgues, J. Portier, J. Bonnet, A. Vekris, and E. Duguet, J. Magn. Magn. Mater. **234**, 409 (2001).

[57] R. Kotitz, P. C. Fannin, and L. J. Trahms, J. Magn. Magn. Mater. **149**, 42 (1995).

[58] R. W. Chantrell, M. El-Hilo, and K. O'Grady, IEEE Trans. Magn. **27**, 3570 (1991).

[59] R. W. Chantrell, A. Lyberatos, M. El-Hilo, and K. O'Grady, J. Appl. Phys. **76**, 6407 (1994).

[60] V. P. Torchilin, Eur. J. Pharm. Sci. **11**, S81 (2000).

[61] R. T. Gordon, J. R. Hines, and D. Gordon, Med. Hypotheses **5**, 83 (1979).

[62] H. Maeda, J. Wu, T. Sawa, Y. Matsumura, and K. J. Hori, J. Control. Release **65**, 271 (2000).

[63] C. C. Berry and A. S. G. Curtis, J. Phys. D **36**, R198 (2003).

[64] U. O. Häfeli, G. J. Pauer, W. K. Roberts, J. L. Humm, and R. M. Macklis, in "Scientific and Clinical Applications of Magnetic Carriers", Eds U. Häfeli *et al.*, Plenum Press, New York 1997, p. 501.

[65] C. Alexiou, W. Arnold, R. J. Klein, F. G. Parak, P. Hulin, C. Bergemann, W. Erhardt, S. Wagenpfeil, and A. S. Lubbe, Cancer Res. **60**, 6641 (2000).

[66] D. Devineni, A. Kleinszanto, and J. M. Gallo, J. Neuro-Oncol. **24**, 143 (1995).

[67] A. S. Lubbe, C. Bergemann, J. Brock, and D. G. McClure, J. Magn. Magn. Mater. **194**, 149 (1999).

[68] A. Jordan, R. Scholz, P. Wust, H. Fahling, and R. Felix, J. Magn. Magn. Mater. **201**, 413 (1999).

[69] K. Widder, R. Morris, G. Poore, D. Howard, and A. Senyei, Eur. J. Cancer Clin. Oncol. **19**, 135 (1983).

[70] A. S. Lübbe, C. Alexiou, and C. Bergemann, J. Surg. Res. **95**, 200 (2001).

[71] Z. G. Forbes, B. B. Yellen, K. A. Barbee, and G. Friedman, IEEE Trans. Magn. **39**, 3372 (2003).

[72] A. D. Ebner, H. J. Ploehn, and J. A. Ritter, Separ. Sci. Technol. **37**, 3727 (2002).

[73] J. A. Ritter, A. D. Ebner, K. D. Daniel, and K. L. Stewart, J. Magn. Magn. Mater. **280**, 184 (2004).

[74] H. Chen, A. D. Ebner, J. A. Ritter, M. D. Kaminski, and A. J. Rosengart, in "Scientific and Clinical Applications of Magnetic Carriers", Lyon, France 2004.

[75] B. B. Yellen, Z. G. Forbes, G. Friedman, and K. A. Barbee, in "Scientific and Clinical Applications of Magnetic Carriers", Lyon, France 2004.

[76] S. Y. Mukhmudov, A. A. Kuznetsov, and V. I. Filippov, in "Scientific and Clinical Applications of Magnetic Carriers", Eds U. Häfeli *et al.*, Plenum Press, New York 1997, p. 495.

[77] P. L. Kronick, G. L. Campbell, and K. Joseph, Science **200**, 1074 (1978).

[78] A. Radbruch, B. Mechtold, A. Thiel, S. Miltenyi, and E. Pfluger, Meth. Cell Biol. **42**, 387 (1994).

[79] K. Kriz, J. Gehrke, and D. Kriz, Biosens. Bioelectron. **13**, 817 (1998).

[80] W. S. Prestivik, A. Berge, P. Mørk, P. Stenstad, and J. Ugelstad, in "Scientific and Clinical Applications of Magnetic Carriers", Eds U. Häfeli *et al.*, Plenum Press, New York 1997, p. 11.

[81] I. Šafarik and M. Šafariková, in "Scientific and Clinical Applications of Magnetic Carriers", Eds U. Häfeli *et al.*, Plenum Press, New York 1997, p. 323.

[82] S. E. Matthews, P. Parzuchowski, A. Garcia-Carrera, C. Grüttner, J. F. Dozol, and V. Bohmer, Chem. Commun., 417 (2001).

[83] M. Kaminski, S. Landsberger, L. Nunez, and G. F. Vandegrift, Separ. Sci. Technol. **32**, 115 (1997).

[84] A. S. Bahaj, P. A. B. James, and F. D. Moeschler, J. Magn. Magn. Mater. **177**, 1453 (1998).

[85] C. Weber and D. Falkerhagen, in "Scientific and Clinical Applications of Magnetic Carriers", Eds U. Häfeli *et al.*, Plenum Press, New York 1997, p. 371.

[86] H. W. Gu, P. L. Ho, K. W. T. Tsang, C. W. Yu, and B. Xu, Chem. Commun., 1966 (2003).

[87] H. W. Gu, P. L. Ho, K. W. T. Tsang, L. Wang, and B. Xu, J. Am. Chem. Soc **125**, 15702 (2003).

[88] J. Ugelstad, W. S. Prestivik, P. Stenstad, L. Kilaas, and G. Kvalheim, Eds W. Andrä and H. Nowak (New York, Berlin, 1998), p. 471.

[89] G. Blankenstein, in "Scientific and Clinical Applications of Magnetic Carriers", Eds U. Häfeli *et al.*, Plenum Press, New York 1997, p. 233.

[90] S. Miltenyi, W. Muller, W. Weichel, and A. Radbruch, Cytometry **11**, 231 (1990).

[91] J. J. Chalmers, Y. Zhao, M. Nakamura, K. Melnik, L. Lasky, L. Moore, and M. Zborowski, J. Magn. Magn. Mater. **194**, 231 (1999).

[92] D. L. Graham, H. A. Ferreira, and P. P. Freitas, Trends Biotechnol. **22**, 455 (2004).

[93] L. Lagae, R. Wirix-Speetjens, J. Das, D. Graham, H. Ferreira, P. P. F. Freitas, G. Borghs, and J. De Boeck, J. Appl. Phys. **91**, 7445 (2002).

[94] T. Deng, G. M. Whitesides, M. Radhakrishnan, G. Zabow, and M. Prentiss, Appl. Phys. Lett. **78**, 1775 (2001).

[95] A. Rida, V. Fernandez, and M. A. M. Gijs, Appl. Phys. Lett. **83**, 2396 (2003).

[96] G. X. Li, V. Joshi, R. L. White, S. X. Wang, J. T. Kemp, C. Webb, R. W. Davis, and S. H. Sun, J. Appl. Phys. **93**, 7557 (2003).

[97] J. Schotter, P. B. Kamp, A. Becker, A. Puhler, G. Reiss, and H. Bruckl, Biosens. Bioelectron. **19**, 1149 (2004).

[98] J. Schotter, P. B. Kamp, A. Becker, A. Puhler, D. Brinkmann, W. Schepper, H. Bruckl, and G. Reiss, IEEE Trans. Magn. **38**, 3365 (2002).

[99] D. R. Baselt, G. U. Lee, M. Natesan, S. W. Metzger, P. E. Sheehan, and R. J. Colton, Biosens. Bioelectron. **13**, 731 (1998).

[100] R. L. Edelstein, C. R. Tamanaha, P. E. Sheehan, M. M. Miller, D. R. Baselt, L. J. Whitman, and R. J. Colton, Biosens. Bioelectron. **14**, 805 (2000).

[101] M. M. Miller, P. E. Sheehan, R. L. Edelstein, C. R. Tamanaha, L. Zhong, S. Bounnak, L. J. Whitman, and R. J. Colton, J. Magn. Magn. Mater. **225**, 138 (2001).

[102] M. Tondra, M. Granger, R. Fuerst, M. Porter, C. Nordman, J. Taylor, and S. Akou, IEEE Trans. Magn. **37**, 2621 (2001).

[103] M. M. Miller, P. E. Sheehan, R. L. Edelstein, C. R. Tamanaha, L. Zhong, S. Bounnak, L. J. Whitman, and R. J. Colton, J. Magn. Magn. Mater. **225**, 138 (2001).

[104] L. Ejsing, M. F. Hansen, A. K. Menon, H. A. Ferreira, D. L. Graham, and P. P. Freitas, Appl. Phys. Lett. **84**, 4729 (2004).

[105] G. X. Li and S. X. Wang, IEEE Trans. Magn. **39**, 3313 (2003).

[106] J. Connolly and T. G. St. Pierre, J. Magn. Magn. Mater. **225**, 156 (2001).

[107] S. H. Chung, A. Hoffmann, S. D. Bader, C. Liu, B. Kay, L. Makowski, and L. Chen, Appl. Phys. Lett. **85**, 2971 (2004).

[108] C. D. Delgratta, S. Dellapenna, P. Battista, L. Didonato, P. Vitullo, G. L. Romani, and S. Diluzio, Phys. Med. Biol. **40**, 671 (1995).
[109] E. Romanus, M. Huckel, C. Gross, S. Prass, W. Weitschies, R. Brauer, and P. Weber, J. Magn. Magn. Mater. **252**, 387 (2002).
[110] Y. R. Chemla, H. L. Crossman, Y. Poon, R. McDermott, R. Stevens, M. D. Alper, and J. Clarke, P. Natl. Acad. Sci. USA **97**, 14268 (2000).
[111] C. Gosse and V. Croquette, Biophys. J. **82**, 3314 (2002).
[112] S. B. Smith, L. Finzi, and C. Bustamante, Science **258**, 1122 (1992).
[113] T. R. Strick, J.-F. Allemand, D. Bensimon, A. Bensimon, and V. Croquette, Science **271**, 1835 (1996).
[114] T. R. Strick, J.-F. Allemand, D. Bensimon, and V. Croquette, Biophys. J. **74**, 2016 (1998).
[115] S. H. Leuba, M. A. Karymov, M. Tomschik, R. Ramjit, P. Smith, and J. Zlatanova, P. Natl. Acad. Sci. USA **100**, 495 (2003).
[116] B. Fabry, G. N. Maksym, R. D. Hubmayr, J. P. Butler, and J. J. Fredberg, J. Magn. Magn. Mater. **194**, 120 (1999).
[117] S. M. Mijailovich, M. Kojic, M. Zivkovic, B. Fabry, and J. J. Fredberg, J. Appl. Physiol. **93**, 1429 (2002).
[118] R. Fulconis, A. Bancaud, J. F. Allemand, V. Croquette, M. Dutreix, and J. L. Viovy, Biophys. J. **87**, 2552 (2004).
[119] C. Wilhelm, A. Cebers, J. C. Bacri, and F. Gazeau, Eur. Biophys. J. Biophy. **32**, 655 (2003).
[120] C. Wilhelm, F. Gazeau, and J. C. Bacri, Phys. Rev. E **67** (2003).
[121] M. Kumar, J. Pharm. Pharm. Sci. **3**, 234 (2000).
[122] C. Alexiou, R. Jurgons, R. J. Schmid, C. Bergemann, J. Henke, W. Erhardt, E. Huenges, and F. Parak, J. Drug. Target. **11**, 139 (2003).
[123] U. O. Häfeli, Int. J. Pharm. **277**, 19 (2004).
[124] P. K. Gupta and C. T. Hung, Life Sci. **44**, 175 (1989).
[125] P. K. Gupta and C. T. Hung, in "Microspheres and Regional Cancer Therapy", Eds N. Willmott and J. M. Daly, CRC Press, Boca Raton 1993, p. 1.
[126] J. Panyam and V. Labhasetwar, Adv. Drug Deliv. Rev. **55**, 329 (2003).
[127] M. A. Morales, Tapan Kumar Jain, V. Labhasetwar, and D. L. Leslie-Pelecky, J. Appl. Phys. **97**, 10Q905 (2005).
[128] C. Bergemann, D. Muller-Schulte, J. Oster, L. a Brassard, and A. S. Lübbe, J. Magn. Magn. Mater. **194**, 45 (1999).
[129] A. S. Lübbe, C. Bergemann, H. Riess, F. Schriever, P. Reichardt, K. Possinger, M. Matthias, B. Dorken, F. Herrmann, R. Gurtler, P. Hohenberger, N. Haas, R. Sohr, B. Sander, A. J. Lemke, D. Ohlendorf, W. Huhnt, and D. Huhn, Cancer Res. **56**, 4686 (1996).
[130] K. Widder, G. Flouret, and A. Senyei, J Pharm Sci. **68**, 79 (1979).
[131] K. J. Widder, R. M. Morris, G. Poore, D. P. Howard Jr., and A. W. Senyei, P. Natl. Acad. Sci. USA **78**, 579 (1981).
[132] S. Goodwin, C. Peterson, C. Hoh, and C. J. o. M. a. M. M. Bittner, J. Magn. Magn. Mater. **194**, 132 (1999).

[133] L. M. Allen, T. Kent, C. Wolfe, C. Ficco, and J. Johnson, in "Scientific and Clinical Applications of Magnetic Carriers", Eds U. Häfeli *et al.*, Plenum Press, New York 1997, p. 481.
[134] U. O. Häfeli, S. M. Sweeney, B. A. Beresford, J. L. Humm, and R. M. Macklis, Nucl. Med. Biol. **22**, 147 (1995).
[135] U. O. Häfeli, S. M. Sweeney, B. A. Beresford, E. H. Sim, and R. M. Macklis, J. Biomed. Mater. Res. **28**, 901 (1994).
[136] U. Häfeli, G. Pauer, S. Failing, and G. Tapolsky, J. Magn. Magn. Mater. **225**, 73 (2001).
[137] A. S. Lubbe, C. Bergemann, W. Huhnt, T. Fricke, H. Riess, J. W. Brock, and D. Huhn, Cancer Res. **56**, 4694 (1996).
[138] C. Alexiou, A. Schmidt, R. Klein, P. Hulin, C. Bergemann, and W. Arnold, J. Magn. Magn. Mater. **252**, 363 (2002).
[139] S. Päuser, R. Reszka, S. Wagner, K.-J. Wolf, H. J. Buhr, and G. Berger, in "Scientific and Clinical Applications of Magnetic Carriers", Eds U. Häfeli *et al.*, Plenum Press, New York 1997, p. 561.
[140] M. Babincova, D. Leszczynska, P. Sourivong, P. Babinec, and J. Leszczynski, Med. Hypotheses **62**, 375 (2004).
[141] D. Szabo, G. Szeghy, and M. Zrinyi, Macromolecules **31**, 6541 (1998).
[142] P. M. Xulu, G. Filipcsei, and M. Zrínyi, Macromolecules **33**, 1716 (2000).
[143] E. R. Edelman and R. Langer, Biomaterials **14**, 621 (1993).
[144] J. Kost and R. Langer, Adv. Drug Deliver. Rev. **46**, 125 (2001).
[145] H. Richert, O. Surzhenko, S. Wangemann, P. Payer, and P. Görnert, in "Scientific and Clinical Applications of Magnetic Carriers", Lyon, France 2004.
[146] F. Scherer, M. Anton, U. Schillinger, J. Henkel, C. Bergemann, A. Kruger, B. Gansbacher, and C. Plank, Gene Ther. **9**, 102 (2002).
[147] F. Krotz, H. Y. Sohn, T. Gloe, C. Plank, and U. Pohl, J. Vasc. Res. **40**, 425 (2003).
[148] C. Plank, U. Schillinger, F. Scherer, C. Bergemann, J. S. Remy, F. Krotz, M. Anton, J. Lausier, and J. Rosenecker, Biol. Chem. **384**, 737 (2003).
[149] A. S. Arbab, E. K. Jordan, L. B. Wilson, G. T. Yocum, B. K. Lewis, and J. A. Frank, Hum. Gene Ther. **15**, 351 (2004).
[150] E. Odeblad and G. Lindstrom, Acta Radiol. **43**, 469 (1955).
[151] P. C. Lauterbur, Nature **242**, 190 (1973).
[152] A. Oppelt, in "Magnetism in Medicine: A Handbook", Eds W. Andrä and H. Nowak, Wiley, New York - Berlin 1998, p. 305.
[153] C. Chapon, F. Franconi, L. Lemaire, L. Marescaux, P. Legras, J. P. Saint-Andre, B. Denizot, and J. J. Le Jeune, Invest. Radiol. **38**, 141 (2003).
[154] A. K. Fahlvik, E. Holtz, and J. Klaveness, Mag. Reson. Imaging **8**, 363 (1990).
[155] J. Gellissen, C. Axmann, A. Prescher, K. Bohndorf, and K. P. Lodemann, Magn. Reson. Imaging **17**, 557 (1999).
[156] D. K. Kim, Y. Zhang, W. Voit, K. V. Rao, and M. Muhammed, J. Magn. Magn. Mater. **225**, 30 (2001).
[157] C. H. Reynolds, N. Annan, K. Beshah, J. H. Huber, S. H. Shaber, R. E. Lenkinski, and J. A. Wortman, J. Am. Chem. Soc. **122**, 8940 (2000).

[158] S. Flacke, S. Fischer, M. J. Scott, R. J. Fuhrhop, J. S. Allen, M. McLean, P. Winter, G. A. Sicard, P. J. Gaffney, S. A. Wickline, and G. M. Lanza, Circulation **104**, 1280 (2001).

[159] G. M. Lanza, X. Yu, P. M. Winter, D. R. Abendschein, K. K. Karukstis, M. J. Scott, L. K. Chinen, R. W. Fuhrhop, D. E. Scherrer, and S. A. Wickline, Circulation **106**, 2842 (2002).

[160] M. L. Zelivyanskaya, J. A. Nelson, L. Poluektova, M. Uberti, M. Mellon, H. E. Gendelman, and M. D. Boska, J. Neurosci. Res. **73**, 284 (2003).

[161] J. M. Perez, L. Josephson, and R. Weissleder, Chembiochem. **5**, 261 (2004).

[162] M. Zhao, M. F. Kircher, L. Josephson, and R. Weissleder, Bioconjugate Chem. **13**, 840 (2002).

[163] J. M. Perez, L. Josephson, T. O'Loughlin, D. Hogemann, and R. Weissleder, Nat. Biotechnol. **20**, 816 (2002).

[164] R. Weissleder, A. Moore, U. Mahmood, R. Bhorade, H. Benveniste, E. A. Chiocca, and J. P. Basilion, Nat. Med. **6**, 351 (2000).

[165] D. Hogemann, L. Josephson, R. Weissleder, and J. P. Basilion, Bioconjugate Chem. **11**, 941 (2000).

[166] M. D. Chabot and J. Moreland, J. Appl. Phys. **93**, 7897 (2003).

[167] W. Andrä, in "Magnetism in Medicine: A Handbook", Eds W. Andrä and H. Nowak, Wiley, New York - Berlin 1998, p. 455.

[168] M. H. Seegenschmiedt, P. Fessenden, C. C. Vernon, and M. Abe, "Thermoradiotherapy and Thermochemotherapy", London, Berlin 1996.

[169] P. Moroz, S. K. Jones, and B. N. Gray, J. Surg. Oncol. **77**, 259 (2001).

[170] I. Hilger, W. Andra, R. Hergt, R. Hiergeist, H. Schubert, and W. A. Kaiser, Radiology **218**, 570 (2001).

[171] M. Ahmed, W. E. Monsky, G. Girnun, A. Lukyanov, G. D'Ippolito, J. B. Kruskal, K. E. Stuart, V. P. Torchilin, and S. N. Goldberg, Cancer Res. **63**, 6327 (2003).

[172] K. Dowlatshahi, A. K. Bhattacharya, B. Silver, T. Matalon, and J. W. Williams, Surgery **112**, 603 (1992).

[173] A. Jordan, P. Wust, R. Scholz, H. Faehling, J. Krause, and R. Felix, in "Scientific and Clinical Applications of Magnetic Carriers", Eds U. Häfeli *et al.*, Plenum Press, New York 1997, p. 569.

[174] R. J. Myerson, L. Leybovich, B. Emami, P. W. Grigsby, W. Straube, and D. Vongerichten, Int. J. Hypertherm. **7**, 937 (1991).

[175] J. Dobson, FEBS Letters **496**, 1 (2001).

[176] M. G. de Jode, J. A. Vale, and W. M. Gedroyc, J. Magn. Reson. Imaging **10**, 545 (1999).

[177] M. G. de Jode, G. M. Lamb, H. C. Thomas, S. D. Taylor-Robinson, and W. M. Gedroyc, J. Hepatol. **31**, 347 (1999).

[178] C. K. Chou, Int. J. Hypertherm. **6**, 367 (1990).

[179] R. Hergt, W. Andra, C. G. d'Ambly, I. Hilger, W. A. Kaiser, U. Richter, and H. G. Schmidt, IEEE Trans. Magn. **34**, 3745 (1998).

[180] D. C. F. Chan, D. B. Kirpotin, and P. A. Bunn, J. Magn. Magn. Mater. **122**, 374 (1993).

[181] Q. A. Pankhurst, J. Connolly, S. K. Jones, and J. Dobson, J. Phys. D **36**, R167 (2003).

[182] A. Jordan, P. Wust, H. Fahling, W. John, A. Hinz, and R. Felix, Int. J. Hypertherm. **9**, 51 (1993).

[183] I. Hilger, R. Hergt, and W. A. Kaiser, Invest. Radiol. **35**, 170 (2000).

[184] R. E. Rosensweig, J. Magn. Magn. Mater. **252**, 370 (2002).

[185] R. K. Gilchrist, R. Medal, W. D. Shorey, R. C. Hanselman, J. C. Parrott, and C. B. Taylor, Ann. Surg. **146** (1957).

[186] I. Hilger, A. Kiessling, E. Romanus, R. Hiergeist, R. Hergt, W. Andra, M. Roskos, W. Linss, P. Weber, W. Weitschies, and W. Kaiser, Nanotechnology **15**, 1027 (2004).

[187] P. Moroz, S. K. Jones, and B. N. Gray, J. Surg. Oncol. **80**, 149 (2002).

[188] S. K. Jones and J. G. Winter, Phys. Med. Biol. **46**, 385 (2001).

[189] G. A. Flores and J. Liu, Eur. Cells. Mater. **3**, 9 (2002).

[190] A. S. Lübbe and C. Bergemann, in "Scientific and Clinical Applications of Magnetic Carriers", Eds U. Häfeli *et al.*, Plenum Press, New York 1997, p. 457.

[191] A. Jordan, in "Scientific and Clinical Applications of Magnetic Carriers", Lyon, France 2004.

[192] P. Moroz, S. K. Jones, and B. N. Gray, Int. J. Hypertherm. **18**, 129 (2002).

[193] M. Mitsumori, M. Hiraoka, T. Shibata, Y. Okuno, S. Masaunaga, M. Koishi, K. Okajima, Y. Nagata, Y. Nishimura, M. Abe, K. Ohura, M. Hasegawa, H. Nageae, and Y. Ebisawa, Int. J. Hypertherm. **10**, 785 (1994).

[194] J. Giri, A. Ray, S. Dasgupta, D. Datta, and D. Bahadur, Biomed. Mater. Eng. **13**, 387 (2003).

[195] A. A. Kuznetsov, O. A. Shlyakhtin, N. A. Brusentsov, and O. A. Kuznetsov, Eur. Cells Mater. **3**, 75 (2002).

[196] M. Bettge, J. Chatterjee, and Y. Haik, Biomagn. Res. Technol. **2**, 4 (2004).

[197] I. A. Brezovich, M. B. Lilly, R. F. Meredith, B. Weppelmann, R. A. Henderson, W. Brawner, and M. M. Salter, Int. J. Hypertherm. **6**, 117 (1990).

[198] I. A. Brezovich and R. F. Meredith, Radiol. Clin. North Am. **27**, 589 (1989).

[199] T. Kobayashi and Y. Kida, Stereotac.t Funct. Neurosurg. **54-55**, 514 (1990).

[200] T. Kobayashi, Y. Kida, T. Tanaka, N. Kageyama, H. Kobayashi, and Y. Amemiya, J Neuro-Oncology **4**, 175 (1986).

[201] H. Pardoe, P. R. Clark, T. G. St Pierre, P. Moroz, and S. K. Jones, J. Magn. Magn. Mater. **21**, 483 (2003).

[202] S. P. Mulvaney, H. M. Mattoussi, and L. J. Whitman, Biotechniques **36**, 602 (2004).

[203] M. F. Kircher, U. Mahmood, R. S. King, R. Weissleder, and L. Josephson, Cancer Res. **63**, 8122 (2003).

[204] L. Josephson, M. F. Kircher, U. Mahmood, Y. Tang, and R. Weissleder, Bioconjugate Chem. **13**, 554 (2002).

[205] M. Ogiue-Ikeda, Y. Sato, and S. Ueno, IEEE Trans. Nanobiosci. **2**, 262 (2003).

[206] J. P. Stevenson, M. Rutnakornpituk, M. Vadala, A. R. Esker, S. W. Charles, S. Wells, J. P. Dailey, and J. S. Riffle, J. Magn. Magn. Mater. **225**, 47 (2001)

Appendix

MAGNETIC MATERIALS

The behavior of magnetic nanostructures reflects both nanoscale features, such as particle size and geometry, and the intrinsic properties of the magnetic substances. For example, the magnetization reversal in nanodots crucially depends on the anisotropy of the dot material. Furthermore, nanostructures are often used as bulk materials, so that their extrinsic properties must be evaluated from the point of view of bulk materials. This appendix summarizes the characteristics of some important classes of magnetic materials and provides exemplary data.

A.1. CLASSES OF MAGNETIC MATERIALS

Traditionally, magnetic materials are classified by their magnetic coercivity or hardness. The term is historical and refers to iron, where the addition of carbon increases not only the mechanical hardness (steel) but also the coercivity. In descending order, the hardness gives rise to a classification into hard magnets (permanent magnets), recording media, and soft magnets. Note that some classes of magnetic materials, such as ferrofluids, that do not fall in any of these categories.

A.1.1. Permanent Magnets

These materials have a wide range of applications, from toys and refrigerator magnets to everyday electrotechnology—electromotors, loudspeakers, windshield wipers, locks, and microphones—and high-tech devices such as hard-disk drives. Key requirements for permanent magnets are a high remanent magnetization M_r, a high coercivity H_c, and a nearly rectangular hysteresis-loop shape. These properties are epitomized by the *energy product* $(BH)_{max}$ [1, 2], that is, by the magnets' ability to store magnetostatic energy outside the magnet. Energy products never exceed $\mu_o M_r H_c$ or $\mu_o M_r^2/4$, whichever is smaller. Until the mid 20th century, the main limitation was coercivity, as exemplified by the energy product of hard-magnetic steel, about 1 kJ/m^3. Steel magnets are now obsolete, because their low coercivities and energy products made it necessary to resort to cumbersome horseshoe shapes, but the high saturation induction of $Fe_{65}Co_{35}$ (2.43 T) and its pronounced temperature stability continue to be exploited in alnico-type magnets.

Today, the high anisotropy of rare-earth intermetallics makes it relatively easy to create sufficient anisotropy [3], and the main limitation is the magnetization. The best permanent magnets, currently made from $Nd_2Fe_{14}B$ [4], have room-temperature energy products exceeding 450 kJ/m^3 [5, 6]. This means that a compact Nd-Fe-B magnet of about 2 grams is now able to replace a 1-kg horseshoe magnet—a feature of major importance for advanced consumer electronics, car design, and computer technology. Nanostructuring is a major tool in fully exploiting and exceeding the performance of bulk rare-earth transition-metal intermetallics.

A.1.2. Soft Magnets

Soft-magnetic materials are used for a variety of low- and high-frequency applications. Examples are flux guidance in permanent-magnet and other systems, transformer cores, microwave applications, and recording heads. A feature of soft magnets is their low coercivity, which is often several orders of magnitude smaller than those of hard and semi hard materials. Another figure of merit is the initial permeability $\mu_r = (dM/dH)_0$, which exceeds 1000 in soft magnets [7, 8]. High-frequency applications require small hysteresis losses, so that the small coercivities are often more important than a high permeability.

Iron-based metallic magnets, such as Fe-Si, $Fe_{50}Co_{50}$, and permalloy ($Ni_{80}Fe_{20}$) have long been used as soft-magnetic materials. For example, permalloy has an anisotropy of about 0.15 kJ/m^3, an anisotropy field of about 0.4 mT [4 Oe], and coercivities of about 0.04 mT. Oxides, such as simple ferrites (TFe_2O_4, where T = Mn, Fe, Ni, Zn) and garnets ($R_3Fe_5O_{12}$, R = Y, Gd, …) have a ferrimagnetic spin structure and, therefore, a rather low magnetization. However, their comparatively large resistivity suppresses eddy-current losses and makes them suitable for high-frequency applications, for example in antennas and microwave devices [7, 9].

More recently, amorphous and nanostructured metals have attracted much attention as soft-magnetic materials. Essentially, they have the composition $T_{100-x}Z_x$ (T = Fe, Co, Ni and Z = B, C, P, Zr,…) where $x \approx$ 10-20. An example is the Yoshizawa alloy $Fe_{73.5}Si_{13.5}B_9Cu_1Nb_3$], where Fe_3Si grains having the cubic DO_3 structure are embedded in an amorphous matrix [10, 11]. As pointed out by Herzer [12], this random-anisotropy nanostructuring is a powerful tool to reduce the coercivity of soft magnetic materials. It is exploited, for example, in Fe-Si-B-Cu-Nb alloys [10].

An important application of soft materials is inductive and magneto-resistive recording heads [9]. The function of inductive head materials is to realize flux closure for reading and writing on recording media. Typical materials are $Ni_{80}Fe_{20}$ (permalloy, H_c = 0.01 to 0.05 mT), hot-pressed Ni-Zn and Mn-Zn ferrites (H_c = 0.02 mT); Fe-Si-Al (sendust, H_c = 0.025 mT), as well as Fe-Ti-N and Fe-Rh-N alloys [9]. Magnetoresistive read heads exploit the anisotropic magnetoresistance due to the spin-dependent scattering of conduction electrons by magnetic atoms or, more recently, the giant magnetoresistive (GMR) effect exploiting the different Fermi-level spin-up and spin-densities of the involved components. Many ferromagnets exhibit GMR, but soft-magnetic materials—such as permalloy—are easier to switch.

A.1.3. Recording Media

The first magnetic recording medium, invented by Poulsen in 1898, was the magnetic wire. Until the early 1940s, steel wires were used quite extensively for dictation, telephone recording, and radio broadcasting. Tape recording, invented in 1935 [13], has been used for data storage since 1951 and preceded modern PC hard disks. The first data-storage tapes had an areal storage density of order 0.002 Mb/in^2, as compared to 2.4 Mb/in^2 for the 3½" disks the widely used in the 1990s and about 10^5 Mb/in^2 in present-day hard disks.

Magnetic used in magnetic data storage tend to have moderate coercivities of the order of 0.3 T (240 kA/m), although some recently developed materials have higher coercivities, which improves the thermal stability of the stored information but makes the media more difficult to write. Traditional storage media are made using materials such as granular Fe, Fe_2O_3 and CrO_2, whereas advanced high-density recording media are based on materials such as Co-Cr-Pt-B, where the Pt improves the anisotropy. Other classes of materials presently being considered are $L1_0$ magnets, such as FePt, and nanocomposite rare-earth transition-metal films.

The main advantage of magnetic recording is the potential storage density. The bit size of optical and magneto-optical media, such as amorphous $Tb_{22}Fe_{66}Co_{12}$ [9], is limited by the wavelength of the used light, whereas semiconductor devices are difficult to handle on very small length scales. Some limitations in magnetic recording are the present use of longitudinal or in-plane recording media and thermal instabilities at ultrahigh recording densities. These problems can be tackled by suitable nanostructuring, using perpendicular recording and highly anisotropic media to avoid thermal demagnetization. The role of nanostructuring is seen from the quadratic dependence of the areal storage density of the bit size: exceeding 100 Gb/in^2 requires structures with feature size of less than about 50 nm.

A.2. DATA TABLES

Tables A.1 to A.3 show typical intrinsic and extrinsic parameters compiled from various sources [1, 2, 7]. Note that Ba and Sr ferrite (Table A.2) are isostructural and have very similar properties.

Table A.1 Intrinsic properties of some magnetic compounds. Note that γ-Fe_2O_3 crystallizes in a disordered spinel structure

Substance	$\mu_0 M_s$ (T)	T_c (K)	K_1 (MJ/m^3)	Symmetry	Structure
Fe_3O_4	0.60	858	- 0.011	cubic	$MgAl_2O_4$
$MnFe_2O_4$	0.52	573	- 0.0028	cubic	$MgAl_2O_4$
$CoFe_2O_4$	0.50	793	0.270	cubic	$MgAl_2O_4$
$NiFe_2O_4$	0.34	858	- 0.0069	cubic	$MgAl_2O_4$
$CuFe_2O_4$	0.17	728	- 0.0060	cubic	$MgAl_2O_4$
$SrFe_{12}O_{19}$	0.46	733	0.35	hex.	M ferrite
CrO_2	0.56	390	0.025	tetr.	TiO_2
$NiMnO_3$	0.13	437	- 0.26	hex.	$FeTiO_3$
γ-Fe_2O_3	0.47	863	- 0.0046	cubic	(spinel)
$Y_3Fe_5O_{12}$	0.16	560	- 0.00067	cubic	(garnet)
$Sm_3Fe_5O_{12}$	0.17	578	- 0.0025	cubic	(garnet)
$Dy_3Fe_5O_{12}$	0.05	563	- 0.0005	cubic	(garnet)
PtCo	1.00	840	4.9	tetr.	CuAu (I)
PtFe	1.43	750	6.6	tetr.	CuAu (I)
PdFe	1.37	760	1.8	tetr.	CuAu (I)
MnAl	0.62	650	1.7	tetr.	CuAu (I)
MnBi	0.78	630	1.2	hex.	NiAs
$NdCo_5$	1.23	910	0.7	hex.	$CaCu_5$
YCo_5	1.06	987	5.2	hex.	$CaCu_5$
Sm_2Fe_{17}	1.17	389	- 0.8	rhomb.	Th_2Zn_{17}
Y_2Fe_{17}	0.84	320	- 0.4	hex.	Th_2Ni_{17}
$Sm_2Fe_{17}N_3$	1.54	749	8.9	rhomb.	Th_2Zn_{17}
$Y_2Fe_{17}N_3$	1.46	694	- 1.1	hex.	Th_2Ni_{17}
Sm_2Co_{17}	1.20	1190	3.3	rhomb.	Th_2Zn_{17}
Dy_2Co_{17}	0.68	1152	- 2.6	rhomb. or hex.	2:17
Y_2Co_{17}	1.25	1167	- 0.34	rhomb. or hex.	2:17
$Sm(Fe_{11}Ti)$	1.14	584	4.9	tetr.	$ThMn_{12}$
$Y(Fe_{11}Ti)$	1.12	524	0.89	tetr.	$ThMn_{12}$
$Sm_2Fe_{14}B$	1.49	618	- 12.0	tetr.	$Nd_2Fe_{14}B$
$Dy_2Fe_{14}B$	0.67	593	4.5	tetr.	$Nd_2Fe_{14}B$
$Y_2Fe_{14}B$	1.36	571	1.06	tetr.	$Nd_2Fe_{14}B$

Table A.2 Intrinsic and micromagnetic properties of some magnetic materials.

	Fe	Co	Ni	$BaFe_{12}O_{19}$	$SmCo_{15}$	$Nd_2Fe_{14}B$
M_S (T)	2.15	1.76	0.62	0.47	1.07	1.61
T_C (K)	1043	1388	631	723	1003	585
K_1 (MJ/m^3)	0.05	0.53	-0.005	0.33	17	5.0
A (pJ/m)	8.3	10.3	3.4	6.1	22.0	7.7
δ (nm)	40	14	82	14	3.6	3.9
γ (mJ/m^2)	2.6	9.3	0.5	5.7	77	25
l_{ex} (nm)	1.5	2.0	3.4	5.9	4.9	1.9
R_{sd} (nm)	6	34	16	290	764	107
μ_0H_a (T)	0.06	0.76	0.03	1.8	40	7.6

Table A.3 Extrinsic properties of some bulk magnets.

	μ_oM_r [T]	μ_oH_c [T]	$(BH)_{max}$ [kJ/m^3]
Cobalt steel	1.0	0.025	8
Annealed iron	1.0	0.0001	0.04
Sintered Ba-ferrite	0.39	0.30	28
Anisotropic alnico	1.30	0.07	50
Metal-bonded $SmCo_5$	0.92	1.88	175
Polymer-bonded $SmCo_5$	0.58	1.00	60
Sintered Sm-Co	1.08	1.0	225
Sintered Nd-Fe-B	1.33	1.6	400
Polymer-bonded Nd-Fe-B	0.55	0.75	48

References

[1] R. M. Bozorth "Ferromagnetism", van Nostrand, Princeton1951.

[2] R. Skomski and J. M. D. Coey, "Permanent Magnetism", Institute of Physics, Bristol 1999.

[3] K. Kumar, J. Appl. Phys. **63**, R13 (1988).

[4] M. Sagawa, S. Hirosawa, H. Yamamoto, S. Fujimura, and Y. Matsuura, Jpn. J. Appl. Phys. **26**, 785 (1987).

[5] W. Rodewald, B. Wall, M. Katter, K. Üstüner, and S. Steinmetz, in "Rare-Earth Magnets and their Applications" (Proc. REM), Eds. G. C. Hadjipanayis and M. J. Bonder, Rinton Press, Princeton 2002, p. 25.

[6] Y. Kaneko, in: "High Performance Magnets and their Applications" (Proc. REM), Eds. N. M. Dempsey and P. de Rango, CNRS, Grenoble, 2004, p. 40.

[7] R. A. McCurrie, "Ferromagnetic Materials—Structure and Properties", Academic Press, London 1994.

[8] J. E. Evetts (Ed.), "Concise Encyclopedia of Magnetic and Superconducting Materials", Pergamon Press, Oxford 1992.

[9] R. L. Comstock, "Introduction to Magnetism and Magnetic Recording", Wiley, New York 1999.

[10] Y. Yoshizawa, S. Oguma, and K. Yamauchi, J. Appl. Phys. **64**, 6044 (1988).

[11] D. Jiles, "Introduction to Magnetism and Magnetic Materials", Chapman & Hall, London 1998.

[12] G. Herzer, Scripta Metal. **33**, 1741 (1995).

[13] T. Volk, AEG-Mitteilungen **25**, 199 (1935).

Index

MIX
Papier aus verantwortungsvollen Quellen
Paper from responsible sources
FSC® C105338

If you have any concerns about our products,
you can contact us on
ProductSafety@springernature.com

In case Publisher is established outside the EU,
the EU authorized representative is:
Springer Nature Customer Service Center GmbH
Europaplatz 3, 69115 Heidelberg, Germany

Printed by Libri Plureos GmbH
in Hamburg, Germany